MATHEMATICAL MODELING

A Chemical Engineer's Perspective

This is Volume 1 of
PROCESS SYSTEMS ENGINEERING
A Series edited by George Stephanopoulos and John Perkins

MATHEMATICAL MODELING

A Chemical Engineer's Perspective

Rutherford Aris
Department of Chemical Engineering and Materials Science
University of Minnesota
Minneapolis, Minnesota

ACADEMIC PRESS
San Diego London Boston New York Sydney Tokyo Toronto

Front cover photograph: The disposition of the general hexagonal pattern. (For more details, see Chapter 10, Figure C.1.)

This book is printed on acid-free paper. ♾

Academic Press
a division of Harcourt Brace & Company
525 B Street, Suite 1900, San Diego, California 92101-4495, USA
http://www.apnet.com

Academic Press
24-28 Oval Road, London NW1 7DX, UK
http://www.hbuk.co.uk/ap/

Library of Congress Catalog Card Number: 98-89644

International Standard Book Number: 0-12-604585-2

PRINTED IN THE UNITED STATES OF AMERICA
99 00 01 02 03 04 QW 9 8 7 6 5 4 3 2 1

In Memoriam
Albert Moscowitz
1929–1996

. . . the rest is commentary.

CONTENTS

2 Manipulation of Models

3 Solving the Equations

4 Presenting the Model and Its Behavior

5 Maxims for Modelers

6 Style

II Matter

7 Dispersion in Flow

9 STATICS AND DYNAMICS OF CHEMICAL REACTORS

PREFACE

In bygone days, when the full mystery of a volume was declared upon its title page, that page of this book might have read as follows:

CONFESSIONS
OF A
MATHEMATICAL MODELER
WHEREIN IS ADUMBRATED THE ART & CRAFT OF MODELING
DIVERS
CHEMICAL ENGINEERING SYSTEMS
BY THE FABRICATION, SOLUTION, AND PRESENTATION OF
ORDINARY AND PARTIAL DIFFERENTIAL EQUATIONS
COPIOUSLY ILLUSTRATED BY EXAMPLES FROM
THE PAPERS OF
RUTHERFORD ARIS
A SELECTION OF WHICH IS HERE REPRODUCED FROM THE
ORIGINAL SOURCES, SOME HAVING BEEN MUCH SOUGHT AFTER
& OTHERS ILLUSTRATIVE OF THE ERRORS TO WHICH THE
AUTHOR HAS AT TIMES BEEN OBNOXIOUS: TO WHICH IS ADDED
A COMPLETE BIBLIOGRAPHY OF THE AUTHOR'S WORKS AND AN
AUTOBIOGRAPHICAL APPENDIX OF ACKNOWLEDGMENTS
(WITH ASIDES ON SUNDRY PERSONS OF NOTE):
THE WHOLE BEING BOUND TOGETHER BY A PAINFUL INDEX
COMPILED BY THE AUTHOR'S OWN WORD PROCESSOR
AT THE SIGN OF THE LIRIPIPE
NOV. EBOR.

That it should so read today is more questionable. The publishers, quite rightly, point out that it scarcely facilitates either proper indexing or reader recognition. Certainly I would hate to languish among the tomes of pastoral theology in some seminary library, as Hobson's *Spherical and Elliptical Harmonics* slumbered for many years in the Music Library of a university that shall be nameless. But, if this fate is to be avoided, some explanation of the book's unusual composition is necessary. Let me note first, however, that I am concerned with models consisting of ordinary and partial differential equations. Allusions may be made to difference, stochastic, difference-differential, and differential-algebraic equations, but no general theory will be presented for them.

This book can be used in a number of different ways, to explain which I must speak of its history. At the beginning it was to have been a volume of selected papers on the chemical engineering systems that I have been modeling for the past forty years. This suggestion, no doubt a little pretentious on my part, was not totally unsupported by the Citation Index, nor, indeed, by the opinion of a few colleagues to whom I mentioned it. The selected papers would be divided into topics, with each section introduced by a brief paragraph of commentary. Then it occurred to me that I had learned a great deal from my mistakes and that the papers chosen should include some whose shortcomings, or even outright errors, had been instructive. The whole would have been introduced by an autobiographical essay. Though it may sound a little presumptuous, I have always appreciated the context that a little biography provides for understanding the development of a subject or the solution of particular problems. In the evolution of this book the proposed prelude has become a postlude—an autobiographical appendix of acknowledgments with asides. This is the right place for it. What is autobiography if not the acknowledgment of the people who have been influential in one's career?

A publisher is justifiably wary of a volume of *selecteana simpliciter.* In these days of diminishing book budgets, to produce a book nine-tenths of which might be found somewhere on the library shelves is not too wise. Some publishers indeed have a policy of not publishing the selected papers of X.Y.Z. until X.Y.Z. is dead—an option that I was not minded to take up. Several encouraged me to write a book on modeling in general or on the developments in an area to which I had earlier contributed. For various reasons I was unable to do this, but George Stephanopoulos made the creative suggestions I have tried to follow here. First, the text should be more explicitly didactic so that it might "provide a more tutorial exposition of the art of modeling." Second, quotation from, rather than complete reproduction of, the selected papers should be used in the text. To avoid a typographical nightmare, such extended quotations and the several papers reprinted would be reset. (Ultimately the publisher decided to reset all the reprinted papers). Third, the several parts would be linked by a system of indexes designed to give the reader immediate access to any topic.

Method and manner are the themes of the first six chapters and a score of illustrative examples are woven into this part, allowing me to refer to papers that are not suitable for reproduction in full. The matter of chemical engineering is represented in a selection of seventeen papers that constitute the second part, Chapters 7–11. It is, of course, impossible to separate these topics completely, but the first part emphasizes general principles and cites papers that illustrate these, as well as taking note of some pitfalls that beset them. The second commentary in the second part is light, being merely introductory to the different topical groupings. Because Chapters 5 and 6 show the difficulty of dividing up the subject, they might well have gone in a separate, third section. However, it was felt that they deal with questions of method. In them I get a few things off my chest that I hope will provoke my colleagues and be of some value to the tyromathematical modeler.

The reprints have been repaginated so that they can be indexed. These are designated by capital letters for purposes of internal reference and are designated A, B, and so on. The balance of the book required that the number of papers reproduced in full be strictly limited. I was particularly mortified that so many—and they so distinguished a group—of my co-authors are not represented in these reprinted papers. All are listed in the appropriate index, of course, but it would have been pleasant to have had the early papers of such an outstanding group of students represented. They are more justly treated in the scattered allusions to them in the first part of the book. Certain topics—optimization and countercurrency, for example—have been omitted completely, since to have included any one of them would have added another fifty pages to this section. Fortunately six of the most relevant general papers [145, 204, 168, 243, 253, 136] are readily accessible as appendixes to the Dover reprint of *Mathematical Modelling Techniques* [J′].

A word should be said about this reprint, which rejoices in the easy availability of the distinguished reprint house of Dover. In general, my earlier book complements, rather than competes with, the present volume. Of course, the logic of the subject is much the same, for this is dictated by its nature, but the emphases are slightly different. For instance, in the earlier book, much was made of hierarchies of models, which are mentioned here but not developed. Where the topics are the same, different examples are generally chosen, or different parts of a paper used as illustration. Frequent cross-referencing will show that this book in no sense supercedes the other. A reference to page n of *Mathematical Modelling Techniques* will be given as [J′. n], J′ being the *siglum* for MMT in the bibliography (see p. 455); throughout the text, square bracketed numbers, [m], will refer to the main bibliography of my books and papers, p. 455; all other references will be given as footnotes, since an assembled bibliography would convey an unjustified impression of completeness. This book makes no claim to comprehensiveness or to being a systematic treatise, though it has its proper order, as I have just outlined. Rather, it is concerned with the ideas that should be grasped and the tools that should be learned and used until they "curve rightly to the hand" and "build up their plain particular poetry" ([308], p. 2454).

In brief, this book is elementary and idiosyncratic; it is not necessarily easy and idiotic. It is not necessarily easy, for it presumes a basic understanding of ordinary and partial differential equations. I hope it is not idiotic, for I have tried to see beyond my own little world. It is idiosyncratic in the sense that it reflects my preferences and does not attempt to be systematic or comprehensive. It is elementary in that it deals with the elements of the craft of mathematical modeling. However powerful the modern software available for mathematical modeling may be, or become, the basic principles that I have tried to elaborate are still needed. The use of the indefinite article in the title and of "Engineer's" rather than "Engineering" is meant to indicate to the reader that this book is, in some ways, more personal than is usual in a mathematical text.

To illustrate the ways in which this book might be used I will sketch two scenarios. Suppose the reader were tackling a particular problem, say, the burning of a small particle. She might well get the mass and heat balances down but hesitate before getting them into dimensionless form. Either a direct reference to the Contents or by taking a route through the Index, she would find some direct discussion of this manipulation in Chap. 2, Examples 7 and 8, pp. 28–34. At a later stage, she might have a differential equation that could be solved by quadrature, in which case Chap. 3, Example 7, p. 50 might be useful. If her system were nonisothermal, the index to the papers of the main bibliography, A, under "reaction, gas/solid" would lead to [74], a discussion of the transition between two possible steady states during burning, and [113], which raises the question whether a sudden transition to a high rate of reaction will dig a pit in the particle and destroy all semblance of spherical symmetry. If she were to consider the effect of particle shape, the indexes, under "Shape Factor," would give access to Chapter 2, Examples 10, p. 36. If the burning were of a reservoir of solid material, then the Subject Index to the Papers in the Bibliography entry, "Stock piles, spontaneous ignition in" leads to [269].

Or again, suppose the reader were faced with a problem in optimization and were interested in getting into the aspects of mathematical modeling that might be needed. Notice how I have put it. This is a book on mathematical modeling, not a compendium of complete models, though it tries to open up many roads to various areas that I have worked in. This is not, at first blush, a very hopeful task, for he will have read in this preface that, unfortunately, optimization is not represented among the reprints. That should not stop him, however, for in the two subject indexes there are references to "optimization" aplenty. Take the Index to the Papers in the Bibliography: if it is a matter of method, the entry "optimization, methods" reads, "*see under the name of method* PD, Pontryagin, etc.," and specifically mentions Green's function approach. Since this is an index of the Bibliography, he finds that these papers [56–58] are the work of Denn (who, after the manner of the first rank of Ph.D. students, took a simple idea I had adumbrated, and soon ran ahead of me with it) and, by consulting the Index of Graduate Students and Co-authors, he learns that seven papers came from Denn's graduate work. Noticing that one of them, [50], is a derivation of the maximum principle, the reader is led by the cross-referencing to Pontryagin and so to the PMP and more papers that use that approach, notably Siebenthal's [42, 43]. The method of dynamic programming would yield a number of references and lead round to Example 12, where an error into which the unwary might fall is discovered. Consulting the Index, would also make sure that he did not fall into any confusion between "Iterative Optimization," an idea for holding to optimal conditions by correlating random fluctuations [26], and "Iterative Dynamic Programming," Luus' more recent development that has lifted the curse of dimensionality.

The Index would also refer to some anecdotal material and historical material. It is my hope that the indexes will allow the reader to move wittingly through a volume, which, though necessarily diverse, is still coherent.

I am indebted to George Stephanopoulos, an *ancien ami* and the co-editor of this series, for suggesting a format that would bring together the various needs and intentions that it seeks to fulfill. Nicholas Stephanopoulos made some useful comments from an essayist's point of view.

My colleagues at the University of Minnesota have set a standard of excellence that, over forty years, has been as stimulating as it has been exacting.

Outwith the Department of Chemical Engineering and Materials Science I have, and have had, many friends and acquaintances, two of whom I would not fail to mention: Al Moscowitz, whose memory, as John of Salisbury said of another scholar, is held "in pleasantness and blessing" by not a few; and Mischa Penn, with whom I hope to walk and talk for many years to come.

I also thank David Packer, Executive Editor, who has made the experience of publishing my last book with Academic Press as pleasant as it was when I published my first nearly forty years ago. My best thanks to Mercedes Heston, the copy editor, for her excellent work with a more than usually difficult manuscript.

I

METHOD AND MANNER

1 WHAT IS MATHEMATICAL MODELING?

Without definite examples to focus our thinking, it is easy to get entangled in a quasi-philosophical discussion of just what a mathematical model might be. To avoid this, I present a dogmatic statement on modeling and proceed to consider an elementary example, returning later to the philosophical caveats and more general considerations.

A mathematical model is a representation, in mathematical terms, of certain aspects of a nonmathematical system. The arts and crafts of mathematical modeling are exhibited in the construction of models that not only are consistent in themselves and mirror the behavior of their prototype, but also serve some exterior purpose.

A VERY SIMPLE EXAMPLE

Example 1. The Well-Stirred Tank

The simplest type of chemical reactor is a well-stirred vessel, into which a feedstock flows and out of which a product stream comes. We shall assume that the densities of the two streams and the contents of the reactor are all the same, say ρ. If the volumetric flow rates in and out are q_{in} and q_{out}, respectively, and V is the volume in the reactor, a mass balance gives

$$d(\rho V)/dt = \rho(q_{\text{in}} - q_{\text{out}}) \tag{1}$$

and this equation simplifies to

$$dV/dt = (q_{\text{in}} - q_{\text{out}}). \tag{2}$$

If, in addition, $q_{\text{in}} = q_{\text{out}} = q$, then the volume in the reactor is constant. Let us pause and consider just what assumptions we have made. The conservation of mass is a general principle of physics that holds in everyday life and only gives way to the conservation of mass–energy when atomic physics is involved. Equation (1) merely states that the rate of change of mass in the vessel is the difference between the rate at which mass flows in and the rate at which mass flows out. The assumption that the density is the same in all three terms is of a rather different character, for it says something about the constitution of the specific substances involved, not about the constitution of the everyday world. This assumption might be invalid if, for example, the reactants were gaseous and the pressure in the feed line was greater than the pressure in the reactor, which, in turn, was greater than in the product stream. In such a case we might need three different densities or prefer to work in mass flowrates.

The third assumption made is of yet another character, for it is concerned with the operation of the reactor. It is the solution of the differential Eq. (2) when the operating mode is that of the so-called steady-state in which q_{in} and q_{out} are kept at the same value q. Although the solution of the equation tells us that the volume is constant, it does not tell us what that constant is until we add an initial condition, $V(0) = V_0$; then $V(t) = V_0$. When q and V are constant, their quotient, $V/q = \theta$, is called the *residence time* and its reciprocal is called the *space velocity*.

Now suppose the feed contains a reactant A that irreversibly goes to a product B. Let c_{in} be the concentration of A in the feed and c its concentration both in the tank and the effluent (the assumption of good mixing ensures the last two will be the same), and let the reaction be of the first order so that it proceeds at a rate kc, where k is a constant. The concentrations are in moles/unit volume and the rate is in moles/(unit volume · unit time). A is not conserved as mass was, for the whole point of the reactor is to get A to become B, but we must still be able to account for it. Coming in with the feed stream there are $q_{\text{in}}c_{\text{in}}$ moles/unit time; leaving, unreacted, in the product stream there are qc moles/unit time. When the volume is V, the rate of disappearance of A by reaction is Vkc and the number of moles of A in the vessel is Vc. The rate of change of this last must be the difference between the inflow and the total rate of disappearance of A by outflow and reaction. Thus

$$(d/dt)(Vc) = q_{\text{in}}c_{\text{in}} - q_{\text{out}}\,c - Vkc \tag{3}$$

and initially, $c(0) = c_0$. If we multiply Eq. (2) by c and subtract it from Eq. (3), we have

$$V(dc/dt) = q_{\text{in}}(c_{\text{in}} - c) - Vkc. \tag{4}$$

If, in addition, V is constant, we can write this as

$$\theta(dc/dt) = c_{\text{in}} - c - \theta kc. \tag{5}$$

In deriving this equation, we have again used a universal principle (that A must be accounted for), a constitutive relation (the kinetics of the reaction), and an operating mode (constant volume). These three elements will be found in the setting up of any model. The differential equation for c has an initial value $c(0) = c_0$, and the equations are so elementary that the solution could be almost written down at sight. But this will seldom be the case, and it will pay to work the equations into the most transparent form possible.

Consider first the single equation, Eq. (5) and ask what it might be used for. We could, for example, ask what the effect of comparing the behavior for various, but still constant, values of c_{in} might be. In this case, the answer is obvious because, the equations being linear, c is proportional to c_{in}. If we take c_{in} as the characteristic concentration, then

$$u = c/c_{\text{in}} \tag{6}$$

and we have a dimensionless concentration $u(t)$ satisfying

$$\theta(du/dt) = 1 - u - \theta ku, \quad u(0) = c_0/c_{\text{in}}. \tag{7}$$

We have already seen that θ is a characteristic time, and it is evident that

$$\tau = t/\theta. \tag{8}$$

This leaves only the combination θk in the equation. This is clearly a measure of the ratio of the rate of reaction to the rate of flow, for it could be written Vkc/qc, the numerator being the rate of disappearance of A by reaction and the denominator its disappearance by outflow. Just as k is a first-order rate constant, so also can $(1/\theta)$ be regarded as a convective rate constant. The ratio is often written as

$$\text{Da} = \theta k \tag{9}$$

in honor of Gerhardt Damköhler, who was the first to isolate the important dimensionless groups in chemical reaction engineering.[1]

We now have

$$du/d\tau = 1 - u - \text{Da}\cdot u, \quad u(0) = U, \tag{10}$$

where U is the initial value of u. Before committing ourselves to this form, however, we should consider the purpose this model will serve.

If only the steady state is of interest, we set all time derivatives to zero and U becomes irrelevant. There is only one parameter, and the solution that

[1] The dimensionless groups of engineering are commonly designated by two letters, an upper-case letter followed by a lower-case letter. They are the only exceptions to the rule that a single quantity should be denoted by a single letter and precautions may have to be taken to avoid confusion with a product. It is no bad thing to have these reminders of flesh and blood and the practice avoids the typographical torture of N_{da}. In computer programming, names tend to be spelled out.

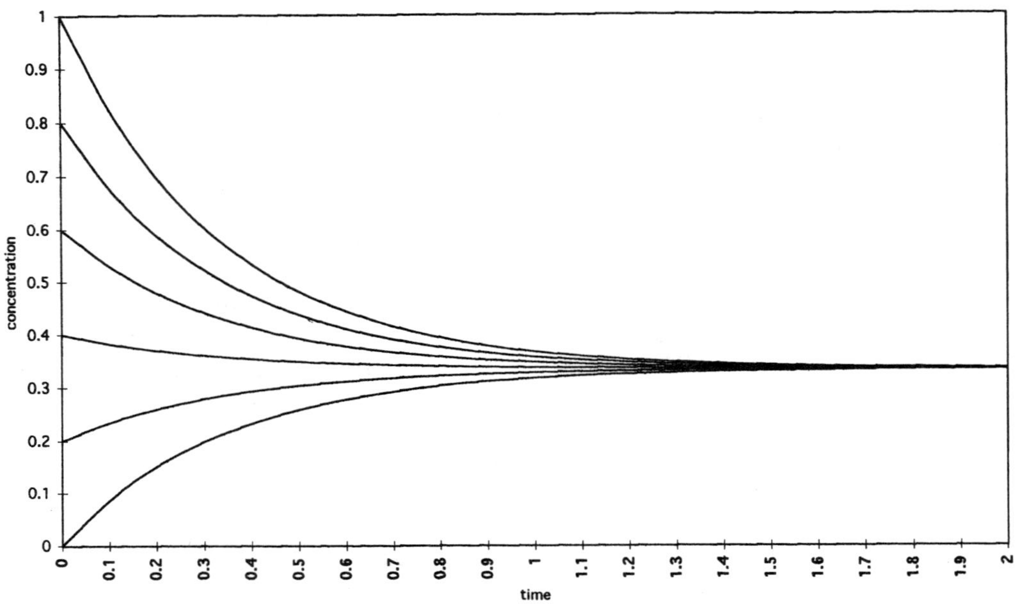

FIGURE 1 The approach to the steady state for $Da = 2$ and various initial conditions.

gives the steady-state output is immediate,

$$u_s = 1/(1 + \mathrm{Da}). \tag{11}$$

We can plot a single curve for the steady-state performance and for any Da we may choose, u_s [or $(1 - u_s)$, the conversion] is immediately calculable or can be read off the graph.

If we want to learn how the steady state is approached and how this depends on the initial value of the concentration, U, we have to solve the differential equation, which in this particular example can be done at sight:

$$u(\tau) = u_s\{1 - \exp(-(1 + \mathrm{Da})\tau)\} + U \exp(-(1 + \mathrm{Da})\tau). \tag{12}$$

A few minutes with a spreadsheet program suffices to run out a set of curves such as those in Fig. 1 drawn for Da = 2 and various U.

Whenever we have derived a formula, it is important to see that all the special cases we can think of are satisfied. Here, we see that if the initial concentration is the steady-state concentration, $U = u_s$, then the concentration stays at the steady-state value, which is just what we would expect it to do. If $U \neq u_s$, we can set

$$v = (u - u_s)/(U - u_s) \tag{13}$$

and v satisfies the even simpler equation

$$dv/d\tau = -(1 + \mathrm{Da})v, \quad v(0) = 1, \tag{14}$$

from which we have the solution almost without thinking, namely

$$v(\tau) = \exp -(1 + \mathrm{Da})\tau. \tag{15}$$

Finally, we see that time is running on a scale of, for instance,

$$(1 + \mathrm{Da})\tau = ((1/\theta) + k)t = \sigma \qquad (16)$$

and

$$v(\sigma) = \exp - \sigma \qquad (17)$$

which is the neatest and most compact form of the solution—so simple, in fact, that it would be an insult to the reader to insert a graph of it. $v = 0$ is the steady state, and v is the difference between the current state from the ultimate steady state in units of the initial value of this difference. Equation (17) shows that this transition takes place exponentially with a rate constant that is the sum of the rate constants for convection and reaction.

This is very satisfying intellectually. We feel that we have grasped the totality of the reactor's behavior. And so we have, although in an example as simple as this, the linearity plays a leading role. For some purposes, however, it may be desirable to unpack this tightly wrapped package. The first stage takes us back through Eq. (16) to Eq. (12), which, using Eq. (11), can be written

$$u(\tau) = \{1 - \exp(-(1 + \mathrm{Da})\tau) + (1 + \mathrm{Da})\, U \exp(-(1 + \mathrm{Da})\tau)\}/(1 + \mathrm{Da}). \qquad (18)$$

By a *parametric study* we mean a presentation of the way in which the performance of the system depends on the choice of a parameter. If we want to make a parametric study of the effect of variations of, say, temperature, which affects k but nothing else, this form is ideal, because k is present in Da but in no other parameter. We perform a quick computation of the conversion (%) and time to reach 99% of steady state (min), for $U = 0$, $\theta = 12$ s, and $k = 6.2 \times 10^{-7} \exp - 14000/T\ \mathrm{s}^{-1}$ (this last datum is a rate constant taken from L. D. Schmidt's *The Engineering of Chemical Reactions* (Oxford, 1997) for a reaction involving butadiene). Any spreadsheet program will immediately give a table of results such as:

Temperature (°C)	**Da**	**Conversion (%)**	**Time to 99% (min)**
500	0.169	14.5	3.94
550	0.508	33.7	3.05
600	1.345	57.4	1.96
650	3.207	76.2	1.09
700	6.992	87.5	0.58
750	14.13	93.4	0.30

However, if q is to be varied, Eq. (12) will not do, because q appears also in the dimensionless time τ. Thus k must be used to make time dimensionless, for instance, $\tau' = kt$. Da is still the Damköhler number, but the equation is now

$$du/d\tau' = (1 - u)/\mathrm{Da} - u. \qquad (19)$$

To unpack the result completely, we go back to the original variables and have

$$c(t) = c_{\mathrm{in}}\{1 - \exp(-(1/\theta + k)t)\}/(1 + k\theta) + c_0 \exp(-(1/\theta + k)t. \qquad (20)$$

REVIEW OF THE SIMPLEST EXAMPLE

Let us go over the leading points that this example has brought out.

1. We first *choose variables* sufficient to describe the situation. This choice is tentative, for we may need to omit some or recruit others at a later stage (e.g., if V is constant, it can be dismissed as a variable). In general, variables fall into two groups: independent (in our example, time) and dependent (volume and concentration) variables. The term *lumped* is applied to variables that are uniform throughout the system, as all are in our simple example because we have assumed perfect mixing. If we had wished to model imperfect mixing, we would have had either to introduce a number of different zones (each of which would then be described by lumped variables) or to introduce spatial coordinates, in which case the variables are said to be *distributed.*[2] Lumped variables lead to ordinary equations; distributed variables lead to partial differential equations.

2. When we make a balance to obtain a differential equation, we are invoking a *natural law,* the conservation of matter in our case. If the net flux of any conserved quantity into a lumped system over its boundaries is F, the rate of generation within the system is G, and the amount contained in it is H, then the balance gives

$$F + G = dH/dt. \tag{21}$$

We shall derive the equations for distributed systems from this equation later.

3. In expressing F, G, and H in terms of the dependent variables, we use certain properties of the materials involved (e.g., that the reaction is first order, so $G = Vkc$). These are sometimes called *constitutive relations* because they invoke the constitutions of the various components. They are not principles applying to everything, as are natural laws, but apply only to the materials in question.

4. We should also distinguish *operating assumptions,* which may be different even when the same materials are involved. For instance, if we were interested in finding out what goes on during the filling of the reactor, we would need both equations. However, if the reactor were operated at constant volume, only one would be needed.

5. A dimensionless variable is the ratio of that variable to a characteristic quantity of the same dimensions (e.g., $\tau = t/\theta$). A dimensionless parameter is the ratio of combinations of characteristic quantities having the same dimensions [$\text{Da} = k/(1/\theta)$]. We shall go into more detail on the question of rendering equations dimensionless later (See pp. 28–33), but two principles are of sufficient importance that they are worth reiterating.[3]

[2] The locutions *lumped parameter system* and *distributed parameter system* are enough "to frighten children in their beds" as Housman might have put it (T. Stoppard, *The Invention of Love,* in Act II, where Pollard and Housman are discussing Postgate's edition of Propertius (p. 56) and Housman says, "*Voces* in verse 33 is an emendation to frighten children in their beds"). Parameters, the quantities that go alongside the problem, cannot but be lumped, because, were they not, they would have to be treated as variables.

[3] The didactic principle of reiteration is a sort of converse to Ockham's Razor and might be called Ior's Ears: "Could I have that again in the other ear?" (Cf. [B, p. viii]).

They are:

A. Characteristic quantities used in making variables dimensionless must be capable of being held constant;

B. If a parametric study of the effect of some quantity is to be done, that quantity should appear in the numerator of one and only one of the dimensionless parameters.

THE SIMPLEST DISTRIBUTED MODEL

Example 2. The Tubular Reactor

Suppose now that the reactor we have been considering is not mixed at all; that is, each little packet of molecules is in the reactor for a time θ and does not have contact with the molecules in any other packet (see Example 2). Then we might as well think of the reactor as a straight tube of length L of cross-sectional area A through which the reactant moves with speed v. Note that $\theta = L/v$. Let $c(z, t)$ be the concentration of the reactant at a distance z from the entrance, where the feed concentration is c_{in}. We can proceed in two ways at this point: applying the general balance $F + G = dH/dt$ either to an infinitesimal slice between the planes at z and $z + dz$, or to the whole section from the entrance to z. In the first case, $F = -vA\{c(z + dz, t) - c(z, t)\} = -v(\partial c/\partial z)A\,dz$, $G = -kc(z, t)A\,dz$ and $H = c(z, t)A\,dz$. Then, applying Eq. (21) and dividing by $A\,dz$, we have

$$(\partial c/\partial t) + v(\partial c/\partial z) + kc = 0. \tag{22}$$

This is a partial differential equation, as we should expect from a plug–flow tubular reactor with a single reaction. We note in passing that the solution requires the specification of an initial distribution and a boundary, or feed, value. These are both functions (the first of z because $t = 0$; the second of t because $z = 0$) in the distributed system. Of the corresponding quantities, c_0 and c_{in}, in the lumped system, the latter is embodied in the ordinary differential equation itself and the former is the initial value.

If we want to avoid the use of infinitesimals, we may certainly do the balance over an arbitrary section of the tube, for example, $a < z < b$. Then

$$\begin{aligned} F &= vA\{c(a, t) - c(b, t)\} \\ G &= -A\int_a^b kc(z, t)\,dz \\ H &= A\int_a^b c(z, t)\,dz. \end{aligned} \tag{23}$$

However, F can be written

$$vA\int_a^b \{\partial c(z, t)/\partial z\}\,dz,$$

and so the balance can be written (using the parts in the order H, F, G)

$$A\int_a^b [\partial c(z, t)/\partial t + v\partial c(z, t)/\partial z + kc(z, t)]\,dz = 0. \tag{24}$$

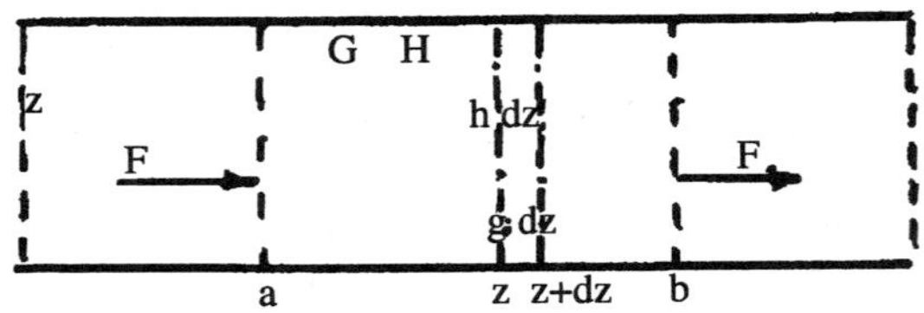

FIGURE 2 Finite and infinitesimal balances on the plug flow reactor P^*.

This is true for an arbitrary section of the reactor, and it follows that, if the integrand is continuous, the integrand must also vanish everywhere (Fig. 2). For, if it did not vanish at a point of continuity, but were, for instance, positive there, then there would have to be a finite interval in which it remained positive and, no matter how small this interval might be, a and b could be chosen within it and Eq. (24) would be violated. Thus, we arrive again at Eq. (22).

We will not take this equation further at this point, but will return to it later. It should be noticed, however, that although $\theta = L/v$ is the natural way of defining the residence time, we may want to consider a "very long" reactor, that is, one for which $L \to \infty$, and in this case it is not suitable to use it in forming the dimensionless variables.

THE GENERAL BALANCE EQUATIONS FOR DISTRIBUTED SYSTEMS

Consider a conserved quantity in a three-dimensional region Ω, with boundary $\partial\Omega$. x, y, and z are the space coordinates and t is the time coordinate. Let $h = h(x, y, z, t)$ be its concentration, that is, the amount per unit volume, so that the total amount in a subregion ω of Ω is

$$H(t) = \iiint h(x, y, z, t)\, dV. \tag{25}$$

Similarly, $g(x, y, z, t)$ is the rate of generation per unit volume, so that

$$G(t) = \iiint g(x, y, z, t)\, dV, \tag{26}$$

both integrals being taken over the region ω. The flux is defined by a vector $\mathbf{f}$ with three components, f_x, f_y, and f_z, and the flux across a plane whose normal is in the direction (l, m, n) is $lf_x + mf_y + nf_z = \mathbf{f}\cdot\mathbf{n}$. Thus the net flux into the region ω is

$$F(t) = \iint \mathbf{f}(x, y, z, t) \cdot \mathbf{n}\, dS, \tag{27}$$

where $\mathbf{n}$ is the outwardly directed normal to the surface of ω, of which dS is the surface element. This surface integral can be written as a volume integral by the application of Green's theorem $\iiint \mathbf{f}\cdot\mathbf{n}\, dS = \iiint \mathbf{\Delta}\cdot\mathbf{f}\, dV$, where $\mathbf{\Delta f} = (\partial f_x/\partial x + f_y/dy + f_z/\partial z)$. Then, bringing all terms to the same side of the equation,

$$\iiint [\partial h/\partial t + \mathbf{\Delta f} - g]\, dV = 0,$$

and the same argument can be used to show that the integrand is zero at every point of continuity. We therefore have at every point of continuity in Ω

$$\partial h/\partial t + (\partial f_x/\partial x + \partial f_y/\partial y + \partial f_z/\partial z) = g. \tag{28}$$

Of course, we have to express f, g, and h in terms of a common variable, u for example, by means of a constitutive relation for the material under study; often $u = h$.

At a surface of discontinuity in a three-dimensional space (or a line in a two-dimensional space, or a point in a one-dimensional space), the flux normal to the surface must be continuous, for the surface has no capacity to hold anything or volume to generate anything. Because there can be no accumulation in the surface, the flux up to it from one side must equal the flux away from it on the other. Thus,

$$[\mathbf{f}]\cdot\mathbf{n} = 0,$$

where

$$[\mathbf{f}] = (f_{x+} - f_{x-}, f_{y+} - f_{y-}, f_{z+} - f_{z-}). \tag{29}$$

Let us apply Eq. (28) to an important generalization of P* in which the z-coordinate is parallel to the axis of a straight tube, and x and y are coordinates in the cross-section. First, let there be no diffusion but merely a first-order reaction; then $h = c(x, y, z, t)$, $f_z = c(x, y, z, t)v(x, y)$, $f_y = f_x = 0$, $g = -kc(x, y, z, t)$, and so

$$\partial c/\partial t + v(\partial c/\partial z) + kc = 0. \tag{22 bis}$$

In the simplest P*, when v is a constant and $c(z, 0)$ and $c(0, t)$ are given, the solution of this equation is

$$\begin{aligned} c(z, t) &= c(z - t/v, 0)\cdot\exp(-kt),\ vt < z, \\ &= c(0, t - z/v)\cdot\exp(-kz/v),\ z < vt. \end{aligned} \tag{30}$$

When any quantity a varies over the cross-section, we can define an average across the tube in the usual way as

$$\langle a\rangle = (1/A)\iint a\cdot dx\cdot dy.$$

Then Eq. (22) can be averaged to give

$$\partial\langle c\rangle/\partial t + \langle v\rangle(\partial\langle c\rangle/\partial z) + \partial\langle j\rangle\partial/z + k\langle c\rangle = 0, \tag{31}$$

where

$$\langle j\rangle = (1/A)\iint c\cdot(v - \langle v\rangle)dx\cdot dy \tag{32}$$

is the net flux across a plane moving with the mean speed of flow. In the case of a flat profile, $v = \langle v\rangle$ and $\langle j\rangle = 0$. If there is a variation of flow profile or longitudinal mixing, $\langle j\rangle$ is not zero and might be called the *dispersive flux* because it acts to disperse material in the direction of the tube. If there is longitudinal diffusion and Fick's law is valid for the averaged quantities,

$\langle j \rangle = -D(\partial\langle c \rangle \partial / z)$ and we have (dropping the averaging signs, which are not needed)

$$\partial c/\partial t + v(\partial c/\partial z) + kc = D(\partial^2 c/\partial z^2). \tag{33}$$

This is the standard dispersion model for P*, and the aim of investigating more complex situations has often been to reduce them to this form with $D = D_e$, an effective dispersion coefficient that wraps up the complexities of the underlying situation in a single quantity. Whether this is wise is another matter. For example, in a packed bed the flow is obviously very complex, but both theory [4] and experiment can be invoked to justify an effective Peclet number, UR/D_e, of about 2. The question that hangs over the use of Eq. (33) is that it is a parabolic equation, with infinite signal speed and controversial boundary conditions.

We shall look at the boundary conditions for this equation in the next section. I want to mention here a very important model introduced by Westerterp and his collaborators[4] and to do so will revert to the classical form of the Taylor problem, in which there is no reaction, the tube is circular (radius R), and the flow is laminar (average velocity U). Also, we ignore the effect of molecular diffusion in the longitudinal direction and are concerned only with the effects of the lateral diffusion across the flow profile. Thus, we have

$$\partial c/\partial t + 2U(1 - r^2)(\partial c/\partial z) = D[\partial^2 c/\partial r^2 + (1/r)(\partial c/\partial r)] \tag{34}$$

or, briefly,

$$L[c] = M[c].$$

To get an approximation for the variation of c across the tube, we argue that it must be the average $\langle c \rangle$ plus a deviation c_1 and, because the Laplacian of a constant is zero,

$$L[\langle c_0 \rangle] = M[c_1]. \tag{35}$$

However, the solubility condition (see the section entitled "Observing Conditions" in this chapter for more detail) that must be imposed is easily seen to be that the average of $L[\langle c_0 \rangle]$ must be zero and

$$\partial\langle c_0 \rangle/\partial t + U(\partial\langle c_0 \rangle/\partial z) = 0, \tag{36}$$

which is just the P* equation. Had we been considering reaction, there would obviously be a term $k\langle c_0 \rangle$ in this equation. Subtracting (36) from (35) gives an equation that can be solved for c_1 and written in the form

$$c_1 = \langle c_1 \rangle - (UR^2/24D)\{2 - 6(r/R)^2 + 3(r/R)^4\}(\partial\langle c_0 \rangle/\partial z).$$

Substituting this into Eq. (32), we get Taylor's result,

$$j_1 = \langle j_1 \rangle = -D_e(\partial\langle c_0 \rangle/\partial z),\ D_e = R^2U^2/48D. \tag{37}$$

By going one more step in approximating the concentration and flux profiles, a consistent second equation is obtained, yielding the pair of first-order partial

[4] Westerterp, K. R., V. V. Dil'man, and A. E. Kronberg. Wave model for longitudinal dispersion: Development of the model. *A.I.Ch.E. J.* **41,** 2013–2028 (1995).

differential equations

$$\begin{aligned} &\partial c/\partial t + U(\partial c/\partial z) + (\partial j/\partial z) = 0 \\ &D_e(\partial c/\partial z) + (R^2/15D)\{\partial j/\partial t + (5/4)U(\partial j/\partial z)\} + j = 0. \end{aligned} \tag{38}$$

A similar treatment of the case of first-order reaction gives

$$\begin{aligned} &\partial c/\partial t + U(\partial c/\partial z) + (\partial j/\partial z) + kc = 0 \\ &D_e(\partial c/\partial z) + \tau\{\partial j/\partial t + (1 + \lambda)U(\partial j/\partial z)\} + k\tau j = 0, \end{aligned} \tag{39}$$

where τ is a relaxation time and λ the ratio of a drift velocity to the mean velocity U.

It is possible to combine the two equations in one hyperbolic second-order PDE. This has the property of finite wave speed, both boundary conditions at the entrance are easily calculable, and it accounts for some of the phenomena of unmixing. This is not the place to treat this model in detail and, indeed, it is still finding fruitful applications.[5] Another method for a hyperbolic model is to be found in [173].

BOUNDARY CONDITIONS

When Amundson taught the graduate course in mathematics for chemical engineering, he always insisted that "all boundary conditions arise from nature." He meant, I think, that a lot of simplification and imagination goes into the model itself, but the boundary conditions have to mirror the links between the system and its environment very faithfully. Thus if we have no doubt that the feed does get into the reactor, then we must have a condition that ensures this in the model. We probably do not wish to model the hydrodynamics of the entrance region, but the inlet must be an inlet. One merit of the wave model we have looked at briefly is that both boundary conditions apply to the inlet.

Example 3. The Danckwerts Boundary Conditions

That notorious pair, the Danckwerts boundary conditions for the tubular reactor, provides a good illustration of boundary conditions arising from nature. Much ink has been spilt over these, particularly the exit condition that Danckwerts based on his (perfectly correct, but intuitive) engineering insight. If we take the steady-state case of the simplest distributed example given previously but make the flux depend on dispersion as well as on convection, then, because there is only one-space dimension, $f = vAc - DA(dc/dz)$, where D is a dispersion coefficient. Then, as the assumption of steady state eliminates

[5] Westerterp, K. R., V. V. Dil'man, A. H. Benneker, and A. E. Kronberg. Wave model for longitudinal dispersion: Analysis and applications. *A.I.Ch.E. J.* **41,** 2029–2039 (1995). Westerterp, K. R., V. V. Dil'man, A. H. Benneker, and A. E. Kronberg. Wave concept in the theory of hydrodynamical dispersion—A Maxwellian type approach. *Trans. I. Ch. E.* **74,** A 944–953 (1996).

the time derivative of h and $g = -kc$, we have

$$D(d^2c/dz^2) - v(dc/dz) - kc = 0. \tag{40}$$

It is assumed that no reaction takes place outside the reactor, as would be the case if the reactor were packed with catalyst, and that the concentrations are therefore uniform outside the reactor. Thus, the inlet condition is merely the continuity of flux, which is $\{vAc_{in}\}$ just before entering and $A\{vc - D(dc/dz)\}_+$ just after. Thus, at the inlet, there is the condition

$$c(0) - (D/v)(dc(0)/dz) = c_{in}. \tag{41}$$

At the exit

$$c(L) - (D/v)(dc(L)/dz) = c_{product}, \tag{42}$$

because the product stream is uniform, and so the derivative of c vanishes outside the reactor. However, we have to ask whether there is discontinuity in c at the exit as there is at the inlet? If $c(L) = c_{product}$, then $dc(L)/dz = 0$, which is the condition Danckwerts had given quite independently of Langmuir, who had used it some years before.

The question can be answered by noting that, as the value of D goes to infinity, the tubular reactor becomes more and more completely mixed until in the limit it is a stirred tank. We should therefore be able to get the equations for the stirred tank as a limiting case. At this point, we should really work in dimensionless variables. $\zeta = z/L$ is a natural way of reducing the length and, because the residence time is L/v, the dimensionless time is $\tau = tv/L$. Note that, by comparing the two models, $\theta = V/q = L/v$, $\mathrm{Da} = k\theta$, and we need the dimensionless dispersion coefficient $\mathrm{Pe} = vL/D$. The limit we want is then $\mathrm{Pe} \to 0$. With $u(\zeta) = c(z)/c_{in}$ and $U = c_{product}/c_{in}$

$$(1/\mathrm{Pe})(d^2u/d\zeta^2) - (du/d\zeta) - \mathrm{Da}\, u = 0 \tag{43}$$

with the boundary conditions

$$u - (1/\mathrm{Pe})(du/d\zeta) = 1 \text{ at } \zeta = 0$$

and

$$u - (1/\mathrm{Pe})(du/d\zeta) = U \text{ at } \zeta = 1. \tag{44}$$

Integrating the differential equation from 0 to 1 and using the two boundary conditions, we get

$$1 - U - \mathrm{Da}\int_0^1 u(\zeta)d\zeta = 0, \tag{45}$$

which corresponds to the steady-state equation if the average value of u is the same as the product U and, hence, with the solution of the equation. But if $u(1) = U$, then, by the second boundary condition, $dc/d\zeta = 0$. This is Danckwerts' exit condition, namely

$$du(1)/d\zeta = 0. \tag{46}$$

The beauty of this argument is that it can be used for the transient case (we only gave up the time derivative in the interests of clarity), for nonlinear kinetic expressions, and for the energy equation in a nonisothermal situation. You might feel that we have only established these boundary conditions for a limiting case, but this is not so. We have established that the only set of boundary conditions that are consistent with the requirement that, in the limit of very great dispersion, the tubular system becomes a stirred tank is the Danckwerts pair.[6]

RESPECTING UNIFORMITY

There is a subtle point that must be understood in setting up the equations for systems that are both lumped and distributed. No one in their right mind would start making infinitesimal balances in lumped systems; they would treat the lumps as lumps. No mathematical modeler of any experience would do a balance over a slice containing both lumped and distributed parts, but a modeler might back into such a situation if parts of the system now lumped had, in a previous version of the model, been distributed. Such was one of the more blatant errors that I have committed in a published paper. The full story is told in my Danckwerts Lecture of 1990 ([254] and [J, Appendix 5]) and I will not retell it here. Suffice it to say that a part of the model, introduced as a distributed phase, was combined with another, also distributed, and a balance was taken over the two at once. When the first phase was subsequently assumed to be uniform, the inappropriate infinitesimal balance that was embedded there gave a wrong answer. The answer was so spectacularly wrong that it defied the second law of thermodynamics, and anyone perpetrating such an error is destined for a special place in the mathematical *purgatorio*. Fortunately, the paper was not a leading runner in the Citation Stakes and the fact that, of the several who pointed it out to me, none could lay their finger on the error gave me some comfort when I did finally find out what was wrong.

Example 4. Two-Phase Reactor, One-Phase Uniform

A simplified example is the best way to explain a two-phase reactor, one-phase uniform. Suppose we have a tubular reactor embedded in a stirred tank and communicating with it (Fig. 3 shows the arrangement). A is the cross-sectional area of both reactors ($A\varepsilon$ for the tubular reactor, P*, and $A(1 - \varepsilon)$ for the tank, C*[7]). Their common length is L, giving volumes of $A\varepsilon L$ and $A(1 - \varepsilon)L$, respectively. The area of interchange is S and k_c the exchange coefficient, so that with $C(z)$ denoting the concentration in P* at z, the amount exchanged over the next distance dz is $(Sdz/L)k_c[C(z) - c]$, where c is the concentration in C*. The first-order reaction takes place only in C*, and there

[6] That the Danckwerts boundary conditions still elicit reasearch is demonstrated by the reference: J. A. Barber, J. D. Perkins, and R. W. H. Sargent. "Boundary conditions for flow with dispersion." *Chem. Eng. Sci.* **53,** 1463–1464 (1998).

[7] C* is a convenient abbreviation for the sesquipedalian *continuous flow stirred tank reactor,* and P* might be coined for the *plug flow tubular reactor.* We might call the family that includes both T* for *tubular* (SDM for *standard dispersion model* is sometimes used).

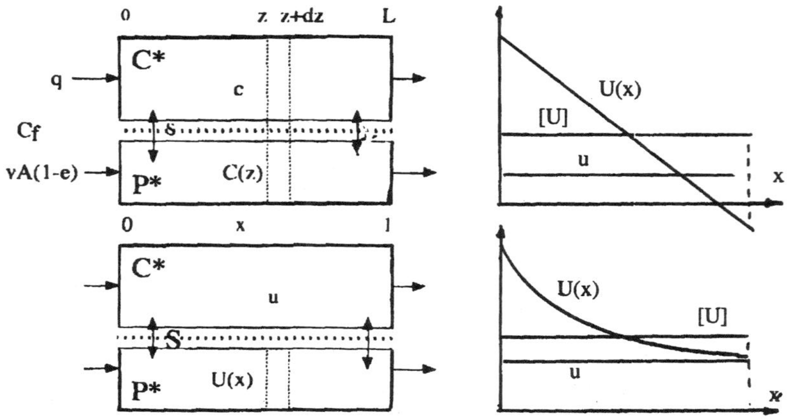

FIGURE 3 (Left) Incorrect and correct ways of making the balance when one phase is uniform. The latter is in dimensionless variables. (Right) Incorrect and correct solutions. (Top) How $U(x)$ can become negative if $B > 1$.

is no reactant in the feed to C*; it all comes through the internal contact with TP*, which it enters with concentration C_f.

We consider the steady state. When a balance is taken over both phases between z and $z + dz$, the transfer term disappears because what leaves P* turns up in C*. Thus, $\varepsilon A[vC(z + dz) - vC(z) - D(dC(z + dz)/dz - dC(z)/dz)$ is the net flux in T*; $qcdz/L$ is the convective flux engendered by the net flow through C*, of which a fraction dz/L must be ascribed to the infinitesimal slice [when integrated, this will give the convective term $q(0)$, 0 being the feed concentration to the C*]; the term $A(1 - \varepsilon)kc$ is the rate of reaction. Thus the balance over the two phases gives

$$\varepsilon A(DC'' - vC') - (q/L)c = (1 - \varepsilon)Akc. \tag{47}$$

The balance over the lumped, or uniform, phase recognizes that there is only one source of the reacting species, namely the transfer from the distributed phase, whereas the chemical disappears by reaction and by washout:

$$(Sk_c/L) \int [C(z) - c]dz = (1 - \varepsilon)Vkc + qc. \tag{48}$$

These equations can be made dimensionless by setting

$$\begin{gathered} U = C/C_f, u = c/C_f, x = z/L, m = Sk_c/L, n = (1 - \varepsilon)Vk/q, \\ P = vL/D, b = qL/V\varepsilon v, [U] = \int_0^1 U(x)dx. \end{gathered} \tag{49}$$

Then the balance over the two phases gives

$$(1/P)U'' - U' = b(1 + n)u \tag{50}$$

and this is subject to

$$U(0) - (1/P)U'(0) = 1 \text{ and } (1/P)U'(1) = 0, \tag{51}$$

whereas the balance over C* gives

$$m[U] = (1 + m + n)u. \tag{52}$$

Because the problem shows up even in the limiting case of no dispersion in T*, i.e., $P \to \infty$, let us simplify further and for U write

$$U' = -b(1 + n)u, \quad U(0) = 1. \tag{53}$$

Then U is clearly a straight line with negative slope, and

$$U(x) = 1 - \{bm(1 + n)/(1 + m + n)\}[U]x \tag{54}$$

$$[U] = 1/\{1 + \}bm(1 + n)/2(1 + m + n)\}, \tag{55}$$

from which we can extract the exit T* concentration

$$U(1) = \{1 - \mathrm{B}\}/\{1 + B\}, \tag{56}$$

where

$$B = \{bm(1 + n)/2(1 + m + n)\}. \tag{57}$$

Now there is nothing to stop B from taking on any positive value whatsoever, yet when $B > 1$, $U(1) < 0$, which is impossible because it is a concentration. Figure 3 shows the three cases $B < 1$, $B = 1$, and $B > 1$. If Eq. (50) had been used, a more complicated solution would have been obtained, but the same difficulty would have been encountered.

The error—the one that I backed into in [28] and corrected in [254]—lies in using a balance that cuts across both phases, Eq. (47), when one is uniform and the other distributed. The balance over the lumped phase, Eq. (38), stands, but we must do the other balance over P* only. This gives

$$\varepsilon A\{DC'' - vC'\} = (Sk_c/L)\,\{C - c\}. \tag{58}$$

Then with the same dimensionless variables as before

$$(1/P)U'' - U' = bm(U - u) \tag{59}$$

$$m[U] = (1 + m + n)u \tag{60}$$

subject to

$$-U'(0)/P + U(0) = 1 \quad \text{and} \quad U'(1) = 0. \tag{61}$$

These equations can be solved easily enough, but again, the point is made by the limiting case of no dispersion. We have immediately

$$U(x) = u + (1 - u)\cdot\exp - bmx, \quad u = m[U]/(1 + m + n). \tag{62}$$

Then, if we calculate the mean $[U]$ from the first of these equations, we have two equations for $[U]$ and u, giving

$$[U] = (1 + m + n)E/(1 + n + mE), \quad u = mE/(1 + n + mE), \tag{63}$$

where E is an abbreviation for $\{1 - \exp(-bm)\}/bm$. To the solution

$$U(x) = \{(1 + n)\exp(-bmx) + mE\}/\{1 + n + mE) \tag{64}$$

no taint of negativity attaches.

The moral of all this seems to be that, if any parts of the system are assumed to be uniform, the balances must be made over the whole of that part, and one must beware of backing into such an assumption *after* making a distributed balance.

EXTENSIVE AND INTENSIVE QUANTITIES

It is of the utmost importance to recognize that balances can only be made on extensive variables or quantities. If you double the system, an extensive measure is doubled, whereas an intensive measure remains the same. Thus the mass of two identical bricks is twice the mass of one, but the density, or mass per unit volume, remains the same in the duplicated system because we have doubled both the mass and the volume. The first (mass) behaves like a homogeneous function of degree one, the second (volume) of degree zero. Thus in the simple example used in Example 1, we did not make our balance on the concentration, moles per unit volume $= c$, but on the amount, moles $= Vc$.

A function, $f(x_1, x_2, \ldots x_n)$, of n variables is called *homogeneous of degree m* if $f(\lambda x_1, \lambda x_2, \ldots \lambda x_n) = \lambda^m f(x_1, x_2, \ldots x_n)$. Euler's theorem states that $\Sigma x_i(\partial f/\partial x_i) = nf(x_1, x_2, \ldots x_n)$, as can be easily seen by differentiating the definition and putting $\lambda = 1$.

Example 5. The Nonisothermal Stirred Tank

The *locus classicus* of this important principle for chemical engineers is the nonisothermal stirred tank in which a single reaction takes place (for multiple reactions, see pp. 16–17). Consider the reaction

$$\sum_1^S \alpha_j A_j = 0 \tag{65}$$

between the chemical species $A_1, A_2, \ldots, A_S$, where by convention the products of the reaction have positive stoicheiometric coefficients and the reactants negative; inerts can be included by assigning them the stoichiometric coefficient zero. If this takes place in a stirred tank of constant volume, $V = q\theta$ [as in Eq. (5)], the state can be characterized by the S concentrations c_j (moles per unit volume) and the temperature T. Let $r = r(c_j, T)$ be the reaction rate in moles per unit volume per unit time, then for A_j, in the notation we have been using,

$$F = q(c_{jf} - c_j),\ G = V\alpha_j r(c_j, T),\ H = Vc_j.$$

Hence,

$$(d/dt)(Vc_j) = q(c_{jf} - c_j) + \alpha_j Vr(c_j, T), \tag{66}$$

or, dividing through by the constant V,

$$\theta(d/dt)(c_j) = (c_{jf} - c_j) + \alpha_j \theta r(c_j, T). \tag{67}$$

c_{jf} is, of course, the feed concentration of A_j.

We will leave the further reduction of these S equations until later and turn to the equation for temperature. Again, it is important to realize that an equation for T cannot be written down immediately, but must be derived from the balance of some form of energy, here the enthalpy. Let $h_j(T)$ be the enthalpy per mole at temperature T (we will ignore the dependence on c for the moment), so that the total enthalpy of the reacting mixture is $H = V\Sigma_1^S$

c_jh_j, and the net flux $F = q\Sigma_1^S\{c_{jf}h_j(T_f) - c_jh_j(T)\}$. As for the generation of enthalpy, it is represented only by the external heating or cooling. *One does not put in a term for the rate of generation of heat by reaction;* that falls out naturally from the rearrangement of the enthalpy. Let $G = Q(T)$; then

$$V(d/dt)\left(\sum_1^S c_jh_j\right) = q\sum_1^S \{c_{jf}h_j(T_f) - c_jh_j(T)\} + Q(T) \tag{68}$$

or, dividing by q and expanding the derivative of the product.

$$\begin{aligned}\theta\sum_1^S (dc_j/dt)h_j(T) + \theta\left\{\sum_1^S c_j(dh_j(T)/dT)\right\}(dT/dt) \\ = \sum_1^S \{c_{jf}h_j(T_f) - c_jh_j(T)\} + Q(T)/q.\end{aligned} \tag{69}$$

Now $(dh_j(T)/dT) = c_{pj}$ is the heat capacity per mole of A_j, and the second term on the left-hand side of this equation could be written $\theta C_p(dT/dt)$, where C_p is the total heat capacity per unit volume. Also, by multiplying each of the equations in (57) by the corresponding h_j and subtracting from (59), we have

$$\theta C_p(dT/dt) = \sum_1^S \{h_j(T_f) - h_j(T)\} + \theta\sum_1^S \{\alpha_jh_j\}r + Q(T)/q. \tag{70}$$

Now $\Sigma_1^S\{a_jh_j\}$ is the heat of reaction ΔH because it is the difference between the enthalpy of the products and that of the reagents. For the sake of simplicity, let us assume that the heat capacity is constant so that the first term on the right is $C_p(T_f - T)$. Substituting and dividing through by the heat capacity C_p gives

$$\theta(dT/dt) = T_f - T - \theta(-\Delta H)r(c, T)/C_p + Q(T)/qC_p. \tag{71}$$

We should ask what difference it would make if the dependence of h_j on the concentrations were taken into account. Quite obviously, it would have given us a rather messy term in addition to the two on the left-hand side of Eq. (59),

$$\theta\sum_j\sum_k c_j\{(\partial h_j/\partial c_k)(dc_k/dt)\} = \theta\sum_k\left\{\sum_j c_j\,(\partial h_j/\partial c_k)\right\}(dc_k/dt).$$

Now the summation on j can be written $\Sigma_j\, c_j\,(\partial h_k/c_j)$, because

$$h_j = \partial h/\partial c_j$$

and

$$\partial h_j/\partial c_k = \partial^2 h/\partial c_j\partial c_k = \partial h_k/\partial c_j.$$

But h_k is an intensive variable, i.e., a homogeneous function of degree zero, and this implies (by a well-known result attributed to Euler, quoted previously) that

$$\sum_j c_j(\partial h_k/\partial c_j) = 0. \tag{72}$$

Thus, a potentially messy term vanishes.

GENERAL OBSERVATIONS ON FORMING THE MODEL

We have now had sufficient acquaintance with actual, though very simple, mathematical models so that we can safely return to the definition given at the beginning. It is important to recognize that, although the term *mathematical modeling* has gained considerable currency in recent years, the practice of mathematical modeling has been going on as long as mathematics itself under the guise of applied mathematics (which, in some quarters, has been used pejoratively) and natural philosophy (which avoids the danger of being pretentious as long as it observes its etymology). Penn and I tried to tease out "the mere notion of a model" in a paper in the first issue of the journal *Mathematical Modelling* (now *Mathematical and Computer Modelling*). We felt that the notion of craftsmanship was a useful one as providing a *via media* between the extremes of abstract formalism and pure subjectivism [168 = J, App. 3] at once doing justice to the modeler's mental activity and providing a basis for discussion. Thus a difference in the choice of assumptions generally leads to different, although related, models, and, in the definition given at the beginning of the chapter, the claim is only that *certain aspects* of the non-mathematical situation are being represented. In this, as in many other things, economy, and sometimes even minimalism, is accounted a virtue. The classic statement of this is in Newton's *Principia*, where, in his Rules for Reasoning in Philosophy, he says: "We are to admit no more causes of natural things than such as are both true and sufficient to explain their appearances. To this purpose the philosophers say that Nature does nothing in vain, and more is in vain when less will serve; for Nature is pleased with simplicity and affects not the pomp of superfluous causes."

In several papers I have advanced the notion that there is an empathy between the poet and the mathematical modeler [K′, Ch. 2; 243 = J′, App. 4; 254 = J′, App. 5; 308 = Q]. The tag *Ut pictura poesis* (which, in Dryden's translation of Du Fresnoy's *De Arte Graphica,* becomes *Painting and poesy are two sisters*) has long been a peg on which to hang some theory of literary or artistic criticism, and it would be inappropriate to build a massive critical structure on a twist of it; as the poem, so the mathematical model. Nevertheless there is the shared knowledge of craftsmen who "have held reality down fluttering to a bench,"[8] the common search for the most telling imagery,[9] the same working, "lonely with their tools,"[10] "in the still night,"[11] and the same discovery that "to make an end is to make a beginning."[12] I will not dilate on this further, as the last paper reproduced in this book [308*] goes over

[8] V. Sackville-West. "The Land, Summer," in *Collected Poems.* London: Hogarth Press, 1933.

[9] A spontaneous lecture on the analogy between imagery and the dimensionless parameter is something my friends have learned to endure with a grace for which I am thankful. Because it is to be found in the papers here reproduced [277 = P, 308 = Q], as well as in others [J′, K′], it will only be mentioned in passing.

[10] A few lines later in the poem referred to in footnote 8.

[11] D. Thomas, In *My Craft and Sullen Art.* New York: New Directions, 1957.

[12] T. S. Eliot. "Little Gidding," in *Four Quartets* (New York, 1943). The relationship of ends and beginnings is a recurrent theme in *Four Quartets.*

this ground fairly thoroughly and may be augmented by the fourth and fifth appendixes of J.

Although the model may be worked on in isolation and, as a system of mathematical equations, has a life of its own, it is at all stages part of an ongoing analysis of some other entity. This is sometimes represented as a commutative diagram in which the state of the nonmathematical (let us call it *physical,* for definiteness) system, S, is mirrored in the model, Σ, by the modeling relationship h. Changes in the physical, say, $S_1 \rightarrow S_2 = gS_1$, are mirrored in the model as $\Sigma_1 \rightarrow \Sigma_2 = \gamma\Sigma_1$. The model is satisfactory if the diagram

$$\begin{array}{ccc} S_1 & \xrightarrow{\quad g \quad} & S_2 \\ \Big\downarrow h & & \Big\downarrow h \\ \Sigma_1 & \xrightarrow{\quad \gamma \quad} & \Sigma_2 \end{array}$$

commutes, that is, $hg = \gamma h$.

One can also think of modeling as an ongoing process of understanding in which the preliminary ideas are brought together in a first model. The consequences of this model are worked out and compared with experience. I have used the word *experience* as a broader term than *experiment*, although it often takes that form. I take an experiment to be in some way tailor-made for the situation, and so capable of giving a more precise answer to our questions than the happenstance of undirected experience, although the existence of natural experiments must be recognized. If the agreement is adequate, this first model may be modified and the cycle of formulation, manipulation, solution, and comparison begun again. If these iterations lead to a closer agreement with a particular system or physical situation, we shall follow Maynard Smith[13] in calling it a *simulation.* The word *model* will be used for the product of successive iterations aimed at a getting a broad view of the situation and results that apply to a whole class of objects rather than to one particular object. Of course, the boundary between the two endeavors cannot be drawn rigidly, but the general intents are usefully distinguished. It is easy to see that dimensionless quantities are the natural constituents of the model, whereas it might be quite appropriate to use quantities with dimensions in a simulation. Using dimensional quantities has the advantage that, by checking the consistency of dimensions, one can avoid lapses of algebra, although we shall see that dimensionless numbers have ghostly dimensions that may also serve this end.

In my preliminary definition, I mentioned the exterior purpose, and the use of the words *model* and *simulation* speaks to the respective purposes of conceptual understanding and actual approximation. Different models may be appropriate to the same purpose, and function best in different regions of parameter space. [See J (pp. 18–22) for a further discussion of this point.]

[13] J. M. Smith. *Models in Ecology.* Cambridge: Cambridge Univ. Press. Cambridge, U.K., 1974.

The subject of the value of mathematical modeling in nonscientific disciplines is a broad one, about which we cannot go into detail here. In the social sciences they have to be carefully formulated to avoid being self-fulfilling and are obnoxious to inputs of widely variable accuracy. Statistical models, a subclass of mathematical models, abound. They often have underlying models that serve to give correlations, sensitivities, and the like, but often fails to give any insight into the mechanisms involved. Ancillary questions in humanistic disciplines can often be illuminated by some elementary models; but, not surprisingly, they do not lie at the core of these subjects. In the study of military history, Bachrach has done some interesting things with logistical models and we have worked together on the defense strategy that might have been used by archers [250] and the use of the battering ram [281]. In the latter problem, we regarded the ram as a bifilar pendulum suspended from an A-frame on wheels and asked what regime of battering would deliver the greatest energy to the impinging ram. The initial run-up should stop short of the wall while the operators "pumped" it up and, with the last swing, run the whole frame forward so that the ram would strike when in the lowest position. It is not suggested that medieval soldiers could work this out in detail. However, they might come close to such a *modus operandi* by trial and error. Moreover, the mathematical model is useful in setting the limits of what is possible, a valuable tool in assessing the reliability of some medieval records.

Example 2. The Plug-Flow Tubular Reactor (Reprise)

We have seen that the basic P* model has the form of a first-order partial differential Eq. (22) describing each narrow slice as a little batch reactor being transported through the reactor at constant speed. This equation was so elementary that it could be solved at sight in Eq. (30). When we added a longitudinal dispersion term governed by Fick's law and took the steady state, Eq. (40), we had a second-order o.d.e. with controversial boundary conditions. This is the model with $u(\zeta) = c(z)/c_{\text{in}}$ and $\text{Pe} = vL/D$, $\text{Da} = kL/v$,

$$(1/\text{Pe})(d^2u/d\zeta^2) - (du/d\zeta) - \text{Da}\, u = 0, \qquad \text{(43 bis)}$$

with the boundary conditions

$$u - (1/\text{Pe})(du/d\zeta) = 1 \text{ at } \zeta = 0$$

and

$$(1/\text{Pe})(du/d\zeta) = 0 \text{ at } \zeta = 1. \qquad \text{(44 bis)}$$

We have justified the second set of boundary conditions by the fact that when $\text{Pe} \to 0$ (i.e., the dispersion is so large that the reactor is uniform, like C*), the C* equation is recaptured. If we put $\gamma = 2/(2 + \text{Pe})$ the solution of Eq. (43) can be written

$$\begin{aligned} u(\text{Da}, \gamma) &= 2/\{(\sigma + 1)\exp(\nu) - (\sigma - 1)\exp(-\mu)\} \\ \sigma &= [1 + \text{Da}\,\gamma/(1-\gamma)]/[1 + 2\text{Da}\,\gamma/(1-\gamma)]^{1/2}, \\ \mu &= (1-\gamma)\{[1 + 2\text{Da}\,\gamma/(1-\gamma)]^{1/2} + 1\}/\gamma, \\ \nu &= (1-\gamma)\{[1 + 2\text{Da}\,\gamma/(1-\gamma)]^{1/2} - 1\}/\gamma, \end{aligned} \qquad \text{(T*)}$$

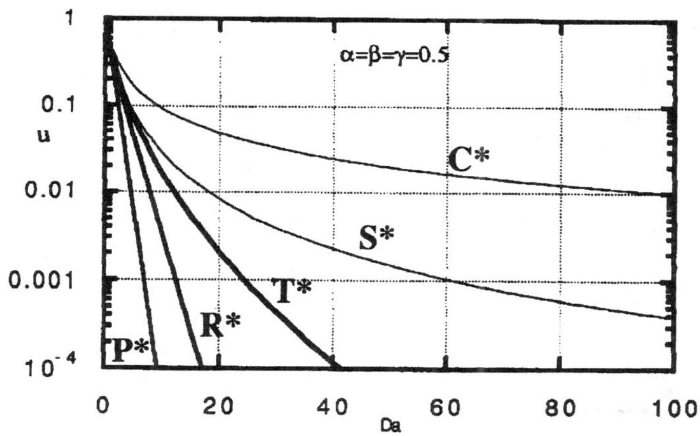

FIGURE 4 Exit concentrations as functions of *Da* for five reactors.

and as γ passes from 0 to 1, the exit concentration goes from exp(−Da) for P* to 1/(1 + Da) for C*. The family of reactors with finite dispersion is thus a homotopy between the two ideal types, P* and C*.[14]

There are other ways to devise homotopic families [270], the simplest being to divide the C* into a series (S*) of N equal reactors and use them in series. Because $u_{n-1} - u_n = (\mathrm{Da}/N)u_n$, we can write

$$u(\mathrm{Da}, \beta) = \{1 + \beta \mathrm{Da}\}^{-\beta}. \qquad \text{(S*)}$$

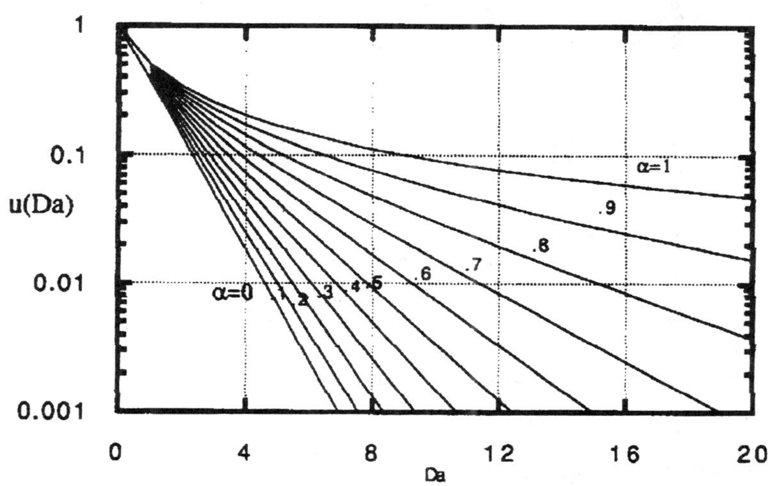

FIGURE 5 Exit concentrations as functions of *Da* for the homotopic family *R**.

[14] V. Balakotaiah has pointed out that $\Gamma = (2/\mathrm{Pe}) - (2/\mathrm{Pe}^2)(1 - \exp(-\mathrm{Pe}))$ would be a better homotopic parameter. It cannot be inverted, so $\mathrm{Pe}(\Gamma)$ would have to be used parametrically.

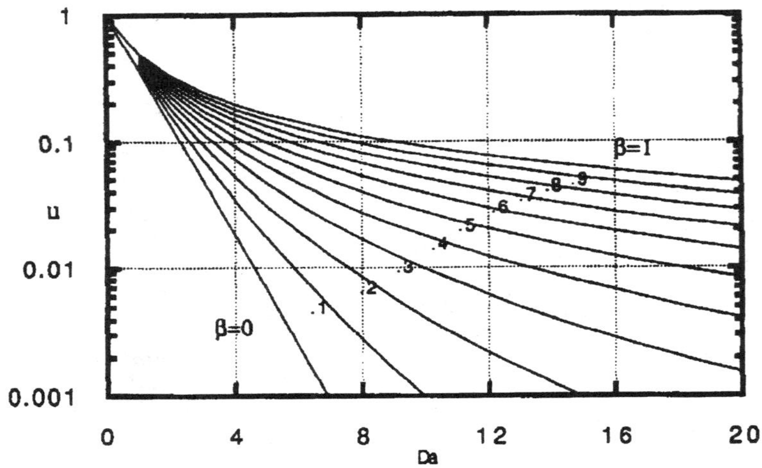

FIGURE 6 Exit concentrations as functions of *Da* for the homotopic family *S**.

A third way is to recycle (R*) some of the product of a P* back through its feed, because no recycle is P* and infinite recirculation is C*. In this case

$$u(\text{Da}, \alpha) = (1 - \alpha)\{\exp - (1 - \alpha)\text{Da}\{/[1 - \alpha\{\exp - (1 - \alpha)\text{Da}\}], (\text{R}^*)$$

where α is the ratio of the recycle rate to the total flow rate of feed and recycle. The curves of exit concentration versus Damköhler number for $\alpha = \beta = \gamma = 0.5$ are given in Fig. 4, and the three families R*, S*, and T* in Figs. 5–7.

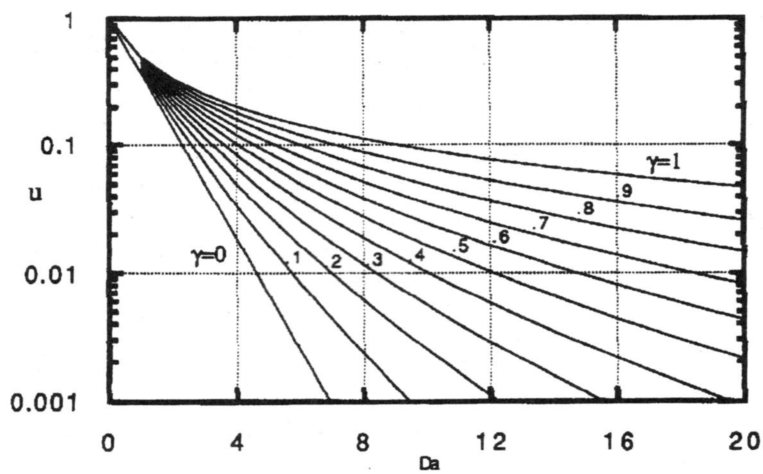

FIGURE 7 Exit concentrations as functions of *Da* for the homotopic family *T**.

2 MANIPULATION OF MODELS

The previous section was concerned with the process of getting the differential equations and their boundary and initial conditions down on paper. These equations are not usually in their simplest form when they are first written down, and it pays to spend some time getting familiar with them and reducing them to the simplest form. Of course, one can go overboard in this respect also, and achieve so tight a wrap-up that the model can only utter platitudes of an oracular type. Such is the case in Example 1, in which Eq. (17) reduces the whole system to a simple exponential. We are not surprised by this, because it is a first-order system, but we have to unwrap it partially before we can get much use out of it. In this section, we consider ways of reducing the number of equations in the model to the useful minimum, by partial solution or algebraic considerations, and the ways of best reducing them to dimensionless form.

Some of the considerations of this section have been discussed by the late Gianni Astarita, whose death in 1997 cut short a tremendously productive career. In a paper dedicated to M. M. Sharma on his sixtieth birthday,[1] Astarita points out that Euclid recognized the need for dimensional homogeneity and Ptolemy wrote a book entitled *On Dimension,* and that their overly strict rules of manipulation were not relaxed until Newton's time. The modern

[1] G. Astarita. Dimensional analysis, scaling, and orders of magnitude. *Chem. Eng. Sci.* **53,** 1461–1469 (1998).

phase of the history of dimensional analysis begins with Rayleigh's *Theory of Sound* and that it is intertwined the development of scale-up in design and the recognition of order of magnitude.[2] Indeed, Astarita claims that "the three techniques of dimensional analysis, appropriate scaling, and intelligent estimation of orders of magnitude, if used jointly, constitute an extremely powerful tool for the chemical engineer. Very often more than nine-tenths of what we can ever hope to know about a problem can be obtained from this tool, without ever solving the problem; the remaining tenth requires painstaking algebra and/or lots of computer time, it adds very little to our understanding of the problem, and if we haven't done the first part right, all that the algebra and the computer will produce will be a lot of nonsense. Of course, when nonsense comes out of a computer people have a lot of respect for it, and that's exactly the problem." This is why the principles of mathematical modeling need to be understood apart from the devices that can greatly ease the way that has to be taken to get the final result. The same engines can ease the way to a slippery slope that leads to totally false results. I have sometimes tried to dramatize this by saying, "Don't start computing until you know the answer," which is, of course, not true if taken too literally, but echoes Astarita's sentiment. This is not to say that the desktop and main-line computers do not have their uses, even at this stage. Such features as the graphing calculator in the Apple computer menu allow one to see the shape of an expression on dull days. They are invaluable as means of checking one's estimates, but it takes half the joy of mathematical modeling away if one resorts to them too readily.

We discuss here various aspects of the manipulation of models, not attempting to be comprehensive, but introducing further examples of systems of chemical engineering significance.

GETTING RID OF UNNECESSARY EQUATIONS

Example 6. Multiple Reactions in a C*

In deriving the equations for the concentrations in a stirred tank we made balances on each of the components and obtained

$$\theta(d/dt)(c_j) = (c_{jf} - c_j) + \alpha_j \theta r(c_j, T), \quad c_j(0) = c_{j0}. \qquad \text{(67 bis)}$$

Here, c_j is the concentration of A_j. A participant in the reaction whose stoichiometry is $\Sigma\alpha_j A_j = 0$. Now it seems a little unnecessary that we should have S equations when there is only one reaction, and one feels instinctively that only one equation is necessary for one reaction. This indeed is the case, for, if we put $c_j = c_{jf} + \alpha_j \xi$, we get

$$\theta(d\xi/dt) = -\xi + r(\xi, T). \qquad (73)$$

[2] I have mentioned the outstanding grasp of order of magnitude on the part of people I have worked with: Bosanquet (p. 426), Ranz and Cussler (p. 452), and Astarita's name should be added to the list.

If the initial condition can be expressed as

$$c_{j0} = c_{jf} + \alpha_j \xi_0, \tag{74}$$

all is well, for the differential Eq. (63) has the initial condition $\xi = \xi_0$ and the solution can be calculated. But, if the initial and feed compositions are incompatible[3] (in the sense that no ξ_0 can be found to satisfy (74), then we must introduce another variable ζ such that

$$c_j = c_{jf} + \alpha_j \xi + (c_{j0} - c_{jf})\zeta, \tag{75}$$

for then $\xi = 0$ and $\zeta = 1$ will correspond to the initial conditions. On substituting this into Eq. (57) and dividing through by α_j, we have

$$\{\theta(d\xi/dt) + \xi - \theta r(\xi, T)\} + [(c_{j0} - c_{jf})/\alpha_j]\{\theta(d\zeta/dt) + \zeta\} = 0. \tag{76}$$

Now not all the factors $[(c_{j0} - c_{jf})/\alpha_j]$ are equal, for otherwise the initial and feed compositions would be compatible. It follows that the only way in which these equations can all be zero is for the two curly brackets to be zero. The first is the equation we had before for the extent of reaction ξ, whereas the second tells us that the incompatibility, ξ, washes out of the reactor as any nonreacting chemical does, $\zeta = \exp(-t/\theta)$. Thus, nothing is added to our understanding of the reactor's behavior by considering the incompatibility in this model. Of course, if this were a simulation, the case might be quite different.

A similar proof goes through if there are R independent chemical reactions taking place. If the suffix i denotes a particular reaction $\Sigma \alpha_{ij} A_j = 0$,

$$c_j = c_{jf} + \Sigma \alpha_{ij} \xi_i + (c_{j0} - c_{jf})\zeta,$$

substituted into Eq. (57) gives an equation like (66). When the independence of the reactions and the incompatibility of the initial state are invoked, we have R equations for the several extents

$$\theta(d\xi_i/dt) = -\xi_i + \theta r_i(\xi_1, \xi_2 \ldots \xi_P, T), i = 1, 2, \ldots R \tag{77}$$

and the same equation for ζ.

Examples of the reduction of the number of equations abound. With adiabatic reactors in the steady state, the temperature is a function of the composition (see [A, D, F] and the introductions to papers under "Adiabatic" in the Index of Subjects in Publications.) This is the case also with the catalyst particle (see I, Ch. 2, and [81, 122] and Example 9). Examples of reduction in the number of equations because one or more can be solved immediately are given in Example 12; see also [280] and [310].

A more sophisticated form of reduction is obtained when the so-called center manifold theorem is invoked. This says essentially that a subspace of lower dimension than the whole state-space gives a true representation of the essential features of the system and one that can be built on to give a yet more accurate picture. We shall not attempt to go into this here; to see the method in action, the reader cannot do better than to read C. Chang and

[3] I well remember R. A. Schmitz, as a graduate student taking my course in kinetics, pointing this out to me. I had overlooked the possibility in my early work on the algebra of stoichiometry.

V. Balakotaiah's treatment of "Dispersion in Chemical Solutes in Chromatographs and Reactors"[4]; Carr's *Applications of Center Manifold Theory*[5] is the standard text.

THE REDUCTION OF THE EQUATIONS TO DIMENSIONLESS FORM[6]

From our earliest example, we saw that it was advantageous to use dimensionless variables and that the characteristic quantities should be capable of being held constant. In addition, if a parametric study on the effect of varying some input quantity is to be performed, that quantity should appear in only the distinguished parameter. This is no restriction, for the others are proportional to powers of the distinguished parameter, and the proportionality constants are themselves dimensionless numbers. For example, if the viscosity is to be varied, the Reynolds and the Schmidt numbers are both functions of ν, but ReSc is not; so, if Sc is chosen as the dimensionless viscosity, Re $= C/\text{Sc}$, where $C = \text{ReSc}$ is independent of ν.

To be effective, the reduction to dimensionless form must be complete. Going halfway will neither add to the understanding of the physical situation nor make the equations any more tractable than they are to begin with.[7]

Another broad principle may be mentioned. It is a good idea in most cases to try to make the dimensionless dependent variables be of order of magnitude 1. Sometimes this is best done by using boundary or initial values, as in the following example.

Example 7. The Dissolving Sphere

Consider the case of a solid sphere falling through a stagnant fluid, in which the sphere is soluble. This is a problem that could be solved on the computer, taking into consideration the change of shape of the sphere due to the differences of mass transfer at different locations. Even if that were the objective, we should do a modeling study before the more detailed analysis and simulation. We assume that the sphere falls under gravity and attains the Stokes velocity at all times; we shall return to examine this assumption later. Thus its downward velocity is

$$u = 2gr^2\,\Delta/9\nu,\ \Delta = (\rho_s - \rho_l)/\rho_l, \tag{78}$$

where r is its radius, ν the viscosity of the fluid, and Δ the density difference of the solid and liquid expressed as a fraction of the liquid density. Because

[4] *Phil. Trans. Roy. Soc.* **A351,** 1–37 (1995).

[5] Springer-Verlag, 1981.

[6] *Dedimensionalization* is an ugly word, fit only to be used in Washington. I am sure I have fallen into using it occasionally, but repent in favor of this title in spite of its length. Where there is no possibility of confusion, I will merely use *reduction.*

[7] I learned this the hard way when teaching a course in the old Humble Lectures at Baytown. When I was dealing with the nonisothermal stirred tank, I had one equation for the concentration of the reactant and one for the temperature. To compare them, I expressed both in moles per unit volume. The notion of a temperature in moles per unit volume was not acceptable to the class.

the radius and the velocity both decrease from their initial values, $x = r/R$ and $v = u/U$, where $U = 2gR^2\,\Delta/9\nu$ is the initial velocity. To achieve this, we would have to drop it from a height of $U^2/2g$ and ignore the splash. The Reynolds number is

$$\mathrm{Re} = 2gr^3\,\Delta/9\nu^2 = P(r/R)^3, \tag{79}$$

where P is the initial Reynolds number. (Note that we avoid using RE, which would be a natural for the initial value of Re, as we have an R and this could easily become detached, lacking, as it does, the ligaturing effect of the upper and lower cases in Re, etc. Lowercase ρ is in use for densities, so an uppercase P(rho) will be used). Let

$$x = r/R,\ v = u/U = x^2,\ U = 2gR^2\,\Delta/9\nu,\ \mathrm{Re} = Px^3. \tag{80}$$

The number of moles of the solid chemical in the sphere is $N = (4/3)\pi r^3\rho_s/m$, and this is diminished by the mass transfer from the falling sphere into the stagnant fluid. At the surface of the sphere the concentration of the solute is S, the saturation solubility, and far from the sphere it is zero. Thus,

$$dN/dt = (4\pi r^2\rho_s/m)(dr/dt) = -k_c 4\pi r^2(S - 0). \tag{81}$$

Let $\Sigma = Sm/\rho_s$, so that

$$dr/dt = -\Sigma k_c. \tag{82}$$

We will use the empirical formula for the mass transfer coefficient (checking later on the range of its validity[8])

$$k_c = (D/r)\{1 + 0.672\ \mathrm{Re}^{1/2}\mathrm{Sc}^{1/3}\} \tag{83}$$

and write it in terms of K_c, its initial value multiplied by a dimensionless function of x,

$$k_c/K_c = \{1 + ax^b\}/x(1 + a) = \kappa(x),\ K_c = (D/R)\{1 + 0.672\,P^{1/2}\,\mathrm{Sc}^{1/3}\}, \tag{84}$$

$$a = 0.672\,P^{1/2}\,\mathrm{Sc}^{1/3},\quad b = 3/2. \tag{85}$$

We now have the single equation

$$dx/dt = -(\Sigma K_c/R)\kappa(x), \tag{86}$$

and we see that the natural dimensionless time is $R/\Sigma K_c$. We therefore set

$$\tau = \Sigma K_c t/R, \tag{87}$$

making Eq. (76)

$$dx/d\tau = -\kappa(x). \tag{88}$$

[8] An alternative form is $k_c = (D/r)\{1 + 0.266\ (\mathrm{Re}(\nu/D)^{1/2})^{0.62}\}$, which differs only in its constants. Both reek of empiricism, but there is no doubt as to their serviceability when used in the right way. In the engineering literature, Re and k_c are commonly based on diameter rather than radius, and forms such as the above may differ by a factor of 2.

Returning to the Stokes equation, we see that if we define a length scale for the depth y ($=\int u\,dt$) of $UR/\Sigma K_c$, we have

$$dz/d\tau = x^2, \quad z = y\Sigma K_c/UR. \tag{89}$$

We shall return to this model when discussing methods of solution, but turn now to the other method of systematically reducing equations to dimensionless form.

AN ALTERNATIVE METHOD OF REDUCTION

In the method outlined previously, we started from the observation that the initial radius of the sphere was a naturally occurring length, and that $x = r/R$ would therefore be a suitable dimensionless radius, conveniently in the range (0, 1). However, if our purpose were to investigate the effect of varying the initial radius, R, over several orders of magnitude, we would have all the parameters varying simultaneously as we varied R. In this case, it is better to build up a set of characteristic dimensions from the physical constants of the process. We need three that span the three dimensions of mass, length, and time. In the previous example, we have ρ, ν, and g, whose dimensionality can be read from the following matrix:

	M	L	T
ρ	1	−3	0
ν	0	2	−1
g	0	1	−2

This tableau merely means that ρ has the dimensions M^1L^{-3}, etc. The matrix is of rank 3, so these three are an independent set and span the three dimensions. The matrix can be inverted to express the basic dimensions in terms of the three parametric quantities:

	ρ	ν	g
M	1	2	−1
L	0	2/3	−1/3
T	0	1/3	−2/3

meaning that $\rho\nu^2/g$ has purely dimensions of mass and so on.

Then, using these combinations, we set

$$x' = \alpha r(g/\nu^2)^{1/3},\ z' = \beta y(g/\nu^2)^{1/3},\ t' = \gamma t(g^2/\nu)^{1/3}, \tag{90}$$

where a prime has been used to distinguish these new dimensionless variables and α, β, and γ are real numbers we can choose later to make the equations simple. If we use Eq. (89) and substitute in the equations, we can choose the other constants to make the coefficients as simple as possible. In addition to the dimensionless (primed) variables of Eq. (89), let

$$v' = \delta(g\nu)^{-1/3}u \tag{91}$$

and

$$K' = \varepsilon\{1 + 0.672\ \mathrm{Re}^{1/2}\mathrm{Sc}^{1/3}\} = \{1 + 0.672\ \mathrm{Sc}^{1/3}\}\kappa(x') = (1 + a)\kappa(x'). \tag{92}$$

Then

$$\text{Re} = v'x' = x'^3,\ dz'/dt' = v' = x'^2,\ dx'/dt' = -\kappa(x')/x' \tag{93}$$

if we set

$$\alpha = [2\ \Delta/9]^{1/3},\ \beta = \alpha\Sigma\text{Sc},\ \gamma = \alpha^2\Sigma\text{Sc},\ \delta = 1/\alpha,\ \varepsilon = 1/(1 + a),\ \text{Sc} = D/\nu. \tag{94}$$

We shall make use of this in discussing families of solutions.

Example 8. The Rising Bubble Problem

Let us look at this method in a cognate situation. Instead of dropping dense particles and allowing them to dissolve, consider a small gas bubble ascending through a stagnant liquid and transferring its contents to that liquid, where it reacts instantaneously. As it ascends, the decreasing hydrostatic pressure allows it to expand, but the transfer of its contents tends to diminish it. We will assume that the bubble is a sphere, that its internal pressure is the external pressure plus the pressure induced by the surface tension, that the gas is a perfect gas, that the motion is governed by the Stokes equation, that the rate of transfer is proportional to the concentration in the bubble (i.e., the concentration of gas in the liquid is zero), and, finally, that the density of the gas is negligible in comparison with that of the liquid. The depth is calculated from a point at which the extrapolated hydrostatic pressure vanishes, i.e., if P is the pressure at the surface of the liquid, h is the depth within the liquid plus $P/\rho g$ (Table I gives the setup). Of the equations listed, that for the volume is purely geometrical; that for the depth is Stokes' law; the hydrostatic and pressure due to surface tension are put into the gas law in the third; and the mass transfer is expressed in the fourth. For the moment, k_c can be regarded as a constant. The fact that the density of the gas is negligible in comparison with that of the liquid means that $\Delta = 1$.

		Equations	M	L	T	ρ_l	ν_l	g	*
t	time				1		1/3	−2/3	α
r	radius			1			2/3	−1/3	β
V	volume	$V = 4\pi r^3/3$		3			2	−1	γ
h	depth	$dh/dt = -2gr^2/9\nu_l$: Stokes' law		1			2/3	−1/3	δ
p	pressure	$\rho_l gh + 2\sigma/r = NRT/V$	1	−1	−2	1	2/3	2/3	ε
N	content	$dN/dt = -3k_c N/r$	1	2	−2	1	8/3	−1/3	
ρ_l	density	liquid density $\gg$ gas density	1	−3		1			
ν_l	viscosity	kinematic viscosity		2	−1		1		
g	gravity			1	−2			1	
k_c	m.t.coeff.	mass transfer coefficient		1	−1		1/3	1/3	
σ	surface	tension; overpressure $= 2\sigma/r$	1		−2	1	4/3	1/3	
τ	t/t^*								
ξ	V/V^*	$\varpi\xi = \zeta = [\delta/\varepsilon]\eta\xi + 3[\Sigma/\beta\varepsilon]\xi^{2/3}$							
η	h/h^*	$d\eta/dt = -[2\alpha\beta/9]\xi^{2/3}$							
ζ	N/N^*	$d\zeta/dt = -[K\alpha/\beta]\zeta\xi^{-1/3}$							
ϖ	p/p^*	$\varpi = [\delta/\varepsilon]\eta + [\Sigma/\beta\varepsilon]3\xi^{-1/3}$							
		$K = \{3k_c/\nu_l^{1/3}g^{1/3}\}$;							
		$\Sigma = \{2\sigma/3\rho_l\nu_l^{4/3}g^{1/3}\}$							

The columns headed M, L, T give the dimensions of the variables with respect to the physical basis of mass, length, and time. We can change the basis to the three constant properties of the liquid and obtain characteristic times, lengths, and so on. For instance, the first line means that a characteristic time is $t^* = \alpha \nu_l^{1/3} g^{-2/3}$, and we can take the dimensionless time to be $t = t/t^*$ and choose a later time to make as many coefficients as possible equal 1. The characteristic magnitudes must be consistent with the equations; i.e., $V^* = 4\pi r^{*3}/3$ or $c = 4\pi b^3/3$ and $N^* = p^* V^*/RT$.

Substituting $t = t^*\tau$, $V = V^*\xi$, ... with $t^* = \alpha \nu^{1/3} g^{-2/3}$, ... from the top right-hand columns, we find that ρ_l, ν_l, and g disappear (a valuable check on the algebra) and leave only the ratios of α, β, ... These equations are given in lines: η, for Stokes' law; ζ, for the mass transfer; ϖ, for the over-pressure due to surface tension; and ξ, for the gas law in the first equality of the line (the second equality is obtained by substituting for ϖ). There are two coefficients composed of physical quantities: K for mass transfer and Σ for surface tension. If the numerical constants are chosen to be

$$\begin{aligned} &\alpha = (9/2K) = (3\nu_l^{1/3} g^{1/3}/2k_c)^{1/2}, \beta = (9K/2) = (27k_c/2\nu_l^{1/3} g^{1/3})^{1/2} \\ &\gamma = 4\pi\beta^3/3, \quad \delta = \varepsilon = \Sigma/\beta = [2/9]^{3/2}(\sigma/\rho_l \nu_l k_c), \end{aligned} \tag{95}$$

all the coefficients become 1, save for a deliberate 3. Thus, we have

$$\zeta = \eta\xi + 3\xi^{2/3} \tag{96}$$

$$\delta\eta/\delta\tau = -\xi^{2/3} \tag{97}$$

$$\delta\zeta/\delta\tau = -\xi^{-1/3}\zeta \tag{98}$$

$$\varpi = \eta + 3\xi^{-1/3}. \tag{99}$$

Differentiate the first of these equations, (96), with respect to time:

$$d\zeta/d\tau = (d\eta/d\tau)\xi + (d\xi/d\tau)\{\eta + 2\xi^{-1/3}\}. \tag{100}$$

Now, from the first equation itself, the last factor is

$$\{\eta + 2\xi^{-1/3}\} = \zeta/\xi - \xi^{-1/3}. = [\zeta - \xi^{2/3}]/\xi.$$

Substituting this and the two differential equations into (96) gives

$$d\xi/d\tau = -\xi^{2/3}\{(\zeta - \xi^2)/(\zeta - \xi^{2/3})\}.$$

Dividing this by the differential equation for η, gives us the pair of autonomous equations

$$d\xi/d\eta = \{(\zeta - \xi^2)/(\zeta - \xi^{2/3})\} \tag{101}$$

$$d\zeta/d\eta = \zeta/\xi. \tag{102}$$

It is very practical to work with the depth as an independent variable, for, if we need it, we can find the time of rise from the solution of the equations by quadrature. The pressure is given by the gas law and is, in fact, equal to either side of the differential equation for ζ. The depth was calculated from the point at which the extrapolated hydrostatic pressure is zero, so that if P is the pressure at the surface, the equations must be integrated until

$\eta = P/p^* = P$. Now P can never be negative, so the least value ϖ can have is $3\xi^{-1/3}$, or the least value that ζ can have is $3\xi^{2/3}$. This is just as well, for the equations are clearly singular in the neighborhood of $\zeta = \xi^{2/3}$, so singular, in fact, that no chemical engineering journal would tolerate their analysis.[9]

SCALING

Example 7. The Dissolving Sphere (Reprise)

To illustrate what is meant by scaling, let us return to the problem of the dissolving sphere and try to make sure that all the important dependent variables are in the interval [0, 1]. The independent variable, time, must be allowed to run its course. We have certainly done this with the radius of the sphere, for r can only diminish; so, if the initial radius is R, $x = r/R$ is obviously the correct choice. With an eye to extending the model later, we define U as the terminal velocity of a sphere of radius R, and because this decreases with decreasing radius, $v = u/U$ is certainly in [0, 1]. The very simple relationship $v = x^2$ holds as long as our assumption of the validity of Stokes' law is true.

Thus far, we are merely confirming what we did before, but when we come to the equation for the flux of solute from the surface, we have probably wrapped too much up in the dimensionless time. The empirical factor $\{1 + 0.672\ \text{Re}^{1/2}\text{Sc}^{1/3}\}$ in k_c accounts for the convective enhancement of the mass transfer and the time the sphere takes to be dissolved in a stagnant fluid is the time for x to decrease from 1 to 0, which is

$$(R^2/D\Sigma)\int_0^1 x\,dx = R^2/2D\Sigma. \tag{103}$$

This suggests that if we incorporate the solubility constant and the factor 2 into the dimensionless time,

$$s = 2\Sigma Dt/R^2. \tag{104}$$

All times would be of magnitude less than 1. If we now write the convective factor as

$$g(x) = \{1 + 0.672\ \text{Re}^{1/2}\text{Sc}^{1/3}\}, \tag{105}$$

we have for the Stokes case (Re < 1)

$$dx/ds = -g(x)/2x, \quad x(0) = 1, \tag{106}$$

of which the immediate solution is (by quadrature)

$$s = \int_{x(s)}^1 [2x/g(x)]\,dx. \tag{107}$$

Again, we will leave the solution of this equation until later.

Scaling is the art of choosing the dimensionless variables so that we know the ranges of magnitude in which they will lie. Clearly, this is an indispensable prelude to calculation. Usually there are some dimensionless parameters that

[9] See, however, pp. 52–54 and [312].

cannot be made equal to 1, and these show the relative importance of the various steps in the process. Take, for instance, the basic problem of diffusion and reaction in an isothermal porous catalyst pellet.

Example 9. The Spherical Catalyst Particle

Take the pellet to be a sphere with r the distance from the center and R the outside radius. If $c(r)$ is the concentration as a function of r (this assumes symmetry), the usual steady state balance gives

$$D(d/dr)(r^2dc/dr) = r^2f(c), \tag{108}$$

where $f(c)$ is the rate of disappearance of the reactant per unit volume of the pellet. At the surface $r = R$ there is a mass transfer resistance and

$$D(dc/dr) + k_c c = k_c c_f, \tag{109}$$

k_c is the mass transfer coefficient and c_f the concentration far from the particle. Obviously, c_f is the concentration to use to make the dependent variable c dimensionless and R will serve the same function for r, so we let

$$x = r/R \text{ and } u(x) = c(Rx). \tag{110}$$

We also want to scale the reaction rate, although we cannot guarantee that it will be less than 1 if auto-catalysis is present.

Let

$$F(u) = f(c_f u)/f(c_f). \tag{111}$$

Then

$$(d/dx)(x^2(du/dx)) = \varphi^2 x^2 F(u), \tag{112}$$

and at $x = 1$

$$du/dx + \nu u = \nu, \tag{113}$$

where

$$\varphi^2 = R^2 f(c_f)/Dc_f, \ \nu = k_c R/D. \tag{114}$$

A second-order equation demands two boundary conditions, and symmetry supplies the other, namely

$$u'(0) = 0. \tag{115}$$

Clearly there are two parameters, the first of which, written as a square in an intelligent anticipation of square roots to come, is the *Thiele modulus.* It measures the intensity of reaction in terms of the potential rate of diffusion, for it may be written

$$\varphi^2 = 3[(4/3)\pi R^3 f(c_f)]/[4\pi R^2 D(c_f/R)]. \tag{116}$$

In this expression, 3 is a purely geometric factor generated from the ratio of the surface area to the volume of the sphere; the first bracket is the maximum total rate of reaction, achieved when the concentration is everywhere at c_f; the second bracket is the total diffusive flux across the surface when the concentration gradient is (c_f/R). Similarly,

$$\nu = [4\pi R^2 k_c(c_f - 0)]/[4\pi R^2 D(c_f/R)] \tag{117}$$

is the ratio of the external flux to the internal diffusive flux. The gradient (c_f/R) and the concentration difference $(c_f - 0)$ are putative but the numerators and denominators are characteristic.

In the nonisothermal case, the reaction rate is a function of both the concentration c and the temperature T and it is not as easy to see what characteristic temperature should be used because with an exothermic reaction the heat may be trapped within the particle. We will not go through the derivation in detail (Chapter 2 of [I] goes into it quite thoroughly), but an energy balance yields

$$(d/dr)(r^2(dc/dr) + (-\Delta H/k)f(c, T) = 0 \tag{118}$$

subject to

$$k(dT/dr) + hT = hT_f, \tag{119}$$

and the symmetry condition

$$T'(0) = 0. \tag{120}$$

k is the effective thermal conductivity of the pellet and h the heat transfer coefficient, while ΔH is the heat of reaction. T_f is the obvious characteristic temperature to use, and we introduce

$$v(x) = T/T_f \quad \text{and} \quad F(u, v) = f(c_f u, T_f v)/f(c_f, T_f), \tag{121}$$

and the parameters

$$\mu = hR/k \quad \text{and} \quad \beta = (-\Delta H)Dc_f/kT_f \tag{122}$$

and retain the Thiele modulus φ^2 to give

$$(d/dx)(x^2(dv/dx)) + \beta\varphi^2 x^2 F(u, v) = 0 \tag{123}$$

at $x = 1$,

$$dv/dx + \mu v = \mu, \tag{124}$$

and at $x = 0$

$$dv/dx = 0. \tag{125}$$

This differential equation, (123), with its boundary conditions, (124) and (125), has to be solved simultaneously with Eq. (112), modified to include the temperature dependence of the reaction rate (i.e., $F(u, v)$ replaces $F(u)$), and its boundary conditions (123) and (125).

Elementary scaling has gotten the equations into a clean form and has revealed that there are a number of parameters[10] whose magnitudes will show the importance of the several processes: φ^2 the Thiele modulus or Damköhler

[10] There will be some parameters in the reaction rate expression, such as the order of an irreversible reaction or an activation energy E/RT_f. The full process of getting the simplest set of equations for a set of independent reactions is covered in great detail in Chapter 2 of [I, Vol. 1].

number, for the intensity of reaction relative to diffusion; μ and ν for the external heat and mass transport; and β a measure of the heat of reaction, often called the *Prater number.* We shall see the importance of the Prater number in Example 12, but first let us say something about shape.

SHAPE FACTORS

Example 10. Diffusion and First-Order Reaction

The result of solving these equations is often expressed as an effectiveness factor, the ratio of the total rate of reaction to the rate that would obtain if the composition and temperature were everywhere equal to the boundary composition and temperature. For the sphere, this would be

$$\eta = 3\int_0^1 x^2 F(u(x), v(x))\, dx. \tag{126}$$

When $\varphi \to 0$, $\eta \to 1$, because, in the absence of reaction, the only solution of Laplace's equation is the constant one. As $\varphi \to \infty$, the reaction rate falls to zero everywhere except in a thin layer just below the surface, in which the rate falls off exponentially with depth. It can be shown that when there is no external resistance, $\eta\varphi$ tends to a constant. This is not surprising, as the reaction rate will be proportional to the surface area and it is being averaged over the volume and so would be proportional to

$$4\pi R^2/(4/3)\pi R^3 = 3/R.$$

The differential operator in Eq. (102) will be recognized as the Laplacian and, taking the isothermal case with $\nu = \infty$ for simplicity, a balance of the same sort as was treated in the section entitled "The General Balance Equations for Distributed Systems" ($\partial/\partial t = 0$, $\mathbf{f} = -D$ grad c, $g = f(c)$) gives

$$\operatorname{div}(\operatorname{grad} c) = f(c) \text{ in } \Omega \text{ and } c = c_f \text{ on } \partial\Omega,$$

where Ω is the space of the pellet and $\partial\Omega$ its surface. Making this equation dimensionless in the same way as before, we have

$$\operatorname{div}(\operatorname{grad} u) = \varphi^2 F(u) \text{ in } \Omega,\ u = 1 \text{ on } \partial\Omega. \tag{127}$$

R is now a characteristic half thickness that we will choose later. The equations for the sphere, cylinder with sealed ends (i.e., the diffusion takes place as in a circle), and a slab with sealed edges (i.e., as in an interval or sphere in one dimension) are included in the equation

$$x^{-n}(d/dx)(x^n(du/dx)) = \varphi^2 F(u), \tag{128}$$

as the cases $n = 3, 2, 1$. In case of a first-order reaction $F(u)$, the equations can be solved explicitly. When the effectiveness factor is calculated, all three can be embraced in a single expression

$$\eta = ((n + 1)/\varphi)(I_{(n+1)/2}(\varphi)/I_{(n-1)/2}(\varphi)). \tag{129}$$

The asymptotic behavior this expression confirms what was said above, that it approaches 1 as φ tends to zero and $(n + 1)/\varphi$ as φ tends to infinity. Consider

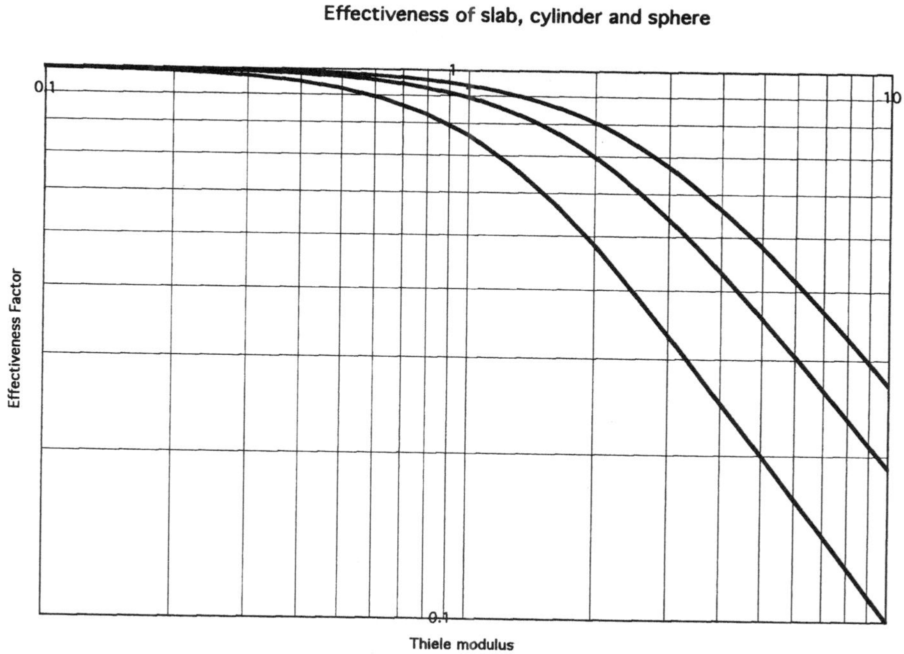

FIGURE 8 Effectiveness factor as a function of the Thiele modulus.

the three shapes with their volumes V and surfaces S (A is the superficial area of a flat plate of thickness $2R$ and L is the length of the cylinder of radius R):

Shape	Plate	Cylinder	Sphere
$n + 1$	1	2	3
S	$2A$	$2\pi RL$	$4\pi R^2$
V	$2AR$	$\pi R^2 L$	$4\pi R^3/3$
SR/V	1	2	3

It follows that, if we used V/S instead of R to define the parameter

$$\Phi^2 = (V/S)^2 f(c_f)/Dc_f, \tag{130}$$

then the three curves share the common asymptote $\eta \to 1/\Phi$ as η tends to infinity. Having thus brought the three curves together at their two ends, we are not surprised to find that they lie in a narrow band and do not differ by more than 13%.[11]

[11] The paper "On Shape Factors for Irregular Particles" is not reproduced in this volume as it is readily available in the special edition of *Chem. Eng. Sci.* **50,** 3897–3903 (1995); [3a]. The companion paper [3b] is directly related to [7 = C] in Chapter 7. See also some relevant comment in the Acknowledgments (p. 432). I was able to return to this topic in the Institute Lecture given in 1997 to the American Institute of Chemical Engineers in Los Angeles [314]. R. J. Wijngaarden, A. Kronberg, and K. R. Westerterp in *Industrial Catalysis* (New York, Wiley-VCH, 1998) use the square of the fully normalized Thiele modulus as An_0 and the limit as $\eta \to 1$ of $1 - \eta^2$ as An_1 (calling these Aris numbers in recognition of [3] and [52]) and find the solution of the cubic $\eta^2\{1 + (1 - \eta)An_0 + \eta An_1\} = 1$ to be a very good approximation.

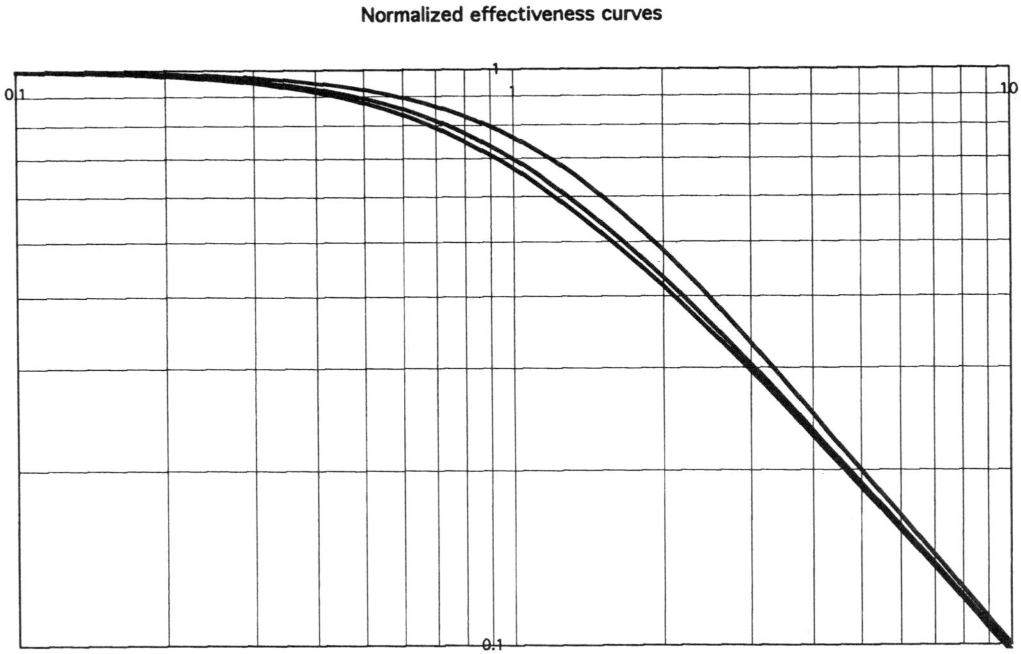

FIGURE 9 Effectiveness factor as a function of the normalized Thiele modulus.

It is not hard to conjecture that of all pellets of the same volume, the sphere must have the least effectiveness for a first-order irreversible reaction. After all, the sphere has the smallest surface area for a given volume and its interior is, in this sense, least accessible. One way of simulating shape is to have a distribution of straight cylindrical pores of such lengths and diameters as will give the same distribution of accessible area as a function of depth [84, 85]. It was not for some years, however, that Luss and Amundson proved it rigorously[12] in an elegant paper using isoperimetric inequalities and Steiner symmetrization.

In treating shape factors, I have gone ahead of the purely preliminary manipulation of the model. But it is useful to know that these shape factors exist as it may be advantageous to incorporate the experience that one has had of them in the way one starts the reduction to dimensionless form. In the light of what we have found—and this has now been common knowledge for some decades—we might start by setting $x = Sr/V = 3r/R$. Various formulae would have the numeral 3 where they did not have them before, but the final result be simpler. Perceptiveness and experience play a role here. If, for example, we had a packing problem, the sphere of the same volume might be more appropriate; if we had a drag problem, the sphere with the same average area normal to the flow would be better.

The discussion of dispersion in linear kinematic waves (Chapter 7, pp. 136, [7] = C; see also [3b] and [314]) brings out another shape factor, this

[12] D. Luss and N. R. Amundson. "On a conjecture of Aris: Proof and remarks" *A.I.Ch.E. J.* **13,** 739–743 (1967).

time for the particles retaining the heat as a temperature wave moves through the bed.[13]

Shape factors of a different sort are involved in the Taylor dispersion problem. With parabolic flow at mean speed U through a cylindrical tube of radius R, Taylor found that the longitudinal dispersion of a solute from the interaction of the flow distribution and transverse diffusion was $R^2U^2/48D$. The number 48 depends on both the geometry of the cross-section and the flow profile. If, however, we insist that the flow should be laminar, then the geometry of the cross-section determines the flow and hence the numerical constant in the Taylor dispersion coefficient.

A PRIORI ESTIMATES

Example 11. The Nonisothermal Catalyst Pellet

For the most adroit scaling we shall need the best estimates of the variables that we can get, and in any case, it is important get as much *a priori* information as possible.[14] Some bounds are physically obvious, as when we have a system in which a substance is disappearing by an irreversible reaction and its concentration cannot exceed the value at the inlet. A nice illustration of a less trivial estimate is provided by the nonisothermal catalyst pellet, the equations of which are here transcribed from (Eqs. 112–114 and 123–125): for $u(x)$, the concentration of the reactant,

$$(d/dx)(x^2(du/dx)) = \varphi^2 x^2 F(u, v), \tag{131}$$

at $x = 1$,

$$du/dx + \nu u = \nu, \tag{132}$$

and at $x = 0$,

$$du/dx = 0, \tag{133}$$

and for the temperature $v(x)$,

$$(d/dx)(x^2(dv/dx)) + \beta\varphi^2 x^2 F(u, v) = 0, \tag{134}$$

at $x = 1$,

$$dv/dx + \mu v = \mu, \tag{135}$$

and at $x = 0$,

$$dv/dx = 0. \tag{136}$$

We shall use these equations bearing in mind that, unless specifically excepted, all the results are valid for any shape of particle.

[13] Actually, the problem was worked out for chromatography when I was working for Imperial Chemical Industries, but the company had been a bit cautious about letting me publish before and, when I came to write the work up, I made the change to avoid having to submit the paper for permission to publish.

[14] Gavalas was the first to recognize the value of *a priori* estimates. See *his "Nonlinear Equations of Chemically Reacting Systems,"* Heidelberg: Springer-Verlag, 1968.

We observe that the combination $w = \beta u + v$ satisfies Laplace's equation. If μ and ν are both very large so that we have $u = v = 1$, then $w = 1 + \beta$ over the whole bounding surface, in this case, $x = 1$. Now, it is well known that a potential function takes its maximum on the boundary of its domain, and so, if it is constant there, it is the same constant value within. It follows that

$$w(x) = \beta u(x) + v(x) = 1 + \beta. \tag{137}$$

Now u is the concentration of a reactant disappearing irreversibly, and it must lie in the interval [0, 1]. Hence

$$0 \le u(x) \le 1, \quad 1 \le v(x) \le 1 + \beta. \tag{138}$$

The relation in Eq. (127) can be used to reduce the two equations to one by making $F(u, v)$ a function of a single variable, by substituting $v = 1 + \beta(1 - u)$ or $u = 1 - (v - 1)/\beta$, and so obtaining $F(u, v)$ as a function of u or v alone. For example, if $F(u, v) = u \cdot \exp(\gamma(v - 1)/v)$ as is the case for a first-order reaction,

$$G(u) = u \exp\{\beta\gamma(1 - u)/(1 + \beta(1 - u))\} \tag{139}$$

expresses the rate in terms of the concentration alone.

When the shape is symmetrical and so is the solution, the surface values are uniform and, for finite μ and ν,

$$v = (1 + \alpha\beta) + \beta(1 - \alpha)v_\sigma - \beta u.$$

Finding the maximum of v is thus an elementary exercise in linear programming, for we must observe the constraints $1 > u_s > u > 0$. It is so simple that the results may be obtained by inspection and tabulated.

	$\alpha < 1: \nu < \mu$	$\alpha > 1: \nu > \mu$
$\beta > 0$	$1 = v = 1 + \beta$	$1 = v = 1 + \alpha\beta$
$\beta < 0$	$1 + \beta = v = 1$	$1 + \alpha\beta = v = 1$

More detail can be found on this particular problem in [I] (pp. 89–94).

SCALING AND PARTIAL SOLUTION IN LINEAR SYSTEMS

In linear systems, an appropriate scaling can sometimes be used to reduce the number of equations. This is really equivalent to solving some of the equations and substituting in the others to get a smaller set—the very method we learned in kindergarten. It may be illustrated in the nontrivial example of the bubbling fluidized bed, which I used in a talk to a group in an English department [284] on the grounds that they would have seen the mathematics before. It is applied to a more general system of reactions in [280, 292] and generalized in [310]. I present the simplest version here as it shows the principle of the modification most clearly.

Example 12. The Bubbling Fluidized Bed

Consider the reaction A $\rightarrow$ B taking place in the dense, or particulate, phase of a bubbling bed of fluidized catalyst particles (Fig. 10). It is in steady opera-

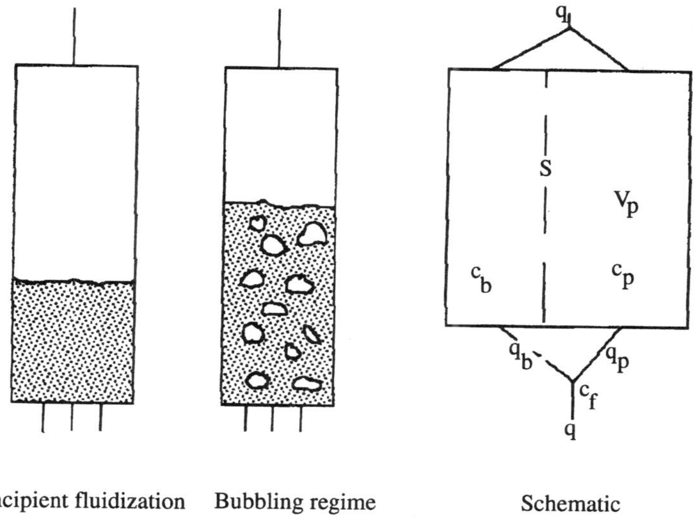

FIGURE 10 Schematic of the bubbling bed.

tion, and suffixes p and b are used to denote quantities in the particulate and bubble phases, respectively. We assume first that both phases are well mixed. This is not a very convincing model for the fluidized bed, but we shall improve on it later; it might be entirely appropriate for a membrane reactor, however. Let β be the split of the feed $(q_b/q) = q_b/(q_p + q_b)$ and c_f the feed concentration, h a mass transfer coefficient, and S the total area of exchange. Then, for the bubble phase

$$\beta q c_f = \beta q c_b + hS(c_b - c_p), \tag{140}$$

and for the dense phase

$$(1 - \beta)qc_f = (1 - \beta)qc_p - hS(c_b - c_p) + Vkc_p. \tag{141}$$

The quantity we are really interested in is

$$C = (1 - \beta)c_p + \beta c_b. \tag{142}$$

The obvious way to render the equations dimensionless is to set

$$u = c_p/c_f,\ v = c_b/c_f,\ w = C/c_f,\ Da = Vk/q,\ Tr = hS/q,^{15} \tag{143}$$

whence

$$(1 - \beta) = (1 - \beta)u + Tr(u - v) + Da\,u, \tag{144}$$

$$\beta = \beta v + Tr(v - u), \tag{145}$$

$$w = (1 - \beta)u + \beta v. \tag{146}$$

We now recognize that even if the reaction rate is extremely large so that u falls to zero, v will not go to zero. It can never be less than $v^* = \beta/(\beta + Tr)$

[15] This dimensionless number might be justly called the *Davidson number* in recognition of John Davidson's leadership in all matters of fluidization. Moreover, in view of his lifelong connection with Trinity College, *Tr* is a peculiarly suitable way of avoiding confusion with *Da*.

and so w can never be less than, for instance, $\beta^2/(\beta + Tr) = w^*$. Let us now use this to rescale w so that it must lie in the interval (0, 1) by setting

$$W = (w - w^*)/(1 - w^*). \tag{147}$$

From Eq. (145), $v = v^* + (1 - v^*)u$, and so, by Eq. (146), $w = (1 - \beta)u + \beta v^* + \beta(1 - v^*)u = w^* + (1 - \beta v^*)u$, or $W = u$. However, if we then add Eqs. (144) and (145) together and divide by $(1 - w^*)$, we have

$$1 - W + Da^*W = 0, \tag{148}$$

where

$$Da^* = Da/(1 - \beta v^*) \tag{149}$$

is a modified Damköhler number. Moreover, this modification is physically sensible, for, if we write

$$\begin{aligned} Da^* &= Vkc_f/qc_f(1 - w^*) = Da/\mathscr{A} \\ \mathscr{A} &= 1 - \beta^2/(\beta + Tr), \end{aligned} \tag{150}$$

we see immediately that, because it is a ratio of reaction rate to advection rate, the denominator has been modified to use only the fraction that is available. Had the word not already a thermodynamic usage, $\mathscr{A}$ might be called the *availability*.

The equations for the two-phase fully mixed system are thus reduced to the equations for a single stirred tank by the physically motivated notion of only using the available fraction of the feed. This has been made possible by the uniformity of the dense phase and the linearity of the transfer process in the bubble. This allows us to see how the rather implausible assumption that the bubble phase is really well mixed can be made more realistic. Let us go to the other extreme, and suppose that the bubbles ascend with uniform velocity U. The surface area per unit length of reactor is S/H, where S is, as before, the total interphase area and H the height of the bed. If h is the transfer coefficient and z the height of a given point, a balance over the interval $(z, z + dz)$ gives the equation for the concentration in the bubble phase $b(z)$

$$U(dc_b/dz) = h(S/H)(c_p - c_b) \quad c_b(0) = c_f. \tag{151}$$

As we showed in Example 4, we must make the balance over the whole of the dense phase because we have assumed it to be uniform. If its volume is V and the volumetric flow q_b, then

$$q_a c_f = q_p c_a + h(S/H) \int_0^H (c_a - c_b)\, dz + Vkc_a. \tag{152}$$

The appropriate dimensionless form is clearly to be achieved by setting

$$x = z/H, u = c_p/c_f, v(x) = c_b/c_f, \beta = q_b/(q_p + q_b), Tr = hS/q. \tag{153}$$

Then

$$dv/dx = (Tr/\beta)(u - v), \quad v(0) = 1 \tag{154}$$

$$(1 - \beta) = (1 - \beta)u + \int_0^1 (u - v(x))\, dx + Da\, u. \tag{155}$$

Because u is constant in Eq. (144),

$$\begin{aligned} v(x) &= u + (1 - u) \exp(-Tr \cdot x) \\ &= \nu^* + (1 - \nu^*)u, \end{aligned} \tag{156}$$

and

$$w^* = \beta \nu^* = \beta \exp(-Tr/\beta). \tag{157}$$

The same argument goes through, and the appropriate availability is

$$\mathscr{A} = 1 - \beta \exp(-Tr/\beta). \tag{158}$$

As with shape factors, the availability can be calculated independently and for a great variety of systems. For example, if the residence time of a group of bubbles is a fraction θ of the mean, and a fraction $f(\theta)\,d\theta$ of the surface area is associated with residence times in the interval $(\theta, \theta + d\theta)$,

$$\mathscr{A} = 1 - \int f(\theta)\beta \exp(-Tr\theta/\beta)\, d\theta, \tag{159}$$

with the normalization

$$\int f(\theta)\, d\theta = 1 = \int \theta f(\theta)\, d\theta. \tag{160}$$

Equation (150) can be recovered in this way by making $f = \exp(-\theta)$, the residence time distribution of the stirred tank. A limited degree of dispersion as would be afforded by a sequence of equally stirred tanks,

$$f(\theta) = n^n \theta^{n-1} e^{-n\theta}/\Gamma(n) \tag{161}$$

gives

$$\mathscr{A} = \beta(1 - \beta(1 + Tr/n\beta)^{-n}\}, \tag{162}$$

from which the two limiting cases from Eqs. (150) and (158) can be recovered by letting $n = 1$ or tend to infinity.

We observe that it is the uniformity of the dense phase and the linearity on the bubble side that allows us the freedom to use any residence time distribution. The reaction term, which we made first order for simplicity, can be nonlinear. A discussion of various generalizations is given in [310].

3 SOLVING THE EQUATIONS

This is not the place for a treatise on the solution of differential equations, ordinary or partial. There are some excellent mathematical texts and some, such as Varma and Morbidelli's *"Mathematical Methods in Chemical Engineering"* (Oxford University Press, 1997), are specifically directed at the chemical engineer. What we shall try to do, however, is to explore some of the *ad hoc* methods that take advantage of peculiar features of particular problems and those that give partial solutions, as well as mentioning a few fall-traps for the unwary.

Although many routine programs for solving differential equations of all sorts are currently available at all levels of computing, from the desk-top to the supercomputer, it is desirable to have as much understanding as possible of the mathematics that lies behind them. Above all, it is essential to get as clear an idea as possible of what the solution looks like, *before* going to the computer.

GETTING A FEEL FOR THE SOLUTION

Example 13. Two Populations Growing in a Chemostat Competing for a Common Nutrient

An example that illustrates this is found in a paper by Gregory Stephanopoulos and Fredrickson on commensual growth of two microbial populations in a chemostat [158]. If x_1 and x_2 are the concentrations of the two organisms and μ_1 and μ_2 their specific growth rates, V the volume of the chemostat and q the flow rate, then the elementary balance gives

$$V(dx_1/dt) = -qx_1 + Vx_1\,\mu_1(s) \tag{163}$$

$$V(dx_2/dt) = -qx_2 + Vx_2\,\mu_2(s). \tag{164}$$

The specific growth rates are functions of the substrate concentration, which itself must satisfy a balance equation

$$V(ds/dt) = q(s_f - s) - V\{x_1\,\mu_1(s)/Y_1 + x_2\,\mu_2(s)/Y_2\}, \tag{165}$$

where s_f is the feed concentration of nutrient and Y is the yield coefficient. There are only two independent populations, so we immediately expect to have only two equations. We notice that the combination $s + x_1/Y_1 + x_1/Y_1 = w$ decays exponentially from its initial value to s_f. It remains constant at s_f, if that is its initial value. Let us keep the third equation for the moment, and make concentrations dimensionless by a characteristic concentration S and time by a characteristic rate M:

$$x = x_1/SY_1,\ y = x_2/SY_2,\ z = s/S,\ \tau = Mt,\ \theta = q/MV \tag{166}$$

with

$$\mu_1(Sz)/M = f(z),\ \mu_2(Sz)/M = g(z) \tag{167}$$

to give

$$\begin{aligned} dx/d\tau &= x\,[f(z) - \theta] \\ dy/d\tau &= y\,[g(z) - \theta] \\ dz/d\tau &= \theta[z_f - z] - x\,f(z) - y\,g(z). \end{aligned} \tag{168}$$

The specific growth rates are of the form

$$f(z) = Az/(B + z),\ g(z) = Cz/(D + z). \tag{169}$$

Let us assume that $x + y + z = z_f$ so that we work in the triangular region

$$0 = x,\ 0 = y,\ x + y = z_f. \tag{170}$$

There are four possible steady states

$x = y = 0$	Both populations wash out.
$x = 0,\ g(z) = \theta$	X_1 washes out; X_2 grows.
$y = 0,\ f(z) = \theta$	X_2 washes out. X_1 grows.
$f(z) = g(z) = \theta$	The populations coexist.

The various conditions under which each alternative would prevail were worked out in a paper that Humphrey and I wrote in 1977 [144]. Here, we are interested in the last case where the two populations coexist. We note that they can be present in any proportion whatsoever, for the condition is merely that the growth curves should cross and the dilution rate θ should be their common value. Moreover, the points on the line of coexistence are all neutrally stable, and if some perturbation moves the state away from the line, it will move back to the line, but not necessarily to the point from which it was disturbed. Thus if the dilution rate were to fluctuate, as in practice it most certainly would, the state might drift along the line.

Let us now choose the characteristic rate and concentration to make the point of intersection of f and g the point (1, 1). We are going to put a random fluctuation on the flow rate of zero mean so we shall be particularly interested in the dynamics in the immediate neighborhood of the line $x + y = z_f - 1$. Now, in this neighborhood, the linearization of the growth rates gives

$$f(z) = 1 + \alpha(z - 1),\ g(z) = 1 + \beta(z - 1), \tag{171}$$

where

$$\alpha = B/A,\ \beta = D/C.$$

Then the equations can be written

$$d(\ln x)/d\tau = 1 - \theta + \alpha(z_f - 1) - \alpha(x + y) \tag{172}$$

$$d(\ln y)/d\tau = 1 - \theta + b(z_f - 1) - \beta(x + y). \tag{173}$$

Multiplying the first equation by β and the second by α and subtracting gives

$$(d/d\tau)[\beta \cdot \ln x - \alpha \cdot \ln y] = (\beta - \alpha)(\theta - 1),$$

or

$$(x^\beta\ \eta^\alpha)/(\xi^\beta\ y^\alpha) = \exp[(1 - \theta)(\beta - \alpha)\tau], \tag{174}$$

where ξ and η are the starting points in the neighborhood of the neutral line. Now, using the relation $x + y = z_f - 1$ in Eq. (164), we obtain

$$dx/d\tau = (\beta - \alpha)(\theta - 1)\{x(z_f - 1 - x)\}/\{\beta(z_f - 1) + (\alpha - \beta)x\},$$

and, rescaling so that the variable along the curve goes from 0 to 1 (instead of $z_f - 1$) by setting $\chi = x/(z_f - 1)$, we have

$$d\chi/d\tau = -(1 - \theta)\chi(1 - \chi)/(\gamma + \chi)\ \gamma = \beta/(\alpha - \beta). \tag{175}$$

The random fluctuations of flow rate are modeled by setting

$$\theta = \theta_0 + \zeta(\tau), \tag{176}$$

where the random element ζ is normally distributed with zero mean and infinite variance, but no correlation with its value at any other instant, i.e.,

$$E[\zeta(\tau)] = 0,\ E[\zeta(\tau_1)\zeta(\tau_2)] = \sigma^2\delta(\tau_1 - \tau_2). \tag{177}$$

This white noise perturbance can be derived from a Wiener process W, and χ then satisfies the stochastic differential equation

$$d\chi = (\theta_0 - 1)F(\chi)\ d\tau + F(\chi)\ dW \tag{178}$$

with

$$F(\chi) = \{\chi(1 - \chi)/(\gamma + \chi)\}. \tag{179}$$

The most telling form of result from this equation is the conditional probability distribution $p(\chi, \tau; \chi_0, \tau_0)\ d\chi$. This is the probability of the system's being in the interval $(\chi, \chi + d\chi)$ at time τ, given that it was at χ_0 at time τ_0. p satisfies the so-called Fokker-Planck equation

$$\partial p/\partial\tau = (\sigma^2/2)(\partial^2/\partial\chi^2)(F^2(\chi)p) - (\theta_0 - 1)(\partial/\partial\chi)(F(\chi)p). \tag{180}$$

Clearly,

$$p(\chi, \tau_0; \chi_0, \tau_0) = \delta(\chi - \chi_0), \tag{181}$$

because this merely means that the state is certainly χ_0 at time τ_0.

So much for the background. The question now is whether we can visualize the solution before going to the computer. This is particularly necessary as we have a delta function as initial condition. However, the equation looks rather like the equation for diffusion and flow in a straight tube, which, if ϖ were the concentration of a solute, $D(\chi)$ its spatially varying diffusion coefficient, and $V(\chi)$ its spatially varying convective velocity, would be

$$\partial\varpi/\partial\tau = (\partial/\partial\chi)\{D(\chi)(\partial\varpi/\partial\chi) - V(\chi)\varpi\}. \tag{182}$$

After a bit of rearrangement, we see that (Fig. 11)

$$D(\chi) = \sigma^2 F^2(\chi)/2 \text{ and } V(\chi) = (\theta_0 - 1)F(\chi) - \sigma^2 F(\chi)F'(\chi). \tag{183}$$

Now $F(\chi) = \chi(1 - \chi)/(\gamma + \chi)$ is zero at both ends of the interval and has a maximum of $(1 + 2\gamma)/\{\sqrt{\{\gamma(1 + \gamma)\}}\}$ at $\chi^* = \sqrt{\{\gamma(1 + \gamma)\}} - \gamma$. If diffusion were the only influence at work, the initial delta function would quickly turn into a peak that would subside and the build-up of material at the two ends would be exceedingly slow because the diffusion coefficient vanishes there. The velocity term in the equation augments this tendency, for, if we set $\theta_0 = 1$, we see that $V(\chi)$ is positive for $\chi > \chi^*$ and negative for $\chi < \chi^*$. Thus the convective effect is to rend the peak in two parts and build them up into peaks near the ends of the interval.

If we wanted to control about the point at which the two populations can coexist, we should set $\theta_0 = 1$. However, the expression for $V(\chi)$ shows that if the setting for the mean of θ is close to 1 in comparison to the variance of the fluctuations, nothing very different will happen. If, however, $(\theta_0 - 1)/\sigma^2$ is outside the interval between the lowest and highest values of F', V will not

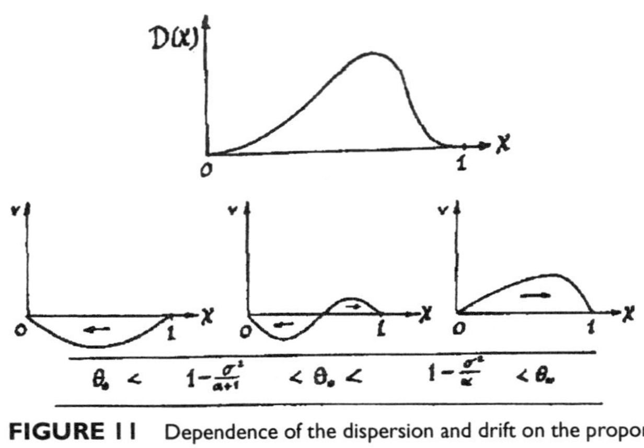

FIGURE 11 Dependence of the dispersion and drift on the proportionality of the species.

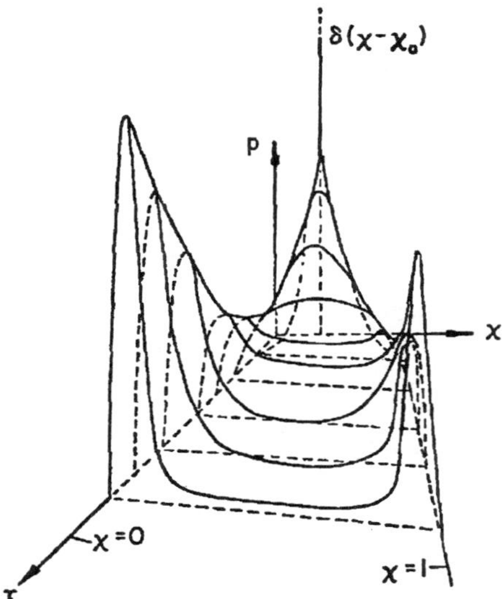

FIGURE 12 Evolution of the probability density of proportionality from certainty of χ_0 to the split between one species and the other.

change sign and the convective influence of V will dominate and drive out the slower-growing species. It is now safe to go to the computer, and the figure shows the expected result. The diminishing peak from the initial-sharp distribution is rent in twain, and two sharp peaks build up at the sides, their relative areas giving the relative probabilities of extinction of one species or the other (Fig. 12).

The further results with this model—time to extinction, stabilization by periodic variation on feed rate, and so on—are not our concern here (see [158] and [164]). Rather, the stress is on getting a feel for the solution before going to its numerical solution.[1]

SPECIAL FORMS

It is surprising how often some of the special forms of easily soluble differential equations turn up. For single equations in which the derivative is a function only of the state, $dx/dt = f(x)$, $x(0) = X$, the solution by quadratures is immediate

$$t = \int_{x(t)}^{X} dx/f(x). \tag{184}$$

Of course, the answer is in implicit form and it may not always be possible to get a formula for $x(t)$ explicitly. However, this does not matter if a graph

[1] We have been able to get a feel for the Fokker–Planck equation by turning it in to a Diffusion–Convection equation. A mathematician on Greg Stephanopoulos' *viva* confessed that he would have had to translate the D–C equation into an F–P equation! *Tot homines, quot aditus.*

is to be the mode of presentation, and interpolation can be used if a table of evenly spaced $x(t)$ is required.

Example 14. Michaelis–Menten Kinetics

A simple example is the so-called Michaelis–Menten kinetics for enzymatic reactions $A + E \leftrightarrow C \rightarrow B + E$, which, when the pseudo–steady-state hypothesis is invoked, gives for the concentration of A, for instance, a,

$$da/dt = -\,ka/(K + a), \text{ whence } kt = a_0 - a + K\ln(a_0/a). \tag{185}$$

Moreover, because $t \rightarrow \infty$ as $a \rightarrow 0$, the last term dominates after a time and so gives an explicit approximation

$$a(t) = a_0 \exp((-kt + a_0)/K). \tag{186}$$

This shows the obvious limits of first-order reaction with rate constant k/K if $K \gg a_0$ and zero-order as $K \rightarrow 0$.

Example 7. The Dissolving Sphere (Reprise)

Equation (88) for the dissolving sphere is another example. From Eqs. (84) and (88) we have two constants, $a = 0.672\ \mathrm{P}^{1/2}\,Sc^{1/3}$ and $b = 3/2$, in

$$dx/d\tau = -\kappa(x) = -\,\{1 + ax^b\}/x(1 + a), \tag{187}$$

where $x(0) = 1$. This equation can be solved by quadratures and expressed in terms of incomplete Beta functions. For

$$\begin{aligned} F(x; a, b, c) &= \int_s^1 w^c \{(1 + a)/(1 + aw^b)\} dw \\ &= M\{B(m; 1 - m; a/(1 + a)) - B(m; 1 - m; ax^b/(1 + ax^b))\}, \end{aligned} \tag{188}$$

with $m = (c + 1)/b$ and $M = (1 + a)/(ba^m)$. $B(\nu, \mu, x)$ is the incomplete Beta function

$$B(\nu, \mu; x) = \int_0^x t^{\nu-1} (1 - t)^{\mu-1}\, dt. \tag{189}$$

Thus,

$$\tau = F(x; 0.475P^{1/2}Sc^{1/3}, 3/2, 1) \tag{190}$$

$$z = F(x; 0.475P^{1/2}Sc^{1/3}, 3/2, 3). \tag{191}$$

The dissolution is complete when $x = 0$ and the second term vanishes, giving

$$T = M\ B(4/3, -1/3, a/(1 + a)) \tag{192}$$

$$Z = M\ B(8/3, -5/3, a/(1 + a)) \tag{193}$$

for the time, T, and the depth, Z, of extinction by solution.

This integration by quadratures works well for a number of transfer problems in burning or dissolution, for example, the burning of a sphere where the burnt ash builds up an insulating layer on the outside. With the various resistances to heat and mass transfer in series and the assumption that the burning front moves slowly, each form of resistance contributes an additive term to the burning time. In the nonisothermal case, interesting questions concerning uniqueness and stability arise [73].

Example 15. Diffusion and Reaction in a Slab

The special form of second-order equation in which the right-hand side is a function only of the dependent variable also turns up in the theory of diffusion and reaction in a slablike particle. Corresponding to equations (123–125) for the sphere, we would have, thanks to the reduction described in Chapter 2 and the example of a first-order nonisothermal reaction given by Eq. (129),

$$d^2u/dx^2 = \varphi^2 G(u) \quad \text{in } (0, 1) \tag{194}$$

$$u'(0) = 0, \; u(1) = 1. \tag{195}$$

Multiplying both sides by $2(du/dx)$ and integrating from 0 to x gives

$$[du/dx]^2 = 2\varphi^2 \int_v^{u(x)} G(w)dw = \varphi^2 [H(u(x), v)]^2, \tag{196}$$

where $v = u(0)$, of which we know that its value is a decreasing function of φ (1 when $\varphi = 0$ and approaching zero as φ goes to infinity). However, the effectiveness factor is the average rate of reaction

$$\eta = \int_0^1 G(u)dx = (du/dx)_1/\varphi^2 = H(v, 1)/\varphi, \tag{197}$$

whereas from Eq. (186)

$$\varphi = \int_v^1 dw/H(w, v). \tag{198}$$

Here, we are treating v as a parameter and calculating η and φ as functions of it, which we may then plot in the φ, η plane with ν as parameter along the curve. We will return to this example later in the chapter.

GETTING THE MOST FROM CALCULATIONS

Example 11. The Nonisothermal Catalyst Pellet (Reprise)

Villadsen and Michelsen[2] have drawn attention to a remarkable property of functions associated with irreversible reactions such as G in equation (194). Let us take the slightly more general case of a pth order reaction in a sphere, for which

$$G(u) = u^p \exp\{\beta\gamma(1 - u)/(1 + \beta(1 - u))\} = G_p(u; \beta, \gamma) \tag{199}$$

and put $u = u_1 U$. Then, after a bit of algebra, we have

$$G_p(u; \beta, \gamma) = G_p(u_1; \beta, \gamma) \cdot G_p(U; \mathrm{B}, \Gamma), \tag{200}$$

[2] Villadsen, J. and M. L. Michelsen. "Diffusion and reaction in spherical catalyst pellets." *Chem. Eng. Sci.* **27**, 751 (1972).

provided that

$$B = \beta u_1/\{1 + \beta(1 - u_1)\} \tag{201}$$

$$\Gamma = \gamma/\{1 + \beta(1 - u_1)\}. \tag{202}$$

We see that (B, Γ) lies on the line that joins $(-1, 0)$ and (β, γ) and traverses this segment as u_1 goes from $-\infty$ to 1.

Now, the equation for the Dirichlet problem can be written

$$d^2u/d\xi^2 + (2/\xi)(du/dx) = G_p(u; \beta, \gamma), \tag{203}$$

where $\xi = \varphi\rho$ and $u(\varphi) = 1$, whereas $u'(0) = 0$. This is a two-point boundary value problem and would probably be tackled by integrating from the center with initial conditions $u'(0) = 0$, $u(0) = v$ until $u(\xi) = 1$. Then, because $\xi = \varphi$ at the surface, we have the value of φ for which the value of v we had chosen to get the integration started is the center concentration corresponding to φ. Also, the effectiveness factor is given by the gradient at the surface

$$\eta = (3/\varphi^2)(du/d\rho)_1 = (3/\varphi)(du/d\xi)_\varphi = (3/\xi)(du/d\xi)_{\xi=\varphi}.$$

This determines η for the chosen v, and thus v acts as a parameter and the η, φ curve is obtained by varying v.

However, if we take the solution up to the point where $u = u_1$, we have solved the problem for the values B and Γ given by Eqs. (201) and (202). To see this, we put

$$u = u_1 U, \ \zeta^2 = \rho^2 \varphi^2 G(u_1; \beta, \gamma)/u_1. \tag{204}$$

Keeping u_1 constant, we have for U the equation

$$d^2U/d\zeta^2 + (2/\zeta)(dU/d\zeta) = G(U; B, \Gamma). \tag{205}$$

If this is integrated until $U = 1$, the value of ζ^2 is $\Phi^2 = \varphi^2\rho^2 G_p(u_1; \beta, \gamma)$ and $\eta = (3/\zeta)(dU/d\zeta)$. Thus, integrating Eq. (203) to the point where $u(\xi_1) = u_1$ gives

$$\Phi = \zeta_1 \sqrt{\{G(u_1; \beta, \gamma)/u_1\}} \tag{206}$$

$$H = \{3/G(u_1; \beta, \gamma)\}[(1/\zeta_1)(du/\zeta)_1] \tag{207}$$

for the B and Γ that are given by Eqs. (201) and (202). (Further details are given in [I], pp. 251–253.)

THE USE OF PARAMETRIC REPRESENTATIONS

Example 8. The Rising Bubble Problem (Reprise)

In Example 15, we saw that v, the concentration at the center, appears naturally as a parameter in terms of which η and φ can be calculated. It is, then, a very natural thing to plot the two against one another, regarding v as a marker along the curve. We shall see later in the chapter that this aids the asymptotic analysis.

An awareness of the possible parameterization of solution curves is helpful in setting up the numerical procedure on the computer, as may be illustrated by referring back to Example 8, which yielded the equations

$$d\zeta/d\eta = \{(\zeta - \xi^2)/(\zeta - \xi^{2/3})\} \qquad \text{(101 bis)}$$

$$d\zeta/d\eta = \zeta/\xi. \qquad \text{(102 bis)}$$

In the first place, we see that the right-hand sides are infinite along the axis $\xi = 0$ and the curve $\zeta = \xi^{2/3}$. However, the solution paths are the same as those for the system

$$d\xi/d\eta = \xi(\zeta - \xi^2) \qquad (208)$$

$$d\zeta/d\eta = \zeta(\zeta - \xi^{2/3}), \qquad (209)$$

and this will at least not fly off the handle when submitted to an ODE computer package. Such a package will probably take care of the choice of method and step size, but if you are doing the integration on a desk-top computer without the benefit of such a program, the arc length transformation is useful. Writing the pair of equations as

$$dx/dt = A(x, y)/B(x, y),\ dy/dt = C(x, y)/D(x, y), \qquad (210)$$

the non-singular isodic[3] equations are

$$dx/dr = \mathrm{A} \cdot \mathrm{D},\ dy/dr = \mathrm{C} \cdot \mathrm{D}. \qquad (211)$$

Let

$$\Delta = \surd\{A^2D^2 + B^2C^2\}. \qquad (212)$$

Then

$$dx/ds = AD/\Delta,\ dy/ds = BC/\Delta,\ dr/ds = 1/\Delta, \qquad (213)$$

and

$$dt/ds = BD/\Delta. \qquad (214)$$

In the ascending bubble problem (Example 8), the requirement that the pressure at the surface should be positive rules out the dramatic behavior near the singular curve $\zeta = \xi^{2/3}$, but, by suitably defining the value of the derivative at the discontinuity, solution paths of perilous stability can be found[4] (Cf. [312]). In Fig. 13, the broken line $A = 0$ is manifestly unstable. However, the broken line $B = 0$ can be reached in a finite time by a solution starting above $A = 0$ in the lower part of the phase plane and below $A = 0$ in the upper part. If the value at the singularity, $B = 0$, which is otherwise undefined, is correctly chosen, the solution turns and follows the path $B = 0$. In practice, it falls victim to the peril, even when stable.

[3] *Isodic* = means "having the same paths." I am not sure if this term, obvious though it is, is in current usage.

[4] They are rather like summit ridge of Mt. Everest but with equally violent winds blowing from both east and west. Along the ridge the winds cancel each other out, but let the climber step into Nepal and he or she will be hurled through the gossamer stagnation surface into the storm from Tibet.

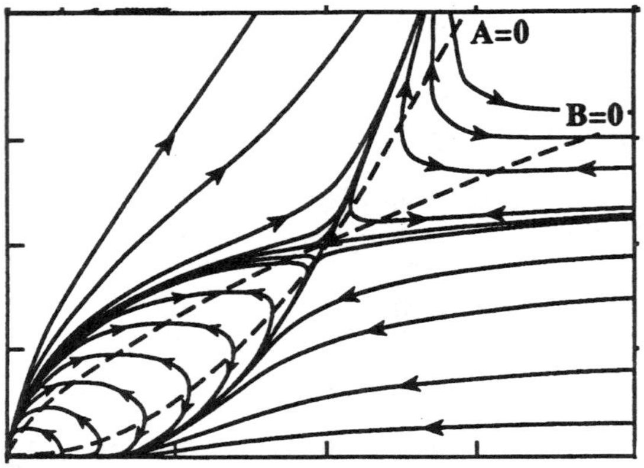

FIGURE 13 The phase-plane perilous.

Example 16. The C* with the Gray–Scott Reactions

The autocatalytic reaction scheme $A + 2B \rightarrow 3B$, $B \rightarrow C$ was introduced in 1983[5] and has proved itself to be fecund of useful applications in the study of reactor stability and chemical oscillations.[6] We shall depart from their notation for we wish to be able to generalize to several species, A_i, and it is not desirable to use the concentration of A as a reference concentration when it is going to be varied. Similarly, the several species will have different rate constants for the several autocatalytic steps and therefore the first-order rate constant of $B \rightarrow C$ is most apt for the time scale.

With θ for the residence time of the stirred tank and an otherwise obvious notation,

$$\theta \cdot da/dt = a_f - a - \theta k_1 ab^2, \; \theta \cdot db/dt = b_f - b - \theta k_1 ab^2 - \theta k_2 b. \quad (215)$$

Let v and w be the two dimensionless concentrations and u the conversion of A;

$$u = 1 - a/a_f, \; v = a/b_f, \; w = b/b_f, \quad (216)$$

and the dimensionless time is

$$\tau = k_2 t. \quad (217)$$

There are three parameters, and they are the space velocity or dilution rate,

$$\lambda = 1/k_2\theta, \quad (218)$$

the feed concentration,

$$\alpha = a_f/b_f, \quad (219)$$

and the Damköhler number,

$$\mu = k_1 b_f^2/k_2. \quad (220)$$

[5] Gray, P. and S. K. Scott. *Chem. Eng. Sci.* **38**, 29 (1983).

[6] Gray, P. and S. K. Scott. *Chemical Oscillations and Instabilities.* Oxford: Clarendon Press, 1990.

Then

$$v' = \lambda(\alpha - v) - \mu v w^2 \tag{221}$$

and

$$w' = \lambda(1 - w) + \mu v w^2 - w, \tag{222}$$

while

$$u = 1 - v/\alpha. \tag{223}$$

We wish to get a comprehensive picture of the way in which the steady state, denoted by capital U, V, W, depends on the parameters. In bifurcation studies of this kind, it is usual to choose a distinguished parameter, here the dilution rate λ, and plot some norm of the steady state against it. The conversion will be taken as this norm and the bifurcation diagram will be a plot of U versus λ. Upper-case letters will be used for the steady state values and the equations connecting U, V, and W are:

$$\lambda\alpha U = \lambda(\alpha - V) = \mu V W^2 = \mu\alpha(1 - U)W^2 = (\lambda + 1)W - \lambda. \tag{224}$$

Using the first and the last of these expressions to establish the linear relationship between U and W, and substituting for W in the fourth term gives an equation in U alone:

$$\Lambda(\lambda) = [(1 + \lambda)^2/\lambda] = \mu(1 - U)(1 + \alpha U)^2/U = \mu f(U; \alpha). \tag{225}$$

This is a cubic and so has one or three real roots, but it is in a very suitable form for a geometric analysis.

In the first place, the parameters have separated themselves nicely into two groups, α and μ, defining a family of cubic curves going from infinity at $U = 0$ to $(1, 0)$ in the U, f-plane, and λ, the distinguished parameter, being the sole parameter on the other side of the equation. Moreover, as a function of the distinguished parameter λ, the value of $\mu f(U)$ that determines U, namely $\Lambda(\lambda) = [(1 + \lambda)^2/\lambda]$, comes from infinity (when $\lambda = 0$) to a minimum of 4 at $\lambda = 1$ and retreats to infinity as λ goes to infinity. For $\alpha > 8$, there are two extrema to the curve $f(U; \alpha)$, as shown in Fig. 14. These are at

$$U = \{\alpha \pm \surd(\alpha(\alpha - 8))\}/4\alpha, \tag{226}$$

for which f has values of

$$\Phi(\alpha), \phi(\alpha) = \{\alpha^2 + 20\alpha - 8 \pm (\alpha - 8)\surd(\alpha(\alpha - 8))\}/8. \tag{227}$$

Thus, if $\mu < 4/\Phi(\alpha)$, the solution is unique, for the quantity on the left of Eq. (225) never has a small enough value for there to be more than the single intersection at a relatively low conversion. If $4/\Phi(\alpha) > \mu > 4/\phi(\alpha)$, the branch of solutions created when the left-hand side of the equation dips below Σ never joins up with the low conversion solution because the function of λ retreats to infinity before its value decreases to ϕ. Thus, an isola of solutions lying above the low conversion solution is formed. It is now easy to see that, if $\mu > 4/\phi(\alpha)$, the bifurcation diagram will be of the type commonly called

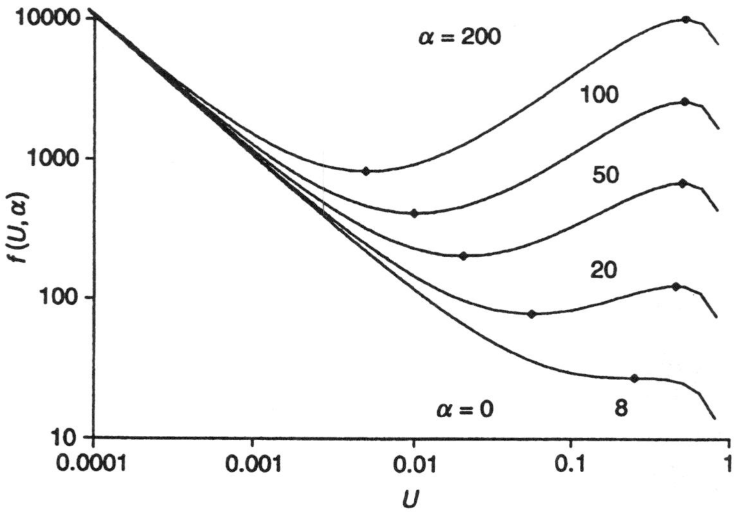

FIGURE 14 Gray/Scott reaction: The curve $f(U; \alpha)$.

mushroom, in which uniqueness obtains for an intermediate range of λ as well as for high and low values.

The neatness of this argument lies in the separation of the parameters and, in particular, the way in which the distinguished parameter, λ, occurs in such a way that its course from zero to infinity is mirrored in the decrease of $[(1 + \lambda)^2/\lambda]$ to the value 4 and its subsequent increase. It might be better to write Eq. (225) in the form

$$[(1 + \lambda)^2/\lambda] = \mu(1 - U)(1 + \alpha U)^2/U = \mu f(U; \alpha, \mu), \qquad (228)$$

for it will prove to be impossible to separate the parameters related to μ from the parameters related to α. Nevertheless, we shall be able to retain the structure of f as the product of two factors, one decreasing with the increase of conversion and the other increasing. Equations (226) and (227) are shown in Fig. 14 as markers at the extrema of $f(U)$, which are shown as functions of U for $\alpha = 8, 10, 20, 50, 100, 200$.

Our method shows that there can be no more than two intersections of a bifurcation curve with a line of constant U for, if $\Lambda(\lambda)$ denote $(1 + \lambda)^2/\lambda$, the two values of λ that satisfy the quadratic $\Lambda = \mu f$ are the only two values possible. Furthermore, because their product is 1, a scale that is logarithmic in λ will make the bifurcation curve symmetric about the line $\lambda = 1$. As $U \to 0$, $\Lambda = \mu f \to \infty$ and one λ goes to zero as $1/\Lambda$, whereas the other goes to infinity as Λ. In fact, the λ are asymptotically μ/U and U/μ. The curve in Fig. 15 is plotted with the U-axis vertical, and there is a symmetric part (not shown) below the λ-axis. At $\alpha = 201$ and $\mu = .005$, it is clearly a mushroom about to become an isola.

We also introduce the parametric representation, which, although not necessary in the case of a single component, is the key to the case of many

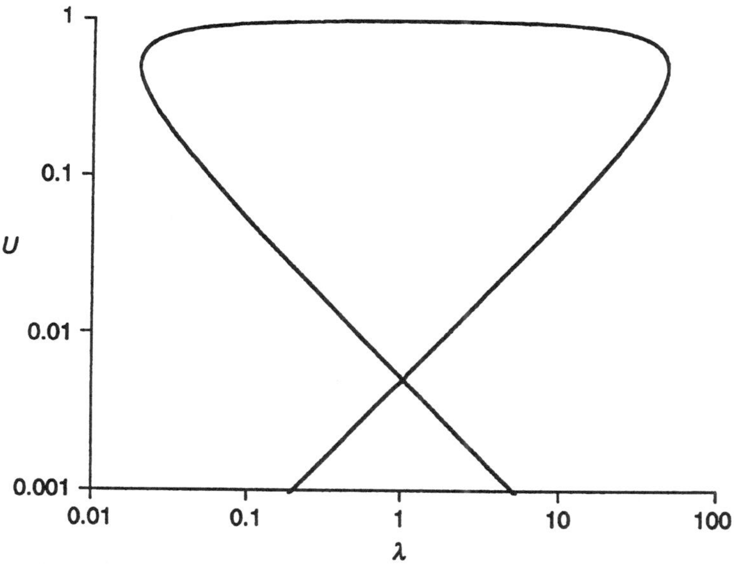

FIGURE 15 Gray/Scott reaction: Bifurcation diagram at the point where *a* mushroom is turning into an isola.

parallel reactions. If we set

$$U = \mu\Omega/(1 + \mu\Omega), \tag{229}$$

then U goes over its full range, 0 to 1, as Ω goes from zero to infinity. Moreover, the relation is monotonic, so there is the inverse function

$$\Omega = U/\mu(1 - U). \tag{230}$$

Equation (218) can be written

$$\Lambda(\lambda) = (1 + \alpha U(\Omega))^2/\Omega. \tag{231}$$

Example 17. Continuous Mixtures and Parallel Gray–Scott Reactions

The parametric approach, which is not strictly needed for a single Gray–Scott reaction, works very well for an arbitrary number of parallel reactions and for continuous mixtures. Figure 16 shows a case of two parallel reactions for which an isola and a mushroom coexist. Because the notions of continuous mixtures and reactions will be treated in Chapter 8, G & H and in the group of papers listed in the "Index of Subjects in Publications" under the heading "Continuous mixtures," we can be very brief and start with the nondimensional equations. Let x be the index of the mixture whose "species" are $A(x)$. The steady-state concentration of the material with index in $(x, x + dx)$ is $V(x)dx$, the feed concentration $a(x)dx$ and the conversion $U(x) = 1 - V(x)/\alpha(x)$, the last being defined only for values of x for which $\alpha(x)$ is not zero. B, the autocatalytic agent, forms itself as an undifferentiated product whose concentration is W. The rate of the first reaction, and hence $\mu(x)$, depends on the

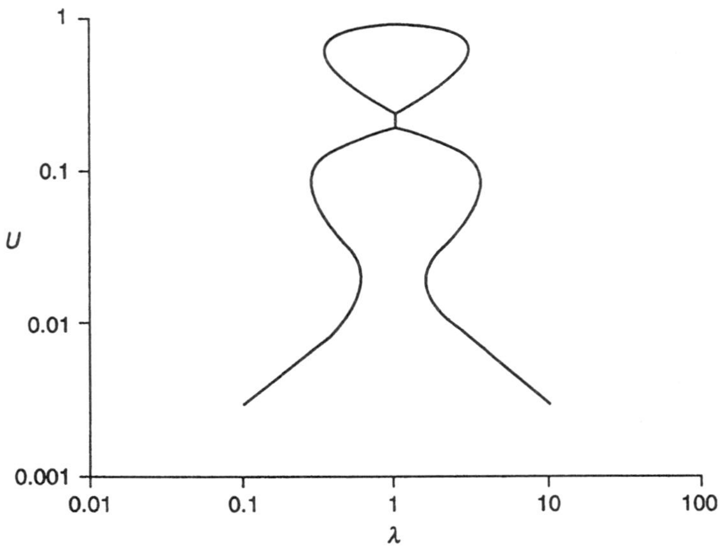

FIGURE 16 Gray/Scott reaction: Bifurcation diagram for two parallel G/S reactions.

index, x. Then the analogs of Eqs. (224) and (225) are

$$\lambda\ \alpha(x)\ U(x) = W^2\ \mu(x)\ \alpha(x)[1 - U(x)] \tag{232}$$

and

$$(\lambda + 1)\ W - 1 = W^2 \int \mu(x)\alpha(x)[1 - U(x)]dx = \lambda \int \alpha(x)U(x)dx. \tag{233}$$

The first of these equations shows that $U(x)/\mu(x)[1 - U(x)]$ is independent of x and may be taken as the parameter Ω, so that

$$U(x) = \mu(x)\Omega/\{1 + \mu(x)\Omega\}, \tag{234}$$

and the same manipulations as before yield

$$\Lambda(\lambda) = (\lambda + 1)^2/\lambda = (1 + \alpha U)^2/\Omega = \mu(1 + \alpha U)^2(1 - \langle \mathrm{U} \rangle)/U, \tag{235}$$

where

$$\begin{aligned} \alpha &= \int \alpha(x)dx, \alpha\mu = \int \alpha(x)\mu(x)dx, \\ \alpha U &= \int a(x)U(x)dx, \quad \alpha\mu\langle U\rangle = \int \alpha(x)\mu(x)U(x)dx. \end{aligned} \tag{236}$$

Thus, the parametric approach works well for continuous mixtures and we have only to calculate integrals, which, except in particular cases, we may have to calculate using finite sums.

The parametric approach lends itself to a spreadsheet type of calculation. The first column can contain a large number of values of Ω spaced logarithmically, i.e., if there are N values between a and b, the nth is $a(b/a)^{n/N}$. Actually, it is best to have three contiguous intervals (a,b), (b,c), and (c,d) each with, for example, 100 values each. Then a and d can be chosen to get a convincing part of the asymptotes and b and c can be used to enlarge the region near $\lambda = 1$ and achieve continuity of the branches. Successive columns can be used to calculate U_i, U, μf, λ (from the quadratic $\Lambda = \mu f$; only one root is needed if only one-half of the logarithmically symmetric diagram is to be plotted), Ω. Then, if the program has an equation solver, succeeding columns can readily be used to compute the Jacobean and the eigenvalues and hence the nature of the steady state. Above all, the parametric method takes care of tracking the isolae, always a problem for continuation methods. (Cf. [302], [306]).

ASYMPTOTICS AND PERTURBATIONS

Whereas a convergent power series gives greater accuracy the more terms one takes, asymptotic approximations have a fixed number of terms and give a closer approximation as the variable tends to some limit, usually zero or infinity. Although, with the widespread availability of powerful computing engines, we are less disposed to use them for evaluating functions, it is still important to have the correct impression of limiting behavior, if only as a check on the plethora of numbers that these machines too easily produce. We have already met various forms of these approximations in several of the examples and will return to two of these. First, however, we look at a regular perturbation that reveals the structure of a shape factor

Example 18. Shape Factors for Particles in Packed Bed Exchange

In Reprint C in Chapter 7, the behavior of a tracer pulse in a stream flowing through a packed bed and exchanging heat or matter with the particles is studied. It is shown that the diffusion in the particles makes a contribution to the apparent dispersion coefficient that is proportional to v^2d^2/D. The constant of proportionality has one part that is a function of the kinematic wave speed μ, but otherwise only a factor that depends on the shape of the particle (see p. 145 and in equation (42) ignore all except the last term and even the suffixes of this; e, being unsuitable as special notation, will be replaced by λ. e is defined in the middle of p. 143 of Chapter 7). In this equation, we should not be surprised to find a term of the same form as the Taylor dispersion coefficient, for it is diffusion across streams of different speeds that causes the dispersion in that case just as it is the diffusion into stationary particles that causes the dispersion in this.[7] What is surprising is that the isothermal diffusion and reaction equation should come up, for λ is defined by

$$\begin{gathered} \lambda = -\Omega'(0), \Omega(\varphi^2) = \langle u \rangle, \\ \operatorname{div}(\operatorname{grad} u) = \varphi^2 u \text{ in } S, u = 1 \text{ on } \partial S, \end{gathered} \tag{237}$$

that is to say, if the effectiveness factor $\eta(\varphi)$ is expanded for small φ, $\eta = 1 - \lambda\varphi^2 + \ldots$, and λ is the negative of the coefficient of φ^2. We can thus

[7] Howard Brenner has generalized the method to a whole class of phenomena in his magisterial paper, "A general theory of Taylor dispersion phenomena. *Physicochem. Hydrodyn.* **1**, 91–123 (1980).

take over the whole body of results for the diffusion and reaction literature (including those with the Robin boundary condition) and use them to calculate these shape factors for particles in the packed bed. Thus the factors for the three canonical shapes in Example 10 are obtained immediately from equation (119) as $1/((n+1)(n+3))$, or 1/3, 1/8, and 1/15 for the slab, cylinder, and sphere, respectively, when there is no external transfer limitation.

The prescription for this shape factor given in Eq. (237) requires only that we find the first coefficient in the expansion and suggests that we might try a solution of the form

$$u = u_0 - \varphi^2 u_1 + \ldots, \tag{238}$$

which is really a perturbation for small φ of the trivial solution $u_0 = 1$ of div(grad u_0) $= 0$ in the interior of the particle S with $u_0 = 1$ on the boundary.

Then, u_1 satisfies div(grad u_1) $= -1$ in S and $u_1 = 0$ on ∂S. We are sure of getting a positive constant, because

$$\lambda = \langle u_1 \rangle = -\langle -1 \cdot u_1 \rangle = -\langle \text{div}(\text{grad } u_1) \cdot u_1 \rangle = \langle \text{grad}^2 \, u_1 \rangle, \tag{239}$$

where the surface integral in Green's formula vanishes because $u_1 = 0$ there.

The unexpected overlap of this problem is more surprising than significant, for the first derivative of the effectiveness factor with respect to the Thiele modulus (Figs. 8 and 9) does not stand out as of particular interest. [Note added in proof (Jan. 99). I could not have been more wrong. See p. 37, fn. 11 and reference then cited.]

Example 15. Diffusion and Reaction in a Slab (Reprise)

While we have contact with this problem, however, we will look at the asymptotics of reactions in a slab of catalyst. In Chapter 3 we used the special form of the equations for diffusion and reaction to get rather general expressions for the Thiele modulus and effectiveness in terms of the center concentration ν

$$\eta = \int_0^1 G(u)dx = (du/dx)_1/\varphi^2 = H(1, \nu)/\varphi \tag{197 bis}$$

$$\varphi = \int_\nu^1 dw/H(w, \nu), \tag{198 bis}$$

where

$$2\int_\nu^{u(x)} G(w)dw = [H(u(x), \nu)]^2. \tag{240}$$

Clearly, $\nu = 1$ corresponds to no reaction, $u = 1$ everywhere, and $\eta = 1$. If the integral of $1/H$ diverges as $\nu \to 0$, then φ will go to infinity as $\nu \to 0$, but $\eta\varphi$ will tend to the finite quantity $H(0,1)$, that is, η is asymptotically proportional to $1/\varphi$ and the constant of proportionality is

$$\surd\{2\int_0^1 G(w)\,dw\}. \tag{241}$$

Take, for example, the easiest case, that of a first-order reaction. Then $G(u) = u$, $H(u, \nu) = \surd\{u^2 - \nu^2\}$ and $\nu = \text{sech}\,\varphi$ by Eq. (240). Then Eq. (197)

gives $\eta = \sqrt{\{1 - \nu^2)}/\varphi = \tanh\varphi/\varphi$. However, $\varphi = \text{sech}^{-1}(\nu)$, and so φ goes to infinity as ν goes to zero. $H(1, 0) = 1$ and $\eta \sim 1/\varphi$. Suppose, however, that the kinetics are those of a zeroth-order reaction, $G(u) = 1$ for $0 \leq u \leq 1$ and zero elsewhere. $H(u, \nu) = \sqrt{2\{u-\nu\}}$ and $\int_\nu^1 dw/\sqrt{2(w - \nu)} = \sqrt{2(1 - \nu)}$, which converges to the square root of 2 as $\nu \to 0$. For values of φ greater than $\sqrt{2}$, which are certainly possible, both the concentration and its derivative go to zero inside the pellet at a point ξ, where $\varphi(1 - \xi) = \sqrt{2}$. We see that $\varphi = H(1, \nu)$ for $\varphi < \sqrt{2}$, so $\eta = 1$. For values of $\eta = \sqrt{2}$, $\eta = 1/\varphi$ and the "curve" and its asymptote coincide.

If we have an external mass transfer resistance at the surface, we need two parameters, $\nu = u(0)$ and $w = u(1)$, $0 \leq \nu \leq w \leq 1$, and then there are various limiting cases according the path that (w, ν) takes to the boundary of its triangular domain. The relevant equations are

$$\eta/\varphi^2 = u'(1) = \varphi H(w, \nu) = \nu(1 - w) \text{ and } \varphi = \int_\nu^w ds/H(s, \nu), \qquad (242)$$

but we shall not go into the detail of their manipulation here. This approach is dealt with rather fully in Chapter 3, Vol. 1 of [I]. We turn instead to the asymptotics of a system that is changed in an essential way in the limit of a small parameter going to zero.

Example 14. Michaelis–Menten[8] Kinetics (Reprise)

In the section entitled "Special Forms" we introduced Michaelis-Menten kinetics rather abruptly as an example of integration by quadrature. The *substrate,* as it is often called in biochemical circles, in the reaction $A \to B$ disappears at a rate

$$da/dt = -ka/(K + a). \qquad (243)$$

This rate law is approximately zero order when $a \gg K$ and first order when $a \ll K$ so that, in some sense, it gets saturated and shows an inhibition of reaction at high concentrations of substrate. It is assumed that the enzyme that catalyses the reaction has a number of sites to which A can attach itself, be transformed to B, and then detach itself. If E denotes the enzyme and e_0 (and e) the concentration of binding sites it initially (and currently) provides, C the complex of the enzyme and substrate and c its concentration, then because sites that are not occupied are available

$$e + c = e_0. \qquad (244)$$

In practice, e_0 is very much smaller than a_0. The mechanism is $A + E \leftrightarrow C \to B + E$, and we assume that these are elementary steps with mass action

[8] This kinetic scheme, of fundamental importance to biochemistry, was first analyzed correctly by Briggs and Haldane. It is often referred to in physicochemical circles as *Langmuir–Hinschelwood* kinetics and chemical engineers encounter it as *Hougen–Watson.* In spite of the injustice it does to Briggs and Haldane, it seems best to leave it as Michaelis–Menten. Certainly, BrH_3LaM_2W is an impossible compound.

kinetics for each. Thus

$$da/dt = -k_1ae + k_2c, db/dt = k_3c \qquad (245)$$
$$dc/dt = -de/dt = k_1ae - (k_2 + k_3)c.$$

At time $t = 0$, $a = a_0$, $e = e_0$, $c = b = 0$. We see that, thanks to Eq. (234), we need only two equations, namely, those for a and c. We will make the equations dimensionless by scaling a and c into the interval [0, 1]. Let

$$a = a_0\, u,\ b = a_0\, w,\ c = e_0\, v,\ \varepsilon = e_0/a_0 \qquad (246)$$

$$\tau = k_1a_0t,\ \mu = (k_2 + k_3)/k_1a_0,\ \kappa = k_3/k_1a_0, \qquad (247)$$

then

$$du/d\tau = -u + (u + \mu - \kappa)v \qquad (248)$$

$$\varepsilon dv/d\tau = u - (u + \mu)v, \qquad (249)$$

to be solved subject to $u(0) = 1$, $v(0) = 0$.

This form of the equations shows clearly that different scales of time are at work in the two equations. Time is τ/ε in the second equation and the concentration $u(\tau)$ changes very slowly in comparison with v. Now if u were constant, the second equation, which has a stable steady state, v would almost instantly reach this so-called pseudo–steady-state $v = u/(u + \mu)$. Then Eq. (248) would be

$$du/d\tau = -\kappa\, u/(\mu + u), \qquad (250)$$

of which we found the solution by separation of variables and quadrature, namely [Eq. (195) in the present notation]

$$\kappa\tau = u_0 - u(\tau) + \mu \ln(u_0/u(\tau)), \qquad (251)$$

where u_0 has been left undetermined for the moment. Notice that the initial conditions cannot be satisfied, for u_0 would have to be zero if v were to be initially zero and the pseudo–steady-state $v = u/(u + \mu)$ were true. Likewise, if u_0 were 1, $v(0)$ would not be zero. The reason for this is that, in the first interval of time of order ε, the variable v has not had time to latch onto the PSSH (pseudo–steady-state hypothesis) relationship.

Evidently, we need another solution for the first interval of time, say $\chi(\varepsilon)$. Let us scale the time in the initial interval and write $\sigma = \tau/\chi$ so that Eqs. (248) and (249) become

$$(\varepsilon/\chi)\ du/d\sigma = \varepsilon\{-u + (u + \mu\ \kappa)v\} \qquad (252)$$

$$(\varepsilon/\chi)\ dv/d\sigma = u - (u + \mu)v. \qquad (253)$$

If we put $\varepsilon = \chi$ and let $\varepsilon \to 0$, the first equation says that u does not really change on this time scale so that $u(\sigma) = 1$ satisfies the initial condition and the equation. Putting $u = 1$ in the second equation and solving it with the initial condition $v = 0$ gives

$$v(\sigma) = \{1\text{-}\exp(-\sigma(1 + \mu))\}/(1 + \mu)\}. \qquad (254)$$

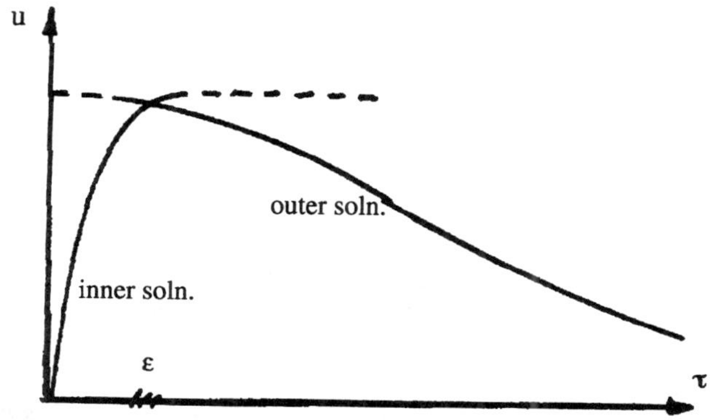

FIGURE 17 The merging of inner and outer solutions in singular perturbation.

This is called the *inner solution,* and is valid in a time interval of order ε; Eq. (241) is the outer solution, valid everywhere else; it will be denoted by capitals U and V (see Fig. 17). The challenge is to see if we can splice it onto the inner solution. We have u_0 at our disposal to do this.

What we want is an intermediate time that runs quickly in comparison with τ but slowly in comparison with σ. Let it be $\rho = \tau/\theta(\varepsilon) = \varepsilon\sigma/\theta(\varepsilon)$ where $\theta(\varepsilon) \to 0$ and $\theta(\varepsilon)/\varepsilon \to \infty$ as $\varepsilon \to 0$. $\varepsilon^{1/2}$ very neatly fills these requirements. If the inner solution is to match up with the outer solution,

$$u_0 = U(\tau) = L_{\varepsilon\to 0}\, U(\rho\, \varepsilon^{1/2}) = u(\sigma) = L_{\varepsilon\to 0}\, u(\rho/\varepsilon^{1/2}) = 1 \qquad (255)$$

$$V(\tau) = L_{\varepsilon\to 0}\, V(\rho\, \varepsilon^{1/2}) = v(\sigma) = L_{\varepsilon\to 0}\, v(\rho/\varepsilon^{1/2}) = 1/(1 + \mu). \qquad (256)$$

The 1, which is the splicing value, is said to be the common part of the inner and outer solutions for u; likewise, the $1/(1 + \mu)$ for v. A uniform approximation can be constructed by adding the inner and outer solutions and subtracting their common part. Thus, with $U(\tau)$ given implicitly by

$$\kappa\tau = 1 - U + \mu \log(1/U), \qquad (257)$$

the uniform approximation is

$$u^*(\tau) = U(\tau),\ v^*(\tau) = U(\tau)/(1 + U(\tau)) - e^{-\tau(1+\mu)/\varepsilon}/(1 + \mu). \qquad (258)$$

Higher order terms can be obtained by writing the inner and outer solutions as expansions in powers of ε and solving the sets of equations obtained by comparing coefficients. This enzymatic example is treated extensively in [73] and a connection with the theory of materials with memory is made in [82]. The essence of the *singular perturbation analysis,* as this method is called, is that there are two (or more in some extensions) time (or spatial) scales involved. If the initial point lies in the domain of attraction of steady states of the fast variables and these are unique and stable, the state of the system will rapidly pass to the stable manifold of the slow variables and, one might hope, be more easily analyzed in the subsystem.[9] An example of a change of stability during the course of an autocatalytic reaction was studied by Schneider [107].

MOMENTS AND GENERATING FUNCTIONS

In chemical engineering, there are many situations in which the response is montiform. Chromatography comes immediately to mind as a process in which sharp peaks are desirable for analysis or preparation. Such peaks can be very roughly characterized by their means and standard deviations, and if, as is quite often the case, we have reason to suppose that a distribution is asymptotically Gaussian, then these two statistics suffice for its description. If $f(r)$ is the distribution in question, its moments about the origin are

$$M_n = \int r^n f(r)\, dr, \tag{259}$$

and the mean and variance are

$$\mu = M_1/M_0 \text{ and } \sigma^2 = \int (r - \mu)^2 f(r) dr \Big/ \int f(r) dr = M_2/M_0 - \mu^2. \tag{260}$$

The integration is over the whole region in which $f(r)$ is defined. The third moment about the mean gives an index of skewness, it being zero for a perfectly symmetric distribution. The fourth moment is an index of the sharpness or flatness of the peak and is rarely called for.

Take the case of flow and diffusion in a long tube with a flat velocity profile but no reaction. The concentration of solute is governed by

$$D(\partial^2 c/\partial z^2) - v(\partial c/\partial z) = \partial c/\partial t. \tag{261}$$

We assume that the tube is very long compared to the region in which significant changes of concentration and as boundary conditions simply take $c \to 0$ as $z \to \pm\infty$. We can imagine the following tracer experiment.[10] The fluid is light sensitive and, at time $t = 0$, we zapp the fluid in a very thin cross-section of the tube centered on $z = 0$. That is about as near as we can get to a delta function in non-mathematical life.[11] Now there is a solution to Eq. (261) with these boundary conditions and the initial condition $c(z, 0) = \delta(z)$, namely the Gaussian distribution

$$c(z, t) = (1/4\pi Dt)^{1/2} \exp(-(z - vt)^2/4Dt), \tag{262}$$

where the factor $\sqrt{\pi}$ has been inserted to make $M_0 = 1$. Then the first spatial

[9] There are any number of good books on this subject, and three come to mind: J. D. Murray, *Asymptotic Analysis*. Oxford: Clarendon Press, 1974; C. C. Lin and L. A. Segal, *Mathematics Applied to Deterministic Problems in the Natural Sciences*. New York: Macmillan, 1974; and R. E. O'Malley. *Introduction to Singular Perturbations*. New York: Academic Press. 1974. See also Varma and Morbidelli. *Mathematical Methods in Chemical Engineering*. New York: Oxford University Press, 1997.

[10] This is not far-fetched, for Adler of Case Western Reserve University did use this method in studies of dispersion in helical tubes.

[11] I do not say *real life,* nor do I call the extrauniversity world *the real world*.

moment is

$$\mu(t) = \int_{-\infty}^{\infty} z\, c(z,t)\, dz = vt. \tag{263}$$

The higher moments are most easily calculated about this mean by using the substitution $z = vt + x\sqrt{(4Dt)}$ in

$$m_n = \int_{-\infty}^{\infty} (z - vt)^n\, c(z,t)\, dz, \tag{264}$$

whence

$$m_n = 0,\ n \text{ odd};\ m_n = (2Dt)^p(2p-1)(2p-3)\cdots 1,\ n = 2p. \tag{265}$$

Thus,

$$\sigma^2 = 2Dt. \tag{266}$$

If we have reason to suppose that a peak is asymptotically Gaussian, then a measurement of the mean gives an average velocity and the rate of growth of $\sigma^2/2$ gives the apparent diffusion coefficient. Only in this simple case of the PFTR in an infinite tube is the peak always Gaussian and the mean and variance suffice for a complete knowledge of the profile.

Let us see how the moments might be computed directly. Take the original equation for c, multiply by z^n, and integrate from $-\infty$ to ∞. The product of z^n and a derivative of c with respect to z can be integrated by parts until only c itself remains, so that

$$(d/dt)M_n = vnM_{n-1} + Dn(n-1)M_{n-2}. \tag{267}$$

The initial values of M is calculated from the initial distribution and the equations can be solved sequentially. Thus for the delta zapp,

$$M_0(0) = 1,\ M_1(0) = M_2(0) = \cdots = 0, \tag{268}$$

so that

$$(d/dt)M_0 = 0, \qquad M_0(t) = 1 \tag{269}$$

$$(d/dt)M_1 = vM_0 \qquad M_1(t) = vt \tag{270}$$

$$(d/dt)M_2 = 2\,vM_1 + 2DM_0 \qquad M_2(t) = v^2t^2 + 2Dt, \tag{271}$$

from which $\mu = vt$ and $\sigma^2 = 2Dt$ as before.

If there were a first-order reaction taking place, M_0 would decay exponentially, but so would M_1 and M_2, so that the mean and variance would be the same. If we had a number of parallel tubes communicating with each other, we would have to develop equations for the moments in each before averaging them to get the overall mean and variance.

It should be stressed that the moment approach works only for systems of linear equations but that, when it is applicable, it is capable of giving insight into some otherwise impossibly complicated situations.

We have looked at spatial moments as functions of time and seen that in the nonreacting PFTR they are especially simple and take their asymptotic form right from the start. The asymptotic behavior of quite a number of systems of importance in process technology is roughly Gaussian, and easily calculable moments give a valuable insight into the factors that affect the flow and dispersion.

We now turn to the temporal moments, i.e., we take a stand at a fixed point z and observe the values of c as the peak passes. The moments are

$$T_n = \int_0^\infty t^n \, c(x, t) \, dt, \tag{272}$$

and they are most easily found by using the Laplace transform as a generating function. Using an asterisk for the Laplace transform with respect to t of any function

$$c^*(z, s) = \int_0^\infty \exp(-st) \, c(z, t) \, dt = \sum_0^\infty (-)^n s^n \, T_n/n!, \tag{273}$$

or

$$T_n = (-)^n \, (d^n c^*/ds^n)_{s=0}. \tag{274}$$

The precise conditions for these formulae to be valid are discussed in Chapter 9. A useful short-cut if the output is of the form of a single exponential, $\exp(-as + bs^2 \cdots)$ is that the mean is a and the variance $2b$.

For temporal moments it is best to consider a semi-infinite tube and let $c(0, t) = \delta(t)$ be the ideal tracer. Initially, $c(z, 0) = 0$ and c must remain finite as $z \to \infty$. These are not the most sophisticated conditions, as we know from Example 3, but they give the swiftest answer. If we take the Laplace transform of Eq. (261), we have

$$D(d^2c^*/dz^2) - v(dc^*/dz) - sc^* = 0 \tag{275}$$

subject to

$$c^*(0, s) = 1, \; c^* \text{ finite as } z \to \infty. \tag{276}$$

Discarding the positive characteristic root because this would not keep c^* finite, we have

$$\begin{aligned} c^*(z, s) &= \exp(-(vz/2D)\{\surd[1 + (4Ds/v^2)] - 1\} \\ &= \exp(-(z/v)s + (Dz/v^3)s^2 - \cdots). \end{aligned} \tag{277}$$

Hence, by the previous remark,

$$\mu(z) = z/v, \; \sigma^2(z) = 2Dz/v^3.$$

We notice that, if we take a length scale vt for the spatial moments and z for the temporal and call them L, the dimensionless ratio $\sigma^2/\mu^2 = 2D/vL$.

This has been the simplest possible introduction to moments and generating functions. More significant use of them will be found in Chapter 7.

Generating functions occur in other contexts than moments and, in particular, the theory of special functions is full of them. Their use in deriving the solution to a whole family of equations is well illustrated in the reprint M, p. 337, Chapter 10. A simple example would be the ideal residence time tracer the chain of identical C*s governed by

$$du_n/dt = (u_{n-1} - u_n)/\theta, \; u_0 = \delta(t), \; u_n(0) = 0.$$

If we multiply each equation by the corresponding power of λ and write $U(\lambda, t) = \Sigma u_n(t)\lambda^n$, where the summation goes from $n = 1$ to infinity, (this is a formal series and no question of convergence arises), we have an equation for U

$$dU/dt = \lambda\delta(t)/\theta + (\lambda - 1)U/\theta,\ U(\lambda, 0) = 0.$$

The solution is

$$U = (\lambda/\theta)\exp((\lambda - 1)t/\theta) = \Sigma\lambda^n\{t^{n-1}\exp(-t/\theta)/(n - 1)!\theta^n\}.$$

We can read off u_n from the exponential expansion. This is the *Gamma distribution,* and it is the natural distribution in many kinetic problems: see G and H, Chapter 8 and [276].

OBSERVING CONDITIONS

Pure mathematicians (if one may be permitted an old-fashioned and largely discredited distinction) love conditions. They work hard at weakening them or strengthening them as the need may be. They prove the equivalence of different formulations. They rejoice in the subtleties that distinguish Abelian and Tauberian theorems.[12] Indeed, the day that notions of the necessity and the sufficiency of conditions are grasped is a red-letter day in one's education. The temptation of the applied mathematician is to be impatient of the restrictions that they sometimes imply. Certainly, one sometimes rushes ahead with an expansion without worrying about its convergence, but it is with the knowledge that one must go back and check it out and, if necessary, modify things.

Conditions must be observed and should be made explicit even at the expense of an air of pedantry.

Example 19. Solvability Conditions

It is quite clear that I did not properly understand solvability conditions for linear partial differential equations in 1955, in spite of having written out in longhand a translation of the second volume of Courant and Hilbert's *Methoden der Mathematische Physik* during my years at Edinburgh. I was fortunate in that the asymptotic properties that were my chief intention were not vitiated by this ignorance. Had I tried to take the moment calculations any further, I might have gotten into real difficulty. It was not for some years that its shortcomings were repaired by N. G. Barton in a paper that is as exemplary for its academic courtesy as it is for its mathematical correctness.[13]

Briefly, the inhomogeneous equation $Au = f$, where A is a linear operator defined on a domain D_A dense in a Hilbert space H, does not always have a solution. f must be orthogonal to the solutions of the homogeneous adjoint

[12] An example of the distinction in connection with moments is to be found in Chapter 7, p. 141.

[13] Barton, N. G. On the method of moments for solute dispersion. *J. Fluid Mech.* **126,** 205–218 (1983).

equations $A^*v = 0$. This follows because $\langle f, v\rangle = \langle Au, v\rangle = \langle u, A^* v\rangle = \langle u, 0\rangle = 0$. To see where I went adrift in the matter of solvability, I must outline the method of attack used in A, pp. 109–120.

The paper grew out of my work on chromatography at I.C.I. as I tried to understand the roles of convection and diffusion in the broadening of chromatographic peaks. Taylor had solved the basic problem a year or two before and had shown that that the longitudinal dispersion was enhanced by a small molecular diffusion coefficient because this inhibited the exchange between the slower and faster streams that would otherwise average out the speed of the cloud and keep it the more compact. I did not see how to generalize Taylor's approach, for it involved some assumptions that clearly came from an insight far greater than mine, but I hit upon a description in terms of the growth of the spatial moments of an initially confined distribution of solute. I transformed the equation for diffusion and flow to coordinates moving with the mean speed of flow, U, scaling lengths by the tube radius, a, and time by a^2/D. There emerged, of course, the Peclet number $aU/D = \mu$ and the (slightly simplified by making D constant from I[4]) equation was

$$\partial C/\partial\tau = (\partial^2 C/\partial\xi^2) - \mu\chi(\eta, \zeta)(\partial C/\partial\xi) + (\partial^2 C/\partial\eta^2) + (\partial^2 C/\partial\zeta^2). \quad (278)$$

ξ is the coordinate in the direction of flow, η and ζ are in the cross-sectional plane, and χ is the velocity relative to the mean, so $\langle\chi\rangle$, its average over the cross section, is zero. There is no flux across the walls of the tube and $\chi = 1$ there. The initial distribution of the solute is known.

One can calculate the moments of the distribution over any line of flow

$$c_p(\eta, \zeta, \tau) = \int \xi^p\, C(\xi, \eta, \zeta, \tau)\, d\xi \quad (279)$$

and

$$m_p(\tau) = \langle c_p\rangle. \quad (280)$$

The moment equations are treated as in Example 19 and satisfy

$$\partial c_p/\partial\tau = (\partial^2 c_p/\partial\eta^2) + (\partial^2 c_p/\partial\zeta^2) + p(p-1)c_{p-2} + p\mu\chi c_{p-1} \quad (281)$$

$$dm_p/d\tau = p(p-1)\langle c_{p-2}\rangle + p\mu\langle\chi c_{p-1}\rangle. \quad (282)$$

To get an apparent dispersion coefficient we need this last equation for $p = 2$. To get this we need m_0, c_0, m_1, c_1 in succession. $m_0 = 1$, reflecting the fact that the total amount of solute is constant; $c_0 = 1 +$ terms that decay to zero reflecting the spread of solute until it is uniform over the cross section (because $\langle\chi\rangle = 0$, χ is orthogonal to the solution of the homogeneous equation, namely 1); $m_1 = 0 +$ terms that decay to a constant, reflecting the fact that although the center of mass ultimately moves with the mean speed of the flow, there will be an offset due to the inequities of the initial distribution. It was with the equation for c_1 that I went astray through ignoring the orthogonality condition. It is necessary to add and subtract certain terms. Fortunately, it did not affect the asymptotic behavior of the second moment on which the extension of Taylor's formula was based.

Example 20. Conditions Easily Overlooked or Hidden

Dick Bellman was a great mathematician with scarcely an equal in celerity of thought, but he did not always make allowance for the relative slowness of his readers' minds or for the fact that a choice that might be instinctive for him would be a matter of deep cogitation for others. In short, he did not always spell out his conditions as completely as he might have done had he not a mind that was already "galloping off in all directions."[14] At any rate, I made the wrong choice in applying dynamic programming that was only corrected when the conditions spelled out in a theorem of Pontryagin's were observed. I found the mistake instructive and hope others will also.

The optimal problem in question was that of determining the cooling policy for a batch reactor in which an exothermic reaction is taking place. A contour of constant reaction rate $R = \mu$ in the plane of extent and temperature are as shown in the Figure 18. At any level of extent (i.e., conversion), the reaction rate goes through a maximum, on the curve Γ_m, before it goes to zero on the equilibrium line, Γ_e. It stands to reason that we want to keep as close as we can to the locus Γ_m, but the trouble is that it goes off to infinity at low conversions and so is not practically attainable. This problem was tackled in a master's thesis by Blakemore and published as [29], using dynamic programming, and, as part of a larger study of optimization of batch reactors, in [43], the doctoral work of Siebenthal using Pontryagin's method. We will simplify the notation and arguments of each here, but give a concordance at the end. The equations can be obtained from those of the C* by omitting the flux terms. All variables will be dimensionless.

Let

u = dimensionless conversion
v = dimensionless temperature
w = dimensionless rate of cooling
t = dimensionless time.

The governing equations[15] would then be

$$du/dt = r(u, v) \qquad u(0) = U \tag{283}$$

$$dv/dt = r(u, v) - w \quad v(0) = V. \tag{284}$$

In these equations there are various kinetic parameters, which we take to be constants. μ is the value of time and we wish to maximize $G = u(\theta) - \mu\theta$, where θ is the duration which also has to be chosen optimally. The reaction rate has the properties: $R(u) = r(u, v_m(u)) \geqq r(u, v)$ for all u; $r(u, v_e(u)) = 0$; $\partial r/u < 0$.

Dynamic programming argues that the maximum, when found, will depend on the initial state U, V, say, $F(U, V)$. If, during an initial period $(0, \tau)$

[14] I seem to recall a review of one Bellman's books in which the reviewer compared him to the cavalry hero who "mounted his horse and galloped off in all directions," but have not been able to locate the reference.

[15] The scaling has been chosen so that the rate expression appears in both equations with unit coefficient.

we choose w optimally we shall attain the same maximum. If we do not then we shall fall short of the maximum.[16] Thus,

$$F(U, V) = \mathcal{M} F(u(\tau), v(\tau)) - \mu\tau, \tag{285}$$

where by $\mathcal{M}$, we mean the maximum attainable by choice of $w(t)$ over the interval $(0, \tau)$. Letting τ become small, expanding F in a Taylor series, using the equations for the rates of change of u and v, and, finally, dividing by τ and setting it to zero gives

$$\mathcal{M}\{F_U r(U, V) + F_V [r(U, V) - w]\} = \mu. \tag{286}$$

There will certainly be a limit to the rate of cooling, so let us write

$$0 \le w \le W^*. \tag{287}$$

The optimal control policy is now obvious and commonsensical. If it is advantageous that the temperature should rise (i.e., $F_V > 0$), then $F_V[r - w]$ will be greatest when w is least and, so, $w = 0$. If F_V is negative, it is advantageous that the temperature should fall and so w should be as large as possible (i.e., $w = W^*$). This is the origin of the name *bang-bang control,* which, I believe, is due to Bellman. Only on the locus $F_V = 0$ does w lie between its limits. It is not of interest to go above the curve $r(u, v) = \mu$, for the reaction rate is there so small that what we want to maximize is decreasing. Moreover we have seen that the most favorable temperature is $v_m(u)$ so that A in Fig. 18, the intersection of $r = \mu$ and Γ_m, is clearly the place to finish.

Γ_m (ATM in Fig. 18) is the locus $F_V = 0$, for on it the Eq. (285) reduces to $\mathcal{M} F_U\, r(U, V) = \mu$, and the maximum is given by $V = v_m(U)$ because F_U is always positive. In Fig. 18, DA is the adiabatic path ($w = 0$) to A and FA is the path under maximum cooling ($w = W^*$), and we are only interested in the region under DAF. To keep on the singular path Γ_m, the cooling rate must be chosen so that

$$dv/du = 1 - w/R(u) = R'(u)$$

or

$$w = \{1 + (-R'(u)\}\, R(u). \tag{288}$$

At some point on Γ_m (for example, J), $w = W^*$, and it will be impossible for the path Γ_m to be followed because the cooling rate required simply is not there. There are two paths arriving at J; the adiabatic path KJ and the path under maximum cooling HJ. Shown as a broken curve, it comes in to J tangentially to Γ_m. Now, although $F(U, V)$ will clearly be a continuous surface, dynamic programming has nothing to say about its derivatives. So it seemed that a nice, clean way to divide up the plane would be to have adiabatic reaction to the left of $ATMJQH$. Any reactor starting below KJ (for example,

[16] Bellman's statement of the Principle of Optimality is, "An optimal policy has the property that whatever the initial state and initial decision may be, the remaining decisions must constitute an optimal policy with regard to the state resulting from the first decision." In other words: If you don't do the best you can with what you happen to have, you'll never do the best you might have done with what you should have had.

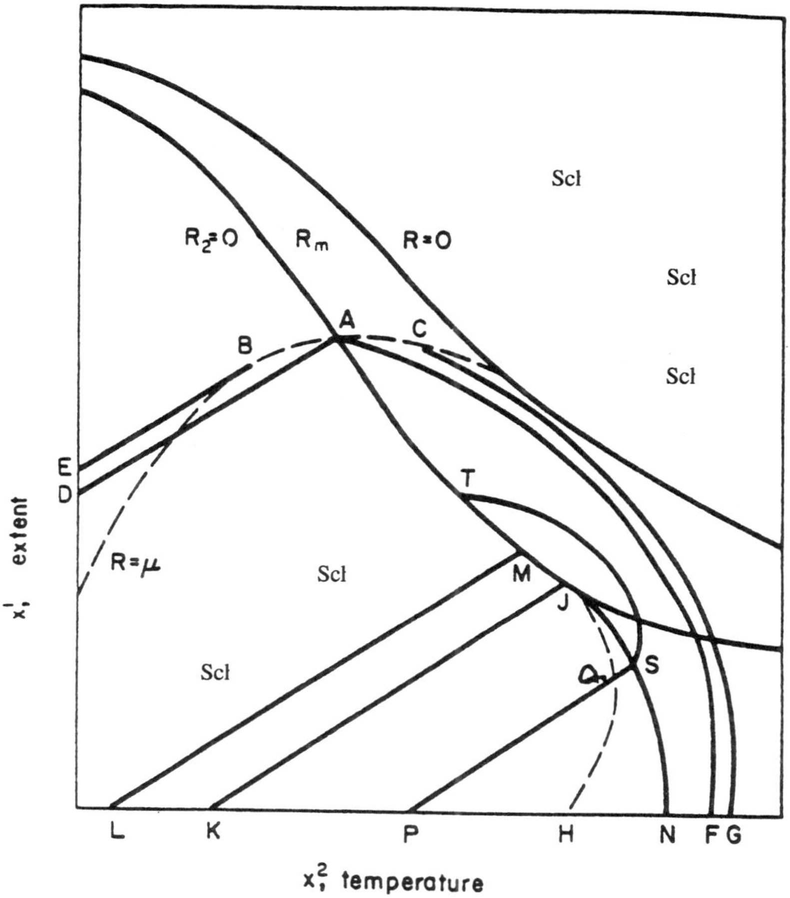

FIGURE 18 Structure of the optimal control of a batch reactor.

at P) would have zero cooling until it reached the curve JQH (at Q), would there jump to full cooling and swing round on a path (QX) that would meet Γ_m (at X) and then drop down to whatever cooling rate was needed to keep it on Γ_m. Such indeed is the structure of the optimal policy, but we were wrong (in [29]) in assuming that the bang-bang transition occurred on the full cooling path JQH. It was only later (in [43]) that the correct transition point was found and that was because of the conditions for optimality were spelled out.

Pontryagin's method can be applied to a number of different cases, but only the form needed for comparison will be given here. Consider an n-dimensional state space vector $\mathbf{x} = (x^1, \dots x^n)$, with r control variables $u = (u^1, \dots u^r)$, related by

$$d\mathbf{x}/dt = \mathbf{f}(\mathbf{x}, \mathbf{u}), \ \mathbf{x}(0) = \mathbf{x}_0, \tag{289}$$

where $\mathbf{f}$ is a vector of n continuously differentiable functions of its $(n + r)$ arguments. Then, if a control $\mathbf{u}^*(t)$ that minimizes $G(\theta) = \mathbf{c} \cdot \mathbf{x}(\theta)$ exists, there exists also an optimal adjoint vector $\psi^*(t)$ satisfying

$$d\psi/dt = -\text{grad}_x\,[\psi \cdot \mathbf{f}(\mathbf{x}, \mathbf{u})],\ \psi(\theta) = -\mathbf{c}, \tag{290}$$

such that at each instant

$$M(t) = H(\mathbf{x}^*, \mathbf{u}^*, \psi^*) = \psi^* \cdot \mathbf{f}(\mathbf{x}, \mathbf{u}) = \text{Max}_u\, H(\mathbf{x}^*, \mathbf{u}, \psi). \tag{291}$$

$M(t)$ is constant and, if the duration of the process, θ, is not specified but allowed to reach its optimal value, this constant is zero. The maximum is over all permitted values of u.[17]

The statement requires an existence theorem and this we are given by Markus and Lee.[18] Translating into the Pontryagin notation [43]: $t = x^0$, $u = x^1$, $v = x^2$, $w = u^1$, $\mathbf{f} = (1, r, r - w)$,

$$H = \psi_0 + (\psi_1 + \psi_2)r - \psi_2 w. \tag{292}$$

In the original formulation, $u(\theta) - \mu\theta$ was to have been maximized, but to conform to the theorem, we minimize its negative. The final values of the adjoint variables are

$$\psi_0 = -\mu,\ \psi_1 = 1,\ \psi_2 = 0, \tag{293}$$

and they satisfy the equations

$$d\psi_0/dt = 0,\ d\psi_1/dt = -(\psi_1 + \psi_2)r_u,\ d\psi_2/dt = -(\psi_1 + \psi_2)r_v. \tag{294}$$

Because we have not specified the time, the maximum value of the Hamiltonian H will be zero, or

$$\text{Max}\{(\psi_1 + \psi_2)r - \psi_2 w\} = \mu. \tag{295}$$

If we insert the final values we get $r = \mu$ (i.e., the path should finish on the locus above which there would be no hope of maximizing the objective function $u - \mu\theta = x^1 - \mu x^0$), because to react is to decrease the objective. Because the quantity to be maximized is linear in w, the rate of cooling, this will be either 0 or W* according as ψ_2 is positive or negative. The singular path that separates the policies is that along which $\psi_2 = 0$ and this, by Eq. (294), is the locus $r_v = 0$. Along this path, ψ_1 satisfies $d\psi_1/dt = \psi_1 r_u$ and so it increases with t (since $r_u < 0$), is 1 at $t = \theta$ and, from the form of the equation, cannot change sign. Its value at J is therefore between 0 and 1. Below the point J the controlled variable w must be either 0 or W*, the former giving KJ and the latter HQJ. It is impossible to come in to J on a full cooling path and maintain the continuity of ψ_2. Because $\psi_1 = \psi_1 + \psi_2 = \mu/r$ [by Eq. (295)] is positive at J and so also in some neighborhood of J. On such a path, $r_v > 0$ (because it is beneath Γ_m), and, near J, ψ_2 is decreasing (by the last Eq. (294)). ψ_2 must be negative and decreasing and so cannot go continuously to zero. No such objection holds for the adiabatic approach KJ. There must therefore be a curve JSN, not a solution of the equations on which ψ_2 changes sign and the adiabatic path (e.g., RS) changes to a full-cooling trajectory (ST). This

[17] Pontryagin, L. S., V. G. Boltyanskii, R. V. Gamkrelidze, and E. F. Mishenko. *The Mathematical Theory of Optimal Processes*. New York: Wiley.

[18] Marcus, L., and E. B. Lee. *Arch. Rat. Mech. Analysis* **8,** 36 (1961).

will swing round and hit some point of Γ_m (T). The whole problem can thus be wrapped up by taking a few points on Γ_m and integrating backwards from each with $w = W^*$ and $\psi_2 = 0$ initially. The locus of points on which ψ_2 changes sign again gives the switching curve.

Pontryagin's adjoint variables ψ are clearly the partial derivatives of Bellman's F and the continuity of the adjoint variables (as solutions of adjoint differential equations) implies the smoothness of the surface that was lacking in the first paper.

4

PRESENTING THE MODEL AND ITS BEHAVIOR[1]

The vitality of the modeler's own work, as well as the impact that a model can have on others, depends critically on its presentation. The ultimate goal may be a table of numbers interpolable to a given accuracy, but it is still probable that an accompanying graph of much lower accuracy will get more attention and convey a more synoptic picture. Both kinds of figure are necessary: the quantitative, if necessary backed up with tables; the qualitative, which can indeed be quantitative, but allows for didactic distortion and magnified sections in its endeavor to convey as much and as vividly as possible. The capabilities of the supercomputer in visualization, its colors, sound effects, stereoscopic three dimensionality, are impressive and should be used to the full, but it is the more immediate graphical presentation that one sketches for oneself before handing over to the computer that we are concerned with here. Fortunately, dynamical systems lend themselves to vivid visualization as Abraham and Shaw have shown in their remarkable books and papers.[2]

[1] I am indebted to the American Physical Society, publishers of the journal *Chaos*, and to the editors of its special issue, G. Kevrekidis and J. L. Hudson, for allowing me to use extensive quotes from [311].

[2] R. H. Abraham and C. D. Shaw. Dynamics: A visual introduction. In F. E. Yates (ed.), *Self-Organizing Systems*. New York: Plenum Publishing Corp. (1987). See also their four-part series, *Dynamics*. Santa Cruz: Ariel Press, 1983.

THE PHASE PLANE

The idea of the phase plane[3] is to let the time be a parametric variable along the curve $(u(t), v(t))$ in the u, v plane. The equations are often autonomous (i.e., the right-hand sides are not functions of t) but, when they are not, they can be made so by adding $w(t) = t$. Then the third differential equation is $w' = 1$. If necessary, a nonvanishing function of F and G, for example $K(u, v)$, can be divided into each equation. This distorts the time that must be recovered from a third equation. A particularly useful transformation of this sort is given by $K^2 = F^2 + G^2$, for then the independent variable is the path length, s, and

$$du/ds = F(u, v)/K(u, v), \; dv/ds = G(u \cdot v)/K(u, v), \; dt/ds = 1/K(u, v). \quad (296)$$

Amundson and Bilous introduced the phase plane to chemical engineering in 1955 using calculations done on an old analog calculator.[4] They took the case of a first-order exothermic reaction and plotted phase planes for situations in which there were one and three steady states. In the case of three, the middle one is always unstable. The value of having a plane diagram on which one could impose one's geometrical imagination was proved in the subsequent problem that Amundson took up just at the time I went to Minnesota. This problem was to try to control the unstable intermediate steady state. I remember sketching the heat removal and generation curves and seeing that controlling by a measurement of the deviation from that state would turn the linear heat removal curve into a parabola, which, with increasing control, would cross the heat generation line ever more steeply until the controlled state was the only steady state and stable. It seemed straightforward enough and I thought it might not be particularly interesting, but when I came to track the phase planes, it became clear that there was more to it than I had realized.

In Fig. 19,[5] the ordinate is extent of reaction and the abscissa the temperature, and we are trying to stabilize the saddle-point B. Local stability is attained when $\mu = 2.25$, but as we shall see, "local" is very local. As we increase the control parameter, μ, the upper steady state, C, moves down, changing its character and passing through B, the state we are trying to control. We were very lucky in our choice of constants, for, when $\mu = 1.125$ in our particular example, there is a homoclinic orbit from A to A, with A as a resting point on it. This immediately breaks away and becomes a limit cycle. Figure 20 shows how this limit cycle shrinks as the single steady state, B, moves toward stability. At first I thought that stability for B would be attained by the limit cycle shrinking to zero. However, this is a subcritical bifurcation and the limit cycle has not vanished when the state becomes stable. An unstable limit cycle grows inside the stable one until they touch and evaporate.[6]

[3] I have not attempted to find the origin of the phase-plane, but it must go back at least to Poincaré, who speaks of "*les courbes definies par* $dx/F(x, y) = dy/G(x, y)$."

[4] N. R. Amundson and O. Bilous. Chemical reactor stability and sensitivity. *A.I.Ch.E.J.* **1,** 513 (1955).

[5] Taken from R. Aris, *Introduction to the Analysis of Chemical Reactors.* Englewood Cliffs, NJ: Prentice Hall. 1965.

[6] N. R. Amundson and R. Aris. An analysis of chemical reactor stability and control. Parts I–III. *Chem. Eng. Sci.* **7,** 121–155 (1958).

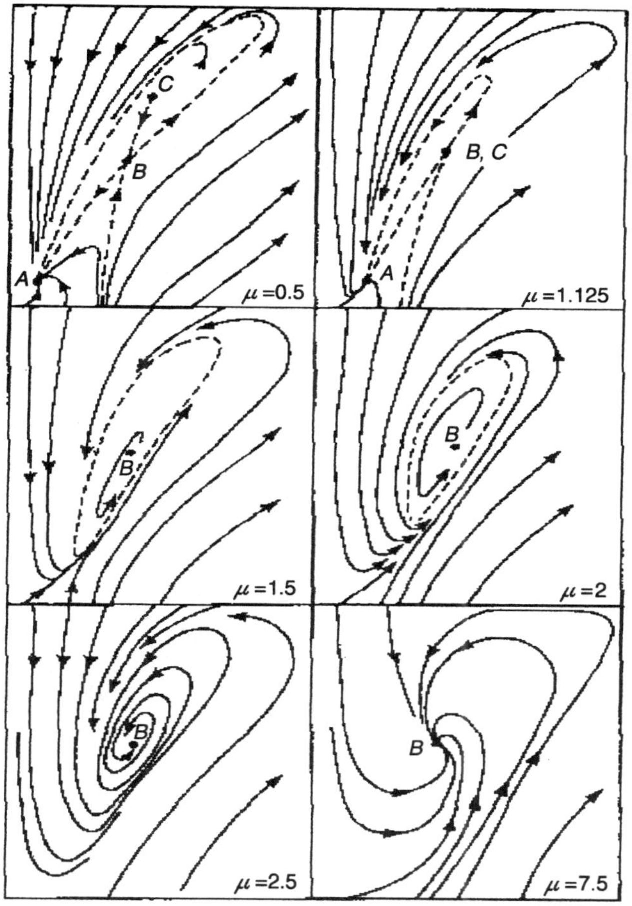

FIGURE 19 Phase-plane of a C^* controlled about the unstable steady state (initially a saddle-point) by cooling proportional to the temperature.

It should be said that the phase plane of itself cannot be guaranteed to give the whole behavior, for, in computing phase planes, one cannot be sure that one has started in the regions of attraction of all the steady or periodic solutions present. In 1956, the theory that is needed for this to be done systematically was 20 years into the future, and the calculating engines were just beginning to become really powerful. Fortran had not been invented, and each cell had to be addressed by its number, in octal![7]

While we are speaking about the history of the subject, it should be said that one of the most influential papers that have been published on the dynamics of the stirred-tank reactor is that by Uppal, Ray, and Poore.[8] They naturally use the phase plane to present their results in a concise and effective

[7] I had tremendous help from a graduate student, Mr. R. A. Wonderly, in the programming and running of this work on a Univac Scientific, the University of Minnesota's first computer.

[8] A. Uppal, W. H. Ray, and A. B. Poore. On the dynamic behavior of continuous stirred tank reactors. *Chem. Eng. Sci.* **29,** 967 (1974).

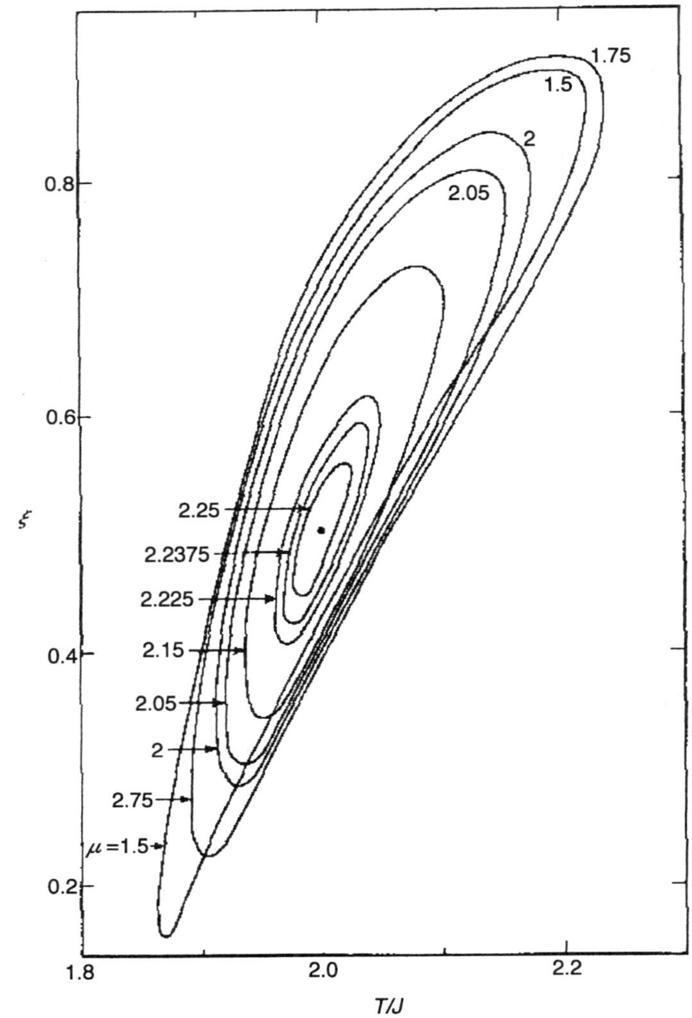

FIGURE 20 Variation of the limit-cycle with increasing control. The steady state is stable when $\mu = 2.25$, but the limit-cycle has not yet shrunk onto the point.

way. However, their presentation has the virtue of putting relevant sections of graphs in close proximity to one another, because the mind most easily takes in information from two related sources when they are close together. Coupled with discreet, but deliberate, didactic distortion,[9] this can pack in a lot of information and, in the simpler cases, cover the whole range of possible parameters. The distortion is necessary because in many cases the areas between the curves of the branch set (across which the bifurcation diagram

[9] An example of didactic distortion is the drawing of the phase-planes in Fig. 6, taken from A. Uppal, W. H. Ray, and A. B. Poore. On the dynamic behavior of continuous stirred tank reactors. *Chem. Eng. Sci.* **29,** 967 (1974).

changes), the intervals of the bifurcation parameter (within which the phase planes are topologically equivalent) and the limit cycles of the phase plane itself are small intervals or slim lenticular regions that are easy to miss completely.

The equations studied by Uppal, Ray, and Poore were written in terms of a dimensionless conversion, x_1, and a dimensionless temperature, x_2, with origin at feed conditions and scaled by the dimensionless activation energy, γ. This left three parameters: Da, the Damköhler number, the ratio of reaction rate to flow rate; B, a dimensionless heat of reaction; β, a dimensionless heat transfer coefficient; and x_{2c}, a dimensionless coolant temperature. The equations were

$$dx_1/dt = -x_1 + \mathrm{Da}(1 - x_1)\exp\{x_2/(1 + x_2/\gamma)\} \tag{297}$$

and

$$dx_2/dt = -x_2 + B\mathrm{Da}(1 - x_1)\exp\{x_2/(1 + x_2/\gamma)\} - \beta(x_2 - x_{2c}). \tag{298}$$

In analyzing the nature of the steady states, a distinguished parameter and characteristic state are chosen, in this case the Damköhler number and the steady-state conversion, respectively. The bifurcation diagram is the plot of the characteristic state versus the distinguished parameter, all other parameters being constant. To each value of the distinguished parameter in the bifurcation diagram there belongs a family of phase planes, but these are indexed by the four other parameters, an immense set. Their first task was to discern the more important parameters. They found that the plane of β and B could be divided into regions (denoted by Roman numerals) such that the points of any one region were associated with topologically equivalent bifurcation diagrams, such as is seen at the center of Fig. 21 (often called the *branch set*). They found further that these were not much influenced by the remaining two parameters, x_{2c} and γ. This allowed them to have good (although, in the nature of the case, not perfect) confidence in the broader validity of the picture that emerged from the special values $x_{2c} = 0$, $\gamma \to \infty$. The boundaries of the regions in the β (actually $1 + \beta$), B-plane are the transitions between different types of bifurcation diagram traced out by keeping track of the various transitions of steady-state stability and Hopf bifurcation. Then a synoptic presentation is achieved in Fig. 21 by surrounding the branch set with the nine possible bifurcation diagrams. They, in turn, have axes that are divided into intervals (designated by Latin capitals) within any one of which the phase plane is topologically equivalent to the sketch of the same letter in Fig. 22.

Uppal, Ray, and Poore did a further study taking residence time as the distinguished parameter[10] and the fine structure of the diagrams was explored by other workers[11] who added many more varieties of phase plane. It should

[10] A. Uppal, W. H. Ray, and A. B. Poore. The classification of the dynamic behavior of stirred tank reactors. Influence of the reactor residence time. *Chem. Eng. Sci.* **31,** 205 (1976).

[11] D. A. Vagonov, N. G. Samoilenko, and V. G. Abramov. Periodic regimes of continuous stirred tank reactors. *Chem. Eng. Sci.* **33,** 1133 (1978); D. C. Williams, and J. M. Calo. Fine structure of the CSTR parameter space. *A.I.Ch.E.J.* **27,** 514 (1981); and T. T. Tsotsis, and V. K. Kwong. Fine structure of the CSTR parameter space. *A.I.Ch.E.J.* **29,** 343 (1983).

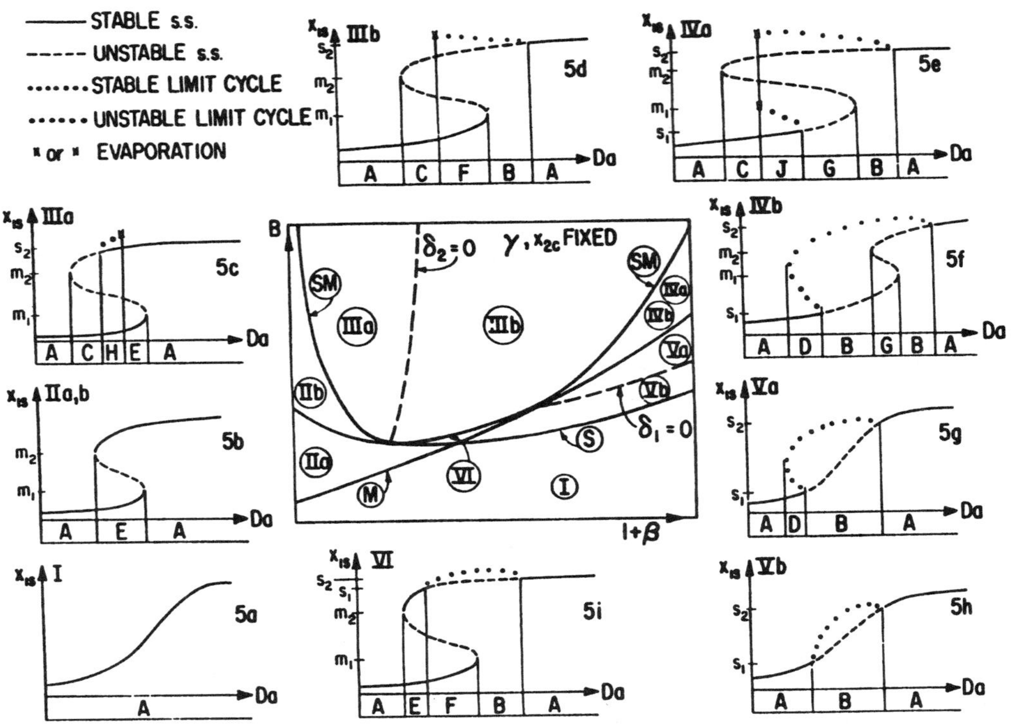

FIGURE 21 Possible bifurcation diagrams surrounding branch set plane of B, the head of reaction, and 1 + β, the cooling capacity of feed and coolant.

also be mentioned that Schmitz had done some experiments that supported the theory and had discovered isolae of steady states; he had even been able to hold their unstable underbellies in a sufficiently stable state to be assured of their existence.[12]

The context of Uppal, Ray, and Poore's work was the revival of interest in Hopf bifurcation at the end of the 1960s and the soon-to-be-transmogrified catastrophe theory of Zeeman. Ray has followed up this type of analysis in studies of polymerization and other important processes.

Meanwhile, singularity theory, which greatly illuminates this type of analysis, was developing both through the interest of some mathematicians[13] and by the insight and ability of a few chemical engineers, notably Luss and Balakotaiah.[14] By a systematic analytical procedure, an organizing center can

[12] I believe I am right in saying that isolae were Schmitz' discovery.

[13] M. Golubitsky and B. L. Keyfitz. A qualitative Study of the continuous flow stirred tank chemical reactor. *SIAM J. Math. Anal.* **11,** 316 (1980).

[14] V. Balakotaiah and D. Luss. Analysis of multiplicity patterns of a CSTR. *Chem. Eng. Commun.* **13,** 111 (1981); **19,** 185 (1982). See also these authors' papers: Structure of the steady state solutions of lumped-parameter chemically reacting systems. *Chem. Eng. Sci.* **37,** 43, 1611 (1982); Multiplicity features of reacting systems. Dependence of the steady-states of a CSTR on the residence time. *Chem. Eng. Sci.* **38,** 1709 (1983); Global analysis of multiplicity features of multi-reaction lumped-parameter systems. *Chem. Eng. Sci.* **39,** 865 (1984).

TYPICAL PHASE PLOTS

CASE		A	B	C	D	E	F	G	H	J
NODE or FOCUS	STABLE	1	0	1	1	2	1	0	2	1
	UNSTABLE	0	1	1	0	0	1	2	0	1
SADDLE POINT		0	0	1	0	1	1	1	1	1
LIMIT CYCLES	STABLE	0	1	0	1	0	1	1	0	1
	UNSTABLE	0	0	0	1	0	0	0	1	1
TOTAL INVARIANTS		1	2	3	3	3	4	4	4	5

ABSCISSA - CONCENTRATION ORDINATE - TEMPERATURE

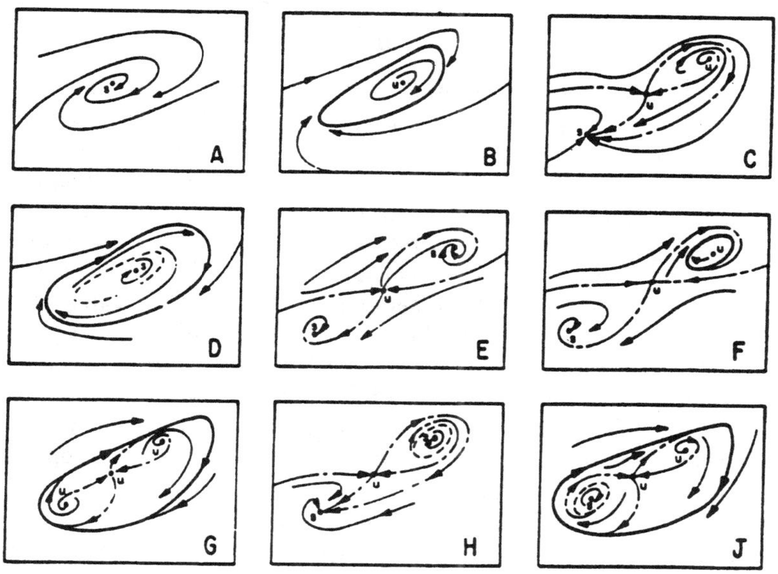

FIGURE 22 Possible types of phase-plane for the nonisothermal C^*.

be found and its character determines the qualitative behavior that unfolds in other parts of parameter space. Golubitsky and Keyfitz[15] recognized the underlying geometry of the C* to be that of a singularity they called the *winged cusp*.

Example 16. The Gray–Scott Reaction (Reprise)

There is merit in considering at a slightly less stiff and more algebraically amenable model that has the same features as the exothermic first-order nonisothermal case. This is the Gray–Scott autocatylator, a pair of coupled reactions

$$A + 2B \rightarrow 3B, \quad B \rightarrow C.$$

[15] *Vide supra* ftn. 14.

In the case of the nonisothermal first-order exothermic reaction heat is auto catalytic, for it raises the temperature and provokes an increase of reaction rate, yet is itself a product of the reaction. In the Gray–Scott scheme, B is plainly autocatalytic and its degeneration by the second reaction plays the role of the direct cooling in the non-isothermal case. This reaction appears in the chemical engineering literature in 1983,[16] and is the keynote reaction in Gray and Scott's 1990 monograph on *Chemical Oscillations and Instabilities.*[17] A justification of the autocatalytic mechanism in terms of successive bimolecular reactions is the subject of Chapter 12.

The equations for this reaction may be written

$$dx/dt = (1 - x)/\theta - xy^2 \tag{299}$$

$$dy/dt = (\beta - y)/\theta + xy^2 - \kappa y, \tag{300}$$

where x is the concentration of A, y that of B, θ is the Damköhler number, β the ratio of B to A in the feed, and κ the relative rate of the second reaction. In addition, we can use $z = 1 - x$, which is simply the conversion of A as the norm of the steady state. The steady states can be found as the roots of a cubic equation and a simple geometric argument can be deployed to show that there are three types of bifurcation diagram (i.e., plots of steady state values of the conversion versus the distinguished parameter θ). Which type will obtain depends on the remaining parameters, and because there are only two of them and they cannot be negative, we can divide the positive quadrant into three regions, as shown in Fig. 23. When we come to consider limit cycles, however, there are several possibilities within each region. Figure 24 shows this elaboration of the branch set into eleven regions, and Fig. 25 shows the corresponding bifurcation diagrams, whereas the possible phase planes are shown in Fig. 26. Thus given β and κ, we can read off a Roman numeral from Fig. 24 and see immediately what kind of bifurcation diagram obtains for this pair of parameters. Then from Figs. 25 and 26 one may read off the sequence of phase planes that would be encountered in any specified variation of the Damköhler number. For example, $\beta = 0.2$, $\kappa = 0.004$ lies in region (iii) of Fig. 24, and so has the bifurcation diagram at the top right-hand corner of the array in Fig. 25. Then an increasing θ would pass through the regions

[16] P. Gray and S. K. Scott. Autocatalytic reactions in the isothermal continuous stirred tank reactor: Isolas and other forms of multiplicity. *Chem. Eng. Sci.* **38,** 29–43 (1983); Autocatalytic reactions in the isothermal continuous stirred tank reactor: Oscillations and instabilities in the system $A + 2B \rightarrow 3B$, $B \rightarrow C$. *Chem Eng. Sci.* **34,** 1087–1097 (1984).

[17] P. Gray and S. K. Scott. *Chemical Oscillations and Instabilities: Nonlinear Chemical Kinetics.* Oxford: Clarendon Press, 1990. See also the Review Lecture by: P. Gray. "Instabilities and oscillations in chemical reactions in closed and open systems." *Proc. Roy. Soc. Lond.* **A 415,** 1–34 (1988).

The literature on this model reaction is already vast and a complete bibliography would be of great use to the mathematical modeler. Of particular interest are: A. d'Anna, P. G. Lignola, and S. K. Scott. The application of singularity theory to isothermal autocatalytic open systems: The elementary scheme $A + mB = (m + 1)\ B$. *Proc. Roy. Soc. Lond.* **A 403,** 341–363 (1986); and S. R. Kay, S. K. Scott, and P. G. Lignola. The application of singularity theory to isothermal autocatalytic open systems: The influence of uncatalyzed reactions. *Proc. Roy. Soc. Lond.* **A 409,** 433–448 (1987).

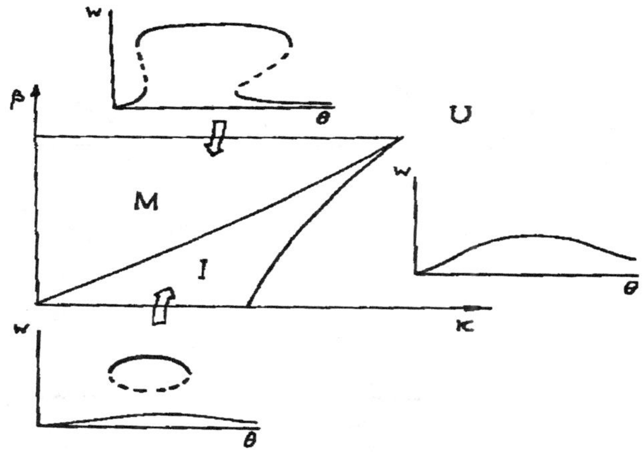

FIGURE 23 Regions of uniqueness, and mushroom and isola multiplicity.

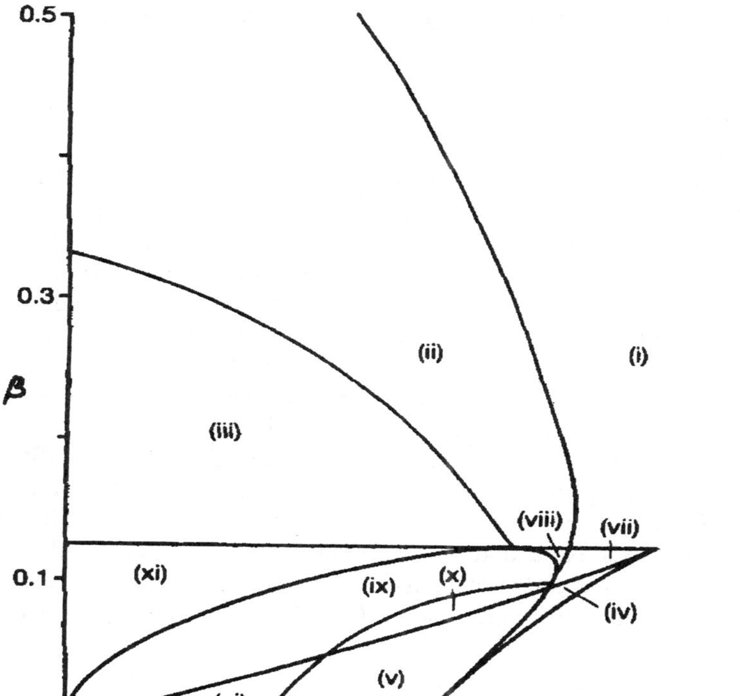

FIGURE 24 Detail of branch set with eleven regions corresponding to the types of bifurcation diagram shown in Fig. 23.

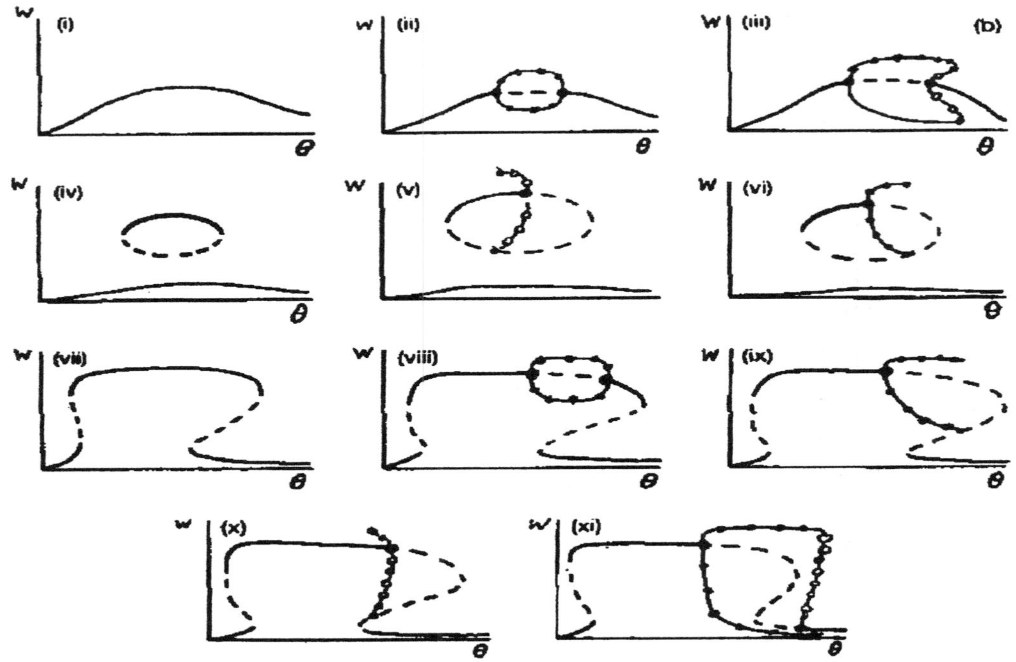

FIGURE 25 Bifurcation diagrams corresponding to the eleven regions of Fig. 24.

corresponding to the phase planes of types a, b, and c and back to a phase-plane of type a.

Figure 27 shows a graphical summary of another region of the branch set in Fig. 24, this time region (ix) represented by the values $\beta = 1/15$, $\kappa = 1/40$. The α-axis is a rescaling of the Damköhler number, $\alpha = \theta/40$. The phase planes that surround the central three-dimensional figure are thumbnails of trajectories in the x, y-plane and may be visualized as stacked vertically, each at the value of α given at the top right-hand corner; their positions are indicated by the numbers and short vertical arrows along the line at the near side of the central figure parallel to the α-axis. The mushroom-like curve $ABC \ldots RST$ is the locus of steady states that lie in the ruled surface $(1 + \beta - x - y)/\theta - \kappa y = 0$ indicated by the dash-dotted line $UVWZ$. The usual bifurcation diagram may be thought of as the projection of the curve $ABC \ldots RST$ onto the back plane and viewed from beyond it in the direction of the x-axis. Note that the five lines between points F and D and the lines between O and Q do not lie in this ruled surface; they are the eigen trajectories that approach the saddle point and form the separatrix between the basins of attraction of the steady states.

Point A represents the situation for all $\alpha < 0.0091$. The unique steady state, A, is a stable node, and the thumbnail phase portrait (1), $\alpha = 0.0075$, shown next is the other extreme case (13), $\alpha = 12.5$, which is a valid phase plane for any greater value of α. [*N.B.* The decimal place in the values of α for thumbnails (1) and (2) should be moved one place to the left]. At $\alpha =$

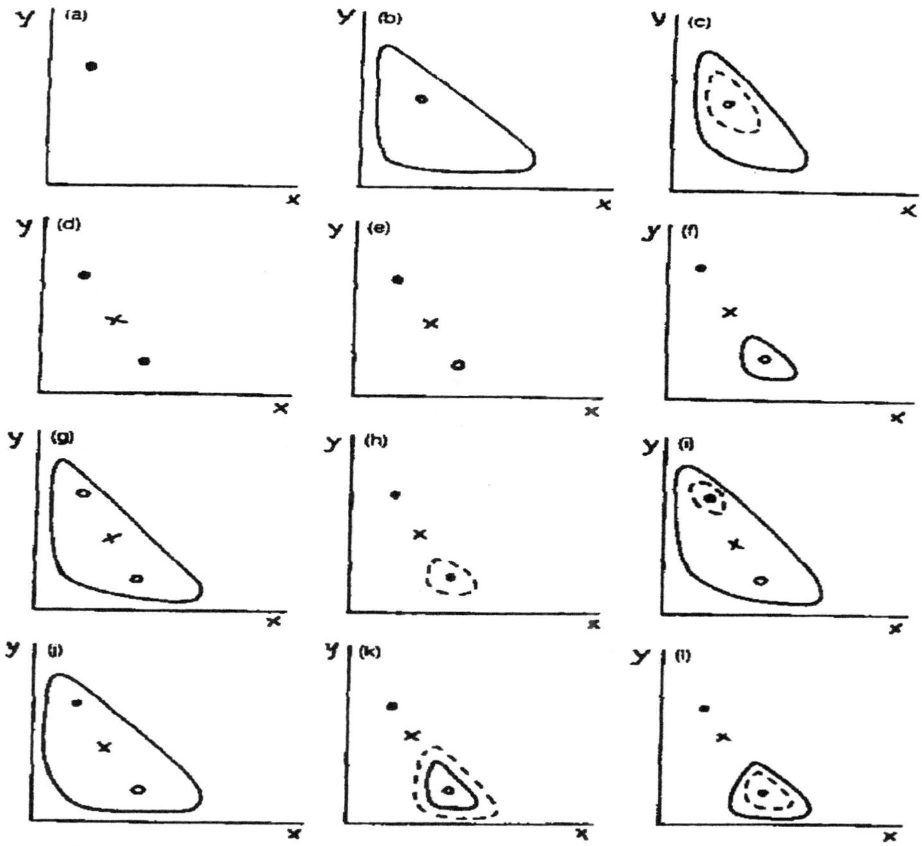

FIGURE 26 Possible types of phase-plane.

0.0091, the node/saddle is born at F, (2) and for $0.0091 < \alpha < 0.129$ the classic pattern of two stable steady states, each with its domain of attraction (their separatrix being the two incoming trajectories of the saddle point), obtains, and the phase plane (3) for $\alpha = 0.01$ is intended to be typical. During this interval, the saddle point moves away from the node with which it was born at F and coalesces with the other node at D. The steady state is unique over the interval HJ, but between the point I' and I it changes its character from a node (5) to a focus (6). This is in preparation for the Hopf bifurcation at J, $\alpha = 5.075$, where a limit cycle is born and the unique steady state becomes unstable. The thumbnails (8) and (9) show phase planes for $\alpha = 6$ and $\alpha = 7$ and the limit cycle is clearly growing and becoming more triangular; it also spends longer and longer in the neighborhood of the right-hand apex. Its period increases until at $\alpha = 7.95$, the point Q (the apex near which the trajectory is very slow) is actually on the trajectory. This means that Q is on a homoclinic trajectory, (10), leaving Q in a northwesterly path, turning to the south and finally arriving back at Q by travelling due east. Because Q is a nascent saddle, the return takes an infinite time. As α increases through the

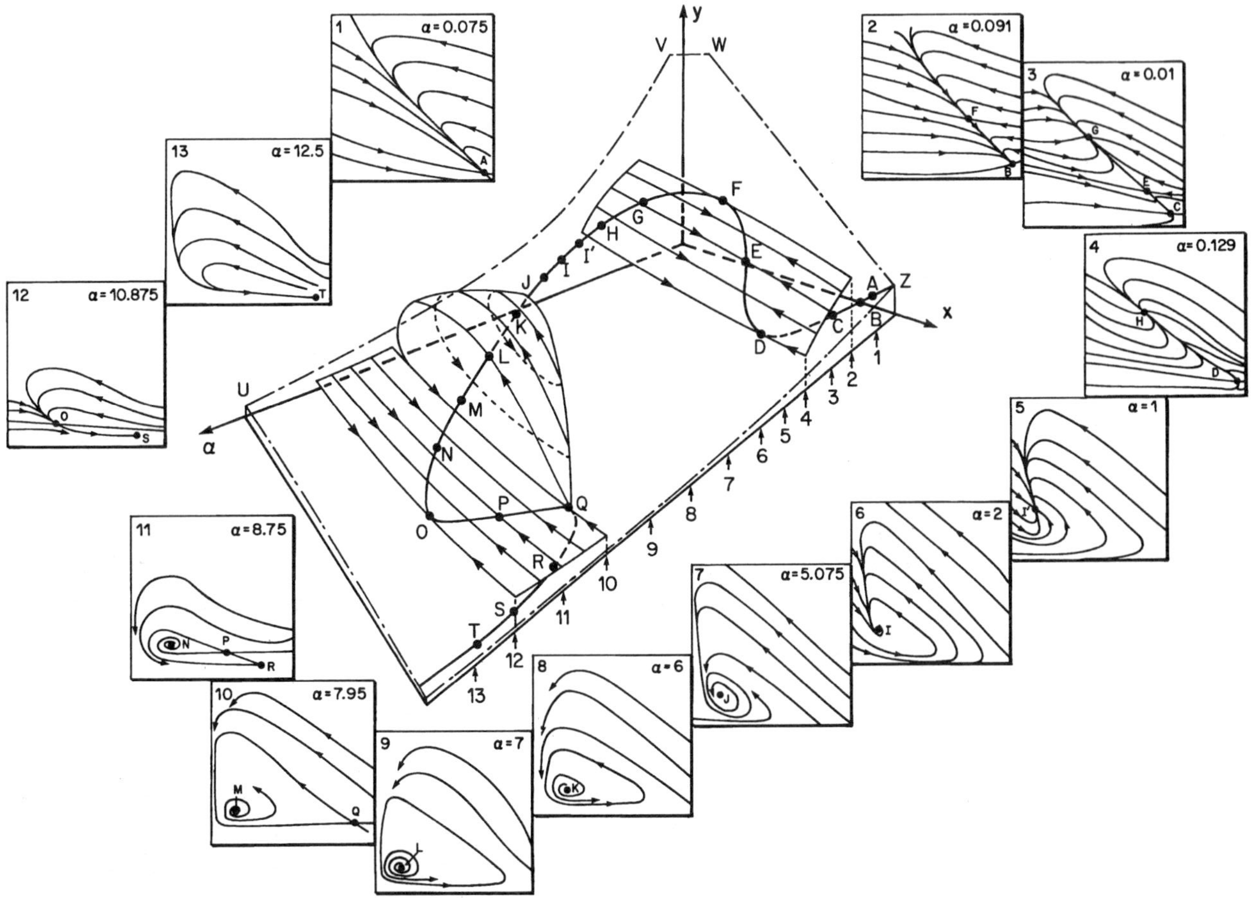

FIGURE 27 The bifurcation curve in three dimensions surrounded by phase-planes.

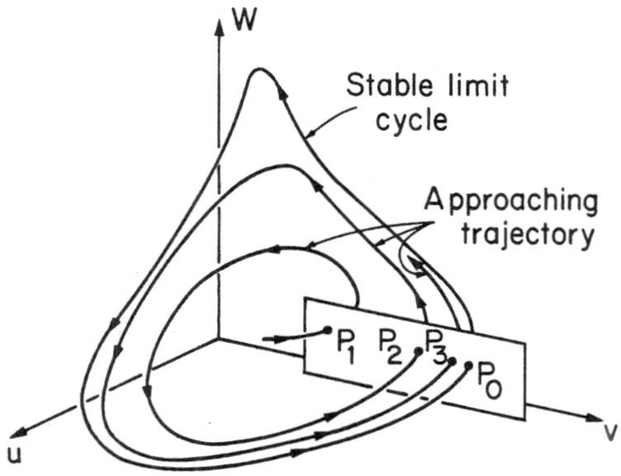

FIGURE 28 A three-dimensional limit-cycle and Poincaŕe map.

interval $7.95 < \alpha < 10.875$, the saddle born at Q moves along QPO, coalescing with the node coming along MNO.

This case has been described in some detail to show how a comprehensive picture of one case can be presented. A gathering of similar presentations, a page for each of the eleven regions of the branch set, would provide a complete portrait of the Gray–Scott autocatalator. Systems of two equations with three parameters are open to such a complete treatment, but even a fourth would lead to a "volume" of such "gatherings," whereas a fifth would demand a "shelf," and a sixth, a "library." A census of low-order dynamical systems to explore how far this might be taken would be an interesting exercise.

When there are three state variables, the autonomous trajectories have to present in perspective drawings that are quite natural extensions of the two-dimensional ones we have been looking at. However, with three dimensions other possibilities emerge in the form of strange attractors and transitions to chaos. There is little hope of covering the totality of such a system's behavior if there are more than three parameters and one must resort to strategic cross-sectioning of parameter space. Alhumaizi[18] explored various possibilities for two related two-phase model reactors.

OSCILLATIONS IN THREE DIMENSIONS

The phase plane has to give place to the three-dimensional state-space when there are three dimensions. If we have a stable periodic solution such as is shown in Fig. 27, all trajectories in its neighborhood tend toward it, but this is harder to show in three dimensions than it is in two and showing that $x(T + t) = x(t)$ when you do not know T is virtually impossible. Poincaré solved this difficulty by setting up a plane transversal to the limit cycle, as in Fig. 28. The limit cycle penetrates this surface at P_0 and, if a nearby trajectory penetrates it at a succession of points that, as shown, converge on P_0, then

[18] K. Alhumaizi and R. Aris. *Surveying a Dynamical System: A Study of the Gray–Scott Reaction in a Two-Phase Reactor.* Longman. Harlow (1995).

the limit cycle is stable and the period is the limiting value of the return times. The mapping that generates successive points of return is called the *Poincaré* or *first-return map.*

FORCED OSCILLATIONS AND THE STROBOSCOPIC PHASE PLANE

If a pair of equations is not autonomous, we must enlarge its representation to three dependent variables to avoid a trajectory crossing itself. The pair $dx/dt = f(x, y, t)$, $dy/dt = g(x, y, t)$ can be made autonomous by writing it as the triple $dx/dt = f(x, y, z)$, $dy/dt = g(x, y, z)$, $dz/dt = 1$. If we think of the time axis as perpendicular to the phase plane, the steady state of an autonomous system becomes a straight line parallel to the t-axis and nearby trajectories spiral into it (if it is a stable focus) or slide in exponentially in an eigen-direction (if a node). A periodic autonomous solution is drawn out into a helix whose pitch is equal to the period. There is, in fact, a whole family whose members are translations of each other in the t-direction and each is associated with a starting point on the limit cycle This family of helices defines a cylinder and, if the periodic solution is stable, nearby trajectories will spiral either from within or from without, approaching the cylinder and one of the helical paths on it. If the system is autonomous, no trajectory can cross the cylinder. If the system is not autonomous, the trajectory can go anywhere but cannot intersect itself because it is always being drawn out in the t-direction.

When the time dependence takes the form of a periodic perturbation of some parameter, we speak of this as *periodic forcing.*[19] The response will obviously not be a steady state, but can be periodic, quasi–periodic, or chaotic. If the response is periodic, it may be with a period that is a multiple of the period of forcing. It is quasi-periodic if the response winds itself onto the cylinder in a helix whose pitch is an irrational multiple of the forcing period, so that it is never quite truly periodic. An example[20] of a forced system is the Gray–Scott autocatalator with the feed concentration sinusoidally perturbed:

$$dx/dt = (1 + \lambda\cos(2\pi t/\tau) - x)/\theta - xy^2 \tag{301}$$

$$dy/dt = (\beta - y)/\theta + xy^2 - \kappa y. \tag{302}$$

Here, the average value of the inlet concentration has been retained as a characteristic quantity and λ is the ratio of the amplitude of the forcing to this mean. (The frequency $\omega = 2\pi/\tau$ will also be used.)

[19] For a good discussion of "the efficacy of concentration forcing," see a paper of that title by R. G. Rinker and R. Yadav. *Chem. Eng. Sci.* **44,** 2191–2195 (1989), and references there given.

[20] A comparative study was done by Kevrekidis and published as: I. G. Kevrekidis, L. D. Schmidt, and R. Aris. Some common features of periodically forced reacting systems. *Chem. Eng. Sci.* **41,** 1263–1276 (1986). See also two papers by the same authors: Resonance in periodically forced processes *Chem. Eng. Sci.* **41,** 905–911 (1986): The stirred tank forced. *Chem. Eng. Sci.* **41,** 1549–1560 (1986). A full study of the Schmidt–Takoudis vacant site mechanism is to be found in: M. A. McKarnin, L. D. Schmidt, and R. Aris. Autonomous bifurcations of a simple bimolecular surface-reaction model. *Proc. R. Soc. Lond.* **A 415,** 363–387 (1988); Forced oscillations of a self-oscillating bimolecular surface reaction model. *Proc. R. Soc. Lond.* **A 415,** 363–388 (1988).

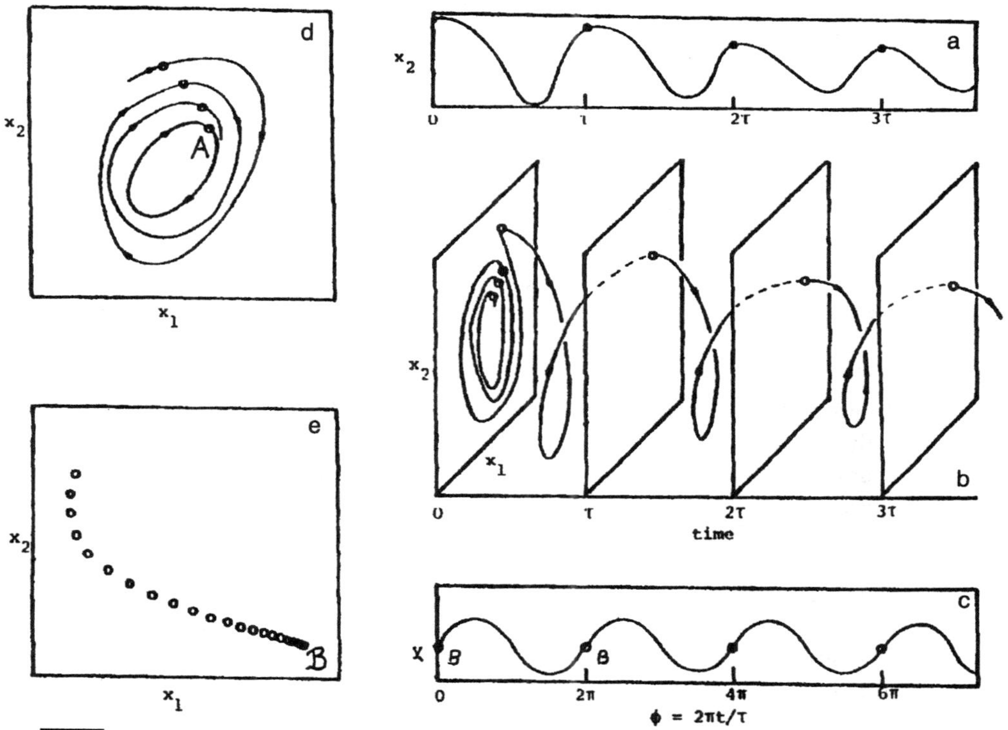

FIGURE 29 The development of the stroboscopic phase-plane. Segments (a) and (d) show the trajectory settling down to a limit-cycle through a sequence of points at times that are multiples of τ. This is drawn out in the time dimension in (b) and shown in its regularity in (c). If there is a unique periodic solution, the stroboscopic plane will show a sequence of states converging on (e).

An obvious map to consider is that which takes the state $(x(t), y(t)$ into the state $(x(t + \tau), y(t + \tau))$, where τ is the period of the forcing function. If we define $x_n = x(n\tau)$ and $y_n = y(n\tau)$, the sequence of points for $n = 0, 1, 2, \ldots$ functions in this so-called stroboscopic phase plane vis-à-vis periodic solutions much as the trajectories function in the ordinary phase plane vis-à-vis the steady states (Fig. 29). Thus if $(x_n, y_n) = (x_{n+1}, y_{n+1})$ and this is not true for any submultiple of τ, then we have a solution of period τ. A sequence of points that converges on a fixed point shows that the periodic solution represented by the fixed point is stable and conversely. Thus the stability of the periodic responses corresponds to that of the stroboscopic map. A quasi-periodic solution gives a sequence of points that drift around a closed curve known as an *invariant circle.* The points of the sequence are often joined by a smooth curve to give them more substance, but it must always be remembered that we are dealing with point maps.

One fixed point, stable or unstable, in the stroboscopic map corresponds to a periodic solution of period τ. If there are two distinct points, each mapping into the other, we have a periodic solution of twice the period. Figure 30, drawn with a continuous fill-in, illustrates the way in which different kinds of periodic solution can coexist. The invariant circle (actually ovoid with a pointed

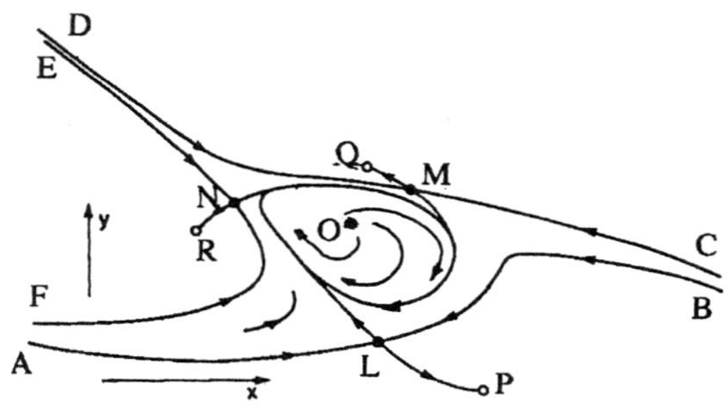

FIGURE 30 Coexistence of forced periodic solutions.

end to the left) is a quasi-periodic solution that is stable and attracts from the triangular region between the three separatrices, AB, CD, and EF. The unlettered dot within it is an unstable solution of the same period as the forcing. A trajectory starting near here would drift away from period τ and with each flash of the strobe get closer to the invariant circle. One starting from near L and above the separatrix would appear first near N, then near M. With no rounding error, L would go to N and then to M and back to L, but this is unstable. A start near P would give a second point nearer R and a third nearer yet to Q as a stable period three solution developed. This example (due to Kevrekidis) shows the complexity of co-existence that obtains under forcing.

The Excitation Diagram

To get a synoptic view of the response of a system to forcing we turn to the so-called excitation diagram.[21] Let us suppose that the parameters of the unperturbed system are such that it has a stable oscillation of frequency ω_0, then it is natural for the dimensionless frequency of the forcing to be $\Omega = \omega/\omega_0$ and the perturbation is $\lambda \cos(\Omega\tau)$. The excitation diagram is the positive quadrant of the Ω, λ-plane divided into regions of qualitatively similar behavior. In practice only a vastly simplified version can ever be drawn for an *Arnol'd tongue,* the name given to a class of these regions, touches down like a tornado at every rational number and, clearly, only a few of these can be drawn. Figure 31 shows just such a simplification with some rational points marked on the Ω-axis and various tongues and regions labeled with the integers 1 to 5. The diagram may best be introduced with reference to the four points A, B, C, D. These have the same value of λ but range from $\Omega = 1$ to about 1.4. A is in a region labeled "1," meaning that the response is of the same period as the excitation. We are not surprised to see this region

[21] I have not attempted to discover who was the first to use this manner of presentation, but the discussion of the Brusselator by: K. Tomita. 'Chaotic response of non-linear oscillators.' *Physics Reports p.* **86,** 113 (1982) was an early work; D. Sincic and J. E. Bailey. *Chem. Eng. Sci.* **32,** 281 (1977). On a mechanical system, see: Y. Ueda, 'Steady motions exhibited by Duffing's equation: A picture book of regular and chaotic motions.' In P. J. Holmes (ed.), *New Approaches to Nonlinear Dynamics.* Philadelphia: SIAM, 311 (1980).

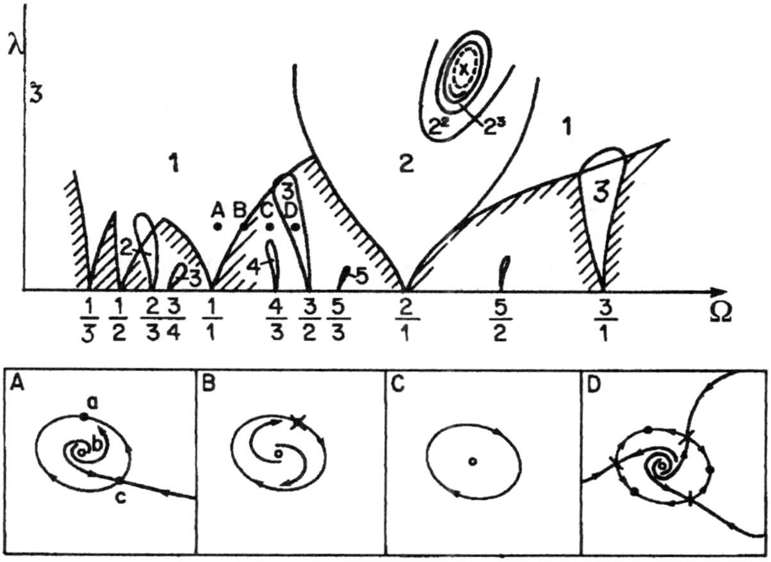

FIGURE 31 Simplified excitation diagram.

broaden out as λ increases, for with increasing amplitude it would seem that the forcing would overcome the innate behavior. At A, the forcing frequency is about the same as the natural frequency, and a in the box below is the period-1 response. This is not the relict of the natural oscillation, for that is still there as the unstable b. Now a is on an invariant circle and cannot attract all points unless there is also a saddle-point c, away from which the stroboscopic point must drift to a round the invariant circle.

As Ω increases from A to B, which is on the boundary of the hatched region, a and c move round the invariant circle and coalesce, evaporating as they do so. We are left with a quasi-periodic response C, in which the stroboscopic point will hop around the circle, never quite returning to the same spot on it. The points will, however, tend to conglobulate at three, eight, and eleven o'clock, because that is where the period-3 solution will spring up as the frequency of forcing pushes the point into the Arnol'd tongue labeled "3" (i.e., from C to D). When Ω has the value Ω_D (about 1.45 by the look of it) there are two period-3 solutions, one stable and one unstable, and the unstable relict of the natural oscillation of period-1. Notice that by traversing at this λ-level we encounter the 3 tongue before the 2.

In the broad region labeled "2," there is a set of concentric ovals labeled with the lowest powers of 2. These mark period doubling transitions whose repetitions lead to chaos. If we calculate the response to the forcing of a small departures from this path (ξ in x and η in y), it will satisfy the linearized equations

$$d\xi/dt = F_x(x, y, t)\xi + F_y(x, y, t)\eta, \quad d\eta/dt = G_x(x, y, t)\xi + G_y(x, y, t)\eta,$$

and integrating them over one cycle will give a linear return map. One eigenvector of the matrix of this map will be tangent to the invariant circle because this is the path on which the representative point moves into fixed point representing a stable periodic solution or out from an unstable one.

If the other (i.e., the transversal) eigenvalue is equal to -1, the infinitesimal vector being transformed is switched from side to side but neither approaches nor departs the fixed point. Thus in this transition the fixed point splits and a pair of points representing a solution with twice the period is formed. This is repeated with Feigenbaumian intensity[22] and leads to a chaotic response, labeled χ.

It is worth repeating that this figure is vastly simplified. Fuller representations are to be found in the papers of Kevrekidis (cf. fn. 62) and others and many ingenious exegeses are to be found in the literature. For more extensive treatments see reprints I and L in Chapter 9.

[22] M. J. Feigenbaum discovered a remarkable regularity in period doublings whereby the ratio of the parameter intervals between successive doublings approaches a universal constant 4.669. See: M. J. Feigenbaum, Quantitative universality for a class of nonlinear transformations. *J. Stat. Phys.* **19,** 25–52 (1978).

5 MAXIMS FOR MODELERS

I said there was to be no overlap with *Mathematical Modelling Techniques* [J] save in general structure and philosophy, but here is a nexus. In "How to get the most out of an equation without really trying" [136 = J, App.6] I set down 13 maxims for mathematical modelers. I will redraft them here in the light of what I have written and done since the late 1970s. This is the appropriate place for them, as in some ways they summarize the craft approach I have tried to take. Moreover, what follows in Parts II and III is less technical and more personal.

1. *Cast the problem in as elegant a form as possible.* At this stage, one cannot expect to achieve the final polish, but it is important to tidy the intellectual bench-top and get rid of irrelevancies (see "Getting Rid of Unnecessary Equations" in Chapter 2).

2. *Choose a sympathetic notation, but be prepared to change it.* Notation is a personal thing and, to some extent, mnemonic. It may not be possible to get it into the best final form at first.

3. *Cast the problem in dimensionless form.* The two approaches to this have been discussed in detail (see "The Reduction of Equations to Dimensionless Form" and "An Alternative Method of Reduction" in Chapter 2). Do not stop half way with only some of the variables dimensionless.

4. *Use any* a priori *estimates you can.* This may involve going to the limit with some parameter or using a maximum principle (see "*A Priori* Estimates" in Chapter 2).

5. *Think geometrically.* A diagram, even a rough one, is seldom useless and often reveals possibilities or rules out impossibilities (see Chapter 4).

6. *Use rough and ready methods but do not carry them beyond their point of usefulness.* A one-point collocation solution, for example, may not be very accurate, but may give an insight. A sketched phase plane may be qualitatively correct even though its numerical value is way off (see "The Phase Plane" in Chapter 4).

7. *Use limiting cases, which may be much more easily soluble, to box the problem.* Perturbation methods, distinguishing between regular and singular, and asymptotics are often useful (see "Perturbations and Asymptotics" in Chapter 3).

8. *Rearrange the problem. Do not get fixed ideas on which variables are dependent and which independent.* The use of parametric representations (see "The Use of Parametric Representations" in Chapter 3) and the hodograph transformation come under this rubric. Even more radically, the shift to a different type of model (e.g., the wave model of Westerterp; see also "General Observations and Forming the Model" in Chapter 1) is a possibility.

9. *Festina lente*[1] is a good motto with regard to computing. Get the very best feel for the solution that you can before computing. Without this, the computer can produce some very expensive waste paper (see "Getting the Most from Calculations" in Chapter 3).

10. *Make your own maxims.* These will self-destruct. They are not intended as an infallible guide, but as a framework within which you can grow in craftsmanship.

[1] Make haste slowly.

6

STYLE

In this brief section I consider some of the peripheral matters of modeling, and I should say at once that my personal views are injected, consciously or unconsciously, to a degree that is not the case in what has gone before. As soon, however, as I make that claim, I have to remember that this book has made no attempt to be comprehensive or systematic, so that the reader has already been subject to my personal predilections in the choice of topics. Once launched into a mathematical topic, however, I am seldom presenting a personal point of view in the same sense as I do when discussing plagiarism, for example, later in this section. These matters are peripheral when the technical content is taken to be central, but they may be central from a different viewpoint. Mathematicians, scientists, and engineers have been exhorted to bring ethical questions into classroom discussion, where this is appropriate, and to recognize that, to some perhaps larger than they realize extent, their vision is shaped by extraneous considerations. Radical deconstruction, such as has scorched the literary earth in its retreat from reason, is very rare in the so-called hard sciences and the best defense against it is a sense of humor.[1] However, although traditional scholars strive for disinterested understanding, they do well to recognize their own limitations and shortcomings.

[1] As in the Sokol affair, wherein a paper, the physics of which could have been detected as spurious by a sophomore but that was in politically correct argot and used the "right" references, was accepted by a journal of so-called cultural studies.

LITERARY STYLE

"English giveth us great occasion to the right use of matter and manner, being indeed capable of any excellent exercising of it." So wrote Sir Philip Sydney in his *Apology for Poetry,* and went on to defend it from the charge of having no grammar: "Nay, truly it hath that praise, that it wanteth not Grammar, for Grammar it might have, but it needs not; being so easy of itself and so void of those cumbersome differences of Cases, Genders, Moods, and Tenses, which I think was a piece of the Tower of Babylon's curse, that a man should be put to school to learn his mother tongue." English encounters the peculiar difficulty that native English speakers are, in the main, largely ignorant of other tongues, and, of those who have some knowledge of another language, most have to labor with dictionary and grammar to compose anything in it. The facility and correctness with which our colleagues in Scandinavia, Germany, or the Netherlands, for example, write and converse in English is a perpetual reproach to our education, and the least we can do is to ensure that we use our own language as well as possible.

Above all, scientific prose must be clear. Science lives by the propagation of its results and the testing of them against the experience of others, so a clear explanation of what one claims to have discovered is of the essence. Lewis is said to have required of his pupils that in their essays they should have reasoning "brief and clear as English sunlight." It goes without saying that scientific prose should be grammatically correct, for otherwise it fails to communicate its substance to a readership that uses standard English.

Ideally we are supposed to gain our knowledge of the rules and manners of our own language from our reading, but, in a day when the very notion of a canon is anathema in most English departments, you will get little encouragement to give your "days and nights to the volumes of Addison," as Johnson advised whomever wished "to attain an English style, familiar but not coarse, elegant but not ostentatious."[2]

In so far as they are motivated by personal courtesy toward our interlocutors, the changes that language saw in the late 1980s may be welcomed, for it is clearly wrong to address anyone in a manner calculated to offend. The range of political correctness runs from a ludicrousness that is too feeble even to be funny ("altitudinally challenged") through careless indifference ("impact" as a verb) to a wanton verbicide ("humankind" for "man"). Many and grave are the offences that male humans have committed against the female of the same species, but rather than seeking to redress them by destroying primary sense of the word "man" [according to the *Oxford English Dictionary,* it is "**I.1.** A human being (irrespective of sex or age)"], I would have thought it simpler to bring back the older form of "werman" for the male "man" and to leave "woman" or bring back "wifman" for the female. Fortunately, in scientific writing these issues seldom arise and the greater danger is the currently fashionable word "impact," "closure," and "concept" at the time of writing.

There are many manuals devoted to style and concerned with grammar,

[2] S. Johnson, S. "Addison." In G. B. Hill (ed.). *Lives of the English Poets.* Oxford: Clarendon Press. §168, p. 150.

syntax, usage, punctuation, and the regularities of format. This is no place to list them, but I cannot resist mentioning two that I have found more than usually helpful. J. M. Williams' *Style* is subtitled *Ten Lessons in Clarity and Grace,* and its scope can be read in its chapters: correctness, clarity in action and characters, concision, cohesion and coherence, emphasis, controlling sprawl, punctuation, and elegance. It is a book to work through and has some useful exercises.[3] After an introductory chapter or two, Robert Graves and Alan Hodge, in *The Reader Over Your Shoulder*, tackle the principles of clear statement and the graces of prose.[4] They summarize the first in 25 principles and the second in 16 principles. In the first set, there are seven leading questions whose answers should be made clear: Who? Which? What ? Where? When? How much? How many? Then the inappropriate, ambiguous, or misplaced word or phrase should be avoided. Banish unintended contrasts, duplication, and self-evident statements; as well as material omissions, unfulfilled promises, and undeveloped themes. Connections should be logical, and their quality, rather than the length of the pause, should be reflected in the punctuation. Confused sequences of ideas, irrelevancy, overemphasis, false contrast, and weak logic are inimical to clarity, as are also changes in standpoint and mixtures of category. The disgraces of style that they list are the mismating and excessive use of metaphors and their confusion with reality. Prose should be prose and not poetry; it should not mismate styles, use overly obscure references, circumlocutions, or variations merely for the sake of elegance. Overlong sentences strain the memory. Words used too often, jingle, and alliteration soon cloy. The same word should not be used in different senses; rhetorical hesitation is not appropriate, nor is an inversion of words that reads unnaturally. Graves and Hodge then examine 18 passages taken from well-known authors, then criticize and rewrite them. Trying to anticipate what they will do with their subjects provides good exercise, but no amount of reading books of this kind or of arguing with friends[5] about these matters can substitute for practice. One must write and rewrite and write again until the external constraints or the internal limitations grant closure.[6] A. N. Whitehead gives Williams a last word on "the most austere of all mental qualities: ... the sense for style. It is an esthetic sense, based on admiration for the direct attainment of a foreseen end, simply and without waste."[7] This stresses directness and simplicity, and one could almost tell that Whitehead had been a mathematician from this passage alone. There are occasions when it is appropriate to use a more expansive style. Such is the publication of a *festschrift,* marking a birthday or a retirement, when an introductory paragraph with an ampler prose style is meet. It must be admitted that a whiff of hyperbole hangs around these

[3] Williams, J. M. *Style* (4th. Ed.) New York: HarperCollins, 1994.

[4] Graves, R., and A. Hodge. *The Reader Over Your Shoulder. A Handbook for Writers of English Prose.* (2nd Ed.) London: Johnathan Cape, 1947.

[5] Many an hour was so spent with the friend to whose memory this book is dedicated.

[6] In the light of what I have said previously, I hope it is clear that I use the word deliberately to deconstruct myself. After all, you must balance on one foot before you can pull your own leg. That writing is an exercise in morality is shown by Graves and Hodge's strictures on recondite quotations. I have seldom been able to resist them, although they are often undiluted show-offs, only partially redeemed by explanatory footnotes; cf. the Watkins number in [145] and [J, p. 195].

[7] Whitehead, A. N. *The Aims of Education and Other Essays.*

encomia, but, provided there is a basis for the reputation, they are not less sincere for being *nil nisi bonum*. I have commented elsewhere on the unfortunate way they pile up, leaving one unable to pay honor to colleagues whom one would have wished to recognize.[8]

GENRE

One of the earliest scientific works, Hesiod's *Works and Days*, was a poem.[9] It is doubtful, however, whether anything would be gained by using verse for our papers, for the constraints of rhythm and rhyme are utterly foreign to scientific discourse.[10] Verse is appropriate enough for special occasions, however, and it is remarkable what a fund of phrases and even usable lines can be found in the bibliographies of distinguished chemical engineers. Co-authors' names can also be used to comic effect, especially if they are linked by sound or meaning to elements in the poem being parodied. I cannot resist quoting one of the score or more I have written for the roasting of the Lacey Lecturer at Caltech. It is a parody of one of Hillaire Belloc's *Cautionary Tales for Children* concerned with Henry King,[11] whose chief delight was chewing little bits of string; in fact, four lines of the original are incorporated in the parody, which is not quite cricket. The suggested procedures have a disjointed actuality about them.

The chief delight of Harmon Ray
Was to polymerize all day.
At breakfast, with his Wheaties, he
Would take his monomer with tea,
For caffeine will initiate
The process at a decent rate.
The very busy morning hours
Displayed him at his highest powers.
Addition prospered, late and soon,
He was quite dimerized by noon.
Corrective intake with his lunch
Allowed his testing of the hunch
That semibatchwise operation
Was fav'rable to propagation.

[8] I have remarked on this in the Acknowledgments (p. 419) and am particularly sorry that I did not have anything for issue of *Ind. Engineering. Chemical Research* dedicated to L. K. Doraiswami, truly a world-leader in chemical engineering and, for many years, director of the National Chemical Laboratory at Pune.

[9] Of which the late D. C. C. Young remarked in his broadest Scots that, "It might be something of an imposition to expect the modern Ph.D. student to write his dissertation in verse, but the results might not be worse than much scientific prose, nor, indeed than much modern verse."

[10] [283] is on the subject of contemporary literary criticism, and nothing could be more appropriate than the heroic couplets of Pope's *Essay on Criticism*, in whose steps it presumed to follow, albeit at a great distance.

[11] This particular *Tale* has served for several victims: the chief delights of Morton Denn and R. Jacckson have been exploited. Like Dr. Chasuble's sermon on the meaning of the manna in the wilderness, it can be adapted to almost any occasion.

This bit of feedback how-de-do
Took him to nearly half past two,
When he'd announce, his cheeks aglow,
"My N-mers are now oligo."
His afternoons were metastable,
And if, perchance, you were unable
Wholely to stifle sneeze or cough,
He and reaction would take off;
The polymer would quickly gel,
Harmon high-tail it home to Nell.

With dinner he'd achieve repletion;
The day's reaction its completion.
Such, with diversions now and then,
Was Harmon's daily regimen.
It could have gone its merry way
Perpetually, but, sad to say,
It was disrupted totally
One afternoon at ten to three
(A time all lovers of the book
Associate with Rupert Brooke).
At this remove it is not poss.
To say exactly what it was.
Some say it was a *Ding am Kopf*,
Others a bifurcation Hopf,
For as he passed through this transition,
This kind of afternoon ignition,
As far as any one could see
His roots were pure imagin'ry
"It was as though right there before us,
He changed his cycle for a torus."
To other friends, concerned and troubled
It seemed his period had doubled.
His loops and whirls grew quite exotic;
His whole behavior quite chaotic.

Physicians of the utmost fame,
Were called upon, but when they came
They murmured, as they took their fees,
"There is no cure for this disease.
Harmon must bifurcate again
Or he'll become a feather-brain."
Since MDs were of no avail
His colleagues hastened to prevail.
Some diagnosed it as big blocks
Of polymer dislodged by shocks–
Such as the sad discovery
That there's no honey still for tea
The Chem. Es were abashed to find
That Harmon did not really mind,
And gathered round him while they pried
All sorts of things for his inside:
Copolymers, test meters wet,
And even a spare Cantor set.
The bio.s sought to save their faces
By trying depolymerases.

His famous pupil, Klavs F. Jensen,
Was summoned to reduce the tension
And gave him (after due permission)
A course of vapor deposition.

Poor Harmon would have heard his knell
But for the commonsense of Nell.
"OK," said she, "Just stop your horsing,
The only thing to try is forcing.
Shake him with force and fervency
At twice his natural frequency,
Slowly reduce the amplitude
To give sufficient spissitude
So that we can just pull him through
The Arnold tongue at p by q.
For only thus, I know, will he
Regain his rationality."

It worked, it worked, at their first try!
Harmon is now a normal guy.
For, once this simple cure was done
One at a time, through minus one,
Lyapounov exponents came
Within the unit cycle's frame.
Where now they sit, so all may see
Harmon's complete tranquility.
Lest his "delight" becomes pandemic
He keeps it strictly academic.

Another variant from the normal plain prose paper is the *dialog*.[12] Some may regard it as something of a conceit in the current pejorative sense of the word (OED, sense 3 of "conceited"). Being a conceit (in the dominant older sense), a dialog can be treated, to some degree, fancifully, although one should avoid the guest-towel[13] effect. The editors of three journals have been kind enough to allow me to publish dialogs on some aspects of mathematical modeling, two of which are appendices to [J′] and the last is P, Chapter 11. The first is set in Caltech and is an introduction to the ideas I was working on when writing *Mathematical Modelling Techniques* [J′]. The logistical equation is adumbrated, the domain of applicability is mentioned, and the problem of diffusion and reaction treated more fully. The shape factor in diffusion and reaction is compared with the calligrapher's treatment of shape in the second dialog, which is in an Oxbridge setting. The third setting is the road that zigzags up to the Benedictine monastery of Montecassino. In 1984, I had received an invitation to give a paper on optimization, a subject I had not thought about for 15 years. However, it came from the University of Cassino, who were opening a new engineering school, and so I instantly revived my interest in the subject. It was a misty December morning when I first approached the Mother-house of Western Monasticism, rebuilt after its desola-

[12] There is an excellent textbook that is a dialog between an older metallurgist and a young scientist, namely W. Hume-Rothery's *Electrons, Atoms, Metals and Alloys*. London. Metal Industry, 1948.

[13] Often defined as a small square of absorbent linen entirely surrounded by waterproof embroidery.

tion just 40 years before. The three friends, whom I imagined conversing on their way up, were obviously Princetonians–a mathematician, an engineer, and a paleographer. The logistical equation is carried much further and then the switch is made to paleography, and some parallels are drawn between the two disciplines. This last and another paper [284] designed for an audience of humanists are the results of Gianni Astarita's challenge to bring my two interests together intelligibly.[14]

PLAGIARISM AND ATTRIBUTION

Plagiarism is the deliberate taking of another person's ideas and the presentation of them as one's own. *Attribution* is the identification and acknowledgment of the source of an idea. In Section III, I have recalled two instances in which I had unintentionally committed a careless plagiarism by completely forgetting that a case had come up in conversation or in a lecture. This might be classed as "venial" rather than "mortal," but its effect is the same: it robs the other person of the full credit for the idea. One can only make reparation as soon as it is discovered and improve the memory or use a notebook. One of the stranger experiences of an editor's work is to find blatant plagiarism flatly denied. By *blatant* I mean a paper that has several passages scattered through it that are word-for-word identical to passages in an earlier paper not cited. The only non-criminal way in which one can imagine this happening is that one of the authors in studying the subject copied out passages without proper attribution into a loose-leaf notebook in which some sheets were later rearranged. Then, when writing the paper, these authors incorporate these passages into their text as if the words were their own first drafts.[15] This is not a very creditable scenario, but at least it does not involve conscious and deliberate theft. Yet there have been instances when, confronted with such evidence, plagiarism is still completely denied. I have not heard of recourse to the monkeys and the British Museum argument!

Attribution gives credit where credit is due, but only in a lumped sense when the paper cited has more than one author.[16] If one knows enough of the way in which the research is being done, one can sometimes credit an individual, although this has its dangers if one's knowledge is imperfect (see p. 435). The order of the authors' names at the head of a paper or on the title page of a book is sometimes a touchy subject. Sometimes the first name is regarded as the most important, sometimes the last, although, in medical journals, the papers have so many names that one suspects the dogcatcher is in there somewhere. I must confess that I have paid so little respect to the

[14] It was accompanied by an invitation to Naples, where he received us with great kindness. It was a little daunting to lecture in the Philosophical Institute of Benedetto Croce, a friend of Gianni's father.

[15] This argues an extraordinary insensitivity to one's own prose, in which one can usually detect changes of ppm.

[16] I must confess that I have only seen the front page of [118], which was published without my knowledge. My name went on because I was a co-advisor and Henry Tsuchiya was punctilious about including me.

order that I have felt free to rearrange them in order to bring the works of the obvious leader of a research project together in the bibliography. In retrospect, this is rather reprehensible, and were I gathering references to the work led by, for example, author C, I should have put the full entry under "C, A, and B," while the entry under the published order "A, B, and C" would have the cross-reference "see C, A, and B."

Even a strict rule of alphabetic order does not work infallibly. Strieder's name comes rightly first on [H], because he drafted three chapters to my one. Alhumaizi's name goes first on [O] for much the same reason, but I embarrassed him by putting it there, since the Arabian tradition puts the teacher's name in the first place regardless of the division of labor.

The Citation Index is a valuable instrument for tracing the influence of individual papers, but each reference has to be examined and evaluated. The crude figures are, of course, not wholly without meaning, and the "classics," based on simple counts, are seldom unworthy of the name. References to earlier work of one's own can be quite legitimate, for they may well be the pegs on which the new fabric is draped, but it is wasteful and misleading to give references that are otiose, and self-serving if they are to work of one's own. That the crude count is misleading can be seen in the fact that there are more references to my paper on Taylor diffusion ([2] = Chapter 7) than there are to his. The only reason for there being more references to my work than to Taylor's is that the major concern has been with the use of moments and not with diffusion and convection.

PUBLISH OR PERISH

I count myself extremely fortunate to have started my academic career before the present tenure hurdles were set up. I really was not too sure of what was expected of me, except that in my case not having a Ph.D. was an impediment that it was only sensible to remove. "It will be harder to get you promoted if you don't get one" were Amundson's words, which seemed to imply his confidence in my ability to grow into the job of teaching, supervising, and getting on with research. Writing proposals came later, and the larger share of the support for graduate students came through the graduate school. Now proposals are sometimes written before a young assistant professor takes his final oral. Jobs are chosen as much by dowry as by departmental reputation. Enormous pressure is imposed on entering faculty and the "dossier," although not due until the sixth year, looms over the first.[17] On the positive side and contrary to the general public's opinion, it is much more usual for the habit of hard work to carry over to the tenured state than it is for the newly tenured faculty member to go to sleep.

[17] R. Byron Bird of Wisconsin has written wisely of the *Plusses and Minuses* of the enormous pressure to get grants and how it has changed the university from a community of scholars to a collection of (admittedly superb) technicians, "but with minimal interest outside their own departments." His essay appeared as an article entitled: "Restore the right priorities." *Chem. Eng. Prog.* **92,** 80–83 (1996).

No doubt the old system could be abused and it may have made prejudice less easy to detect, but that is a poison that can be injected into any system. Certainly it worked best in departments of no great size whose heads were strong, fair-minded, and caring. In the late 1950s the litigious temper was rare; today, the courts are more often of first, rather than of last, resort. Then the business school was peripheral, scarcely more esteemed than the Department of Mortuary Science. Today, universities have grown, and the concepts of Management Science, regardless of whether they be appropriate to academia, are dominant. Research tends to be goal oriented (as it is put) rather than intellectually driven. Industry has "down-sized" its research and seeks more cooperation. All these things are good in their way, but they add up to a Gadarene pace and one cannot help wondering whether this is the best way to nurture the young scientist and to seek understanding of the world in which we live.

My age is showing! It is self-indulgent to bemoan the present, for did not the wisdom of Solomon have it: "Say not the former times were better than these, for thou dost not enquire wisely concerning this matter." The fact is that each generation meets its own challenges, sometimes helped and sometimes hindered by the generations before it. In the small world of mathematical modeling, questions arise from nature or human artefact with a lush penumbra of circumstance, which we try to answer by setting up a parallel system in the sharp-shadowed world of mathematical contrivance. Here we can work in the clean-room of mathematical construction and hope to find that what we have made continues to correspond in a significant way with the "mightier movement" that "sounds and passes" in the natural world. Of one thing we may be sure, and that is that we shall never run out of problems. I have quoted Bridges so often[18] in the context of mathematical modeling that I will turn to his friend Gerard Manley Hopkins for an expression of this plenitude[19]:

> And for all this, Nature is never spent;
> There lives the dearest freshness deep down things;
> And, though the last lights off the black West went
> Oh, morning, at the brown brink eastward, springs—

[18] I have overworked passages from Eliot's *Four Quartets*, Bridges' *Testament of Beauty*, Vita Sackville-West's *The Land*, Keats' *St. Agnes' Eve,* and have elicited a few groans with Horace's "*Omne tulit punctum qui miscuit utile dulce.*" However, I have found them very apt and lack the breadth of reading to vary them elegantly.

[19] G. M. Hopkins, "God's Grandeur." (1877). In Walford Davies (ed.). *The Major Poems of Gerard Manley Hopkins.* New York: Dutton (1979).

II

MATTER

7

DISPERSION IN FLOW

A. On the Dispersion of a Solute in a Fluid Flowing through a Tube
B. On the Dispersion of a Solute by Diffusion, Convection and Exchange between Phases
C. On the Dispersion of Linear Kinematic Waves

INTRODUCTION

The first three reprints in Part II are the outgrowth of an instruction to look into the accuracy of chromatographic analyses, at that time the new and promising method of getting immediate assays of composition at I.C.I. Billingham. I read Martin and Synge's paper and the literature of nonlinear chromatographic theory, including the heated discussion of the Faraday Society, and worked with the plant people on the design of their instruments. However, early in this process I realized I did not understand the interaction of flow profile and diffusion and was very excited when Taylor's paper came out in 1953. Because it was not directly relevant to what I was doing at work, the understanding of Taylor's work and recasting it in the mode of moments were my evenings' occupation. The shortcomings of the first paper are discussed in the section titled "Observing Conditions" in Chapter 3, but, fortunately, they do not affect the first two moments' asymptotic behavior; namely, that the centroid ultimately moves with the mean speed and the solute is dispersed with an apparent longitudinal diffusivity that is the sum of the molecular diffusion coefficient and the Taylor dispersion coefficient. This additivity was splendidly confirmed by the experimental work of Evans and Kenney.[1] Taylor very kindly communicated the paper,[2] and we had some correspondence about his experiments with potassium permanganate.

[1] E. V. Evans and C. N. Kenney, "Gaseous dispersion in laminar flow through a circular tube." *Proc. Roy. Soc.* **A 284,** 540–550 (1965).

[2] Reprint A has been cited 565 times in its first 35 years—Current Contents, Jan. 14, 1991. As I pointed out in [A]: "Some have been kind enough to use the term *Taylor–Aris dispersion.* I take it as a great honor to be so associated, but the honor is really Taylor's, whose name can be attached to so many things that it is useful to hyphenate him with an ordinary mortal." The place of convective dispersion in the range of Taylor's work is well described in G. K. Batchelor's most interesting biography of him: *The Life and Legacy of G. I. Taylor* (1996). New York: Cambridge University Press.

The moment method caught on and, in the hands of Howard Brenner, was a powerful tool both in the generalization of Taylor dispersion and in the analysis of complex situations.[3] The extension to two or more phases, as shown in Reprint B and in [10], although a simple enough extension, was something I did not get to until 1957 when I was back in Edinburgh.[4] The various contributions to the apparent diffusivity are recognizable enough, and it is only the shape factors that require the detailed analysis.[5] The geometry of this paper is that of the coated tube chromatograph.

Reprints A and B deal with the evolution spatial moments in time; Reprint C is concerned with the temporal moments of the dependent quantity at a fixed point—that is, the moments of just such a chart as a chromatograph would produce. It was, in fact, the outgrowth of my early work at I.C.I. Billingham, delayed by my having been switched on to other matters.[6] Meanwhile, Lighthill and Whitham's study of kinematic waves had come out and, although their papers were mainly concerned with nonlinear aspects, their example of a flood wave allowed me forge a link with a really complicated linear model. This was the linear chromatography I had done years before, but transmogrified into a heat storage problem. This avoided any breach of my employment agreement with I.C.I. Billingham, but may have contributed to its not being at all well known.[7] There was a great deal of interest in residence time distributions in the lustrum around 1960, and the Laplace transform was in use as a moment generating function; cf. [9], [11], and [12]. Other papers using moments are [3b], [4], [21], [136], and [145].

[3] H. Brenner, "A general theory of Taylor dispersion phenomena." *Physicochem. Hydrodyn.* **1,** 91–123 (1980).

[4] If evidence were needed to justify the final apology of the Acknowledgements, I would have to confess to working on it while waiting for the taxi to take us to the station for the first stage of a postponed honeymoon!

[5] The notion of a shape factor by which many different situations are represented is a nice one and leads to some difficult variational problems. See Chapter 2 (the section entitled "Shape Factors"), [3], and [314].

[6] See the Acknowledgements, p. 426.

[7] F. J. M. Horn, "Calculation of dispersion coefficients by means of moments." *A.I.Ch.E.J.* **17,** 611–620 (1971) refers to Reprint A, because he is using spatial moment, but, although solving the general chromatographic problem, does not refer to Reprint C.

A.*

ON THE DISPERSION OF A SOLUTE IN A FLUID FLOWING THROUGH A TUBE

R. ARIS

Department of Chemical Engineering, University of Minnesota

(*Communicated by Sir Geoffrey Taylor, F.R.S.—Received 23 September 1955*)

Sir Geoffrey Taylor has recently discussed the dispersion of a solute under the simultaneous action of molecular diffusion and variation of the velocity of the solvent. A new basis for his analysis is presented here which removes the restrictions imposed on some of the parameters at the expense of describing the distribution of solute in terms of its moments in the direction of flow. It is shown that the rate of growth of the variance is proportional to the sum of the molecular diffusion coefficient, D, and the Taylor diffusion coefficient $\kappa a^2U^2/D$, *where U is the mean velocity and a is a dimension characteristic of the cross-section of the tube. An expression for* κ *is given in the most general case, and it is shown that a finite distribution of solute tends to become normally distributed.*

I. INTRODUCTION

In three recent papers, Sir Geoffrey Taylor has discussed the dispersion of soluble matter in a fluid flowing in a straight tube (Taylor 1953, 1954*a*,*b*). In the first of these, viscous flow through a tube of circular cross-section was considered, and it was found that the solute was dispersed about a point moving with the mean velocity of flow, U, with an apparent diffusion coefficient $a^2U^2/48D$. Here D is the molecular diffusion coefficient and a the radius of the tube. In the last of these papers, Taylor showed that the conditions under which this analysis was valid could be expressed as $4L/A \gg Ua/D \gg 6\cdot 9$, where L is the length over which appreciable changes in concentration occur.

It might be hoped that so elegant a result should have some meaning when these restrictions are removed, and it is possible to obtain this by fixing attention on the movement of the centre of gravity of the distribution of solute and the growth of its higher moments. These may be studied in some detail and give a useful picture of the dispersion under the most general conditions.

* Reprinted with permission of the editor of the *Proceedings of the Royal Society*, **A,** volume **235,** pp. 66–77, 1956.

2. THE GENERAL EQUATIONS OF DIFFUSION AND FLOW IN A STRAIGHT TUBE

Consider an infinite tube whose axis is parallel to the axis Ox of a rectangular co-ordinate system $Oxyz$. Let S denote the domain occupied by the interior of the tube in the plane Oyz, and let s be its area and the curve Γ its perimeter. In steady uniform flow the velocity u is everywhere in the direction Ox and is a function of y and z given by

$$u(y, z) = U\{1 + \chi(y, z)\}, \tag{1}$$

where U is the mean velocity and χ defines the velocity relative to the mean. If there is no slip at the wall of the tube $\chi = -1$ on Γ.

Let $C(x, y, z, t)$ denote the concentration of solute at the point x, y, z and time t, and let $D\psi(y, z)$ be its diffusion coefficient. The function ψ defines the variation of the diffusion coefficient and D is its mean value over the cross-section of the tube. The equation governing C is thus

$$\frac{1}{D}\frac{\partial C}{\partial t} = \nabla(\psi\nabla C) - \frac{U}{D}(1 + \chi)\frac{\partial C}{\partial x}. \tag{2}$$

It is convenient to take an origin moving with the mean speed of the stream and to reduce the variables to dimensionless form. Let a be a dimension characteristic of the cross-section S and let

$$\left.\begin{aligned} \xi &= (x - Ut)/a, \\ \eta &= y/a, \\ \zeta &= z/a, \\ \tau &= Dt/a^2, \\ \mu &= Ua/D, \end{aligned}\right\} \tag{3}$$

then the equation for C becomes

$$\frac{\partial C}{\partial \tau} = \psi\frac{\partial^2 C}{\partial \xi^2} - \mu\chi\frac{\partial C}{\partial \xi} + \frac{\partial}{\partial \eta}\left(\psi\frac{\partial C}{\partial \eta}\right) + \frac{\partial}{\partial \zeta}\left(\psi\frac{\partial C}{\partial \xi}\right). \tag{4}$$

The conditions to be applied to the solution are

$$C(\xi, \eta, \zeta, 0) = C_0(\xi, \eta, \zeta), \tag{5}$$

$$\psi\frac{\partial C}{\partial \nu} = 0 \quad \text{on} \quad \Gamma, \tag{6}$$

where $\partial/\partial\nu$ denotes differentiation along the normal to Γ, and sufficient conditions on the behaviour of C as $\xi \rightarrow \pm\infty$. Let

$$c_p(\eta, \zeta, \tau) = \int_{-\infty}^{+\infty} \xi^p C(\xi, \eta, \zeta, \tau)\,\mathrm{d}\xi \tag{7}$$

and

$$m_p(\tau) = \overline{c_p} = \frac{1}{s}\iint_S c_p\,\mathrm{d}\eta\,\mathrm{d}\zeta \tag{8}$$

be the pth moment of the distribution of solute in the filament through η, ζ at time t and the pth moment of the distribution of solute in the tube. The condition to be imposed on C as $\xi \rightarrow \pm\infty$ is thus that these moments should exist and be finite, a condition fulfilled if the solute is originally contained in a finite length of the tube. Multiplying equation (4) by ξ^p and integrating with respect to ξ from $-\infty$ to $+\infty$, we have

$$\frac{\partial c_p}{\partial \tau} = \frac{\partial}{\partial \eta}\left(\psi \frac{\partial c_p}{\partial \eta}\right) + \frac{\partial}{\partial \zeta}\left(\psi \frac{\partial c_p}{\partial \zeta}\right) + p(p-1)\psi c_{p-2} + p\mu\chi c_{p-1}, \tag{9}$$

and the conditions (5) and (6) become

$$c_p(\eta, \zeta, 0) = c_{p0}(\eta, \zeta), \tag{10}$$

$$\psi \frac{\partial c_p}{\partial \nu} = 0 \quad \text{on} \quad \Gamma. \tag{11}$$

If equation (9) is averaged over the cross-section, the use of Green's theorem and the condition (11) reduces it to

$$\frac{\mathrm{d}m_p}{\mathrm{d}\tau} = p(p-1)\overline{\psi c_{p-2}} + p\mu\overline{\chi c_{p-1}} \tag{12}$$

(where, as in equation (8), a bar denotes the average over S) and the condition (10) becomes

$$m_p(0) = m_{p0}. \tag{13}$$

The equations (9) subject to conditions (10) and (11) for $p = 0, 1, 2, \ldots$ now form a sequence of inhomogeneous equations which can be solved for the moments c_p. In principle these might be solved to a sufficiently high value of p for the distribution to be constructed to any degree of accuracy, but a very useful picture of the progress of the dispersion can be obtained from the first three or four moments. Since it will be shown that the distribution ultimately tends to normality the first two moments are ultimately sufficient to describe the distribution.

3. THE TUBE OF CIRCULAR CROSS-SECTION

Before proceeding with the discussion of the general case, it is interesting to consider the case of a tube of circular cross-section and to note how the present treatment agrees with Taylor's. In this case we take Ox to be the axis and a to be the radius of the tube and transform to polar co-ordinates

$$\eta = \rho\cos\theta, \quad \zeta = \rho\sin\theta.$$

In the case of viscous flow $\chi = (1 - 2\rho^2)$, $\psi = 1$ and D is the molecular diffusion coefficient.

Then $$\frac{\partial c_p}{\partial \tau} = \frac{1}{\rho}\frac{\partial}{\partial \rho}\left(\rho\frac{\partial c_p}{\partial \rho}\right) + \frac{1}{\rho^2}\frac{\partial^2 c_p}{\partial \theta^2} + p(p-1)\,c_{p-2} + \mu p(1-2\rho^2)\,c_{p-1}, \tag{14}$$

and of this equation we seek solutions of period 2π in θ and satisfying

$$c_p(\rho, \theta, 0) = c_{p0}(\rho, \theta), \tag{15}$$

$$\frac{\partial c_p}{\partial \rho} = 0, \quad \rho = 1. \tag{16}$$

Equation (12) becomes

$$\frac{dm_p}{d\tau} = p(p-1)\, m_{p-2} + (p\mu/\pi) \int_0^{2\pi} d\theta \int_0^1 \rho(1-2\rho^2)\, c_{p-1}\, d\rho. \tag{17}$$

In particular, $dm_0/d\tau = 0$ so that m_0 is a constant which we may take as unity. This of course merely states that the total quantity of solute is constant.

For $p = 0$, equations (14) and (16) give

$$c_0(\rho, \theta, \tau) = 1 + \sum_{m=0}^{\infty} \sum_{n=1}^{\infty} \{A_{mn}^0 \cos m\theta + B_{mn}^0 \sin m\theta\}\, J_m(\alpha_{mn}\rho) \exp(-\alpha_{mn}^2 \tau), \tag{18}$$

where α_{mn} is the nth zero of $dJ_m(\rho)/d\rho$. The constants A and B are chosen to satisfy equation (15), i.e.

$$\frac{\pi}{2}\left(1 - \frac{m^2}{\alpha^2}\right) \{J_m(\alpha)\}^2 \begin{matrix} A_{mn}^0 \\ B_{mn}^0 \end{matrix} = \int_0^{2\pi} d\theta \int_0^1 \begin{matrix} \cos m\theta \\ \sin m\theta \end{matrix}\, \rho J_m(\alpha\rho)\, c_{00}(\rho, \theta)\, d\rho, \tag{19}$$

the suffixes m and n to α being understood. (It will generally cause no confusion to suppress the suffixes mn on α, A^p and B^p.)

For $p = 1$ equation (17) gives

$$\frac{dm_1}{d\tau} = \frac{\mu}{\pi} \int_0^{2\pi} d\theta \int_0^1 \rho(1-2\rho^2)\, c_0\, d\rho.$$

In this integral the constant term vanishes since the mean value of $(1 - 2\rho^2)$ is zero and the terms with $m > 0$ vanish by integration round the circle. Hence

$$\begin{aligned} \frac{dm_1}{d\tau} &= \mu \Sigma A_{0n}^0\, e^{-\alpha^2\tau} \int_0^1 2\rho(1-2\rho^2)\, J_0(\alpha\rho)\, d\rho \\ &= -8\mu \Sigma A_{0n}^0\, \alpha_{0n}^{-2}\, e^{-\alpha^2\tau} J_0(\alpha_{0n}). \end{aligned}$$

Thus as $\tau \to \infty$, $dm_1/d\tau \to 0$ and the centre of gravity ultimately moves with the mean speed of the stream. Choosing the origin in the original plane of the centre of gravity, $m_{10} = 0$ and

$$m_1 = -8\mu \sum_{n=1}^{\infty} A_{0n}^0 \alpha_{0n}^{-4} J_0(\alpha_{0n})\{1 - \exp(-\alpha_{0n}^2\tau)\}. \tag{20}$$

Thus the centre of gravity ultimately moves to a position

$$m_{1\infty} = -8\mu \Sigma A_{0n}^0 \alpha_{0n}^{-4} J_0(\alpha_{0n}) \tag{21}$$

relative to its original position. The actual distance moved is proportional to a^2U/D, the constant of proportionality depending on the initial distribution of solute. If the mean speed of the stream is being measured by the time taken for the centre of gravity of the distribution to move between two points there will be a slight error, though in practice it will usually be negligible.

For a more detailed picture of the variation in position of centre of gravity across the tube, equation (14) must be solved. For $p = 1$ this becomes

$$\frac{\partial c_1}{\partial \tau} - \frac{1}{\rho}\frac{\partial}{\partial \rho}\left(\rho \frac{\partial c_1}{\partial \rho}\right) - \frac{1}{\rho^2}\frac{\partial^2 c_1}{\partial \theta^2} = \mu(1 - 2\rho^2)\, c_0, \tag{22}$$

of which the complete solution consists of three parts: (i) the term in the particular integral arising from the constant part of c_0; (ii) the rest of the particular integral which may be written

$$\mu\Sigma(A^0 \cos m\theta + B^0 \sin m\theta)\, \phi(\rho) \exp(-\alpha^2\tau),$$

with ϕ_{mn} satisfying the equation

$$\frac{1}{\rho}\frac{\partial}{\partial \rho}\left(\rho \frac{\partial \phi}{\partial \rho}\right) + \left(\alpha^2 - \frac{m^2}{\rho^2}\right)\phi = -J_m(\alpha\rho)(1 - 2\,\rho^2);$$

and (iii) the complementary function

$$\Sigma(A^1 \cos m\theta + B^1 \sin m\theta) J_m(\alpha\rho) \exp(-\alpha^2\tau),$$

with A^1 and B^1 chosen to fit the initial value $c_{10}(\rho, \theta)$.

The first part is independent of τ and θ, being $-\frac{1}{4}\mu\rho^2(1 - \frac{1}{2}\rho^2)$, and choosing A_0^1 so that $m_{10} = 0$, the expression for c_1 is

$$\begin{aligned} c_1 = \tfrac{1}{4}\mu(\tfrac{1}{3} - \rho^2 + \tfrac{1}{2}\rho^4) &+ \mu\Sigma(A^0 \cos m\theta + B^0 \sin m\theta)\phi(\rho) \exp(-\alpha^2\tau) \\ &- \mu\Sigma A_{0n}^0 \overline{\phi}_{0n} + \Sigma(A^1 \cos m\theta + B^1 \sin m\theta) J_m(\alpha\rho) \exp(-\alpha^2\tau). \end{aligned} \tag{23}$$

Thus ultimately the centre of gravity is distributed across the tube according to the formula

$$c_1 \sim m_{1\infty} + \tfrac{1}{4}\mu(\tfrac{1}{3} - \rho^2 + \tfrac{1}{2}\rho^4). \tag{24}$$

It may readily be seen by averaging the equation for ϕ_{0n} that the form given for $m_{1\infty}$ in equation (23) agrees with that previously derived in equation (21). It is interesting that the ultimate distribution of c_1 is independent of θ; physically this is because circumferential diffusion brings a molecule into a stream of precisely the same speed and so contributes nothing to the movement of the centre of gravity.

In Taylor's analysis (1953), 1954*b*) an expression is derived for the variation of C with ρ when $\partial C_m/\partial x$ is constant. Equation (6) of Taylor (1954*b*) may be written in the notation of this paper

$$(C_m - C) \Big/ \frac{\partial C_m}{\partial \xi} = \tfrac{1}{4}\mu(\tfrac{1}{3} - \rho^2 + \tfrac{1}{2}\rho^4), \tag{25}$$

which with $\partial C_m/\partial \xi$ constant means that the distance from any plane to the point at which the concentration $C = C_m$, the mean concentration in the plane, is given by the expression on the right-hand side of (25), i.e. this is the shape of the surfaces of constant concentration. It follows that the centres of gravity would then be distributed on the same surface. Equation (24) shows that this last statement is true even though $\partial C_m/\partial \xi$ is not constant.

Substituting the value of c_1 in equation (17) with $p = 2$ and neglecting terms which tend to zero we have

$$\frac{\mathrm{d}m_2}{\mathrm{d}\tau} = 2 + \mu^2 \int_0^1 \rho(1 - 2\rho^2)(\tfrac{1}{3} - \rho^2 + \tfrac{1}{2}\rho^4)\,\mathrm{d}\rho + O\{\exp(-\alpha^2\tau)\}$$
$$\sim 2(1 + \mu^2/48).$$

Thus the rate of growth of the variance rapidly becomes constant and

$$m_2 \sim 2(1 + \mu^2/48)\,\tau + \text{a constant}$$

which becomes negligible by comparison.

If V is the variance of the distribution of solute about the moving origin, i.e.

$$V = \frac{1}{S}\iint_S \mathrm{d}y\,\mathrm{d}z \int_{-\infty}^{+\infty} (x - Ut)^2 C(x, y, z, t)\,\mathrm{d}x,$$

then

$$\lim_{t\to\infty} \frac{1}{2}\frac{\mathrm{d}V}{\mathrm{d}t} = D + \frac{a^2U^2}{48D} = K. \qquad (26)$$

It is not unreasonable to use the left-hand side of this equation as the definition of the effective diffusion constant K, the more so as it will be shown that any distribution tends to normality. With this definition K is the sum of the molecular diffusion coefficient, D, and the apparent diffusion coefficient $k = \alpha^2U^2/48D$, which was discovered by Taylor in his first paper (Taylor 1953, equation (25)). Equation (26), however, is true without any restriction on the value of μ, or on the distribution of solute. The constant 1/48 is a function of the profile of flow, and for so-called piston flow with $\chi \equiv 0$ this constant is zero and $K = D$ as it should.

In the full expression for c_2 the terms which do not vanish as $\tau \to \infty$ are $m_2(\tau)$, a constant function of ρ and a constant depending on the initial value of c_2. The constant function of ρ must satisfy the equation

$$\frac{1}{\rho}\frac{\mathrm{d}}{\mathrm{d}\rho}\left(\rho\frac{\mathrm{d}f}{\mathrm{d}\rho}\right) = \tfrac{1}{4}\mu^2(-\tfrac{1}{3} + \tfrac{5}{3}\rho^2 - \tfrac{5}{2}\rho^4 + \rho^6) + \tfrac{1}{48}\mu^2 + 2\mu m_{1\infty}(1 - 2\rho^2),$$

i.e.

$$f = g(\rho) + \tfrac{1}{2}\mu m_{1\infty}(-\tfrac{1}{3} + \rho^2 - \tfrac{1}{2}\rho^4),$$

where

$$g(\rho) = \frac{\mu^2}{16}\left\{\tfrac{1}{16}\rho^8 - \tfrac{5}{18}\rho^6 + \tfrac{5}{12}\rho^4 - \tfrac{1}{4}\rho^2 + \frac{31}{5 \times 9 \times 16}\right\}.$$

This is precisely the polynomial obtained by Taylor under the assumption that $\partial C_m/\partial\xi$ is not quite constant but that the second derivative $\partial^2 C_m/\partial\xi^2$ is (Taylor 1954*b*, equation (24)).

Thus, apart from a constant, c_2 is ultimately given by

$$c_2 \sim 2(1 + \mu^2/48)\tau + \tfrac{1}{2}\mu m_{1\infty}(-\tfrac{1}{3} + \rho^2 - \tfrac{1}{2}\rho^4) + g(\rho)$$

and

$$\frac{\mathrm{d}m_3}{\mathrm{d}\tau} \sim 6m_{1\infty} + 3\mu \int_0^1 2\rho(1 - 2\rho^2)c_2\,\mathrm{d}\rho$$
$$= 6m_{1\infty}(1 + \mu^2/48) + 7\mu^3/2^{10}.$$

Neglecting constant terms which become negligible in comparison to the dominant term proportional to τ, this equation gives

$$m_3 - 3m_{1\infty}m_2 \sim 7\mu^3\tau/2^{10}. \tag{27}$$

Apart from a term m_1^3, which is a constant, the expression on the left-hand side of (27) is the third moment about the mean m_1 so that the absolute skewness $\beta_1 = m_3^2/m_2^3$ is

$$\beta_1 = 49\mu^6/2^{20}(1 + \mu^2/48)^3\tau,$$

which tends to zero as $\tau \rightarrow \infty$. Thus any distribution of solute tends to become more symmetrical.

An examination of the equation for c_3 shows that the dominant terms will be proportional to τ and will be $m_3(\tau)$ and the product of τ and a function of ρ. In general,

$$m_{2n} = M_{2n}\tau^n, \quad m_{2n+1} = M_{2n+1}\tau^n, \tag{28}$$

where M_{2n} and M_{2n+1} are even and odd polynomials in μ of degree $2n$ and $2n + 1$ respectively, and

$$c_{2n} = M_{2n}\tau^n + G_{2n}(\rho)\tau^{n-1}, \quad c_{2n+1} = M_{2n+1}\tau^n + G_{2n+1}(\rho)\tau^n, \tag{29}$$

where the mean values of the functions $G_{2n}(\rho)$ and $G_{2n+1}(\rho)$ are zero. This remark may be used to show that the distribution tends to normality but the proof of this will be left for the general case.

4. SOME SPECIAL INITIAL DISTRIBUTIONS OF SOLUTE

The case of a distribution of solute initially constant over any cross-section of the tube is particularly simple for $A^0 = B^0 = 0$, and so by (21) $m_{1\infty} = 0$. When $\tau = 0$, c_1 is independent of ρ and so $A^1 = B^1 = 0$ for all $m > 0$ and

$$\Sigma A_{0n}^1 J_0(\alpha_{0n}\rho) = -\tfrac{1}{4}\mu(\tfrac{1}{3} - \rho^2 + \tfrac{1}{2}\rho^4),$$

i.e.

$$A_{0n}^1 = -8\mu/\alpha_{0n}^4 J_0(\alpha_{0n}) \quad (n > 1)$$

and

$$c_1 = \frac{\mu}{4}(\tfrac{1}{3} - \rho^2 + \tfrac{1}{2}\rho^4) - 8\mu \sum_{n=1}^{\infty} \alpha_{0n}^{-4}\frac{J_0(\alpha_{0n}\rho)}{J(\alpha_{0n})}\exp(-\alpha_{0n}^2\tau). \tag{30}$$

Inserting this value in the equation for m_2 and integrating

$$\begin{aligned} m_2 &= m_{20} + 2(1 + \mu^2/48)\tau + 128\mu^2\Sigma\alpha_{0n}^{-8}\{1 - \exp(-\alpha_{0n}^2\tau)\} \\ &\sim m_{20} + 2(1 + \mu^2/48)\tau + \mu^2/360. \end{aligned}$$

The smallness of the last term may be judged by considering the case $\mu^2 \gg 48$. Then $m_2 - m_{20} = (\mu^2/24)(\tau + 1/15)$, so that this last term is equivalent to an additional time of $a^2/15D$.

The case of an instantaneous point source at $\rho = \rho_0$, $\theta = 0$ is of interest. All the moments are initially zero and $c_{00} = 0$ for $\rho \neq \rho_0$, $\theta \neq 0$ and $m_0 = 1$. Then

$$c_0 = 1 + 2\Sigma\left(1 - \frac{m^2}{\alpha^2}\right)^{-1} e^{-\alpha^2\tau} \cos m\theta J_m(\alpha\rho)J_m(\alpha\rho_0)/\{J_m(\alpha)\}^2, \tag{31}$$

and $$m_{1\infty} = -16\mu\Sigma\alpha^{-4}J_0(\alpha\rho_0)/J_0(\alpha) = \mu(\tfrac{1}{6} - \tfrac{1}{2}\rho_0^2 + \tfrac{1}{4}\rho_0^4). \tag{32}$$

The dependence of $m_{1\infty}$ on the distance of the point source from the axis is the same as the dependence of the centre of gravity of the solute in any filament on its distance from the axis.

5. THE GENERAL CASE

The general case of a tube of arbitrary cross-section, flow profile and variation of diffusion coefficient, for which the equations were set up in §1, may be considered in a similar way. Again, m_0 must be a constant, which may be taken as unity, and for c_0 we have the equations

$$\left.\begin{aligned} \nabla(\psi\nabla c_0) &= \frac{\partial c_0}{\partial \tau}, \\ c_0(\eta, \zeta, 0) &= c_{00}(\eta, \zeta), \\ \psi\frac{\partial c_0}{\partial \nu} &= 0 \quad (\text{on } \Gamma). \end{aligned}\right\} \tag{33}$$

Under very general conditions on Γ and ψ there will exist a positive increasing sequence of eigenvalues λ_n and a complete set of ortho-normal eigenfunctions v_n satisfying the equation

$$\nabla(\psi\nabla v_n) + \lambda_n v_n = 0 \tag{34}$$

in S and the boundary conditions

$$\psi\frac{\partial v_n}{\partial \nu} = 0 \tag{35}$$

on Γ. Any function satisfying the boundary conditions and with continuous derivatives up to the second order can be expanded in a series of these functions (see, for example, Courant & Hilbert 1937). If the constants A_n are such that

$$c_{00}(\eta, \zeta) = 1 + \sum_{n=1}^{\infty} A_n v_n(\eta, \zeta), \tag{36}$$

then the solution of the equations (33) is

$$c_0(\eta, \zeta, \tau) = 1 + \sum_{n=1}^{\infty} A_n v_n(\eta, \zeta)\exp(-\lambda_n\tau). \tag{37}$$

As before, the density of solute in any streamline rapidly becomes uniform across the tube, the relaxation time being of the order of λ_1^{-1}, where λ_1 is the smallest of the eigenvalues.

Inserting this value of c_0 in equation (12) with $p = 1$, it is again evident that the centre of gravity ultimately moves with the mean speed of the stream. Its final position relative to a moving origin originally at the centre of gravity is

$$m_{1\infty} = \mu \sum_{n=1}^{\infty} A_n \lambda_n^{-1} \overline{\chi v_n}, \tag{38}$$

where the bar denotes an average over the cross-section of the tube.

Apart from $m_{1\infty}$ and terms which vanish as $\tau \to \infty$, c_1 will contain a constant function of η and ζ whose mean value is zero. This function arises from the constant term 1 in c_0 when this is inserted in equation (9) with $p = 1$ and so must satisfy the equations

$$\left.\begin{aligned} \nabla(\psi\nabla\phi) &= -\chi, \\ \psi\frac{\partial\phi}{\partial\nu} &= 0 \quad (\text{on } \Gamma). \end{aligned}\right\} \tag{39}$$

Then $c_1 \sim m_{1\infty} + \mu\phi(\eta, \zeta)$ for all other terms will vanish at least as rapidly as $\exp(-\lambda_1\tau)$.

From equation (12) with $p = 2$

$$\frac{1}{2}\frac{\mathrm{d}m_2}{\mathrm{d}\tau} \sim \overline{\psi} + \mu^2\overline{\chi\phi}, \tag{40}$$

where again the vanishing terms have been neglected. But by definition $\overline{\psi} = 1$ and $\overline{\chi\phi} = \kappa$, a pure number dependent only on the geometry of the cross-section and the functions ψ and χ. Thus again defining the effective diffusion constant K as one-half the rate of growth of the variance

$$K = D + \kappa a^2U^2/D, \tag{41}$$

showing how a Taylor diffusion coefficient can be found in the general case. By Green's theorem and equations (39) we have $\kappa = \overline{\phi\chi} = -\overline{\phi\nabla(\psi\nabla\phi)} = \overline{\psi(\nabla\phi)^2}$.

Consideration of the dominant terms in the successive moments again shows that c_{2n} and c_{2n+1} are of the form

$$\left.\begin{aligned} c_{2n} &= M_{2n}\tau^n + G_{2n}(\eta, \zeta)\tau^{n-1}, \\ c_{2n+1} &= M_{2n+1}\tau^n + G_{2n+1}(\eta, \zeta)\tau^n, \end{aligned}\right\} \tag{42}$$

where the mean values of the functions G are zero.

Substituting c_{2n+1} in equation (9) with $p = 2n + 1$ and neglecting all except the dominant term

$$\nabla(\psi\nabla G_{2n+1}) + (2n+1)\mu M_{2n}\chi = 0,$$

so that

$$G_{2n+1}(\eta, \zeta) = (2n+1)\mu M_{2n}\phi(\eta, \zeta),$$

where ϕ has already been defined by the equations (39). Substituting this value of G_{2n+1} in equation (12) with $p = (2n + 2)$ and again neglecting all save the dominant terms

$$\begin{aligned} \frac{\mathrm{d}m_{2n+2}}{\mathrm{d}\tau} &= (2n+2)(2n+1)M_{2n}\tau^n + (2n+2)(2n+1)\mu^2M_{2n}\tau^n\overline{\phi\chi} \\ &= (2n+1)M_{2n}2(1+\kappa\mu^2)(n+1)\tau^n, \end{aligned}$$

and the dominant term of m_{2n+2} is

$$m_{2n+2} \sim (2n+1)M_{2n}\tau^n 2(1+\kappa\mu^2)\tau = (2n+1)m_{2n}m_2.$$

It follows that as $\tau \to \infty$

$$(m_{2n+1})^2/(m_2)^{2n+1} \to 0,$$
$$m_{2n}/(m_2)^n \to (2n-1)(2n-3)\ldots 3.1.$$

These are the relations which exist between the moments of the normal distribution and in this sense the mean concentration is ultimately distributed about a point moving with the mean speed of the stream according to the normal law of error, the variance being $2(1+\kappa\mu^2)\tau$. It should be noted that the approach to normality is as τ^{-1}, a very much slower process than the vanishing of terms in the expressions for the moments, which is as $\exp(-\lambda_1\tau)$.

6. TURBULENT FLOW IN A TUBE OF CIRCULAR CROSS-SECTION

The case of turbulent flow and diffusion which was treated in Taylor's second paper (Taylor 1954*a*) clearly comes under the general case. If ψ and χ are functions of $\rho = (\eta^2 + \zeta^2)^{1/2}$ only and α is the radius of the tube equation (9) becomes

$$\frac{\partial c_p}{\partial \tau} = \frac{1}{\rho}\frac{\partial}{\partial \rho}\left\{\rho\psi(\rho)\frac{\partial c_p}{\partial \rho}\right\} + p(p-1)\psi(\rho)c_{p-2} + p\mu\chi(\rho)c_{p-1}$$

and the equation for ϕ is

$$\frac{\mathrm{d}}{\mathrm{d}\rho}\left(\rho\psi(\rho)\frac{\mathrm{d}\phi}{\mathrm{d}\rho}\right) + \rho\chi(\rho) = 0,$$

of which the solution is

$$\phi(\rho) = -\int_0^{\rho}\frac{\mathrm{d}\rho'}{\rho'\psi(\rho')}\int_0^{\rho'}\rho''\chi(\rho'')\,\mathrm{d}\rho''.$$

This is the function tabulated in column (8) of table 1 (Taylor 1954*a*), and the final column (11) of this table is

$$\kappa = \int_0^1 2\rho\chi(\rho)\phi(\rho)\,\mathrm{d}\rho.$$

In §4 of this paper the effect of longitudinal diffusion is estimated and is added to diffusion coefficient $\kappa\mu^2$. Equation (41) of the present paper supplies a rigorous justification for this.

7. VISCOUS FLOW IN A TUBE OF ARBITRARY CROSS-SECTION

It is well known that the velocity $u(\eta, \zeta)$ for viscous flow in a straight pipe satisfies the equations

$$\left.\begin{aligned}\nabla^2 u &= -a^2 \frac{\mathrm{d}P}{\mathrm{d}x}\bigg/\mu \quad (\text{in } S),\\ u &= 0 \quad (\text{on } \Gamma),\end{aligned}\right\} \tag{43}$$

where in this equation P is the pressure and μ the viscosity, and it follows that χ satisfies the equations

$$\begin{aligned}\nabla^2\chi &= -\alpha \quad (\text{in } S)\\ \chi &= -1 \quad (\text{on } \Gamma),\end{aligned}$$

where $\alpha = a^2 \dfrac{\mathrm{d}P}{\mathrm{d}x}\bigg/ \mu U$ is a constant.

Combining this equation with equation (39), and putting $\psi = 1$, we have the equations for ϕ,

$$\left.\begin{aligned}\nabla^4\phi &= \alpha \quad (\text{in } S),\\ \left.\begin{aligned}\nabla^2\phi &= 1\\ \frac{\partial\phi}{\partial\nu} &= 0\end{aligned}\right\} &\quad (\text{on } \Gamma),\end{aligned}\right\} \tag{44}$$

which enables ϕ and so κ to be calculated for any cross-section.

As an illustration of this a tube of elliptical cross-section may be considered. If the major and minor semi-axes are a and b respectively

$$\chi(y, z) = 1 - 2(y^2/a^2) - 2(z^2/b^2). \tag{45}$$

Taking a as the characteristic length it may be shown, after a certain amount of algebraic labour, that

$$\kappa = \frac{1}{48}\frac{24 - 24e^2 + 5e^4}{24 - 12e^2}, \tag{46}$$

where $e = (1 - b^2/a^2)^{1/2}$ is the eccentricity of the cross-section. For $e = 0$, κ is 1/48 as it should be for a circle and when $e = 1$, κ is 5/12 of this value. A very small value of $(1 - e)$ represents a very narrow elliptical slit and in such a slit diffusion across the slit renders the concentration very nearly constant in this direction, so there are effectively only two spatial dimensions.*

Let $O\zeta$ be in the direction of the narrow dimension of the slit, $2a$ be its major dimension, and $\beta(\eta)$ be proportional to the breadth of the slit at any point. Then the equation for C is

$$\frac{\partial C}{\partial\tau} = \frac{\partial^2 C}{\partial\xi^2} + \frac{1}{\beta}\frac{\partial}{\partial\eta}\left(\beta\frac{\partial C}{\partial\eta}\right) - \mu\chi(\eta)\frac{\partial C}{\partial\xi}.$$

The function $\chi(\eta)$ is here the mean value of $\chi(\eta, \zeta)$ across the slit, so that

$$\int_{-1}^{+1}\beta(\eta)\chi(\eta)\,\mathrm{d}\eta = 0.$$

* I am indebted to Mr. C. H. Bosanquet for pointing this out.

In fact all averages must be taken with $\beta(\eta)$ as a weighting factor. An equation similar to equation (9) obtains for the moment c_p, namely,

$$\frac{\partial c_p}{\partial \tau} = \frac{1}{\beta}\frac{\partial}{\partial \eta}\left(\beta \frac{\partial c_p}{\partial \eta}\right) + p(p-1)c_{p-2} + \mu p c_{p-1}$$

and $$\frac{\mathrm{d}m_p}{\mathrm{d}\tau} = p(p-1)\, m_{p-2} + \mu p \int_{-1}^{+1} \beta \chi c_{p-1}\,\mathrm{d}\eta \bigg/ \int_{-1}^{+1} \beta\,\mathrm{d}\eta.$$

In precisely the same way as before we derive the value of κ,

$$\kappa = \int_{-1}^{+1} \phi(\eta)\chi(\eta)\beta(\eta)\,\mathrm{d}\eta \bigg/ \int_{-1}^{+1} \beta(\eta)\,\mathrm{d}\eta,$$

where $$\frac{\mathrm{d}}{\mathrm{d}\eta}\left(\beta(\eta)\frac{\mathrm{d}\phi}{\mathrm{d}\eta}\right) = -\chi(\eta)\beta(\eta)$$

and $$\frac{\mathrm{d}\phi}{\mathrm{d}\eta} = 0 \quad \text{at} \quad \eta = \pm 1.$$

In the case of an elliptical slit

$$\beta(\eta) = (1-\eta^2)^{1/2},$$
$$\chi(\eta) = \tfrac{1}{3}(1-4\eta^2),$$
$$\phi(\eta) = -\tfrac{1}{6}\eta^2(1-\tfrac{1}{2}\eta^2),$$

and $\kappa = 5/12 \times 48$, which confirms the limiting value in equation (46).

It is well known that for a given pressure drop, the flow is greater in a circular tube than in an elliptical one of the same area, and if in the Taylor diffusion coefficient a^2 is replaced by the area of cross-section (πa^2 for the circle and πab for the ellipse) the constant κ is least for $e = 0$. Thus the dispersion in a circular tube is less than in an elliptical tube of the same area.

REFERENCES

Courant, R. & Hilbert, D. 1937 *Methoden der mathematischen Physik,* **1.** Berlin: Springer.
Taylor, Sir Geoffrey 1953 *Proc. Roy. Soc.* A, **219,** 186.
Taylor, Sir Geoffrey 1954*a* *Proc. Roy. Soc.* A, **223,** 446.
Taylor, Sir Geoffrey 1954*b* *Proc. Roy. Soc.* A, **225,** 473.

B.*

ON THE DISPERSION OF A SOLUTE BY DIFFUSION, CONVECTION, AND EXCHANGE BETWEEN PHASES

R. ARIS

Department of Chemical Engineering, University of Minnesota

(Communicated by Sir Geoffrey Taylor, F.R.S.—Received 25 March 1959)

The method used in an earlier paper (Aris 1956) to discuss the dispersion of a solute in a fluid flowing through a tube is applied here to the case in which the solute can also pass into another fluid phase flowing in an annular region around the first. The apparent diffusion coefficient is the sum of the molecular and Taylor diffusion coefficients in the two phases and a term due to the finite rate of partition between them. It is shown how the Taylor diffusion coefficients depend on the ratio of amounts of solute held in the two phases and how this gives a connexion between the coefficient $a^2U^2/48D$ found by Taylor (1953) for viscous flow in a circular tube and the 11 $a^2U^2/48D$ found by Westhaver (1942) in his analysis of the distillation column. the use of these apparent diffusion coefficients in distillation and partition chromatography is illustrated.

1. INTRODUCTION

If a solute diffuses lin a stream of varying velocity the effect of the lateral diffusion of a solute into a region of different speed is to change its rate of forward motion and hence induce a longitudinal diffusion. The complexity of this situation was first illuminated by Sir Geoffrey Taylor (1953) who showed that this effect could be described by an apparent diffusion coefficient $\kappa a^2U^2/D$, where U = mean speed of stream, a = radius of circular tube, D = molecular diffusion coefficient, and κ is a constant depending on the velocity profile, ($\kappa = \frac{1}{48}$ for viscous flow). The mean concentration across the tube then appears to be dispersing about a plane moving at a speed U with diffusion coefficient $D + \kappa a^2U^2/D$. Taylor later considered turbulent flow profiles and diffusion coefficients (Taylor 1954*a*, *b*) and the turbulent coefficients have recently been calculated in some detail by Tichacek, Barkelew & Baron (1957). An interpretation of the Taylor coefficient which freed it from the restrictions under which it was derived and a method of calculating κ for general flow profiles and cross-sections was given in 1956 (Aris 1956).

* Reprinted with permission of the editor of the *Proceedings of the Royal Society*, **A,** volume 252, pp. 538–550, 1959.

In his analysis of the effect of diffusion on an open-tube distillation column Westhaver (1942) came up with the apparent diffusion coefficient $11a^2U^2/48D$, and since he assumes a parabolic profile it is at first surprising that this should differ by a factor of 11 from Taylor's result. It appears, however, if the more general problem in which the solute can be retained on the wall be considered, that the value of κ varies continuously from $\frac{11}{48}$ to $\frac{1}{48}$ as the fraction of solute held on the wall varies from 1 to 0. This result is implicit in Golay's analysis of the tubular chromatographic column (Golay 1958). He considers the stationary phase of the column as a very thin retentive layer held on the wall and derives an expression for the dispersion by arguments entirely analogous to Taylor's. He has also discussed the effect of diffusion in the retentive layer.

We shall consider in this paper a more general problem which embraces the foregoing as special cases. Since the language of chromatography is convenient and it is a field in which these results can be usefully applied, it will be used here, but it should not be overlooked that the analysis is valid for similar problems in heat and mass transfer. In the next paragraph we consider the two phases to flow in coaxial annular regions with velocities and diffusion coefficients varying only with radial distance from the common axis. The inner and outer fluids are referred to as the gas and liquid phases, respectively, and the distribution of solute concentration as the chromatogram. Numerical results are given for a number of useful special cases. The problem for more general cross-sections is then discussed and it is shown how κ may be calculated. The application of these results to distillation and chromatography is then illustrated.

2. DISPERSION IN COAXIAL CYLINDRICAL ANNULI

The problem to be solved in this paragraph is to determine the rate of spread of the chromatogram under the following conditions. The gas and liquid phases flow in the annular space between two coaxial cylinders of radii r_0 and r_2, the interface being a cylinder with the same axis and radius r_1 ($0 \leq r_0 < r_1 < r_2$). Both phases may be in motion with linear velocity a function of radial distance from the axis, r, and the solute diffuses in both phases with a diffusion coefficient which may also be a function of r. At equilibrium the concentration of solute in the liquid, c_2, is a constant multiple of that in the gas, c_1($c_2 = \alpha c_1$) and at any instant the rate of transfer across the interface is proportional to the distance from equilibrium there, i.e. the value of $(c_2 - \alpha c_1)$. The dispersion of the solute is due to three processes: (i) the combined effect of diffusion and convection in the gas phase, (ii) the finite rate of transfer at the interface, (iii) the combined effect of diffusion and convection in the liquid phase. In what follows the equations will often be in sets of five, labelled (*a*), . . . , (*e*); the differential equations expression the three processes (i), (ii) (iii) above are always (*b*), (*c*) and (*d*), respectively; equations (*a*) and (*e*) represent the condition that there is no flow over the boundaries at $r = r_0$ and $r = r_2$.

The following notation is introduced and unless otherwise stated the suffixes 1 and 2 refer to quantities in the gas and liquid phases respectively:

$$\begin{aligned} c_i &= \text{concentration of solute,} \\ U_i &= \text{mean velocity,} \\ U_i\phi_i(r) &= \text{linear velocity at radius } r, \\ D_i &= \text{mean diffusion coefficient,} \\ D_i\psi_i(r) &= \text{diffusion coefficient at radius } r. \end{aligned}$$

In all these $i = 1$ or 2 and by definition ϕ and ψ satisfy

$$\int_{r_{i-1}}^{r_i} 2r\phi_i(r)\,\mathrm{d}r = (r_i^2 - r_{i-1}^2) = \int_{r_{i-1}}^{r_i} 2r\psi_i(r)\,\mathrm{d}r.$$

Finally

$$\begin{aligned} x &= \text{co-ordinate along common axis,} \\ r &= \text{radial co-ordinate,} \\ t &= \text{time.} \end{aligned}$$

The standard equations for diffusion and flow with the boundary conditions referred to above are now:

$$\left.\begin{aligned}
&D_1\psi_1(r)\frac{\partial c_1}{\partial r} = 0 \quad (\text{at } r = r_0), && (a)\\
&\frac{\partial c_1}{\partial t} = D_1\frac{1}{r}\frac{\partial}{\partial r}\left(r\psi_1(r)\frac{\partial c_1}{\partial r}\right) + D_1\psi_1(r)\frac{\partial^2 c_1}{\partial x^2} - U_1\phi_1(r)\frac{\partial c_1}{\partial x} \quad (\text{in } r_0 < r < r_1), && (b)\\
&D_1\psi_1(r)\frac{\partial c_1}{\partial r} = D_2\psi_2(r)\frac{\partial c_2}{\partial r} = k(c_2 - \alpha c_1) \quad (\text{at } r = r_1), && (c)\\
&\frac{\partial c_2}{\partial t} = D_2\frac{1}{r}\frac{\partial}{\partial r}\left(r\psi_2(r)\frac{\partial c_2}{\partial r}\right) + D_2\psi_2(r)\frac{\partial^2 c_2}{\partial x^2} - U_2\phi_2(r)\frac{\partial c_2}{\partial x} \quad (\text{in } r_1 < r < r_2) && (d)\\
&D_2\psi_2(r)\frac{\partial c_2}{\partial r} = 0 \quad (\text{at } r = r_2). && (e)
\end{aligned}\right\} \quad (1)$$

It is assumed that the cylinders are infinitely long and that c_1 and c_2 drop to zero as $x \to \pm\infty$ so fast that $x^p c_i \to 0$ for all p (in the nomenclature of Lighthill (1958) they are 'good functions' of x). This condition is not so artificial as it sounds for if the solute is initially confined to a small part of the tube its concentration will fall off as $\exp(-kx^2)$ for some positive k so that $x^p c_i \to 0$ for all p. Moreover, in practice the chromatogram occupies a small fraction of the tube length which is effectively infinite. The concentrations c_1 and c_2 initially satisfy some specified distribution of concentration and to simplify the analysis we assume that at $t = 0$ a unit quantity of solute is located at the plane $x = 0$ and equilibrium is attained between the two phases. This again is a justifiable idealization of the initial chromatogram, and the results for more general conditions will be discussed later. Using Dirac's delta function which has the properties $\int_{-\infty}^{\infty} \delta(x)\,\mathrm{d}x = 1$, $\delta(x) = 0$, $x \neq 0$ this may be expressed as

$$\alpha c_1(x, r, 0) = c_2(x, r, 0) = \alpha\delta(x)/s^2, \quad (2)$$

where

$$s^2 = (r_1^2 - r_0^2) + \alpha(r_2^2 - r_1^2),$$

since a unit quantity of solute is taken.

It is convenient at this stage to change the variable of length along the axis so that its origin moves with constant speed V. This quantity V will not be specified at this stage but chosen later to move at the speed of the centre of gravity of the solute. Let

$$y = x - Vt,$$

then $\partial c/\partial t$ must be replaced by $\partial c/\partial t - V(\partial c/\partial y)$ and equations (1) (b) and (d) become

$$\frac{\partial c_i}{\partial t} = D_i \frac{1}{r}\frac{\partial}{\partial r}\left(r\psi_i(r)\frac{\partial c_i}{\partial r}\right) + D_i\psi_i(r)\frac{\partial^2 c_i}{\partial y^2} - \{U_i\phi_i(r) - V\}\frac{\partial c_i}{\partial y} \quad (\text{in } r_{i-1} < r < r_i), \tag{3}$$

with $i = 1$ for (b) and $i = 2$ for (d); equations (a), (c) and (e) are the same.

The moments of c_i are defined as follows:

$$c_i^{(p)}(r, t) = \int_{-\infty}^{\infty} y^p c_i(y, r, t)\, \mathrm{d}y, \tag{4}$$

$$m^{(p)}(t) = \int_{r_0}^{r_1} 2rc_1^{(p)}(r, t)\, \mathrm{d}r + \int_{r_1}^{r_2} 2rc_2^{(p)}(r, t)\, \mathrm{d}r. \tag{5}$$

To obtain equations for them the equations (3) are multiplied by y^p and integrated from $-\infty$ to ∞, giving, after a reduction by partial integration,

$$\left.\begin{aligned}
&D_1\psi_1(r)\frac{\partial c_1^{(p)}}{\partial r} = 0 \quad (\text{at } r = r_0), && (a)\\
&\frac{\partial c_1^{(p)}}{\partial t} = D_1\frac{1}{r}\frac{\partial}{\partial r}\left(r\psi_1(r)\frac{\partial c_1^{(p)}}{\partial r}\right) + D_1\psi_1(r)p(p-1)c_1^{(p-2)} \\
&\qquad + (U_1\phi_1(r) - V)pc_1^{(p-1)} \quad (\text{in } r_0 < r < r_1), && (b)\\
&D_1\psi_1(r)\frac{\partial c_1^{(p)}}{\partial r} = D_2\psi_2(r)\frac{\partial c_2^{(p)}}{\partial r} = k(c_2^{(p)} - \alpha c_1^{(p)}) \quad (\text{at } r = r_1), && (c)\\
&\frac{\partial c_2^{(p)}}{\partial t} = D_2\frac{1}{r}\frac{\partial}{\partial r}\left(r\psi_2(r)\frac{\partial c_2^{(p)}}{\partial r}\right) + D_2\psi_2(r)p(p-1)c_2^{(p-2)} \\
&\qquad + (U_2\phi_2(r) - V)pc_2^{(p-1)} \quad (\text{in } r_1 < r < r_2) && (d)\\
&D_2\psi_2(r)\frac{\partial c_2^{(p)}}{\partial r} = 0 \quad (\text{at } r = r_2). && (e)
\end{aligned}\right\} \tag{6}$$

Finally, multiplying equations (b) and (c) by $2r$ and integrating with respect to r, the first over (r_0, r_1) and the second over (r_1, r_2) and adding, using the conditions (a), (c) and (e) we have

$$\begin{aligned}\frac{\mathrm{d}m^{(p)}}{\mathrm{d}t} = {} & p(p-1)D_1\int_{r_0}^{r_1} 2r\psi_1(r)c_1^{(p-2)}\, \mathrm{d}r + p(p-1)D_2\int_{r_1}^{r_2} 2r\psi_2(r)c_2^{(p-2)}\, \mathrm{d}r \\
&+ p\int_{r_0}^{r_1} 2r(U_1\phi_1(r) - V)c_1^{(p-1)}\, \mathrm{d}r + p\int_{r_1}^{r_2} 2r(U_2\phi_2(r) - V)c_2^{(p-1)}\, \mathrm{d}r.\end{aligned} \tag{7}$$

For the particularly simple initial conditions assumed,

$$\alpha c_1^{(0)} = c_2^{(0)} = \alpha/s^2 \quad (m^{(0)} = 1), \tag{8}$$

and

$$c_1^{(p)} = c_2^{(p)} = m^{(p)} = 0 \quad (p > 0 \quad \text{when} \quad t = 0). \tag{9}$$

The method is to solve equations (6) and (7) step by step. Since the radial variation of $c^{(p-1)}$ is needed for $m^{(p)}$, to obtain $m^{(2)}$ as a function of t, it is only necessary to solve equations (6) for $p = 0, 1$. Now it may be readily seen that with $p = 0$ (7) gives

$$\mathrm{d}m^{(0)}/\mathrm{d}t = 0$$

and so $m^{(0)}$ = a constant, namely its initial value 1. This expresses the fact that the total amount of solute is constant.

Furthermore, it is evident that if $c_1^{(0)}$ and $c_2^{(0)}$ are constants equations (6) are satisfied, so that (8) gives the complete solutions for $p = 0$. Inserting these in equation (7) with $p = 1$ and remembering that the average values of ϕ_i are unity

$$\frac{\mathrm{d}m^{(1)}}{\mathrm{d}t} = (U_1 - V)\frac{r_1^2 - r_0^2}{s^2} + (U_2 - V)\frac{\alpha(r_2^2 - r_1^2)}{s^2}. \tag{10}$$

If therefore V is chosen so that

$$V = \frac{U_1(r_1^2 - r_0^2) + U_2\alpha(r_2^2 - r_1^2)}{(r_1^2 - r_0^2) + \alpha(r_2^2 - r_1^2)}, \tag{11}$$

then $m^{(1)}(t) = 0$ and the first moment or centre of gravity of the solute stays at the origin of the moving axes, i.e. it moves with velocity V. In chromatography the motion of the solute is often described by the ratio of its speed to the speed of the solvent, this ratio being called the R_f value. To ascribe an R_f value, R, to the solute it is convenient to take $V = R(U_1 - U_2)$, for then if the liquid phase is stationary the conventional R_f value is obtained. The following abbreviations are convenient:

$\gamma = U_1/(U_1 - U_2)$ = ratio of gas velocity to the relative velocity,
$\beta = (r_1^2 - r_0^2)/s^2$ = fraction of solute in gas phase at equilibrium.

Then

$$R = \frac{V}{U_1 - U_2} = \beta + \gamma - 1. \tag{12}$$

If the liquid phase is stationary $\gamma = 1$ and if the two phases flow countercurrent $0 < \gamma < 1$. Since β can vary between 0 and 1, R varies between $\gamma - 1$ and γ.

To obtain $m^{(2)}$ and thus a picture of the way the dispersion of the chromatogram grows, it is necessary to find the $c_i^{(1)}$. Putting $p = 1$ in (6) (*b*) and (*d*) we obtain an equation of the following form:

$$\frac{\partial c_i^{(1)}}{\partial t} - D_i \frac{1}{r}\frac{\partial}{\partial r}\left(r\psi_i(r)\frac{\partial c_i^{(1)}}{\partial r}\right) = (U_i\phi_i(r) - V)c_i^{(0)}. \tag{13}$$

Now the solution of this will consist of two parts: (i) a complementary function satisfying the equation with the right-hand side equal to zero, and (ii) a particular integral satisfying the complete equation. It may be seen, following the reasoning of an earlier paper (Aris 1956), that complementary function, which will arise because the initial conditions must be satisfied, can only contribute a constant to the value of $m^{(2)}$ and since this increases as the first power of t, such a constant becomes negligible. Moreover, the particular integral is independent of t and may be obtained by quadratures. In fact if

$$c_1^{(1)}(r) = \frac{1}{2s^2}\left[\frac{1}{D_1}\int_{r_0}^{r}\frac{dr'}{r'\psi_1(r')}\int_{r_0}^{r'}(V - U_1\phi_1(r''))2r''\,dr'' + A\right], \tag{14}$$

$$c_2^{(1)}(r) = \frac{\alpha}{2s^2}\left[\frac{1}{D_2}\int_{r}^{r_2}\frac{dr'}{r'\psi_2(r')}\int_{r}^{r_2}(V - U_2\phi_2(r''))2r''\,dr'' + B\right], \tag{15}$$

equations (6) (a), (b), (d) and (e) and the first part of (c) are satisfied and it remains only to determine the constants A and B, by satisfying the second part of (c) and the condition $m^{(1)} = 0$. The first gives an equation for $B - A$, the second a linearly independent combination of A and B and this determines the constants; however, in the expression for $m^{(2)}$ only the combination $B - A$ is required, so that it is not necessary to find them separately. From the last part of (c),

$$B - A = -\frac{\beta(1-\beta)s^2}{k\alpha r_1}(U_1 - U_2) + A_1 - B_1, \tag{16}$$

where A_1 and B_1 denote the value of the first terms in the brackets of (14) and (15) when $r = r_1$.

Inserting (14) and (15) into (7) with $p = 2$, gives

$$\frac{1}{2}\frac{dm^{(2)}}{dt} = \beta D_1 + (1-\beta)D_2 + (r_1^2 - r_0^2)(U_1 - V)A/2s^2 + \alpha(r_2^2 - r_1^2)$$
$$+ (U_2 - V)B/2s^2 + \frac{1}{2s^2D_1}\int_{r_0}^{r_1} 2r(U_1\phi_1 - V)\,dr\int_{r_0}^{r}\frac{dr'}{r'\psi_1}\int_{r_0}^{r'}$$
$$+ (V - U_1\phi_1)2r''\,dr'' + \frac{1}{2s^2D_2}\int_{r_1}^{r_2} 2r(U_2\phi_2 - V)\,dr\int_{r}^{r_2}$$
$$+ \frac{dr'}{r'\psi_2}\int_{r'}^{r_2}(V - U_2\phi_2)2r''\,dr''.$$

Remembering that $(r_1^2 - r_0^2)(U_1 - V)/s^2 = -\alpha(r_2^2 - r_1^2)(U_2 - V)/s^2 = \beta(1 - \beta)(U_1 - U_2)$ it is clear that the third and fourth terms give a multiple of $B - A$ and so bring in the expression (16). But the integrals may be transformed by partial integration and it will then be found that the A_1 and B_1 vanish to give

$$\frac{1}{2}\frac{dm^{(2)}}{dt} = \beta\left\{D_1 + \kappa_1\frac{U_1^2(r_1^2 - r_0^2)}{D_1}\right\} + (1-\beta)\left\{D_2 + \kappa_2\frac{U_2^2(r_2^2 - r_1^2)}{D_2}\right\}$$
$$+ \frac{\{\beta(1-\beta)s(U_1 - U_2)\}^2}{2k\alpha r_1}, \tag{17}$$

where the κ_i come from the transformed integrals.

The κ_i may be expressed more neatly if two further functions are introduced, namely

$$\Phi_1(r) = \int_{r_0}^{r} 2r'\phi_1(r')\,dr'/(r_1^2 - r_0^2) \quad \text{and}$$
$$\Phi_2(r) = \int_{r}^{r_2} 2r'\phi_2(r')\,dr'/(r_2^2 - r_1^2). \tag{18}$$

They are equal to the fraction of the flow between radii r_0 and r and radii r and r_2, respectively, and hence $\Phi_1(r_0) = \Phi_2(r_2) = 0$ whilst $\Phi_1(r_1) = \Phi_2(r_1) = 1$. Then

$$\kappa_1 = \int_{r_0}^{r_1} \frac{dr}{2r\psi_1(r)}\left\{\Phi_1(r) - \frac{R}{\gamma}\frac{r^2 - r_0^2}{r_1^2 - r_0^2}\right\}^2 = \kappa_{11} - 2\kappa_{12}(R/\gamma) + \kappa_{13}(R/\gamma)^2, \tag{19}$$

$$\kappa_2 = \int_{r_1}^{r_2} \frac{dr}{2r\psi_2(r)}\left\{\Phi_2(r) - \frac{R}{\gamma - 1}\frac{r_2^2 - r^2}{r_2^2 - r_1^2}\right\}^2 = \kappa_{21} - 2\kappa_{22}(R/\gamma - 1) + \kappa_{23}(R/\gamma - 1)^2. \tag{20}$$

Thus each factor κ is a quadratic expression in R whose coefficients may be evaluated from a knowledge of the flow conditions and the variation of the diffusion coefficient alone.

In the case $U_2 = 0$, when $\gamma = 1$ and $U_1 = \text{Lt}\ U_2/(\gamma - 1)$ the Taylor diffusion coefficient in the second bracket of (17) must be written $\kappa_2 U_1^2(r_2^2 - r_1^2)/D_2$ and $\kappa_2 = \kappa_{23}R^2$. Such is the case with a coated tube chromatograph when the liquid phase is stationary. In this case $r_0 = 0$ and

$$\kappa_1 = \tfrac{1}{48}(11 - 16R + 6R^2), \tag{21}$$

$$\kappa_2 = R^2[2\rho^4P^{-1} - (3\rho^2 - 1)]/8(\rho^2 - 1), \tag{22}$$

where $\rho = r_2/r_1$ and $P = (\rho^2 - 1)/\ln \rho^2$. When the coating is very thin, $\rho \doteqdot 1$, $\kappa_2 \doteqdot \frac{1}{12}(\rho^2 - 1)$ and (17) becomes

$$\frac{1}{2}\frac{dm^{(2)}}{dt} = R\left\{D_1 + \frac{U_1^2 r_1^2}{48D_1}(11 - 16R + 6R^2)\right\} + (1 - R)\left\{D_2 + R^2\frac{U_1^2(r_2 - r_1)^2}{3D_2}\right\} + \{R(1 - R)sU_1\}^2/2k\alpha r_1. \tag{23}$$

This is a slight generalization of Golay's result, he being the first to obtain the rather interesting first bracket (Golay 1958). The interest of this lies in the second term which has two limiting cases. The first when $R = 1$, i.e., there is no retentive layer, reduces to Taylor's result $U_1^2r_1^2/48D_1$. The second when $R = 0$ gives Westhaver's coefficient $\frac{11}{48}$ which he obtained from an analysis of the open-tube distillation column (Westhaver 1942). This is to be expected for in distillation the concentration in the liquid layer is very many times greater than the concentration in the vapour corresponding to a very large α and hence a very small R. We shall return to this point in the discussion.

3. CERTAIN SPECIAL CASES

It is useful to list certain special results for the case of viscous flow and molecular diffusion. Results for turbulent flow profiles and diffusion coefficients can be obtained by numerical integration as has been done by Taylor (1954) and more recently by Tichacek and others (1958). For molecular diffusion $\psi_i \equiv 1$, and for the problem of diffusion only in the absence of a second phase, $R = \beta = \gamma = 1$ giving a single factor $\kappa = \kappa_{11} - 2\kappa_{12} + \kappa_{13}$.

The results for diffusion only and for coated tube chromatography are contained in (21) and (22) of the preceding paragraph. We give here the results for diffusion and chromatography when the gas phase flows through the annulus. If the liquid phase is stationary (22) may be used in conjunction with these.

Let $r_0 = \sigma r_1$, $0 < \sigma < 1$, $\Sigma = (1 - \sigma^2)/\ln \sigma^2$, $S = (1 + \sigma^2 + 2\Sigma)$ and $\tau = r^2/r_1^2$. Then

$$\phi_1 = 2\{(1 - \tau) - \Sigma \ln \tau\}/S, \tag{24}$$

$$\Phi_1 = \{2(1 + \Sigma)(\tau - \sigma^2) - (\tau^2 - \sigma^4) - 2\Sigma(\tau \ln \tau - \sigma^2 \ln \sigma^2)\}/(1 - \sigma^2)S. \tag{25}$$

From (19),

$$\kappa_{11} = \{36(5 - 11\sigma^2)\Sigma^2 + 4(38 - 16\sigma^2 - 124\sigma^4)\Sigma + 3(11 + 11\sigma^2 - 25\sigma^4 - 73\sigma^6) - 36\sigma^8\Sigma^{-1}\}/144(1 - \sigma^2)S^2, \tag{26}$$

$$\kappa_{12} = \{9(1 - 3\sigma^2)\Sigma + (4 - 5\sigma^2 - 23\sigma^4) - 6\sigma^6\Sigma^{-1}\}/24(1 - \sigma^2)S, \tag{27}$$

$$\kappa_{13} = \{(1 - 3\sigma^2) - 2\sigma^4\Sigma^{-1}\}/8(1 - \sigma^2), \tag{28}$$

$$\kappa = \{\tfrac{1}{4}(1 + \sigma^2)\Sigma^2 + \tfrac{1}{36}(5 + 32\sigma^2 + 5\sigma^4) + \tfrac{1}{48}(1 + \sigma^2)(1 + 8\sigma^2 + \sigma^4) - \tfrac{1}{4}\sigma^4\Sigma^{-1}\}/(1 - \sigma^2)S^2 \tag{29}$$

As $\sigma \to 0$ these reduce to the values implicit in (21) for the open tube. As $\sigma \to 1$

$$\kappa \to \frac{1}{210}\frac{1 - \sigma}{1 + \sigma} \quad \text{and so} \quad \kappa\frac{U_1^2(r_1^2 - r_0^2)}{D_1} \to \frac{1}{210}\frac{U_1^2(r_1 - r_0)^2}{D_1},$$

the correct result for a parallel slit.

More generally we have as $\sigma \to 1$

$$\kappa_{11} = \tfrac{13}{140}\tau^* + \tfrac{5}{168}\tau^{*2} + \tfrac{3}{200}\tau^{*3} + \dots, \tag{30}$$

$$\kappa_{12} = \tfrac{7}{80}\tau^* + \tfrac{1}{40}\tau^{*2} + \tfrac{2}{175}\tau^{*3} + \dots, \tag{31}$$

$$\kappa_{13} = \tfrac{1}{12}\tau^* + \tfrac{1}{48}\tau^{*2} + \tfrac{1}{120}\tau^{*3} + \dots, \tag{32}$$

$$\kappa_1 = \tfrac{1}{840}\tau^* + \tfrac{1}{1680}\tau^{*2} + \tfrac{1}{2100}\tau^{*3} + \dots, \tag{33}$$

where $\tau^* = 1 - \sigma^2$.

Thus for the parallel walled channel the Taylor diffusion coefficient is

$$\frac{U^2 d^2}{210D}[78 - 147R + 70R^2], \tag{34}$$

where d is the distance between the walls.

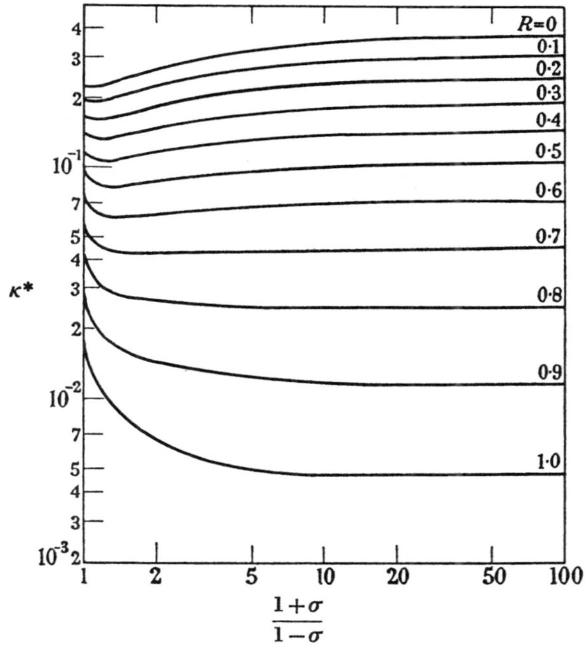

FIGURE 1 κ^* as a function of σ for various R.

For calculation near $\sigma = 1$ formulae (30) to (33) are the easiest, for small errors in (26) to (29) are almost unavoidable. Figure 1 shows how these coefficients vary with R and σ. There are quantity $\kappa^* = (1+\sigma)\kappa/(1-\sigma)$ is plotted against $(1+\sigma)/(1-\sigma)$ logarithmically. The Taylor diffusion coefficient as we have written it is

$$\kappa U_1^2 r_1^2 (1-\sigma^2)/D_1 = \kappa^* U_1^2 r_1^2 (1-\sigma)^2/D_1;$$

in the latter form it passes smoothly to the case of the parallel sided film for $r_1 \to \infty$, $\sigma \to 1$ but $r_1(1-\sigma) \to d$, the film thickness. Thus in the figure the left edge gives the values for the open tube $\sigma = 0$, whilst the asymptotic values on the right are for the parallel film.

4. DISPERSION IN COAXIAL STREAMS OF ARBITRARY CROSS-SECTION

It is useful to derive a general method of determining the coefficients κ when the cross-sections of the regions are of arbitrary shape. This we will outline here omitting as before the transient terms which arise from satisfying the initial conditions; they may be considered on the lines of the earlier paper (Aris 1956, § 5). The three concentric circles of radii r_0, r_1 and r_2 are replaced in the general case by three simple closed curves Γ_0, Γ_1 and Γ_2 such that Γ_0 lies within Γ_1 lies within Γ_2. The regions lying between Γ_0 and Γ_1 and Γ_1 and

Γ_2 are denoted by R_1 and R_2, respectively. Then the equations analogous to equations (1) are

$$\left.\begin{aligned}
&D_1\,\psi_1\frac{\partial c_1}{\partial n}=0 \quad \text{on} \quad \Gamma_0, && (a)\\
&\frac{\partial c_1}{\partial t}=D_1\,\nabla(\psi_1\,\nabla c_1)-U_1\phi_1\frac{\partial c_1}{\partial x} \quad (\text{in } R_1), && (b)\\
&D_1\,\psi_1\frac{\partial c_1}{\partial n}=D_2\psi_2\frac{\partial c_2}{\partial n}=k(c_2-\alpha c_1) \quad (\text{on } \Gamma_1) && (c)\\
&\frac{\partial c_2}{\partial t}=D_2\nabla(\psi_2\,\nabla c_2)-U_2\phi_2\frac{\partial c_2}{\partial n} \quad (\text{in } R_2), && (d)\\
&D_2\,\psi_2\frac{\partial c_2}{\partial n}=0 \quad (\text{on } \Gamma_2). && (e)
\end{aligned}\right\} \tag{35}$$

Here x is the co-ordinate in the direction of flow and the diffusion coefficients $D\psi$ and flow velocities $U\phi$ are taken to be independent of x. Taking

$$\begin{aligned}
&a = \text{some characteristic dimension,} &&\\
&\xi=(x-Vt)/a, && \mu_i=U_ia/D_i,\\
&\eta=y/a, && \delta_i=D/D_i,\\
&\zeta=z/a && \gamma_i=V/U_i,\\
&\tau=Dt/a^2, && \lambda=ka/D,
\end{aligned}$$

and using ∇ to denote the two-dimensional operator $(\partial/\partial\eta,\ \partial/\partial\zeta)$ for the rest of this section and $\partial/\partial\nu$ to be differentiation normal to a curve Γ in the dimensionless co-ordinates, we obtain

$$\left.\begin{aligned}
&\psi_1\frac{\partial c_1}{\partial \nu}=0 \quad (\text{on } \Gamma_0), && (a)\\
&\delta_1\frac{\partial c_1}{\partial \tau}=\psi_1\frac{\partial^2 c_1}{\partial \xi}-\mu_1(\phi_1-\gamma_1)\frac{\partial c_1}{\partial \xi}+\nabla(\psi_1\nabla c_1) \quad (\text{in } R_1), && (b)\\
&\frac{\psi_1}{\delta_1}\frac{\partial c_1}{\partial \nu}=\frac{\psi_2}{\delta_2}\frac{\partial c_2}{\partial \nu}=\lambda(c_2-\alpha c_1) \quad (\text{on } \Gamma_1), && (c)\\
&\delta_2\frac{\partial c_2}{\partial \tau}=\psi_2\frac{\partial^2 c_2}{\partial \xi^2}-\mu_2(\phi_2-\gamma_2)\frac{\partial c_2}{\partial \xi}+\nabla(\psi_2\nabla c_2) \quad (\text{in } R_2), && (d)\\
&\psi_2\frac{\partial c_2}{\partial \nu}=0 \quad (\text{on } \Gamma_2). && (e)
\end{aligned}\right\} \tag{36}$$

We note that the normal differentiation on Γ_1 is from R_1 to R_2. Moments $c_i^{(p)}$ and $m^{(p)}$ are again defined by

$$c_i^{(p)}=\int_{-\infty}^{\infty}\xi^p c_i(\xi,\eta,\zeta,\tau)\,\mathrm{d}\xi=c_i^{(p)}(\eta,\zeta,\tau) \tag{37}$$

and

$$m^{(p)}=\iint_{R_1}c_1^{(p)}\,\mathrm{d}\eta\,\mathrm{d}\zeta+\iint_{R_2}c_2^{(p)}\,\mathrm{d}\eta\,\mathrm{d}\zeta=m^{(p)}(\tau). \tag{38}$$

Proceeding exactly as before we see that $c_i^{(p)}$ and $m^{(p)}$ satisfy

$$\delta_i \frac{\partial c_i^{(p)}}{\partial \tau} = p(p-1)\psi_i c_i^{(p-2)} + p\mu_i(\phi_i - \gamma_i)c_i^{(p-1)} + \nabla(\psi_i \nabla c_i^{(p-1)}), \tag{39}$$

$$\frac{dm^{(p)}}{d\tau} = p(p-1)\sum_{i=1}^{2}\frac{1}{\delta_i}\iint_{R_i}\psi_i c_i^{(p-2)}\,d\eta\,d\zeta + p\sum_{i=1}^{2}\frac{\mu_i}{\delta_i}\iint_{R_i}(\phi_i - \gamma_i)c_i^{(p-1)}\,d\eta\,d\zeta. \tag{40}$$

In this last equation we have used Green's theorem and the first part of condition (c) which together make the integration of the last term in (39) vanish. Again we set $m^{(0)} = 1$ and find that (39) and its associated conditions vanish if

$$\alpha c_1^{(0)} = c_2^{(0)} = \alpha/s^2, \tag{41}$$

where $s^2 = s_1^2 + \alpha s_2^2$ and the areas of R_1 and R_2 are $(s_1 a)^2$ and $(s_2 a)^2$, respectively. Substituting in (40) $dm^{(1)}/d\tau = 0$ if

$$\sum_{1}^{2}\frac{\mu_i}{\delta_i}c_i^{(0)}s_i^2(1-\gamma_i) = 0, \quad \text{i.e. } V = (U_1 s_1^2 + \alpha U_2 s_2^2)/s^2, \tag{42}$$

which shows that the centre of gravity moves with a speed that is the weighted mean of the two stream speeds, the weights being the equilibrium amounts in each stream.

To get the dominant part of $c_i^{(1)}$ for use in determining $m^{(2)}$ we need only consider the time independent part of the solutions of (39) with $p = 1$, namely

$$\nabla(\psi_i \nabla c_i^{(1)}) = -\mu_i(\phi_i - \gamma_i)c_i^{(0)}.$$

Let $c_i^{(1)} = \mu_i c_i^{(0)}(\chi_i + \varepsilon_i)$ where ε_i are constants; then

$$\nabla(\psi_i \nabla \chi_i) = -(\phi_i - \gamma_i) \quad (\text{in } R_i) \tag{43}$$

and

$$c_1^{(0)}\frac{\mu_1}{\delta_1}\psi_1\frac{\partial \chi_1}{\partial \nu} = c_2^{(0)}\frac{\mu_2}{\delta_2}\psi_2\frac{\partial \chi_2}{\partial \nu} = \lambda\{\mu_2 c_2^{(0)}(\chi_2 + \varepsilon_2) - \mu_1 \alpha c_1^{(0)}(\chi_1 + \varepsilon_1)\}$$
$$= \lambda c_2^{(0)}(\mu_2\chi_2 - \mu_1\chi_1) + \lambda c_2^{(0)}(\mu_2\varepsilon_2 - \mu_1\varepsilon_1) \quad (\text{on } \Gamma_1). \tag{44}$$

Now the first part of the condition (44), together with the conditions on Γ_0 and Γ_2, determine the functions χ_1 and χ_2. The constants ε_1 and ε_2 could be evaluated from the latter part of (44) and the condition $m^{(1)} = 0$. This, however, is unnecessary as it will appear that only the combination $(\mu_2\varepsilon_2 - \mu_1\varepsilon_1)$ is required in $m^{(2)}$. To evaluate this we integrate (44) around the contour Γ_1 using Green's theorem and the differential equations to obtain

$$\mu_2\varepsilon_2 - \mu_1\varepsilon_1 = \frac{s_1^2 s_2^2}{s^2}\frac{(U_2 - U_1)}{k\omega}, \tag{45}$$

where ω is the length of Γ_1.

For $m^{(2)}$ we have

$$\begin{aligned}\frac{1}{2}\frac{\mathrm{d}m^{(2)}}{\mathrm{d}\tau} &= \sum_1^2 \frac{1}{\delta_i}\iint_{R_i} \psi_i c_i^{(0)} + \sum_1^2 \frac{\mu_i}{\delta_i}\iint_{R_i} (\phi_i - \gamma_i)c_i^{(1)} \\ &= \sum_1^2 \frac{s_i^2 c_i^{(0)}}{\delta_i} + \sum_1^2 \frac{\mu_i^2 c_i^{(0)}}{\delta_i}\iint_{R_i} (\phi_i - \gamma_i)(\chi_i + \varepsilon_i) \\ &= \sum_1^2 \frac{s_i^2 c_i^{(0)}}{\delta_i} - \sum_1^2 \frac{\mu_i^2 c_i^{(0)}}{\delta_i}\iint_{R_i} \chi_i \nabla(\psi_i \nabla\chi_i) + \sum_1^2 \frac{\mu_i^2 c_i^{(0)}}{\delta_i}\varepsilon_i(1-\gamma_i)s_i^2 \\ &= \sum_1^2 \frac{s_i^2 c_i^{(0)}}{\delta_i} + \sum_1^2 \frac{\mu_i^2 c_i^{(0)} s_i^2}{\delta_i}\left[\frac{1}{s_i^2}\iint_{R_i} \psi_i(\nabla\chi_i)^2\right] \\ &\quad + \left\{\frac{\alpha s_1^2 s_2^2}{s^3}\right\}^2 \frac{a(U_2 - U_1)^2}{Dk\omega\alpha}.\end{aligned}$$

Setting

$$\begin{aligned}\beta &= s_1^2 c_1^{(0)} = s_1^2/(s_1^2 + \alpha s_2^2) = \text{the fraction of solute in } R_1 \text{ at equilibrium} \\ &= 1 - s_2^2 c_2^{(0)},\end{aligned}$$

we have

$$\begin{aligned}\frac{D}{2}\frac{\mathrm{d}m^{(2)}}{\mathrm{d}\tau} = \frac{1}{2}\frac{\mathrm{d}(a^2 m^{(2)})}{\mathrm{d}t} &= \beta\left(D_1 + \kappa_1 \frac{U_1^2 a^2}{D_1}\right) \\ &\quad + (1-\beta)\left(D_2 + \kappa_2\frac{U_2^2 a^2}{D_2}\right) + \beta^2(1-\beta)^2\frac{a(U_2 - U_1)^2 s^2}{k\omega\alpha},\end{aligned} \tag{46}$$

where

$$\kappa_i = \left[\iint_{R_i} \psi_i(\nabla\chi_i)^2 \,\mathrm{d}\eta\,\mathrm{d}\zeta\right] \Big/ \left[\iint_{R_i} \mathrm{d}\eta\,\mathrm{d}\zeta\right] \tag{47}$$

and $a^2 m^{(2)}$ is the variance in terms of the original dimensional co-ordinates.

The additive nature of these terms is again evident, a consequence of the stochastic nature of diffusion. The structure of the constants κ_i, which are functions only of the geometry of the cross-section and distributions of velocity and diffusivity, can be seen better if we put

$$\chi_i = f_i - (1 - \gamma_i)g_i,$$

so that $\qquad \nabla(\psi_i \nabla f_i) = 1 - \phi_i \quad \text{and} \quad \nabla(\psi_i \nabla g_i) = -1.$

Then

$$\kappa_i = \frac{1}{s_i^2}\iint_{R_i} \psi_i(\nabla f_i)^2 - 2(1-\gamma_i)\frac{1}{s_i^2}\iint_{R_i} \psi_i \nabla f_i \nabla g_i + (1-\gamma_i)^2\frac{1}{s_i^2}\iint_{R_i} \psi_i(\nabla g_i)^2.$$

This is the dependence upon the relative concentration in the two phases that has been observed earlier.

5. APPLICATION TO THE THEORY OF CHROMATOGRAPHY

The tubular chromatographic column with the stationary phase held as a thin layer on the inside of the wall has been considered by Golay (1958), who arrives at a formula for the dispersion by an analysis similar to Taylor's. It is desired only to point out here that the practical results of chromatography show that the exit concentration from a column is often well represented by the normal distribution and that this allows the preceding formulae to be used to predict the degree of separation.

Suppose that a mixture of two solutes is moving in the system described and that the partition coefficients of the solutes are α_1 and α_2, respectively. Then their centres of gravity will move with speeds V_1 and V_2 where

$$V_i = \frac{s_1^2 U_1 + \alpha_i s_2^2 U_2}{s_1^2 + \alpha_i s_2^2}.$$

Thus
$$V_1 - V_2 = (\alpha_1 - \alpha_2)(U_2 - U_1)\frac{s_1^2 s_2^2}{(s_1^2 + \alpha_1 s_2^2)(s_1^2 + \alpha_2 s_2^2)}$$

is proportional to $(\alpha_1 - \alpha_2)$ if α_1 and α_2 differ only slightly; it is in these difficult separations that we are interested. In a time of flow t the centres of gravity will be separated by a distance $(V_1 - V_2)t$. In the same time the variances of the two distributions will have grown by $M_1^2 t$ and $M_2^2 t$ where $\frac{1}{2}M_i^2$ is the right-hand side of equation (46) evaluated for the ith solute. The separation may be said to be of degree θ if the distance between the centroids is θ (the sum of the variances), i.e. if $(V_1 - V_2)t^{1/2} = \theta(M_1 + M_2)$, assuming that the variance of the initial distribution was negligible. If both U_1 and U_2 are in the same direction the separation will require an approximate length $\frac{1}{2}(V_1 + V_2)t$ of column, i.e. a length $\frac{1}{2}\theta^2(V_1 + V_2)(M_1 + M_2)^2/(V_1 - V_2)^2$. This length is directly proportional to the square of the degree of separation and inversely to the square of $(U_1 - U_2)$ and $(\alpha_1 - \alpha_2)$. If U_1 and U_2 have opposite signs it is always possible to choose the ratio s_1^2/s_2^2 so that V_1 and V_2 have opposite signs. Then the length of column required is approximately $|V_1 - V_2|t = \theta^2(M_1 + M_2)^2/|V_1 - V_2|$, and this is inversely as $(\alpha_1 - \alpha_2)$ instead of $(\alpha_1 - \alpha_2)^2$. The complicated interaction of the ratios s_1^2/s_2^2, U_1/U_2 in the problem of finding the optimum separation conditions will be considered elsewhere.

6. APPLICATION TO A SIMPLIFIED THEORY OF DISTILLATION

In his analysis of the open tube distillation column Westhaver (1942) goes into a detailed consideration of radial concentration gradients which is very similar to Taylor's approach. His final formula, however, is the same as if he had assumed a constant velocity profile and an effective diffusion coefficient $(D_1 + 11U_1^2 r_1^2/48D_1)$. This is just the diffusion coefficient that we have found for viscous flow in the presence of a film on the tube wall in which the solute concentration is infinitely greater than in the fluid. This is clearly the case for distillation where the pure condensate is much more highly concentrated than the vapour. In a later paper (Westhaver 1947*b*) he states formulae for the height of theoretical plate which imply diffusion coefficients which could be deduced from (19) and (20). Formulae even more directly related to (19) and (20) are obtained in a paper on electromigration (Westhaver 1947*a*).

It is therefore suggested that provided the diffusion coefficient is rightly chosen the following simple theory of distillation will prove quite servicable. In the simplified theory we assume that the velocity of the vapour is everywhere constant and equal to v, that diffusion takes place in the direction of the vapour flow only with diffusion coefficient D and that concentrations are constant on planes perpendicular to the flow. If X is the mole fraction in the vapour phase of the more volatile component of a binary mixture and Y its mole fraction in the condensate, then at total reflux a mass balance over a section of a column gives

$$vX - D\,\mathrm{d}X/\mathrm{d}z = vY.$$

But for a relative volatility α close to 1 (this is of interest since it applies to difficult separations) the equilibrium relationship between X and Y may be written

$$X - Y = (\ln \alpha)X(1 - X),$$

and substituting this in the mass balance,

$$v(\ln \alpha)X(1 - X) = D\,\mathrm{d}X/\mathrm{d}z.$$

If L is the length of the column then integration gives

$$\ln\left\{\left(\frac{X}{1-X}\right)_{z=L}\Big/\left(\frac{X}{1-X}\right)_{z=0}\right\} = \frac{vL}{D}\ln \alpha.$$

If the separation were done in N equilibrium stages the quantity on the left side of this last equation would be $N(\ln \alpha)$. Hence the length L is equivalent to vL/D theoretical plates or the height equivalent to one theoretical plate is D/v. Inserting for D the expression $D_1 + \kappa_{11}U_1^2(r_1^2 - r_0^2)/D_1$, since the ratio of the concentrations in vapour and condensate is very large, we obtain the formula given by Westhaver (1942, equation (6)). A slightly more detailed analysis with longitudinal diffusion also in the condensate shows that D should be the mean of the diffusion coefficients in the two phases weighted according to the area of each. Here taking κ_{11} and κ_{21} for the Taylor diffusion coefficients we obtain in the appropriate special case formulae (1) and (2) of Westhaver (1947*b*).

I am indebted to M. J. E. Golay for allowing me to see his contribution to the Amsterdam Gas Chromatography Symposium whilst it was still in typescript, and to J. W. Westhaver for correspondence on his work. Fruitful discussions of the distillation problem were had with N. Macleod of the Chemical Engineering Department, Edinburgh University. The formulae in § 3 were checked by Mr. Kung-You Lee, who also did the calculation for the figure.

REFERENCES

Aris, R. 1956 *Proc. Roy. Soc.* A, **235,** 67.
Golay, M. J. E. 1958 Contribution to the Gas Chromatography Symposium, Amsterdam.
Lighthill, M. J. 1958 *Fourier analysis and generalized functions.* Cambridge University Press.
Taylor, Sir Geoffrey 1953 *Proc. Roy. Soc.* A, **219,** 186.
Taylor, Sir Geoffrey 1954*a Proc. Roy. Soc.* A, **223,** 446.
Taylor, Sir Geoffrey 1954*b Proc. Roy. Soc.* A, **225,** 473.
Tichacek, L. J., Barkelew, C. H. & Baron, T. 1957 *A.I.Ch.E. Journal,* **3,** 439.
Westhaver, J. W. 1942 *Industr. Engng Chem.* **34,** 126.
Westhaver, J. W. 1947*a J. Res. Nat. Bur. Stand.* **38,** 169.
Westhaver, J. W. 1947*b Industr. Engng Chem.* **39,** 706.

C.*

ON THE DISPERSION OF LINEAR KINEMATIC WAVES

RUTHERFORD ARIS

The Mathematical Institute, University of Edinburgh

(*Communicated by M. J. Lighthill, F.R.S.—Received* 31 *December* 1957)

The theory of kinematic waves, initiated by Lighthill & Whitham, is taken up for the case when the concentration k and flow q are related by a series of linear equations. If the initial disturbance is hump-like it is shown that the resulting kinematic wave can be usually described by the growth of its mean and variance, the former moving with the kinematic wave velocity and the latter increasing proportionally to the distance travelled. Conditions for these moments to be calculated from the Laplace transform of the solution, without the need of inversion, are obtained and it is shown that for a large class of waves, the ultimate wave form is Gaussian. The power of the method is shown in the analysis of a kinematic temperature wave, where the Laplace transform of the solution cannot be inverted.

1. INTRODUCTION

In 1955 Lighthill & Whitman drew attention to a class of wave motion which arises from the equation of continuity of a flow q and a concentration k and a functional relation between q and k (Lighthill & Whitham 1955*a*,*b*). In contrast to the motions governed by the classical wave equation they called these kinematic waves and considered the movement of floods and of traffic as examples. Other examples of kinematic wave motion have been studied in other contexts, notably heat transfer in cross-flow heat exchangers and packed beds (Klinkenberg 1954) and exchange processes in fixed columns (Goldstein 1953). The latter includes the theory of chromatography, already possessed of a considerable literature, and having many features in common with the theory of flood movement.

A kinematic wave may be called linear if the relationship between the flow and the concentration can be expressed by one or more linear equations, algebraic or differential. The term linear may also be applied when a diffusion term is included in the continuity equation as is done in § 3 of Lighthill &

* Reprinted with permission of the editor of the *Proceedings of the Royal Society*, **A**, volume 245, pp. 268–277, 1958.

Whitham (1955*a*). Except in the trivial case when $q = ak + b$ and the simple continuity equation

$$\frac{\partial k}{\partial t} + \frac{\partial q}{\partial x} = 0$$

is used, the kinematic wave changes shape during its motion, and a hump-like disturbance is gradually dispersed as it moves along. Our object here is to show how a simple description of this process may be given by considering the moments of a hump distribution, even in cases where a full solution of the equations is virtually impossible. To exhibit the method, Lighthill & Whitham's example will be considered in the next paragraph. The mathematical justification is given in § 3 and the method applied to the temperature wave in a packed bed, a case where the full solution cannot be obtained, in § 4. The final paragraph considers the asymptotic form of a linear kinematic wave. It should be emphasized that the method presented here, making use as it does of the Laplace transformation is only applicable to linear kinematic waves. In general the elucidation of the asymptotic form of a non-linear kinematic wave requires the knowledge of an exact solution; such is the case in Lighthill's study (1956) of sound waves of finite amplitude, where the Gaussian form is lost as the degree of non-linearity increases.

2. THE DISPERSION OF A FLOOD WAVE

In § 3 of Lighthill & Whitham (1955*a*) the following problem is considered. At $t = 0$ the river downstream of a certain point $x = 0$ is undisturbed, and a disturbance begins to pass this point, the height η of the water above its normal level being specified as a function of t. The equation governing η is derived in Lighthill & Whitham's paper, and for our purposes need only be quoted:

$$(gh_0 - v_0^2)\eta_{xx} - 2v_0\eta_{xt} - \eta_{tt} - 2\lambda\eta_t - 3\lambda v_0\eta_x = 0, \tag{1}$$

$$\eta = f(t) \quad \text{at} \quad x = 0, t > 0, \tag{2}$$

$$\eta = 0 \quad \text{at} \quad t = 0, x > 0, \tag{3}$$

where v_0 = undisturbed stream speed, h_0 = undisturbed water height, S_0 = slope of river bed, $\lambda = gS_0/v_0$, $F = v_0/(gh_0)^{\frac{1}{2}}$, the Froude number. The equation has been made linear by neglecting higher powers of the disturbances, and the second-order terms are introduced by elimination of the flow between the continuity equation and its relation to the concentration.

If $\overline{\eta}$ the Laplace transform of η is introduced by

$$\overline{\eta}(x, p) = \int_0^\infty e^{-pt}\eta(x, t)\,dt, \tag{4}$$

and if

$$A(p) = \int_0^\infty e^{-pt}f(t)\,dt \tag{5}$$

we derive exactly the same form for the Laplace transform of the solution as is given by equation (31) of Lighthill & Whitham, namely

$$\bar{\eta}(x, p) = A(p) \exp\{xP(p)/(gh_0)^{\frac{1}{2}}\}, \tag{6}$$

where $P(p)(1 - F^2) = (p + \tfrac{3}{2}\lambda)F - \{p^2 + \lambda p(2 + F^2) + \tfrac{9}{4}\lambda^2 F^2\}^{\frac{1}{2}}$. (7)

The inversion of this transform gives a somewhat cumbersome integral, of which the physical meaning is far from obvious, and Lighthill & Whitham naturally prefer to elucidate this form the asymptotic behaviour of the transform, by the method of steepest descents. The method presented here also uses the transform without the need for inversion and obtains a description of the wave in terms of its moments.

If the disturbance $f(t)$ is hump-like, falling to zero either at some finite f or sufficiently fast as $t \to \infty$, it will possess a set of moments

$$a_n = \int_0^\infty t^n f(t)\, dt. \tag{8}$$

a_0 is the area under the curve $f(t)$ and is a measure of the total disturbance. a_1/a_0 is a quantity of the dimensions of time being the distance of the centre of gravity of the area under the curve $f(t)$ from $t = 0$. It may be thought of as the mean time of the disturbance. $(a_z a_0 - a_1^2)/a_0^2$ is the variance of the distribution $f(t)$ about its mean time and is clearly a measure of its spread or dispersion. The third moment about the mean is a measure of the skewness and the higher moments can be given increasingly recondite interpretations.

If the disturbance is observed at a station downstream the corresponding moments

$$\mu_n(x) = \int_0^\infty t^n \eta(x, t)\, dt \tag{9}$$

describe its shape as it passes the point x. $\mu_1(x)$ will be the mean time of passing, $(\mu_2\mu_0 - \mu_1^2)/\mu_0^2$ the spread of the disturbance as it passes this station. Thus the general behaviour of the wave can be described quite well in terms of the first two or three moments.

Under conditions which will be discussed in the next paragraph these moments are given by

$$\mu_n(x) = (-)^n \left[\frac{d^n}{dp^n}\bar{\eta}(x, p)\right]_{p=0}. \tag{10}$$

Applying this formula to equation (6) and writing $\xi = x/(gh_0)^{\frac{1}{2}}$,

$$\begin{aligned} \mu_0 &= a_0, \\ \mu_1 &= a_1 + a_0\,\xi[-P'(0)], \\ \mu_2 &= a_2 + 2a_1\xi[-P'(0)] + a_0\{[-P'(0)]^2 + P''(0)\}, \end{aligned} \tag{11}$$

whence, using (7),

$$m_1 = \frac{\mu_1}{\mu_0} - \frac{a_1}{a_0} = [P'(0)]\xi = \frac{2\xi}{3F} = \frac{2x}{3v_0}, \tag{12}$$

$$m_2 = \frac{\mu_0\mu_2 - \mu_1^2}{\mu_0^2} - \frac{a_0a_2 - a_1^2}{a_0^2} = P''(0)\xi = \frac{1}{\lambda}(1 - \tfrac{1}{4}F^2)\left(\frac{2}{3F}\right)^2 \frac{2\xi}{3F}. \tag{13}$$

Equation (11) simply means that the total disturbance $\int_0^\infty \eta(x, t)\, dt$ is the same at any point x as it is at $x = 0$. We could without loss of generality put $a_0 = \mu_0 = 1$. Equation (12) shows that the mean time of the disturbance at x differs from that at $x = 0$ by a quantity strictly proportional to x; moreover, if the wave velocity is judged by the progress of this mean time the wave moves with a constant velocity $3v_0/2$, the kinematic wave velocity. Equation (13) shows that the increase of the variance is also proportional to x and the constant of proportionality is a measure of the rate of dispersion.

We may compare this with the dispersion of a solute diffusing, with diffusion coefficient D, in a stream of velocity V. Here

$$\frac{\partial c}{\partial t} = D\frac{\partial^2 c}{\partial x^2} - V\frac{\partial c}{\partial x},$$

$$c = f(t), \quad x = 0,$$

$$c \to 0, \quad x \to \infty,$$

$$c = 0, \quad t = 0,$$

and hence $$\bar{c}(x, p) = A(p)\exp\{xP^*(p)\},$$

where $$P^*(p) = \frac{V}{2D} - \left(\frac{V^2}{4D^2} + \frac{p}{D}\right)^{\frac{1}{2}}.$$

In this case $m_1 = x/V$, $m_2 = Dx/V^3$, so that $D = V^2 m_2/m_1$. By comparison we may define the apparent diffusion coefficient for flood movement to be

$$D_f = \frac{v_0^2}{F^2\lambda}(1 - \tfrac{1}{4}F^2) = \frac{h_0 v_0}{S_0}\left(1 - \frac{v_0^2}{4gh_0}\right). \tag{14}$$

Similar coefficients have been introduced by Taylor (1953) for combined diffusion and convection, and by Amundson & Aris (1957) for the packed bed. Their value is enhanced by the fact that ultimately the disturbance tends to have the shape of a Gaussian distribution, as is shown in § 5. It is easy to see, however, that the change in the third moment m_3 is also proportional to x, and hence the absolute skewness m_3^2/m_2^3 is $O(x^{-1})$ for large x, so that the disturbance tends to become more symmetrical.

This is fully in accord with the results of Lighthill & Whitham though presented from a different angle. They find that the position of maximum depth ultimately travels with the kinematic wave velocity, and that the maximum depth falls off as $t^{-\frac{1}{2}}$ the constant of proportionality being just what would be calculated for a Gaussian peak of area a_0 and variance m_2. The difference in viewpoint shows up in the fact that their results only hold asymptotically and thus their formula (35) for $\eta_{\max}$. makes no allowance for the initial spread in the disturbance. When the moments μ_1 and μ_2 are used in a Gaussian distribution the accuracy is much greater for smaller values of t. Our treatment lacks precision in that only the moments can be discussed but has the advantage of a slightly easier method and yields results valid for all x. The discussion of the approach to normality in § 5 does something towards removing the imprecision.

3. GENERAL THEOREMS

It is convenient at this point to note the general theorems which are required to justify the proposed method of determining the moments. We require the conditions under which it is true that

$$(-)^n \phi^{(n)}(0) = \int_0^\infty t^n f(t)\, \mathrm{d}t \qquad (15)$$

when

$$\phi(p) = \int_0^\infty \mathrm{e}^{-pt} f(t)\, \mathrm{d}t.$$

In many physical situations the strongest of these conditions will be satisfied, but it is convenient to have weaker conditions in reserve.

If the Laplace transform converges in any half-plane $\mathscr{R}p > \sigma_c$, where σ_c is finite, then $\phi(p)$ is an analytic function of p in this half-plane and

$$\phi^{(n)}(p) = \int_0^\infty \mathrm{e}^{-pt}(-t)^n f(t)\, \mathrm{d}t \quad (\mathscr{R}p > \sigma_c) \qquad (16)$$

(see Widder 1946, p. 57). Thus if $\sigma_c < 0$, $\phi(p)$ is analytic in the neighbourhood of the origin and (15) holds. This will be the case if $f(t) = 0$ for sufficiently large t or if $f(t) \sim A\mathrm{e}^{-\lambda t}(\lambda > 0)$. Conversely, if $\phi(p)$ is analytic in the neighbourhood of the origin $\sigma_c < 0$ and (15) holds, whilst if $\sigma_c > 0$ all the moments are infinite. The case which requires further examination is thus the one where $\sigma_c = 0$ and so $\phi(p)$ is only analytic in a right half neighbourhood of the origin.

The Abelian theorems given in Widder's treatise (p. 180 *et seq.*) cover this case when the moments are known to exist, for it is there shown that if

$$\int_0^t f(t)\, \mathrm{d}t \to A \quad \text{as} \quad t \to \infty$$

then

$$\phi(p) \to A \quad \text{as} \quad p \to 0+.$$

For the nth moment we replace $f(t)$ by $t^n f(t)$, of which the transform is $(-)^n \phi^{(n)}(p)$. Thus if the nth moment is known to exist it is given by $\lim_{p \to 0_+} (-)^n \phi^{(n)}(p)$.

The converse of this theorem, that if $\phi(p) \to A$ as $p \to 0+$ then $\int_0^t f(t)\, \mathrm{d}t \to A$ as $t \to \infty$ is known to be false and further, Tauberian, conditions on $f(t)$ are required. Thus the Tauberian theorem proved by Widder (p. 195) may be adapted for our needs as follows: If $\phi(p)$ is analytic for $p > 0$ and if $(-)^n \phi^{(n)}(p) \to A$ as $p \to 0+$ and if there exist a K such that $Kt + \int_0^t u^{n+1} f(u)\, \mathrm{d}u$ is a non-decreasing function of t in $(0, \infty)$ then $\int_0^t t^n f(t)\, \mathrm{d}t = A$.

4. A KINEMATIC TEMPERATURE WAVE

To show the power of the proposed method we consider the following problem. A cylindrical bed of particles is thermally insulated on its curved surface and a fluid flows uniformly through it. The particles may be of different shapes (denoted by a suffix m) and of different sizes of any one shape (denoted by suffix n), but they are sufficiently small and well mixed that in any sample from the bed the same fraction f_{mn} of the heat capacity is due to particles of the mth shape and nth size. The bed and fluid flowing through it are originally at a constant temperature (taken to be the zero of temperature) and for time $t > 0$ the temperature of the fluid entering the bed is a prescribed function of time. This function will be taken to be hump-like as before, either zero after some finite t or falling to zero sufficiently quickly for the moments to exist. As Lighthill & Whitham remark, the case where the influent temperature does not fall to zero but tends to a constant can be derived from this case, as then the derivative of the temperature is hump-like. The temperature wave will move down the bed as a kinematic wave and at the same time be dispersed by diffusive processes. Four distinct processes will be taken into account as contributing to this:

(a) diffusion of heat within each particle,
(b) diffusion of heat through a thin film of different conductivity on the surface of each particle (e.g. a deposit of 'fur'),
(c) diffusion of heat through a thin film of stagnant fluid surrounding each particle (this differs from (b) in that the thickness of the film is a function of the fluid velocity),
(d) diffusion of heat in the moving fluid in the direction of flow.

Assuming the particles to be sufficiently small for the surrounding temperature to be sensibly constant, the following equations describe the variation of temperature.

(a) Within the m, nth particle denote the space variables by ξ and the temperature by $T^*_{mn}(\xi, x, t)$, where x is the depth of the particle in the bed. Let the surface temperature beneath the deposit of fur be $T'_{mn}(x, t)$. Then

$$\frac{\partial T^*_{mn}}{\partial t} = D_{mn} \nabla^2_\xi T^*_{mn}, \tag{17}$$

$$T^*_{mn} = 0 \quad \text{at} \quad t = 0, \tag{18}$$

$$T^*_{mn} = T'_{mn} \quad \text{on the surface of the particle.} \tag{19}$$

where D_{mn} is the thermal diffusivity in the m, nth species of particle and ∇^2_ξ denotes the Laplace operator in the variables ξ.

(b) For simplicity we assume the thin layer of fur is so thin that it may be considered flat and that heat transfer through it can be described by a single heat transfer coefficient k'_{mn} and the equations

$$\frac{\partial T'_{mn}}{\partial t} = k'_{mn}(T''_{mn} - T'_{mn}), \tag{20}$$

$$T'_{mn} = 0, \quad t = 0, \tag{21}$$

where $T''_{mn}(x, t)$ is the temperature of the outer surface of the fur. Clearly, a more detailed analysis could be carried through but it would only add to the complexities.

(*c*) Similarly, we take a single heat transfer coefficient k_{mn} for the stagnant layer of fluid and write

$$\frac{\partial T''_{mn}}{\partial t} = k_{mn}(T - T''_{mn}), \tag{22}$$

$$T''_{mn} = 0, \quad t = 0, \tag{23}$$

where $T(x, t)$ is the temperature in the fluid at depth x and time t.

(*d*) In deriving the equation for the fluid temperature we shall assume (i) that the bed is effectively semi-infinite, and (ii) that the heat capacity of the stagnant film and fur are negligible in comparison with that of the fluid and particles. Both these assumptions and the one made earlier could be dispensed with, but let 'Ossa' suffice.

Let h_s = heat capacity of the particles per unit volume of the bed,
f_{mn} = fraction of this heat capacity contributed by the m, nth kind of particle,
$T_{mn}(x, t)$ = the mean temperature in the m, nth particle,
= $\iiint T^*_{mn}\, \mathrm{d}v_\xi / \iiint \mathrm{d}v_\xi$, where $\mathrm{d}v_\xi$ is the element of volume within the particle,
h_f = heat capacity of fluid per unit volume of the bed,
K = thermal conductivity of the fluid,
α = voidage of the bed,
v = linear velocity of fluid.
Then $H_f = h_f T$ = heat content of fluid per unit volume of bed,
$H_s = \Sigma f_{mn} h_s T_{mn}$ = heat content of solids per unit volume of bed.

Then taking a heat balance over a small section of the bed gives

$$K\alpha \frac{\partial^2 T}{\partial x^2} - v h_f \frac{\partial T}{\partial x} = \frac{\partial}{\partial t}(H_f + H_s) = h_f \frac{\partial T}{\partial t} + h_s \Sigma f_{mn} \frac{\partial T_{mn}}{\partial t}, \tag{24}$$

$$T = 0, \quad t = 0, \quad x > 0, \tag{25}$$

$$T = f(t), \quad t > 0, \quad x = 0, \tag{26}$$

$$T \to 0 \quad \text{as} \quad x \to \infty. \tag{27}$$

That this is essentially a kinematic wave is seen by dropping the conductivity term in (24) and writing $k = H_f + H_s$, the concentration of heat and $q = v h_f T$, the flow of heat. We then recover the kinematic wave equation given by Lighthill & Whitham. If thermal equilibrium were instantaneously attained so that

$$T = T''_{mn} = T'_{mn} = T_{mn},$$

then
$$k = (h_f + \Sigma f_{mn} h_s)T = (h_f + h_s)T = \{(h_s + h_f)/v h_f\} q$$

since $\Sigma f_{mn} = 1$. Thus in the absence of any dispersion ($D_{mn} = k_{mn} = k'_{mn} = \infty$, $K = 0$) we have a kinematic wave velocity

$$c = \frac{\mathrm{d}q}{\mathrm{d}k} = v \frac{h_f}{h_f + h_s}. \tag{28}$$

We shall see that the mean of the temperature wave travels with just this velocity even in the presence of these dispersive effects and obtain an expression for the growth of the variance.

Let the space variables within the particle be made dimensionless by expressing them as a fraction of some characteristic dimension of the particle d_{mn}. Also let a bar over any function denote its Laplace transform with respect to t, e.g.

$$\overline{T}(x,p) = \int_0^\infty e^{-pt} T(x,t)\,dt.$$

Then equations (17) to (19) give

$$\nabla_\xi^2 \overline{T}^*_{mn} = \frac{pd^2_{mn}}{D_{mn}} \overline{T}^*_{mn}, \tag{29}$$

$$\overline{T}^*_{mn} = \overline{T}'_{mn} \quad \text{on the surface, } S. \tag{30}$$

The theory of the equation

$$\nabla^2 u = \varpi u \quad (u = 1 \text{ on } S),$$

has received considerable attention, and it is known that a solution exists for a large class of closed surfaces S. Moreover, the mean value $\Omega(\varpi)$ of the solution u over the interior of S is a function of the shape of S and ϖ only and $\Omega(0) = 1$, $-\Omega'(0) = e$, a constant for any given shape of S. The evaluation of e has been discussed elsewhere (Aris 1957). Thus if we denote the dependence on shape by the suffix m

$$\overline{T}_{mn} = \overline{T}'_{mn}\Omega_m(pd^2_{mn}/D_{mn}). \tag{31}$$

The transforms of equations (20) and (22), with their associated conditions, yield

$$(p + k'_{mn})\overline{T}'_{mn} = k'_{mn}\overline{T}''_{mn}, \tag{32}$$

$$(p + k_{mn})\overline{T}''_{mn} = k_{mn}\overline{T}, \tag{33}$$

so that
$$\Sigma f_{mn}\overline{T}_{mn} = \overline{T}\Sigma f_{mn}\frac{\Omega_m(pd^2_{mn}/D_{mn})}{(1 + p/k'_{mn})(1 + p/k_{mn})} = \overline{T}\omega(p). \tag{34}$$

Substituting this in the transform of equation (24) we have

$$\frac{K\alpha}{h_f}\frac{d^2\overline{T}}{dx^2} - v\frac{d\overline{T}}{dx} = p\overline{T}\left\{1 + \frac{h_s}{h_f}\omega(p)\right\}.$$

Let $x = v\tau$, $\kappa = K\alpha/v^2 h_f$, $\mu = h_s/h_f$ then

$$\kappa\frac{d^2\overline{T}}{d\tau^2} - \frac{d\overline{T}}{d\tau} - p\{1 + \mu\omega(p)\}\overline{T} = 0, \tag{35}$$

$$\overline{T} = \overline{f}(p), \quad \tau = 0. \tag{36}$$

$$\overline{T} \to 0 \quad \text{as} \quad \tau \to 0. \tag{37}$$

The solution of these equations is

$$\overline{T}(x,p) = \overline{f}(p)\exp\{P(p)\tau\}, \tag{38}$$

where

$$P(p) = [1 - \{1 + 4\kappa p(1 + \mu\omega(p))\}^{\frac{1}{2}}]/2\kappa. \tag{39}$$

This function is analytic in the neighbourhood of the origin if $\overline{f}(p)$ is, so that the strongest condition of § 3 is satisfied. Then we have for the change in the first two moments,

$$m_1 = -P'(0)\tau = x(1 + \mu)/v = x/c, \tag{40}$$

$$m_2 = P''(0)\tau = 2\frac{x}{v}\left[\kappa(1+\mu)^2 + \mu\,\Sigma f_{mn}\left\{\frac{1}{k_{mn}} + \frac{1}{k'_{mn}} + \frac{e_m d^2_{mn}}{D_{mn}}\right\}\right]. \tag{41}$$

If an apparent diffusion coefficient be defined as above by $D = c^2 m_2/m_1$ then

$$D = 2\frac{\kappa v^2}{1+\mu}\left[1 + \frac{\mu}{(1+\mu)^2\kappa}\Sigma f_{mn}\left\{\frac{1}{k_{mn}} + \frac{1}{k'_{mn}} + \frac{e_m d^2_{mn}}{D_{mn}}\right\}\right]. \tag{42}$$

This shows that the mean of the temperature wave moves with the kinematic wave velocity and that an apparent diffusion coefficient may be defined to describe the dispersion. This coefficient is the sum of the diffusion coefficients which would be obtained if each effect were considered independently. Such an additivity has been demonstrated by the author for the molecular and Taylor diffusion coefficients elsewhere (Aris 1956) and is assumed in a paper by Klinkenberg and others (van Deemter, Zuiderweg & Klinkenberg 1956) in their analysis of the dispersion of a chromatogram.

5. THE ULTIMATE FORM OF A KINEMATIC WAVE

It might be expected from a general knowledge of diffusive processes and their stochastic nature that the Gaussian distribution plays a central role in the ultimate behaviour of a disturbance. If it can be shown that the ultimate form of the kinematic wave is Gaussian then the mean and variance suffice to determine the complete disturbance, since all the higher moments are functions of these. This indeed has been the endeavour in the study of kinematic waves. Lighthill & Whitham use this for the flood wave, and the full solution given by (93) of their appendix is doubtless amenable to the type of uniform approximation of Bessel functions by the Gaussian function which has been used by Peiser (1950). Similar approximations have been used by Thomas (1948), van Deemter *et al.* (1956) and many others in the theory of chromatography. The asymptotic behaviour of solute dispersed by the combined effect of diffusion and convection has been discussed elsewhere by the author (Aris 1956) and Gaussian distributions of holding time in a number of systems have been treated by Klinkenberg & Sjenitzer (1956). We shall show here that the Gaussian curve is the asymptotic form of a large class of linear kinematic waves and give some further indication as to how this form is reached. Again it might be mentioned that this analysis will apply only to linear equations. In a recent study (Lighthill 1956) the asymptotic form of a non-linear sound wave was discussed for a wide range of the Reynolds number R. R measures the relative importance of non-linearity and diffusion in the wave, and it was found that only for very small values of R is the disturbance Gaussian in shape.

We prove the following proposition:

If $\bar{f}(p)$ and $xP(p) = -m_1p + \frac{1}{2}m_2p^2 + \varpi(p)$ are analytic in the neighbourhood of the origin and $\bar{f}(p)\exp\{xP(p)\}$ is the Laplace transform of $F(x, t)$ then

$$F(x,t) \sim \left[\frac{1}{(2\pi m_2)^{\frac{1}{2}}}\exp\left\{-\frac{(t-m_1)^2}{2m_2}\right\}\right]\int_0^\infty f(t)\,\mathrm{d}t, \tag{43}$$

as $x \to \infty$ and $\liminf(t - m_1)/m_1)/m_2^{\frac{1}{2}} > 0$, $\limsup(t - m_1)/m_2^{\frac{1}{2}} < \infty$.

Since $\bar{f}(p)\exp\{xP(p)\}$ is a Laplace transform, analytic in the neighbourhood of the origin, the imaginary axis is within its half plane of convergence and may be taken as the contour in the inversion integral. Thus

$$F(x,t) = \frac{1}{2\pi}\int_{-\infty}^{\infty}\exp\{\mathrm{i}\omega t + xP(\mathrm{i}\omega)\}\bar{f}(\mathrm{i}\omega)\,\mathrm{d}\omega. \tag{44}$$

Let $t = m_1 + \xi m_1)/m_2^{\frac{1}{2}}$, then

$$F(x,t) = \frac{1}{2\pi}\int_{-\infty}^{\infty}\exp\{\mathrm{i}\omega\xi m_2^{\frac{1}{2}} - \tfrac{1}{2}\,\omega^2(1+\varpi)\,m_2\}\bar{f}(\mathrm{i}\omega)\,\mathrm{d}\omega$$

and $\varpi(\mathrm{i}\omega)$ contains cubes and higher powers of ω. The asymptotic behaviour of this type of integral where two parameters h and k tend to infinity has been studied by Fulks (1951). Putting $h = m_2$, which tends to infinity as x, and $k = \mathrm{i}\xi m_2^{\frac{1}{2}}$, which if ξ is bounded is proportional to $x^{\frac{1}{2}}$ and so $o(h)$, we find the conditions of Fulks's theorem 2 are fulfilled (Fulks, p. 615). It follows immediately from this theorem that as $h \to \infty$

$$F(x,t) \sim \bar{f}(0)(2\pi m_2)^{-\frac{1}{2}}\exp\{-\tfrac{1}{2}\xi^2\},$$

which on substituting for ξ and putting $\bar{f}(0) = \int_0^\infty f(t)\,\mathrm{d}t$ yields (43).

Roughly speaking, to require that the Laplace transform of a function should be analytic at the origin is to require that the function should tend to zero as $e^{-\lambda t}$, where $\lambda > 0$. Such a function possesses all its moments and this is clearly necessary if it is to be asymptotically Gaussian. The Gaussian distribution, however, tends to zero as $e^{-\lambda t^2}$ with $\lambda > 0$, and if the function is similarly rapid in its decay it may be possible to carry the approximation still further by means of a type A series (Kendall 1943, p. 147). We shall merely outline the way in which this may be possible.

If as above we put

$$xP(p) = m_1p + \tfrac{1}{2}m_2p^2 + \varpi(p) \quad\text{and}\quad \varpi(p) = -\frac{1}{3!}m_3p^3 + \frac{1}{4!}m_4p^4 - \dots,$$

$$\text{then } e^{xP(p)} = \left\{1 - \frac{1}{3!}m_3p^3 + \frac{1}{4!}m_4p^4 - \cdots\right\}\exp\{-m_1p + \tfrac{1}{2}m_2p^2\}. \tag{45}$$

In this series the fourth term is $-m_5p^5/5!$ but for the terms above this the coefficients become increasingly complex combinations of the coefficients m_r in the expansion of ϖ. Now the formal interpretation of $p^3\bar{g}(p)$ is that it is the Laplace transform of $g'''(p)$ so that writing

$$\phi(t) = (2\pi m_2)^{-\frac{1}{2}} \exp\left\{-\frac{(t-m_1)^2}{2m_2}\right\},$$

the formal interpretation of (45) is

$$\phi(t) - \frac{1}{3!} m_3\phi'''(t) + \frac{1}{4!} m_4\phi^{\text{iv}}(t) - \ldots,$$

and this is the so-called Gram–Charlier series of type A. m_3 is in fact the third moment about the mean and m_4 the fourth moment less $3m_2^2$. That this is purely formal is clear from the fact that we have neglected a number of terms involving the value of $\phi(t)$ and its derivatives at $t = 0$. These are of the order of $\exp\{-m_1^2/2m_2\}$ and so vanish as $x \to \infty$. This neglect is equivalent to ignoring the fact that for any finite t, $\phi(t)$ must be truncated at $t = 0$.

It does, however, indicate the way in which the normal distribution is attained. The absolute skewness m_3^2/m_2^3 and the kurtosis m_4/m_2^2 both tend to zero as x^{-1}. The difficulty in the practical application lies in the truncation error and the fact that the higher terms decrease in a somewhat irregular manner. It may often be of value, however, to estimate the skewness from the third moment and even in such a complicated example as that of § 4 this is not impracticable.

REFERENCES

Amundson, N. R. & Aris, R. 1957 *A.I.Ch.E. Journal* **3,** 280.

Aris, R. 1956 *Proc. Roy. Soc.* A, **235,** 67.

Aris, R. 1957 *Chem. Engng Sci.* **7,** 8.

van Deemter, J. J., Zuiderweg, F. J. & Klinkenberg, A. 1956 *Chem. Engng Sci.* **5,** 271.

Fulks, W. 1951 *Proc. Amer. Math. Soc.* **2,** 613.

Goldstein, S. 1953 *Proc. Roy. Soc.* A, **219,** 151.

Kendall, M. G. 1943 *The advanced theory of statistics.* London: Griffin and Co.

Klinkenberg, A. 1954 *Industr. Engng Chem.* **46,** 2285.

Klinkenberg, A. & Sjenitzer, F. 1956 *Chem. Engng Sci.* **5,** 258.

Lighthill, M. J. 1956 *Viscosity effects in sound waves of finite amplitude. Surveys in mechanics.* Cambridge University Press.

Lighthill, M. J. & Whitham, G. B. 1955*a* *Proc. Roy. Soc.* A, **229,** 281.

Lighthill, M. J. & Whitham, G. B. 1955*b* *Proc. Roy. Soc.* A, **229,** 317.

Peiser, A. M. 1950 *Proc. Amer. Math. Soc.* **1,** 650.

Taylor, G. I. 1953 *Proc. Roy. Soc.* A, **219,** 186.

Thomas, H. C. 1948 *Ann. N.Y. Acad. Sci.* **49,** 161.

Widder, D. V. 1946 *The Laplace transform.* Princeton University Press.

8

FORMAL KINETICS

INTRODUCTION

There is an aside in the Acknowledgments (pp. 438–442) in which I have tried to sketch the development of what might be called *formal chemical kinetics*. I take this to be the exploration of the mathematical structure that is engendered by the physics and chemistry of reaction without regard to the chemical peculiarities or physical particulars. Thus, *mass-action kinetics* denotes a relationship between stoichiometry and kinetics that may or may not hold between any given ingredients. Wei's axiomatization[1] and the recognition of the utility of continuous mixtures were what aroused my interest. Reprint D was an attempt to lay out the basic structure, and Reprint E gives some corrections and additions.

I am sorry that there is not space for the paper [62], which was based on Gavalas' dissertation, for it laid the foundations of continuous mixture kinetics of the first order. When Astarita and Ocone revived the subject again in the late 1980s, the emphasis was on underlying kinetics other than first order. Reprint G is a paper from the second period, the first being represented only by the last section of Reprint E. Papers on continuous mixtures may be traced through the Index of Subjects in Publications.

Reprint H is a late paper written as a *festschrift* paper in honor of Davidson's retirement. It combines the continuous mixture techniques with the model reduction method given in Chapter 2, the section entitled "Scaling and Partial Solution in Linear Systems." A closely similar paper was requested

[1] James Wei, "Axiomatic treatment of chemical reaction systems." *J. Chem. Phys.* **36**, 1578 (1962).

for *Advances in Engineering Fluid Mechanics,* and there is much overlap (with copyright permission) of this paper with [292].

Reprint F is an example of analyzing a reaction in formal kinetics. Gray and Scott introduced the autocatalytic A + 2B = 3B as a simple model reaction that proved to have a rich behavior, much richer than the Brusselator for example. However, A + 2B smacks of a three-body interaction, which is a sufficiently rare occurrence as to be avoided. I had done a pseudo–steady-state analysis before I visited Leeds at Gray's invitation, and the chance of working with the *fons et origo* of this reaction, so to speak, was an opportunity to make sure that the limiting behavior was not lost when certain parameters were small, but not actually zero. For another analysis of autocatalytic behavior, see [107].

D.*

PROLEGOMENA TO THE RATIONAL ANALYSIS OF SYSTEMS OF CHEMICAL REACTIONS

RUTHERFORD ARIS

Communicated by C. TRUESDELL

Contents

1. INTRODUCTION

We wish, in this paper, to examine the foundations of a rational analysis of chemical systems and their reactions. Though it should not be necessary, it may help to remove any misapprehension if we say explicitly that the word "rational" is here used in the sense of the title of this journal and implies no derogation of the conventional treatment to the subject. But the immense access of clarity and order that has accrued to continuum mechanics during the last few years encourages us to think that there may be much to be gained in approaching some aspects of chemistry in the same spirit. For just as the algebra and calculus of tensors lies at the basis of continuum mechanics, so it will be found that the algebra of finite dimensional vector spaces lies at the basis of formal reaction kinetics of the simplest representation of chemical species.

It is with the representation of chemical species by their so-called "empirical formula" (*i.e.* with no reference to structure) that we are here concerned. In such a representation ethyl alcohol is C_2H_6O, and not C_2H_5OH or CH_3CH_2OH as more structured representations would have it. They will be referred to, where necessary, as representations of Class I. A formal theory of a simple structured representation has indeed been adumbrated [*1*, *2*], but it is not yet clear what the algebraic structure of the reaction system may be.

* Reprinted with the permission of Springer-Verlag from *Archive for Rational Mechanics and Analysis,* volume **19,** number 2, pp. 81–99, 1965.

By "stoicheiometry" we understand the calculus of changes in composition that take place by reaction; it corresponds to kinematics in the analogy with continuum mechanics for its provides the framework within which chemical "motions" must take place, irrespective of the forces that bring them about. By "kinetics" we understand the relations that govern the speed of the composition changes and this bears some analogy to the dynamics of continua. Just as the latter can only be built on a proper understanding of the kinematics, so the analysis of stoicheiometry must precede that of kinetics.

Although we are concerned with linear algebra it will be convenient to use a notation slightly different from the usual matrix notation and to adopt the summation convention from tensor analysis. Thus α_r^s will denote the element in the rth row and s^{th} column of the matrix $\boldsymbol{\alpha} = (\alpha_r^s)$. The product $\boldsymbol{\alpha} \cdot \boldsymbol{\beta}$ of $\boldsymbol{\alpha}$ with a matrix $\boldsymbol{\beta} = (\beta_s^t)$ is the matrix $(\alpha_r^s \beta_s^t)$, where summation over $s = 1, 2, \dots, \mathrm{S}$ is implied. The range of a particular affix (*i.e.*, $s = 1, 2, \dots, \mathrm{S}$) is given in the table of nomenclature. In this notation the order of any matrix is apparent and the use of partitioned matrices can be represented rather easily.

2. THE REPRESENTATION OF MOLECULAR SPECIES AND REACTIONS BETWEEN THEM

We shall commonly abuse the language to the extent of abbreviating the phrase "the Class I representation of a molecular species $\mathscr{A}_s''$ to read "the molecular species $\mathscr{A}_s''$. Since we are only concerned with Class I representations this will not cause any confusion.

Definition 1. A molecular species, $\mathscr{A}_s$, is an entity of the form

$$\mathscr{A}_s = \beta_s^t \mathscr{B}_t,$$

where β_s^t are integers. The $\mathscr{B}_t$, $t = 1, 2, \dots, \mathrm{T}$, are symbols for the chemical elements or atomic species, T being the number of such elements in the universe of discourse.

Remark 1. If all the atomic coefficients β_s^t are positive we speak (where necessary) of a "proper" molecular species. If all the β_s^t are zero we have the "null" or "trivial" species, and if any are negative a "pseudo-species."

Remark 2. The elements are themselves molecular species with $\beta_s^t = \delta_s^t$ (Kronecker delta).

Remark 3. Although the number of elements known to exist in the universe is finite, it is seldom necessary to consider more than a few of them in a particular universe of discourse. Thus large tracts of organic chemistry require only the three elements, carbon, hydrogen and oxygen.

Rule of Composition. Two species $\mathscr{A}_s$ and $\mathscr{A}_{s'}$ may be combined by a rule of composition, denoted by "+",

$$\mathscr{A}_s + \mathscr{A}_{s'} = (\beta_s^t + \beta_{s'}^t)\mathscr{B}_t. \tag{1}$$

The + on the right hand side denotes the ordinary addition of integers. If γ is any integer $\gamma\mathscr{A}_s$ denotes the species $(\gamma\beta_s^t)\mathscr{B}_t$.

With this definition and rule of composition we have the following:

THEOREM 1. $\mathfrak{A}_{\mathrm{T}}$, *the totality of molecular species formed from the atomic species* $\mathscr{B}_1, \ldots, \mathscr{B}_{\mathrm{T}}$, *constitutes a linear vector space of dimension* T *over the integral domain of integers.*

Proof. $\mathfrak{A}_{\mathrm{T}}$ is a module: for if $\mathscr{A}_s$ and $\mathscr{A}_{s'}$ belong to $\mathfrak{A}_{\mathrm{T}}$

(i) $\mathscr{A}_s + \mathscr{A}_{s'}$ belongs to $\mathfrak{A}_{\mathrm{T}}$,

(ii) $\mathscr{A}_s + \mathscr{A}_{s'} = \mathscr{A}_{s'} + \mathscr{A}_s$,

(iii) $(\mathscr{A}_s + \mathscr{A}_{s'}) + \mathscr{A}_{s''} = \mathscr{A}_s + (\mathscr{A}_{s'} + \mathscr{A}_{s''})$,

(iv) $\mathscr{A}_s - \mathscr{A}_{s'}$ belongs to $\mathfrak{A}_{\mathrm{T}}$.

Also if γ and γ' are any integers

(v) $\gamma\mathscr{A}_s = 0, \quad \mathscr{A}_s \neq 0 \quad$ only if $\gamma = 0$,

$\gamma\mathscr{A}_s = \mathscr{A}_s, \quad \mathscr{A}_s \neq 0 \quad$ only if $\gamma = 1$,

(vi) $\gamma(\mathscr{A}_s + \mathscr{A}_{s'}) = \gamma\mathscr{A}_s + \gamma\mathscr{A}_{s'}$,

(vii) $(\gamma + \gamma')\mathscr{A}_s = \gamma\mathscr{A}_s + \gamma'\mathscr{A}_s$,

(viii) $\gamma(\gamma'\mathscr{A}_s) = (\gamma\gamma')\mathscr{A}_s$.

These properties define a linear vector space over the integers. Also since $\mathscr{B}_1, \ldots, \mathscr{B}_{\mathrm{T}}$ belong to $\mathfrak{A}_{\mathrm{T}}$ and are independent, they form a basis for the space which is therefore of dimension T. ∎

We shall often be concerned with a subset of species in $\mathfrak{A}_{\mathrm{T}}$ consisting of a given set of molecular species $\{\mathscr{A}_s\}$, $s = 1, 2, \ldots, \mathrm{S}$, and it is important to define the subspace within which this lies. Concerning this we have:

THEOREM 2. *The species* $\{\mathscr{A}_s\}$, $s = 1, 2, \ldots, \mathrm{S}$, *lie in a subspace of* $\mathfrak{A}_{\mathrm{T}}$ *of dimension* T′, *where* T′ *is the rank of the matrix* $\boldsymbol{\beta} = (\beta_s^t)$.

Proof. If T′ is the rank of the matrix (β_s^t), then, without loss of generality, we can take the first T′ columns to be independent and express the last (T − T′) in terms of them,

$$\beta_s^{t''} = \beta_s^{t'}\,\nu_{t'}^{t''}, \begin{cases} t' = 1, 2, \ldots, \mathrm{T}', \\ t'' = \mathrm{T}' + 1, \mathrm{T}' + 2, \ldots, \mathrm{T}. \end{cases}$$

Thus,

$$\begin{aligned} \mathscr{A}_s &= \beta_s^{t'}\,\mathscr{B}_{t'} + \beta_s^{t''}\,\mathscr{B}_{t''} \\ &= \beta_s^{t'}\,(\mathscr{B}_{t'} + \nu_{t'}^{t''}\,\mathscr{B}_{t''}) = \beta_s^{t'}\,\overline{\mathscr{B}}_{t'}'. \end{aligned}$$

The species $\{\overline{\mathscr{B}}_{t'}\}$ therefore constitute a basis for the subspace in which the $\{\mathscr{A}_s\}$ lie and this is therefore of dimension T′. ∎

COROLLARY. T′ *cannot be greater than* S.

We can now discuss the representation of reactions between the molecular species of a given set.

Definition 2. A reaction over the set of s species $\{\mathscr{A}_s\}$ is the linear functional

$$\mathscr{R} \equiv \alpha^s \mathscr{A}_s \tag{2}$$

where the α^s are real numbers called stoicheiometric coefficients. The reaction is said to be trivial if all the stoicheiometric coefficients are zero. It is said to be proper if

$$\alpha^s \beta_s^t = 0, \quad t = 1, 2, \ldots, \text{T}. \tag{3}$$

Remark 4. The condition for a reaction to be proper is known in chemistry as "balancing the equation" and it implies that

$$\alpha^s \mathscr{A}_s = 0. \tag{4}$$

This is often written

$$-\alpha^{s'} \mathscr{A}_{s'} = \alpha^{s''} \mathscr{A}_{s''}, \quad s' = 1, 2, \ldots, \text{s}', \quad s'' = \text{s}' + 1, \ldots, \text{s},$$

where the $-\alpha^{s'}$ and $\alpha^{s''}$ are all positive (*e.g.* $2H_2 + O_2 = 2H_2O$). We shall adopt the convention that any particular species that may be regarded as the chemical product of the reaction should have a positive stoicheiometric coefficient. The species with negative coefficients are known as "reactants." This convention implies a certain sense to the reaction and the reactants and products will be interchanged if the reaction is reversed.

Remark 5. The proper reactions $\alpha^s \mathscr{A}_s = 0$ and $\gamma \alpha^s \mathscr{A}_s = 0$ are not distinguished and hence it is only the ratios of the stoicheiometric coefficients that are important. The condition (3) shows that these ratios are rational numbers and that the α^s can always be taken to be integers. Various conventions have been suggested [*3*, *4*], none of which has any claim to authority. The simplest is to make the α^s mutually prime integers; alternatively a particular species may be singled out and its coefficient made unity, but this is arbitrary (see however Rem. 9 *infra*). If the α^s are so chosen that $\sum (\alpha^s)^2 = 1$ we shall say that the stoicheiometry of the reaction has been normalized, and this will prove to have some computational advantage in spite of the chemical incongruity of surds.

Remark 6. Simultaneous reactions over the set of species $\{\mathscr{A}_s\}$ may be written

$$\mathscr{R}_r = \alpha_r^s \mathscr{A}_s, \quad r = 1, 2, \ldots, \text{R}. \tag{5}$$

They are all proper if $\alpha_r^s \beta_s^t = 0$ for all r and t. The matrix $\boldsymbol{\alpha} = (\alpha_r^s)$ is called the matrix of stoicheiometric coefficients of the set of reactions $\{\mathscr{R}_r\}$.

Rule of Composition of Reactions. The sum of two reactions over the same set of species and the product of a scalar and a reaction are defined by

$$\mathscr{R}_r + \mathscr{R}_{r'} = (\alpha_r^s + \alpha_{r'}^s)\mathscr{A}_s,$$
$$\gamma \mathscr{R} = (\gamma \alpha^s)\mathscr{A}_s.$$

DEFINITION 3. The reactions $\mathscr{R}_r$ of a set of reactions $\{\mathscr{R}_r\}$ are said to be independent if there are no non-trivial multipliers γ^r such that

$$\gamma^r \mathscr{R}_r \equiv 0.$$

Remark 7. The number of independent reactions in a given set $\{\mathscr{R}_r\}$ is the rank of the stoicheiometric matrix and may be determined by elementary row operations.

Remark 8. The set of independent, proper reactions $\{\mathscr{R}_r\}$ will be said to be orthonormalized if

$$\sum \alpha_r^s \alpha_{r'}^{s*} = \delta_r^{r'}, \quad \text{or} \quad \boldsymbol{\alpha} \cdot \boldsymbol{\alpha}^* = \boldsymbol{I}. \tag{6}$$

Here $\alpha_{r'}^{s*}$ denotes the typical element of the transposed matrix $\boldsymbol{\alpha}^*$.

Remark 9. An alternative convention has been proposed by HOOYMAN [5] as follows. Since the reactions are independent we may choose R of the S species (without loss of generality these may be $\mathscr{A}_1, \ldots, \mathscr{A}_R$) and transform the stoicheiometric matrix to the form

$$\boldsymbol{\alpha} = (\boldsymbol{I} : \boldsymbol{\alpha}'),$$

where $\boldsymbol{I}$ is the R × R identity matrix and $\boldsymbol{\alpha}'$ an R × (S − R) matrix. A given reaction, say $\mathscr{R}_{r'}$, can then be referred to as the reaction for $\mathscr{A}_{r'}$ with respect to the chemistry involving $\mathscr{A}_1, \ldots, \mathscr{A}_{r'-1}, \mathscr{A}_{r'+1}, \ldots, \mathscr{A}_R$. Again the selection of the R species makes this somewhat arbitrary and the convention does not have the computational advantage of orthonormalization.

We shall show later the sense in which these sets of reactions are equivalent.

THEOREM 3. *The totality of reactions* R *over a given set of species* $\{\mathscr{A}_s\}$, $s = 1, 2, \ldots,$ S, *forms a linear vector space* $\mathfrak{B}_S$ *of dimension* S *over the field of real numbers.*

Proof. $\mathfrak{B}_S$ is a module by the rule of composition given and conditions similar to those in Theorem 1 may easily be verified. ■

Remark 10. $\mathfrak{B}_S$ is formed with respect to a given set of species $\{\mathscr{A}_s\}$ and is not to be confused with the dual vector-space to $\mathfrak{A}_T$.

THEOREM 4. *The proper reactions over a set of species* $\{\mathscr{A}_s\}$ *lie in a subspace of* $\mathfrak{B}_S$ *of dimensions* (S − T′), *where* T′ *is the dimension of the subspace of* $\mathfrak{A}_T$ *within which the set* $\{\mathscr{A}_s\}$ *lies.*

Proof. Consider the dual vector space $\mathfrak{B}_S'$. This is isomorphic to an S dimensional vector space, $\mathfrak{A}_S'$, in which $\mathfrak{A}_T$ is embedded. Let $\mathfrak{b}'$ be the subspace of $\mathfrak{B}_S'$ which is isomorphic to the subspace of $\mathfrak{U}_S'$ within which the set $\{\mathscr{A}_s\}$ lies. By Theorem 2 the dimension of $\mathfrak{b}'$ is T′. Now the proper reactions of $\mathfrak{B}_S$ are the subspace $\mathfrak{b}$ of annihilators of $\mathfrak{b}'$. By a well-known theorem on dual vector spaces [6], $\mathfrak{b}$ must have dimension (S − T′). ■

COROLLARY. *There can be at most* S − 1 *reactions between* S *nontrivial species, for* T′ > 1 *unless all the* β_s^t *are zero.*

3. THE REPRESENTATION AND CALCULUS OF COMPOSITION CHANGES

To each atomic species there belongs a real number called its atomic weight, a_t. The molecular weight of the species $\mathscr{A}_s$ is given by

$$m_s = \beta_s^t a_t. \tag{7}$$

If M_s is the mass of $\mathscr{A}_s$ present in a given system the ratio

$$n^s = M_s/m_s \tag{8}$$

is the number of moles present, and if V is the volume of the system

$$c^s = n^s/V \tag{9}$$

is the (molar) concentration of $\mathscr{A}_s$. Other measures of quantity and concentration may be defined and related to these [7], but this will suffice here.

The chemical meaning of the equation $\alpha^s \mathscr{A}_s = 0$ is that the changes in the number of moles during the reaction $\mathscr{R}$ are in the ratios of the stoicheiometric coefficients, *i.e.* $\Delta n^s/\alpha^s$ is independent of s. Therefore we may define a continuous variable belonging to a reaction as follows:

DEFINITION 4. The extent of the reaction $\mathscr{R}$ is a variable ζ, such that the composition changes by reaction are $\Delta n^s = \alpha^s \zeta$. The extent per unit volume, ξ, is similarly defined by

$$\Delta c^s = \alpha^s \xi.$$

Remark 11. For simultaneous reactions an extent can be defined for each reaction and

$$\Delta n^s = \zeta^r \alpha_r^s, \qquad \Delta c^s = \xi^r \alpha_r^s.$$

THEOREM 5. *The composition changes of a system under the set of reactions* $\{\mathscr{R}_r\}$, $r = 1, 2, \ldots,$ R, *can be fully described by the subset of independent reactions.*

Proof. Let $\{\mathscr{R}_{r'}\}$, $r' = 1, 2, \ldots,$ R′ be the subset of independent reactions and let r'' have the range R′ + 1, ..., R. Then

$$\alpha_{r''}^s = \nu_{r''}^{r'} \alpha_{r'}^s \tag{10}$$

and

$$\begin{aligned} \Delta c^s &= \xi^r \alpha_r^s = \xi^{r'} \alpha_{r'}^s + \xi^{r''} \alpha_{r''}^s \\ &= (\xi^{r'} + \xi^{r''} \nu_{r''}^{r'}) \alpha_{r'}^s \\ &= \bar{\xi}^{r'} \alpha_{r'}^s, \end{aligned}$$

where

$$\overline{\xi}^{r'} = \xi^{r'} + \xi^{r''}\nu^{r'}_{r''} \tag{11}$$

is the modification of the extent of $\mathscr{R}_{r'}$ required to accommodate the dependent reactions.

DEFINITION 5. Two sets of independent reactions $\{\mathscr{R}_r\}$ and $\{\overline{\mathscr{R}}_r\}$, $r, \bar{r} = 1, 2, \ldots, \mathrm{R}$ will be said to be equivalent if there is a non-singular transformation with matrix $\boldsymbol{\gamma} = (\gamma^r_{\bar{r}})$ such that

$$\overline{\mathscr{R}}_{\bar{r}} = \gamma^r_{\bar{r}}\mathscr{R}_r.$$

THEOREM 6. *The composition changes of a system under the set of independent reactions* $\{\mathscr{R}_r\}$, $r = 1, 2, \ldots, \mathrm{R}$, *can be described equally well by any equivalent set.*

Proof. If $\overline{\mathscr{R}}_{\bar{r}} = \gamma^r_{\bar{r}}\mathscr{R}_r$, then $\overline{\alpha}^s_{\bar{r}} = \gamma^r_{\bar{r}}\alpha^s_r$. But

$$\Delta c^s = \xi^r\alpha^s_r = \overline{\xi}^{\bar{r}}\alpha^s_{\bar{r}} = \overline{\xi}^{\bar{r}}\gamma^r_{\bar{r}}\alpha^s_r,$$

so that

$$\xi^r = \overline{\xi}^{\bar{r}}\gamma^r_{\bar{r}},$$

i.e.

$$\boldsymbol{\xi} = \overline{\boldsymbol{\xi}}\boldsymbol{\gamma} \quad \text{or} \quad \overline{\boldsymbol{\xi}} = \boldsymbol{\xi}\boldsymbol{\gamma}^{-1}. \tag{12}$$

This gives an equivalent set of extents. ■

COROLLARY. *The orthonormalized set of reactions* (Rem. 8) *or Hooyman's normal form* (Rem. 9) *may be used when convenient. This follows from the fact that they are non-singular transformations of* $\{\mathscr{R}_r\}$.

THEOREM 7. *The total mass of a system is conserved under a set of proper reactions.*

Proof. If the reactions $\{\mathscr{R}_r\}$ are proper $\alpha^s_r\beta^t_s = 0$. But the total mass $M = \sum_{s=1}^{S} M_s = n^s m_s$. Hence

$$\Delta M = \Delta n^s m_s = \zeta^r\alpha^s_r\beta^t_s a_t = 0. \quad ■ \tag{13}$$

THEOREM 8. *If* R *independent reactions* $\{\mathscr{R}_r\}$ *take place between* S *species* $\{\mathscr{A}_s\}$, *there are* (S − R) *independent linear combinations of the concentrations* c^s *that are invariant under reaction.*

Proof. Let the reactions be written in orthonormal form $\alpha^s_r\mathscr{A}_s = 0$, $r = 1, 2, \ldots, \mathrm{R}$. If ϱ is an index with range $\mathrm{R} + 1, \ldots, \mathrm{S}$, we can find (S − R) independent row vectors α^s_ϱ which are themselves orthonormal and orthogonal to each of the reaction vectors α_r^s, *i.e.* $\alpha_r^s(\alpha_\varrho^s)^* = 0$. Then the S × S matrix

$$A = \begin{bmatrix} \alpha_r^s \\ \cdots \\ \alpha_\varrho^s \end{bmatrix}$$

is an orthogonal matrix, $A^*A = AA^* = I$.
Let

$$\eta^\varrho = c^s(\alpha_\varrho^s)^*. \tag{14}$$

Then

$$\Delta\eta^\varrho = \Delta c^s(\alpha_\varrho^s)^* = \xi^r\alpha_r^s(\alpha_\varrho^s)^* = 0$$

since $r \neq \varrho$. This constructs the invariants of the reaction system. ■

THEOREM 9. *The composition may be expressed in terms of extents and invariants.*

Proof. Let

$$c^s = \xi^r\alpha_r^s + \eta^\varrho\alpha_\varrho^s. \tag{15}$$

This will represent the change in composition and will be satisfactory provided the initial composition can be so expressed. Let $x = (x^1, \ldots, x^s) = (\xi^1, \ldots, \xi^R, \eta^{R+1}, \ldots, \eta^S)$,

$$c = xA, \qquad x = cA^*. \tag{16}$$

If c_0^s, $s = 1, \ldots, S$, are the initial concentrations then the initial extents and invariants are

$$x_0 = c_0A^*. \tag{17}$$

COROLLARY. *The composition space is the direct sum of a reaction subspace and an invariant subspace.*

In the S dimensional space of concentrations the state of the system can be represented by the point $c = (c^1, \ldots, c^S)$. Evidently this point can only move in an R dimensional subspace if the composition changes are due to only R independent reactions. The orientation of this subspace is given by the matrix of stoicheiometric coefficients, but its position is fixed by the initial composition. Indeed the stoicheiometry of the possible reactions can be determined from the atomic matrix by the following theorem: this does not of course imply that all, or even any, of the possible reactions do take place.

THEOREM 10. *The orthonormalized matrix $\boldsymbol{\alpha}$ may be determined from the matrix $\boldsymbol{\beta}$ to within an arbitrary orthogonal transformation that leaves the reaction subspace invariant.*

Proof. Let $\boldsymbol{\beta} = (\beta_s^t)$ be of rank T′ and $\boldsymbol{\beta}' = (\beta_s^{t'})$ be an S × T′ matrix of rank T′ formed from $\boldsymbol{\beta}$ by the deletion of (T − T′) dependent columns. Without loss of generality these can be taken to be the last T − T′ columns. By Theorem 4 we know that there are R = (S − T′) independent reactions $\alpha_r^s\mathcal{A}_s = 0$ and the conditions for these to be proper are

$$\alpha_r^s \beta_s^{t'} = 0.$$

Let these be written

$$\beta_1^1 \alpha_r^1 + \cdots + \beta_{T'}^1 \alpha_r^{T'} = -(\beta_{T'+1}^1 \alpha_r^{T'+1} + \cdots + \beta_S^1 \alpha_r^S),$$

$$\cdots\cdots\cdots\cdots\cdots\cdots\cdots\cdots\cdots$$

$$\beta_1^{T'} \alpha_r^1 + \cdots + \beta_{T'}^{T'} \alpha_r^{T'} = -(\beta_{T'+1}^{T'} \alpha_r^{T'+1} + \cdots + \beta_S^{T'} \alpha_r^S)$$

The $T' \times T'$ matrix of coefficients β on the left is non-singular and there are $(S - T') = R$ independent ways of assigning arbitrary values to the α's on the right hand side. The simplest way would be to set $\alpha_r^{T'+t} = \delta_r^t$, which would give the Hooyman form (Rem. 9). Alternatively the resulting row vectors may be orthonormalized by the Gramm-Schmidt process and this will give the required stoicheiometric matrix $\boldsymbol{\alpha}$. Clearly this can only be defined to within an arbitrary rotation that does not confound the subspaces of reaction and invariance. ■

4. EQUILIBRIUM IN SYSTEMS OF REACTIONS

Since this is not the place to go into the niceties of the thermostatics of nonideal substances we will consider only the case of ideal behaviour.* The proper reaction $\alpha^s \mathscr{A}_s = 0$ will come to equilibrium when

$$\sum_1^S \alpha^s \ln c^s = \varkappa \tag{18}$$

where $\varkappa$ is a constant depending only on the thermodynamic variables, *e.g.* temperature and pressure. Similarly a set of reactions $\{\mathscr{R}_r\}$ is at equilibrium when

$$\sum_1^S \alpha_r^s \ln c^s = \varkappa_r, \qquad r = 1, 2, \ldots, R. \tag{19}$$

By the principle of microscopic reversibility all the reactions of the set $\{\mathscr{R}_r\}$ must be severally at equilibrium if the set as a whole is at equilibrium. The constants $\varkappa$ are the logarithms of the conventional equilibrium constants.

THEOREM 11. *If* $\{\mathscr{R}_r\}$ *and* $\{\overline{\mathscr{R}}_{\bar{r}}\}$ *are two equivalent sets of reactions, related by* $\overline{\mathscr{R}}_{\bar{r}} = \gamma_{\bar{r}}^r \mathscr{R}_r$, *their sets of equilibrium constants are related by* $\bar{\varkappa}_{\bar{r}} = \gamma_{\bar{r}}^r \varkappa_r$.

Proof. The equations $\Sigma \alpha_r^s \ln c^s = \varkappa_r$ and $\Sigma\, \bar{\alpha}_{\bar{r}}^s \ln c^s = \bar{\varkappa}_{\bar{r}}$ must both hold for the equilibrium composition since they are equivalent descriptions of the equilibrium state. Multiplying the first set of equations by the matrix γ gives immediately

$$\bar{\varkappa}_{\bar{r}} = \gamma_{\bar{r}}^r \varkappa_r. \tag{20}$$

* A more complete model of equilibrium has been studied by J. H. HANCOCK & T. S. MOTZKIN [25], who prove some valuable existence and uniqueness theorems.

Since each concentration c^s may be related to its initial value c_0^s by $c^s = c_0^s + \xi^r \alpha_r^s$, the problem of determining the equilibrium composition is that of solving the R simultaneous non-linear equations*

$$\sum_1^S \alpha_r^s \ln(c_0^s + \xi^r \alpha_r^s) = \varkappa_r. \tag{21}$$

The uniqueness of the solution will be assured if the Jacobian, $J = |\partial\varkappa_r/\partial\xi^{r'}|$, does not vanish and this will be shown in the following theorem.

THEOREM 12. *The Jacobian* $J = \partial(\varkappa_1, \dots, \varkappa_R)/\partial(\xi^1, \dots, \xi^R)$ *of a set of independent reactions does not vanish.*

Proof. The Jacobian in question has as its i, j^{th} term

$$\sum_{s=1}^S \frac{\alpha_i^s \alpha_j^s}{c^s}.$$

Let $\boldsymbol{\lambda}_{(i)}^*$ be a row vector with elements $\alpha_i^s/(c^s)^{\frac{1}{2}}$; then the typical term is scalar product $\boldsymbol{\lambda}_{(i)}^* \cdot \boldsymbol{\lambda}_{(j)}$, the second factor being the transpose of $\boldsymbol{\lambda}_{(j)}^*$ *i.e.* a column vector. Thus the Jacobian

$$J = \det[(\boldsymbol{\lambda}_{(i)}^* \cdot \boldsymbol{\lambda}_{(j)})]$$

is a Gramian and, if we can show that the vectors $\boldsymbol{\lambda}_{(i)}$ are independent, it must be positive (see *e.g.* [8]). But the matrix $\boldsymbol{\Lambda}$ with rows $\boldsymbol{\lambda}_{(i)}^*$ is related to the matrix $\boldsymbol{\alpha}$ by $\boldsymbol{\Lambda} = \boldsymbol{\alpha} \cdot \boldsymbol{c}$ where $\boldsymbol{c} = \operatorname{diag}((c^1)^{-\frac{1}{2}}, \dots, (c^R)^{-1\frac{1}{2}})$. Now none of the c^R is zero, since otherwise the $\varkappa_r$ would not be finite, hence $\boldsymbol{c}$ is non-singular and the rank of $\boldsymbol{\Lambda}$ is the same as that of $[\boldsymbol{\alpha}$, namely R. Hence the $\boldsymbol{\lambda}_{(i)}$ are independent vectors and the Jacobian is always positive.

A minor result, which scarcely bears the dignity of a theorem, is the relation between the Jacobian of two equivalent systems. Since the Jacobian matrix J has typical term $\partial\varkappa_r/\partial\xi^{r'}$, the typical term of $\bar{\boldsymbol{J}}$ will be $\partial\bar{\varkappa}_{\bar{r}}/\partial\bar{\xi}^{\bar{r}'}$.

* I owe to Mr. H. P. HUTCHISON the observation that the equilibrium relations may be formulated somewhat differently giving S − R equations to be solved. If free atoms are hypothesised to be present and hence in equilibrium with the molecular species, we have

$$\ln c^s - \sum_{t=1}^{T'} \beta_s^t \ln \gamma^t = \kappa_s'$$

where γ^t is the concentration of the atomic species $\mathscr{B}_t$. The reaction equilibrium constants χ_r are related to the molecular equilibrium constants $\varkappa_s'$ by

$$\varkappa_r = \alpha_r^s \varkappa_s'.$$

If these relations are used to calculate the S concentrations in terms of the T′ atomic concentrations γ^t, the equilibrium relationships will be automatically satisfied but the mass balances not necessarily so. Thus, substituting in the S − R = T′ invariants

$$\eta^\varrho = (\alpha_\varrho^s)^* c^s = \sum_{s=1}^S (\alpha_\varrho^s)^* \exp \varkappa_s' \prod_{t=1}^{T'} (\gamma^t)^{\beta_s^t},$$

there are now only S − R equations to solve for the γ^t and this may on occasion be easier than to solve for the R extents.

Combining the results of Theorems 6 and 11 we have

$$\frac{\partial \bar{\varkappa}_{\bar{r}}}{\partial \bar{\xi}^{\bar{r}'}} = \gamma_{\bar{r}}^{r} \frac{\partial \varkappa_{r}}{\partial \xi^{r'}} \gamma_{\bar{r}'}^{r'}.$$

Hence $\bar{\boldsymbol{J}} = \boldsymbol{\gamma J \gamma}$ and for the determinants $\bar{J} = \gamma^2 J$. ■

5. KINETICS OF REACTIONS

DEFINITION 6. By the kinetics of a set of reactions $\{\mathscr{R}_r\}$, $r = 1, 2, \ldots$, R, over the species $\{\mathscr{A}_s\}$, $s = 1, 2, \ldots$, S, we mean the set of first order differential equations*

$$\dot{\xi}^r = f^r(\boldsymbol{c}, Z) \tag{22}$$

or the set

$$\dot{c}^s = g^s(\boldsymbol{c}, Z), \tag{23}$$

where Z denotes the thermodynamic variables and $\boldsymbol{c}$ is a set of concentrations. If $\boldsymbol{c}$ is contained in the set of concentrations $c^1, \ldots, c^s$ of the species $\{\mathscr{A}_s\}$ so that $c^s = c_0^s + \xi^r \alpha_r^s$ and hence

$$\dot{\xi}^r = f^r(\boldsymbol{\xi}, Z; \boldsymbol{c}_0) \tag{24}$$

the kinetics of the set of reactions are said to be entire. If $g^s \geqq 0$ whenever $c^s = 0$, $s = 1, 2, \ldots$, S and there exists a function $V(\boldsymbol{c})$ such that $V\dot{V} \leqq 0$ then the kinetics are called *proper*. If $g^s > 0$ whenever $c^s = 0$ the kinetics are reversible.

Remark 12. The introduction of the Liapunov function V appears somewhat arbitrary but accords with the axioms proposed by WEI [*9*]; see the discussion at the end of the paper.

Remark 13. The concepts of entirety and independence are quite distinct. We shall assume henceforth that all reactions are proper and will not add this qualification each time. If variable concentrations of species other than the $\mathscr{A}_s$ have to be introduced, or if the f^r, g^s depend explicitly on time, the kinetics are not entire.

Remark 14. Reaction kinetics are commonly classified by their order, the following cases being typical:

(i) If any f^r is linear in the c^s, the reaction R_r is said to be a first order reaction.

(ii) If any f^r is independent of the c^s, the reaction R_r is said to be of the zeroth order.

(iii) If f^r is a polynomial with terms such as $k_{ss'\ldots}^{r} c^s c^{s'} \ldots$, the reaction is said to be of polynomial order. If n is the highest degree of such a polynomial term the reactions are some times said to be of the nth order.

* A dot will be used to denote the derivative with respect to time throughout.

(iv) Many of the properties of reactions of polynomial order are retained if

$$f^r(\boldsymbol{c}, Z) = \chi(\boldsymbol{c}, Z)\varphi^r(\boldsymbol{c}, Z) \tag{25}$$

where φ^r is of polynomial order and χ is independent of r.

(v) If $\beta^s = \frac{1}{2}(|\alpha^s| - \alpha^s)$, $\gamma^s = \frac{1}{2}(|\alpha^s| + \alpha^s)$ and

$$f^r(\boldsymbol{c}, Z) = k(Z)\prod_{s=1}^{\mathrm{S}} (c^s)^{\beta^s} - k'(Z)\prod_{s=1}^{\mathrm{S}} (c^s)^{\gamma^s} \tag{26}$$

the reaction is said to be simple.

Remark 15. It is worth emphasizing that the kinetics is the *set* of functions $\{f^r\}$ and belongs as a set to the set of reactions $\{\mathscr{R}_r\}$. It is not generally possible to associate a particular f^r with a particular $\mathscr{R}_r$ independently of the other reactions, unless $\mathscr{R}_r$ represents a single molecular event. If it is necessary to be precise f^r can be described as the rate of $\mathscr{R}_r$ in the set of reactions $\{\mathscr{R}_r\}$.

Remark 16. There is an ineluctable complication in the formulation of kinetic equations. The expressions $f^r(\boldsymbol{c}, Z)$ are often relatively simple functions of the $\boldsymbol{c}$ involving only two or three of the S concentrations c^s. However when $c^s = c_0^s + \xi^r\alpha_r^s$ is substituted to give $\dot{\xi}^r = f^r(\boldsymbol{\xi}, Z)$ each equation may involve all the $\xi^1, \ldots, \xi^{\mathrm{R}}$. On the other hand by differentiating $c^s = c_0^s + \xi^r\alpha_r^s$ we see that

$$\dot{c}^s = f^r(\boldsymbol{c}, Z)\alpha_r^s = g^s(\boldsymbol{c}, Z). \tag{27}$$

But now the right hand sides of these equations may be complicated because they contain several f^r.

We wish to show next that the stoicheiometry may be deduced from a given set of kinetics to within the usual arbitrary rotation. Whilst the deduction of the stoicheiometry from the atomic matrix can only give the maximum number of reactions, the deduction from a given kinetics may show that not all the possible reactions are taking place.

THEOREM 13. *If the kinetics of any of the types listed in Remark 13 are given in the form*

$$\dot{c}^s = g^s(\boldsymbol{c}, Z),$$

the reaction and invariant subspace may be determined each to within an arbitrary rotation.

Proof. Let

$$g^s = k_0^s + k_{1s_1}^s c^{s_1} + k_{2s_1s_2}^s c^{s_1}c^{s_2} + \cdots k_{ps_1\cdots s_p}^s c^{s_1}\ldots c^{s_p},$$

where p is the highest degree of any term and summation over the $s_1, \ldots, s_p$ is implied. The k's are functions of the thermodynamic variables. If there are R independent reactions between the S species, there are (S − R) invariants $\eta^\varrho = c^s A_s^\varrho$, and hence an S × (S − R) matrix (A_s^ϱ) such that $g^s A_s^\varrho = \dot{\eta}^\varrho = 0$. If the coefficients of the equation are arranged as rows of a matrix of s columns

$$\begin{bmatrix} k_0^1 & k_0^2 & \dots & k_0^{\mathrm{S}} \\ k_{11}^1 & k_{11}^2 & \dots & k_{11}^s \\ \cdot\cdot & \cdot\cdot & \dots & \cdot \\ k_{1\mathrm{S}}^1 & k_{1\mathrm{S}}^2 & \dots & k_{1\mathrm{S}}^{\mathrm{S}} \\ k_{211}^1 & k_{211}^2 & \dots & k_{211}^{\mathrm{S}} \\ \cdot\cdot & \cdot\cdot & \dots & \cdot \end{bmatrix}$$

the rank of the matrix may be determined by conventional methods. This involves reducing the matrix to a form in which the last (S − R) columns are identically zero and the leading R × R submatrix is a triangular one with a unit diagonal. The postmultiplying S × S matrix which performs the reduction can be determined at the same time and its last (S − R) columns constitute the matrix A_s^ϱ. Let these columns be orthonormalized and the matrix completed to an S × S orthogonal matrix. This is the transpose of the matrix $\boldsymbol{A}$ of Theorem 8, which defines the stoicheiometric matrix. Moreover from Eq. (16) $\dot{\boldsymbol{x}} = \dot{\boldsymbol{c}}\boldsymbol{A}^*$ or $f^r = g^s(\alpha_r^s)^*$, $0 = g^s(\alpha_\varrho^S)^*$. Thus the associated set of reaction rates f^r is determined.

THEOREM 14. *If $\bar{c}^{\bar{s}} = c^s\gamma_s^{\bar{s}}$ is any linear transformation of concentration space, the kinetics $\dot{\bar{c}}^{\bar{s}} = \bar{g}^{\bar{s}}$ describe the same set of reaction rates f^r as do the kinetics $\dot{c}^s = g^s = f^r\alpha_r^s$, provided the stoicheiometry is given by $\bar{\alpha}_r^{\bar{s}} = \alpha_r^s\gamma_s^{\bar{s}}$.*

Proof. If $\bar{c}^{\bar{s}} = c^s\gamma_s^{\bar{s}}$, then

$$\dot{\bar{c}}^{\bar{s}} = \bar{g}^{\bar{s}} = \dot{c}^s\gamma_s^{\bar{s}} = f^r\alpha_r^s\gamma_s^{\bar{s}} = f^r\bar{\alpha}_r^{\bar{s}}.$$

THEOREM 15. *If $\{\mathscr{R}_r\}$ and $\{\overline{\mathscr{R}}_{\bar{r}}\} = \{\gamma_{\bar{r}}^r\mathscr{R}_r\}$ are two equivalent sets of reactions both capable of describing the kinetics $\dot{c}^s = g^s(\boldsymbol{c}, Z)$, their sets of reaction rates f^r and $\bar{f}^{\bar{r}}$ are related by*

$$f^r = \bar{f}^{\bar{r}}\gamma_{\bar{r}}^r, \qquad \bar{f}^{\bar{r}} = f^r\Gamma_r^{\bar{r}}$$

where $\gamma_{\bar{r}}^r\Gamma_{r'}^{\bar{r}} = \delta_{r'}^r$.

Proof. Since both sets of reactions are capable of describing the kinetics we have

$$\dot{c}^s = g^s = f^r\alpha_r^s = \bar{f}^{\bar{r}}\bar{\alpha}_{\bar{r}}^s.$$

But $\bar{\alpha}_{\bar{r}}^s = \gamma_{\bar{r}}^r\alpha_r^s$, so $(f^r - \bar{f}^{\bar{r}}\gamma_{\bar{r}}^r)\alpha_r^s = 0$. Since the reactions $\mathscr{R}_r$ are independent $f^r = \bar{f}^{\bar{r}}\gamma_{\bar{r}}^r$. Also, since the systems of reactions are equivalent, the matrix $\boldsymbol{\gamma}$ possesses an inverse $\boldsymbol{\Gamma}$ such that $\mathscr{R}_r = \Gamma_r^{\bar{r}}\overline{\mathscr{R}}_{\bar{r}}$ and $\bar{f}^{\bar{r}} = f^r\Gamma_r^{\bar{r}}$.

6. REACTION MECHANISMS

A simple illustration may serve to introduce this important aspect of chemical kinetics. The reaction $\mathscr{A}_2 \rightarrow \mathscr{A}_1$ may actually take place in two steps when a catalyst $\mathscr{A}_3$ is present. Thus $\mathscr{A}_2 + \mathscr{A}_3 \rightleftharpoons \mathscr{A}_4$, $\mathscr{A}_4 \to \mathscr{A}_1 + \mathscr{A}_3$. Written in our standard notation the overall reaction is

$$\mathscr{R} \equiv \mathscr{A}_1 - \mathscr{A}_2$$

and the two steps

$$\begin{aligned} \overline{\mathscr{R}}_1 &\equiv \qquad - \mathscr{A}_2 - \mathscr{A}_3 + \mathscr{A}_4, \\ \overline{\mathscr{R}}_2 &\equiv \mathscr{A}_1 \qquad + \mathscr{A}_3 - \mathscr{A}_4. \end{aligned}$$

Clearly $\mathscr{R} = \overline{\mathscr{R}}_1 + \overline{\mathscr{R}}_2$. The set of reactions $\{\overline{\mathscr{R}}_{\bar{r}}\} = \{\overline{\mathscr{R}}_1, \overline{\mathscr{R}}_2\}$ is said to be a mechanism for the reaction $\mathscr{R}$ which is a linear combination of them. All reference to $\mathscr{A}_3$ and $\mathscr{A}_4$, the catalyst and its complex, has of course disappeared in making this combination and we are interested in seeing whether we can obtain consistent kinetics which have no reference to c^3 and c^4.

For the kinetics we have

$$\begin{aligned} &\mathscr{R}\colon\ g^1 = f, \quad g^2 = -f, \\ &\{\overline{\mathscr{R}}\}\colon\ \bar{g}^1 = \bar{f}^2, \quad \bar{g}^2 = -\bar{f}^1, \quad \bar{g}^3 = -\bar{f}^1 + \bar{f}^2, \quad \bar{g}^4 = \bar{f}^1 - \bar{f}^2. \end{aligned}$$

If the kinetics of the mechanism is to be the same as that of the reaction $g^1 = \bar{g}^1$, $g^2 = \bar{g}^2$, *i.e.* $\bar{f}^1 = \bar{f}^2 = f$. But this implies that $\bar{g}^3 = \bar{g}^4 = 0$ or c^3 and c^4 are constant.

For definiteness let the two reactions be simple but the second irreversible, so that

$$\begin{aligned} \bar{f}^1 &= k_1 c^2 c^3 - k_2 c^4, \\ \bar{f}^2 &= k_3 c^4. \end{aligned}$$

Then if $\bar{f}^1 = \bar{f}^2$ we have

$$k_1 c^2 c^3 = (k_2 + k_3) c^4.$$

But $\eta^3 = c^3 + c^4$ is an invariant of the system $\{\overline{\mathscr{R}}\}$ so that we may write this

$$c^4(k_2 + k_3 + k_1 c^2) = k_1 \eta^3 c^2.$$

Then

$$f = \bar{f}^1 = \bar{f}^2 = \frac{k_1 k_3 \eta^3 c^2}{k_2 + k_3 + k_1 c^2}, \tag{28}$$

which is an entire kinetic expression for the overall reaction $\mathscr{A}_1 - \mathscr{A}_2 = 0$.

However if c^3 and c^4 are constant then $c^2 = (k_2 + k_3)c^4/k_1 c^3$ must be constant, and no reaction takes place. There is therefore a basic inconsistency in the attempt to make the mechanism $\{\overline{\mathscr{R}}\}$ account strictly for the reaction $\mathscr{R}$. In spite of this, such kinetic equations as (28) have been found to be extremely useful and quite accurate in kinetic studies. The chemical kineticist therefore claims that over an important part of the course of reaction c^3 and c^4 are approximately constant, or often that they are both small and slowly varying. This is called a pseudo-steady-state hypothesis and however "pseudo" it must appear to the mathematician it is sufficiently important to merit formalization. We shall therefore propound a formal definition and illustrate further how it may be used.

DEFINITION 7. The set of independent reactions $\{\overline{\mathscr{R}}_{\bar{r}}\}$ over the species $\{\overline{\mathscr{A}}_{\bar{s}}\}$ is called a mechanism for the reaction $\mathscr{R}$ over the species $\{\mathscr{A}_s\}$, if

(i) $\mathscr{A}_s = \{\overline{\mathscr{A}}_s\}$, $s = 1, 2, \ldots, \mathrm{s} < \bar{\mathrm{s}}$,
(ii) $\mathscr{R}_r = \mu^{\bar{r}}\overline{\mathscr{R}}_{\bar{r}}$ i.e. $\alpha^s = \mu^{\bar{r}}\bar{\alpha}^{\bar{s}}_{\bar{r}}$, $s = 1, 2, \ldots, \mathrm{s}$, $0 = \mu^{\bar{r}}\bar{\alpha}^{\sigma}_{\bar{r}}$, $\alpha = \mathrm{s} + 1, \ldots, \bar{\mathrm{s}}$,
(iii) none of the μ^r is zero.

The vector $\boldsymbol{\mu}$ is called the vector of stoicheiometric numbers (to be distinguished from the stoicheiometric coefficients $\boldsymbol{\alpha}$) associated with the mechanism [*10*].

Remark 17. In this definition (i) is a convention that involves no loss of generality and (iii) is a condition that implies that all the reactions in the mechanism are needed. The species $\mathscr{A}_\sigma$ have been called Bodenstein-products by FRANK-KAMENETSKY [*11*].

This stoicheiometric definition of mechanism is quite precise, but in formulating the kinetic implications we shall avoid the position that this can be tied down in precise terms. To do this would be to stultify the imagination of the practitioner and make absolute a procedure whose validity can only be established by experiment. Nevertheless we shall show that under the above conditions it is possible to construct an entire kinetic expression for $\mathscr{R}$, by the introduction of certain hypotheses. Such an expression is not necessarily unique and only experiment can establish one form of kinetics as a sufficiently accurate description of the course of the reaction.

Without loss of generality we may suppose that $\mathscr{R}$ and $\{\overline{\mathscr{R}}_{\bar{r}}\}$ are normalized and orthonormalized respectively. Then if the mechanism $\{\overline{\mathscr{R}}_{\bar{r}}\}$ is to describe the reaction $\mathscr{R}$ the kinetics $\bar{f}^{\bar{r}}$ and f should give the same composition changes. Thus

$$\dot{c}^s = f\alpha^s = \bar{f}^{\bar{r}}\bar{\alpha}^s_{\bar{r}}, \qquad s = 1, 2, \ldots, \mathrm{s}, \tag{29}$$

$$\dot{c}^\sigma = \bar{f}^{\bar{r}}\bar{\alpha}^\sigma_{\bar{r}}, \qquad \sigma = \mathrm{s} + 1, \ldots, \bar{\mathrm{s}}. \tag{30}$$

The hypothetical method now claims that we may ignore these last $(\bar{\mathrm{s}} - \mathrm{s})$ relations, (30), and disregard any contradictions they may imply. However the relations (29) imply certain restrictions between the $\bar{f}^{\bar{r}}$. Let the vector $\boldsymbol{\alpha} = (\alpha^1, \ldots, \alpha^s)$ be completed to an $\mathrm{s} \times \mathrm{s}$ orthogonal matrix by the addition of $(\mathrm{s} - 1)$ rows. Then postmultiplying by $\boldsymbol{\alpha}^*$ (whose elements we write as $(\alpha^s_r)^* = (\alpha^s)^*$, $r = 1$ and $(\alpha^s_\varrho)^*$, $\varrho = 2, 3, \ldots, \mathrm{s}$) we have

$$f = \bar{f}^{\bar{r}}\bar{\alpha}^s_{\bar{r}}(\alpha^s_r)^*, \qquad r = 1, \tag{31}$$

$$0 = \bar{f}^{\bar{r}}\bar{\alpha}^s_{\bar{r}}(\bar{\alpha}^s_\varrho)^*, \qquad \varrho = 2, 3, \ldots, \mathrm{s}. \tag{32}$$

But since $\alpha^s_r = \mu^{\bar{r}}\bar{\alpha}^s_{\bar{r}}$, $0 = \mu^r\bar{\alpha}^\sigma_{\bar{r}}$ we have for $r = 1$,

$$\begin{aligned}(\bar{\alpha}^s_{\bar{r}})(\alpha^s_r)^* &= (\bar{\alpha}^s_{\bar{r}} : \bar{\alpha}^\sigma_{\bar{r}})(\alpha^s_r : 0)^* \\ &= (\bar{\alpha}^s_{\bar{r}} : \bar{\alpha}^\sigma_{\bar{r}})(\gamma^{\bar{r}}\bar{\alpha}^s_{\bar{r}} : 0)^* = (\gamma^{\bar{r}})^*\end{aligned}$$

by the orthonormality of $\bar{\boldsymbol{\alpha}}$. Thus equation (31) is

$$f = \bar{f}^{\bar{r}}(\mu^{\bar{r}})^*. \tag{33}$$

Eqns. (32) give (s − 1) conditions that must be satisfied and though they are generally non-linear in concentration they can in principle be solved to give as many as (s − 1) of the concentrations c^σ, $\sigma = \mathrm{s} + 1, \ldots, \bar{\mathrm{s}}$, in terms of the others. In addition there are the ($\bar{\mathrm{s}} - \bar{\mathrm{R}}$) invariants of the mechanism which again may yield as many as ($\bar{\mathrm{s}} - \bar{\mathrm{R}}$) of the c^σ in terms of the c^s, but it is impossible to say in general how many of such relations are independent. Between them the invariants and Eqns. (31) might determine as many as ($\bar{\mathrm{s}} - \bar{\mathrm{R}} + \mathrm{s} - 1$) of the ($\bar{\mathrm{s}} - \mathrm{s}$) c^σ's and if the actual number so determined is not sufficient existential hypotheses must be introduced to provide further equations. The kinetics of the reaction $\mathscr{R}$ can then be made entire by substitution for the c^σ in Eqn. (33) which then becomes a function of the c^s.

Although each case must be examined on its own merits three principal situations do arise from this vagueness. In the first place, Eqns. (32) may have physical significance and be eligible as hypotheses, as in the introductory example above. In the second, they may be insufficient and further hypotheses must be introduced. These are usually of the kind that certain of the c^σ are constant and are known generally as pseudo-steady-state hypotheses. In the third place it is sometimes assumed that one reaction is very much slower than the others and so its rate controls the rate of the whole process. The remaining reactions are "at equilibrium" in the sense that the relationship between their concentrations is very close to the equilibrium relationship. They cannot be at equilibrium since then the whole reaction would stop, but their finite rate is really the product of a very large rate constant and a very small driving force, namely, the departure from equilibrium. The indefiniteness of this product of "infinity" and "zero" allows the relations (32) to be satisfied. An example will clarify this third case.

The synthesis of ammonia, $N_2 + 3H_2 = 2NH_3$, can proceed in the following sequence of reactions, $N_2 = 2N^*$, $H_2 = 2H^*$, $N^* + H^* = NY^*$, $NY^* + H^* = NH_2^*$, $NH_2^* + H^* = NH_3$, where the asterisk denotes the adsorbed state [*10*]. Translating this into our notation:

$$\{\mathscr{A}_s\} = \{NH_3, N_2, H_2\}, \qquad \{\overline{\mathscr{A}}_s\} = \{NH_3, N_2, H_2, N^*, H^*, NH^*, NH_2^*\}$$

$$\mathscr{R} = \alpha^s \mathscr{A}_s : (\alpha^s) = [2, -1, -3];$$

$$\{\overline{\mathscr{R}}_{\bar{r}}\} = \overline{\alpha}_{\bar{r}}^{\bar{s}}\, \overline{\mathscr{A}}_{\bar{s}} : \overline{\alpha}_{\bar{r}}^{\bar{s}} = \begin{bmatrix} \cdot & -1 & \cdot & 2 & \cdot & \cdot & \cdot \\ \cdot & \cdot & -1 & \cdot & 2 & \cdot & \cdot \\ \cdot & \cdot & \cdot & -1 & -1 & 1 & \cdot \\ \cdot & \cdot & \cdot & \cdot & -1 & -1 & 1 \\ 1 & \cdot & \cdot & \cdot & -1 & \cdot & -1 \end{bmatrix}$$

$$(\mu^{\bar{r}}) = [1, 3, 2, 2, 2]; \qquad s = 2, \quad \bar{\mathrm{s}} = 7, \quad \bar{\mathrm{R}} = 5.$$

(These are not in orthonormal form but this is unimportant). The equations (30) and (31) give

$$\begin{aligned} f &= \tfrac{1}{14}(\bar{f}^1 + 3\bar{f}^2 + 2\bar{f}^5), \\ 0 &= -\bar{f}^1 - 3\bar{f}^2 + 5\bar{f}^5, \\ 0 &= 3\bar{f}^1 - 5\bar{f}^2 + 6\bar{f}^5, \end{aligned} \tag{34}$$

or more compactly $f = \bar{f}^1 = \frac{1}{3}\bar{f}^2 = \frac{1}{2}\bar{f}^5$. Nothing can be said about $\bar{f}^3$ and $\bar{f}^4$ and some hypothesis needs to be made. The hypothesis chosen (and it is not the only one) is that the rate of $\overline{\mathscr{R}}_2$ is limiting. For $\bar{f}^{\bar{r}}$ we have the expressions

$$\begin{aligned}
\bar{f}^1 &= k^1\{K_1c^2 - (c^4)^2\},\\
\bar{f}^2 &= k^2\{K_2c^3 - (c^5)^2\},\\
\bar{f}^3 &= k^3\{K_3c^4c^5 - c^6\},\\
\bar{f}^4 &= k^4\{K_4c^5c^6 - c^7\},\\
\bar{f}^5 &= k^5\{K_5c^5c^7 - c^1\},
\end{aligned} \tag{35}$$

and the hypothesis is that all the $k^{\bar{r}}$ except k^2 tend to infinity. If the $\bar{f}^{\bar{r}}$ are to remain finite the expressions inside the brackets for $\bar{r} = 1, 3, 4, 5$ must vanish and this gives four equations for $c^4, \ldots, c^7$. They may be written ($\varkappa_i = \ln K_i$)

$$\begin{aligned}
-\ln c^2 + 2\ln c^4 \qquad\qquad\qquad\qquad\qquad &= \varkappa_1,\\
-\ln c^4 - \ln c^5 + \ln c^6 \qquad\qquad &= \varkappa_3,\\
-\ln c^5 - \ln c^6 + \ln c^7 &= \varkappa_4,\\
\ln c^1 \qquad\qquad\qquad -\ln c^5 \qquad\qquad - \ln c^7 &= \varkappa_5,
\end{aligned}$$

whence

$$2\ln c^1 - \ln c^2 - 6\ln c^5 = \varkappa_1 + 2\varkappa_3 + 2\varkappa_4 + 2\varkappa_5$$

or

$$(c^5)^2 = K'(c^1)^{2/3}(c^2)^{-1/3} \tag{36}$$

where $-3\ln K^1 = \varkappa_1 + 2\varkappa_3 + 2\varkappa_4 + 2\varkappa_5$. Substituting into the expression for $\bar{f}^2$ we have

$$f = \tfrac{1}{3}\bar{f}^2 = \tfrac{1}{3}k^3\{K_2c^3 - K'(c^1)^{2/3}(c^2)^{-1/3}\} \tag{37}$$

which is an entire kinetic expression. Its equilibrium condition is when

$$(c^1)^2(c^2)^{-1}(c^3)^{-3} = (K_2/K')^3 = K$$

i.e. when

$$\begin{aligned}
2\ln c^1 - \ln c^2 - 3\ln c^3 &= \varkappa,\\
\varkappa = \varkappa_1 + 3\varkappa_2 + 2\varkappa_3 &+ 2\varkappa_4 + 2\varkappa_5.
\end{aligned}$$

Thus the ratio of the forward and back rate constants, (K_2/K') is not the equilibrium constant for the overall reaction, but $K^{\frac{1}{3}}$, the exponent being the reciprocal of the stoicheiometric number of the limiting reaction. This significance and its exploitation have been examined by HORIUTI [*10*]. FRANK-KAMENETSKY [*11*] has developed some safe conditions for the applicability of certain forms of the hypothetical method, in the form of rather strong inequalities on the partial derivatives of the kinetic expressions.

7. ENTROPY PRODUCTION

The rate of entropy production per unit volume, σ, by the reaction $\alpha^s \mathscr{A}_s = 0$, is

$$\sigma = -T^{-1}\dot{c}^s\mu_s$$

where T is the absolute temperature and μ_s the chemical potential of $\mathscr{A}_s$. Since $\dot{c}^s = f\alpha^s$ we have

$$\sigma T = -f\alpha^s \mu_s = f\mathfrak{a},$$

where

$$\mathfrak{a} = -\alpha^s \mu_s$$

is called the affinity of the reaction. Similarly the entropy production of R simultaneous reactions is given by

$$\sigma T = -\dot{c}^s \mu_s = -f^r \alpha_r^s \mu_s = f^r \mathfrak{a}_r,$$

where $\mathfrak{a}_r$ is the affinity of the rth reaction.

It might be thought that some principle concerning entropy production, such as the minimum principle invoked in irreversible thermodynamics, might distinguish between equivalent sets of reactions, but the next theorem shows that this is not so.

THEOREM 16. *If $\{\mathscr{R}_r\}$ and $\{\overline{\mathscr{R}}_{\bar{r}}\} = \{\gamma_{\bar{r}}^r \mathscr{R}_r\}$ are two equivalent sets of reactions over the same species, their rates of entropy production are the same.*

Proof. Since $\bar{\alpha}_{\bar{r}}^s = \gamma_{\bar{r}}^r \alpha_r^s$, $\mathfrak{a}_{\bar{r}} = -\bar{\alpha}_{\bar{r}}^s \mu_s = -\gamma_{\bar{r}}^r \alpha_r^s \mu_s = \gamma_{\bar{r}}^r \mathfrak{a}_r$. By Theorem 15, $\bar{f}^{\bar{r}} = f^r \Gamma_r^{\bar{r}}$ so that

$$\bar{\sigma} = \bar{f}^{\bar{r}} \bar{\mathfrak{a}}_{\bar{r}} = f^r \Gamma_r^{\bar{r}} \gamma_{\bar{r}}^{r'} \mathfrak{a}_{r'} = f^r \mathfrak{a}_r = \sigma.$$

In a set of coupled reactions $\{\mathscr{R}_r\}$ the signs of all the terms in the sum $f^r \mathfrak{a}_r$ need not be the same although the sum, σ, will be positive. A particular reaction can be called positive or negative according as its contribution to σ is positive or negative. HOOYMAN [*5*] has defended the use of this terminology in the context of Rem. 9 above. ■

8. DISCUSSION

If the foregoing observations are indeed prolegomena, it is desirable to indicate how far they are related to what has been said elsewhere and what may be said after them.

The linear algebraic character of this representation of chemical species and their reactions has been recognised elsewhere by SCHAY & PETHÖ [*12*] and the rule for the number of independent reactions is related to the phase rule of GIBBS and has been associated with his name [*12, 13, 14*], though there does not appear to be an explicit reference to it in GIBBS' writings. PETHÖ in a later paper [*15*] has formulated a scheme for generating the "best" set of reactions, in the sense that the matrix of stoicheiometric coefficients has as many zeros as possible. The term "best" is, of course, relative to the context of its employment and as we have stressed elsewhere [*16*] it is probably more important to recognise the equivalence of reactions.

We have not attempted to lay out a formal axiom scheme, but it will be well to observe that the definitions given here are consistent with the axioms given in WEI's treatment [*9*]. Indeed, though his first axiom (the conservation of mass) is here a theorem, his others have been built into the definition of the kinetics of a set of reactions.

The concept of a mechanism for a reaction, whilst well known and much used by the chemist, is not as yet clearly formulated in a rational analysis. The precise classification of the methods adverted to in Section 6 above, and their extension to sets of reactions are steps that need to be taken. In this connection the work of HORIUTI [*10*] and CHRISTIANSEN [*13*], as well as the vast chemical literature (see for example KONDRAT'EV [*17*]) will provide much material. It is not to be expected however that all vagueness can be removed for the hypothetical method is intrinsically self-contradictory.

HORIUTI & NAKAMURA [*18*] have considered the possibility that there may be more than one set of stoicheiometric numbers leading to an overall reaction between the species $\{\mathscr{A}_s\}$. They find that the number of such "reaction routes" is the row nullity of the matrix $[\overline{\alpha}_r^g]$. NAKAMURA has also considered the affinity of the overall reaction near equilibrium [*19*] and with YAMAZAKI [*20*] has related this to the theorem of minimum entropy production.

The algebra of the kinetic differential equations has already received some attention. WEI & PRATER have given a very complete description of first order systems [21] and a start has been made on systems of the second and higher orders [*22, 23*]. An interesting approach via set theory has been put forward by BARTHOLOMAY [*24*] and deserves to be pursued.

Thus although this formalization and approach to chemical kinetics is in its infancy, there is, as with continuum mechanics, a vast subsoil of prior knowledge and, it is to be hoped, of like fertility.

NOMENCLATURE

$\mathbf{A}$	partitioned, augmented stoicheiometric matrix
a_t	atomic weight
c^s	concentration, n^s/V
f^r	kinetic expression for rate of change of ξ^r
g^s	kinetic expression for rate of change of c^s
J	Jacobian of transformation to equivalent reactions
K_i	equilibrium constants (sec. 6)
$k_{s..}^r$	rate constants
M	total mass
M_s	mass of $\mathscr{A}_s$ present
m_s	molecular weight
n^s	number of moles of $\mathscr{A}_s$ present
T	temperature
V	volume
Z	thermodynamic variables
$\boldsymbol{\alpha}$	matrix of stoicheiometric coefficients
$\boldsymbol{\beta}$	atomic matrix
γ, γ^r	arbitrary scalar multipliers
$\boldsymbol{\Gamma}, \boldsymbol{\gamma}$	arbitrary nonsingular transformation and inverse
ζ	extent in number of moles

η^{ϱ}	invariant
$\varkappa$, $\varkappa_r \bar{\varkappa}_{\bar{r}}$,	logarithms of equilibrium constants
$\mu^{\bar{r}}$	stoicheiometric numbers
ξ	extent in moles per unit volume
σ	rate of entropy production
φ^r, χ	factors of f^r in a kinetic expression
$\mathscr{A}_s$	molecular species
$\mathscr{B}_t$	atomic species
$\mathscr{R}_r$	reaction
$\mathfrak{a}_r$	affinity of reaction
$\mathfrak{A}_T$	vector space of species
$\mathfrak{B}_S$	vector space of reactions

Suffixes and their ranges

r, ϱ	denotes a reaction; ranges: r, (1, R); r', (1, R′); r'', (R′ + 1, R); $\bar{r}$, (1, $\bar{\text{R}}$); ϱ, (R + 1, S).
s, σ	denotes a molecular species; ranges: s, (1, S); s', (1, S′); s'', (S′ + 1, S); $\bar{s}$, (1, $\bar{\text{S}}$); σ, (S + 1, $\bar{\text{S}}$).
t	denotes an atomic species; ranges: t, (1, T); t', (1, T′); t'', (T′ + 1, T).
o	denotes an initial value.

Small capitals

R, $\bar{\text{R}}$	number of reactions
R′	rank of $\boldsymbol{\alpha}$
S, $\bar{\text{S}}$	number of molecular species
T	number of atomic species
T′	rank of $\boldsymbol{\beta}$.

REFERENCES

[1] Mulkhuyse, J. J., Molecules and Models. In: The concept and role of the model in mathematics and natural and social sciences, ed. H. Freudenthal. Dordrecht 1961.

[2] Spialter, L., J. Am. Chem. Soc. **85,** 2012 (1963).

[3] Groot, S. R. de, Thermodynamics of Irreversible Processes. Amsterdam: North Holland Pub. Co. 1951.

[4] Pings, C. J., Chem. Eng. Sci. **17,** 573 (1962).

[5] Hooyman, G. J., Proc. Nat. Acad. Sci. **47,** 1169 (1961).

[6] Halmos, P. R., Finite-dimensional Vector Spaces, 2nd Edn. Princeton: Van Nostrand 1958.

[7] Aris, R., Introduction to the Analysis of Chemical Reactors, Ch. 2. Englewood Cliffs: Prentice-Hall, 1965.

[8] Bechenbach, E. F., & R. Bellman, Inequalities, Ch. 2. § 6. Berlin-Göttingen-Heidelberg: Springer 1961.

[9] Wei, J., J. Chem. Phys. **36,** 1578 (1962).

[10] Horiuti, J., J. of Catalysis **1,** 199 (1962).

[11] Frank-Kamenetsky, D. A., J. Phys. Chem. (USSR) **14,** 695 (1940).

[12] Schay, G., & Á. Pethö, Acta Chim. Acad. Sci. Hungaricae **32,** 59 (1962).

[13] Christiansen, J. A., Advances in Catalysis **5,** 311 (1953).

[14] Aris, R., & R. H. S. Mah, Ind. Eng. Chem. Fundamentals **2,** 90 (1963).

[15] Pethö, Á., Wiss. Zeits. der Tech. Hochschule für Chemie Leuna-Merseburg **6,** 13 (1964).
[16] Aris, R., Chem. Engng. Sci. **18,** 554 (1963).
[17] Kondrat'ev, V. N., Chemical Kinetics of Gas Reactions. Reading: Addison Wesley Pub. Co. 1964.
[18] Horiuti, J., & T. Nakamura, Z. für Phys. Chem. **11,** 358 (1957).
[19] Nakamura, T., J. Res. Inst. for Catalysis, Hokkaido Univ. **6,** 20 (1958).
[20] Nakamura, T., & H. Yamazaki, J. Res. Inst. for Catalysis, Hokkaido Univ. **5,** 98 (1957).
[21] Wei, J., & C. D. Prater, Advances in Catalysis **13,** 203 (1962).
[22] Aris, R., Ind. Eng. Chem. Fundamentals **3,** 28 (1964).
[23] Wei, J., Ind. Eng. Chem. Fundamentals (in the press).
[24] Bartholomay, A. F., Bull. Math. Biophys. **22,** 285 (1960).
[25] Hancock, J. H., & T. S. Motzkin, Analysis of the mathematical model for chemical equilibrium. In: G. S. Bahn & E. D. Zukoski (Eds.), Kinetics, Equilibria and Performance of High Temperature Systems. London: Butterworths 1960.

Department of Chemical Engineering
University of Minnesota
Minneapolis

E.*

PROLEGOMENA TO THE RATIONAL ANALYSIS OF SYSTEMS OF CHEMICAL REACTIONS II. SOME ADDENDA

RUTHERFORD ARIS

1. INTRODUCTION

Since the publication of some prolegomena to the rational analysis of systems of chemical reaction [1] other cognate work has come to light and some earlier statements have been made more precise and comprehensive. I would like first to advert to an earlier work previously overlooked and to mention some recent publications that partially fill some of the undeveloped areas noticed before. Secondly, I wish to extend the theorem on the uniqueness of equilibrium to a more general case and to establish the conditions for the consistency of the kinetic and equilibrium expressions (§§ 2, 3). Thirdly, the conception of a reaction mechanism is to be reformulated in a more general way and the metrical connection between the kinetics of the mechanism and those of the ostensible reactions clarified. The notation of the earlier paper ([1], hereinafter referred to as P) will be followed and augmented where necessary. In particular the reader is reminded that the range of each affix is carefully specified and the summation convention of tensor analysis is employed.

As noticed in a review of P [2]†, an important early reference had been overlooked, namely the two "notes" of Jouguet [3], "Observations sur les principes et les théorèmes généraux de la statique chimique" (pp. 61–180) and "Sur les lois de la dynamique chimique relatives aux sens des réactions irréversibles" (pp. 181–194). In these papers (of which I was culpably unaware when writing P) Jouguet carefully establishes the notion of independence of reactions and the invariants of a system of reactions and considers the ramifications of the phase rule. In particular he is quite clear as to the way in which stoicheiometry lays a foundation for thermodynamics and in going on to discuss the nature of equilibrium he shows that its uniqueness is a consequence of its stability. Thus much of the content of §§ 2–4 of P is to be found in Jouguet. His second note, however, is concerned with the direction

* Reprinted with the permission of Springer-Verlag from *Archive for Rational Mechanics and Analysis,* volume **27,** number 5, pp. 356–364, 1968.

† I am grateful to the reviewer for drawing my attention to these papers and thankful that a more careful reading of them and of my paper does not substantiate the later strictures of his review.

of a reaction provoked by a charge of conditions at equilibrium and not with the kinetics of chemical change as now understood.

In a recent paper SHAPIRO & SHAPLEY [4] have considered the problem of the uniqueness of equilibrium of systems of reactions in several phases in great detail. The computation of equilibrium compositions by direct minimization of the GIBBS free energy function has proved a valuable tool in the discussion of very complex systems and it is important to show that this minimum is unique and achieved under the same conditions that satisfy the mass action laws. This is what SHAPIRO & SHAPLEY have done; SELLERS has suggested some improvements [5].

The most important advance in the algebra of chemical systems has been the work of SELLERS [6, 7] for it is capable of taking into account the structure of a chemical species. Thus JO_1K and JO_2K might represent two isomers, O_1 and O_2 being distinct structural symbols, and the isomerization reaction $JO_1K - JO_2K = 0$ would not be a trivial one, as it is in our structureless representation. The ideas involved in this elegant work centre around chain complexes of free Abelian groups and the resulting homologies: they are fully presented in [7].

2. THE UNIQUENESS OF EQUILIBRIUM UNDER ADIABATIC CONDITIONS

In § 4 of P it was shown that the equilibrium conditions for a set of R independent reactions, namely.

$$\sum_{1}^{S} \alpha_r^s \ln c^s = \kappa_r, \quad r = 1, 2, \dots, \mathrm{R}, \tag{1}$$

have a unique solution for any set of constants κ_r. These constants are fixed by fixing the temperature and pressure and in fact κ_r has the form

$$\kappa_r = K_r^* - (\Delta H)_r/RT, \tag{2}$$

where

T	is the absolute temperature,
$(\Delta H)_r$	the heat of reaction,

and

K_r^* a constant

Suppose now that the reaction takes place adiabatically at constant pressure from an initial composition c_0^s and temperature T_0. Then the concentrations will be governed by

$$c^s = c_0^s + \xi^r \alpha_r^s \tag{3}$$

and the enthalpy

$$H = \Sigma\, c^s h_s(T, c) \tag{4}$$

will be constant. But the constancy of the enthalpy implies that

$$\frac{\partial H}{\partial \xi^r} = \sum c^s \frac{\partial h_s}{\partial T}\frac{\partial T}{\partial \xi^r} + \sum \alpha_r^s h_s = 0$$

or

$$\frac{\partial T}{\partial \xi^r} = \frac{-(\Delta H)_r}{C_p} \tag{5}$$

since $C_p = \Sigma\, c^s(\partial h_s/\partial T)$ is the heat capacity, $(\Delta H)_r = \Sigma\, \alpha_r^s h_s$ and the other terms are identically zero in virtue of h being an intensive variable (*cf.* [9], p. 36).

Let

$$\varphi_r \equiv \sum_{s=1}^{S} \alpha_r^s \ln c^s - \kappa_r,$$

the Jacobian $\partial(\varphi_1, \ldots, \varphi_R)/\partial(\xi^1, \ldots, \xi^R)$ has as its i, j^{th} term

$$\sum_{s=1}^{S} \frac{\alpha_i^s \alpha_j^s}{c^s} + \frac{(\Delta H)_i(\Delta H)_j}{RT^2 C_p}. \tag{6}$$

In theorem 12 of P it was shown that the first term is that of a positive definite Gramian matrix. The second is clearly the general term of a positive semi-definite matrix; we cannot assert that it is definite since one or more of the ΔH_i might be zero. But the sum of a positive definite and a positive semi-definite matrix is definite and so the Jacobian nowhere vanishes.

It is evident from the method of proof that a range of semi-adiabatic cases in which $(\partial T/\partial \xi^r) = \theta(-\Delta H)_r/C_p$ will also yield unique equilibria; it is only necessary that θ be independent of r.

A theorem of this kind is needed to establish the existence and properties of the solution of the equations for a tubular reactor, as in the work of Amundson & Luss [8].

3. THE CONSISTENCY OF CERTAIN KINETIC AND EQUILIBRIUM EXPRESSIONS

A common expression for the rate of a homogeneous reaction is

$$f(c, Z) = k(Z) \prod_{s=1}^{S} (c^s)^{\beta^s} - k'(Z) \prod_{s=1}^{S} (c^s)^{\gamma^s}, \tag{7}$$

in which the β^s and γ^s are called the orders of the forward and backward reaction with respect to $\mathscr{A}_s$ respectively and Z denotes a pair of thermodynamic variables. The reaction rate is zero when

$$\prod_{s=1}^{S} (c^s)^{\gamma^s - \beta^s} = k(Z)/k'(Z). \tag{8}$$

This condition must define the same surface in ($\boldsymbol{c}$, Z)-space as does the equilibrium condition

$$\prod_{s=1}^{S} (c^s)^{\alpha^s} = K_c(Z). \tag{9}$$

It is clear that this implies that

$$k/k' = F(K_c), \tag{10}$$

$$\prod_{s=1}^{S} (c^s)^{\gamma^s-\beta^s} = F\left[\prod_{s=1}^{S} (c^s)^{\alpha^s}\right] \tag{11}$$

for some suitable function F. It is obvious that $F(x) = x^\nu$, $\nu > 0$, is a satisfactory function and implies that

$$\gamma^s - \beta^s = \nu\alpha^s \tag{12}$$

and

$$k/k' = K_c^\nu. \tag{13}$$

We wish to show that this is the only function for which the relation is true; the proofs given elsewhere [9, 10] are in error since they assume too large a number of degrees of freedom*.

By the phase rule of Gibbs the system has s degrees of freedom so that, if the two thermodynamic variables are chosen, only s − 2 of the remaining concentrations may be selected, and the other two are fixed by (9) and

$$\sum_{s=1}^{S} c^s = C. \tag{14}$$

Let $c^3 = c^4 = \cdots = c^s = \gamma < C/(\mathrm{s} - 2)$, then

$$c^2 = (C - (\mathrm{s} - 2)\gamma - c^1) \tag{15}$$

and

$$(c^1)^{\alpha^1}(C - (\mathrm{s} - 2)\gamma - c^1)^{\alpha^2}\gamma^{\alpha'} = K_c, \tag{16}$$

where

$$\alpha' = \sum_{s=3}^{S} \alpha^s. \tag{17}$$

Further, let

$$\nu^s = (\gamma^s - \beta^s)/\alpha^s \tag{18}$$

and

$$\nu'\alpha' = \sum_{s=3}^{S} (\gamma^s - \beta^s). \tag{19}$$

* I am indebted to Professor M. Boudart of Stanford University for pointing this out to me.

Substituting from (15) and (19) into (8)

$$(c^1)^{\eta^1\alpha^1}(C - (s - 2)\gamma - c^1)^{\nu^2\alpha^2}\gamma^{\nu'\alpha'} = (k/k'). \qquad (20)$$

Now if the same manifold is defined by both equations (8) and (9), and therefore also by (16) and (20), we should be able to solve (16) for c^1 and, on substituting it in the left side of (20), find that the resulting expression was independent of γ. This will be so if $\partial(k/k')/\partial\gamma = 0$ when $\partial c^1/\partial\gamma$ is calculated from (16). But, differentiating the two equations logarithmically, we have

$$\left(\frac{\alpha^1}{c^1} - \frac{\alpha^2}{c^2}\right)\frac{\partial c^1}{\partial\gamma} + \left(\frac{\alpha'}{\gamma} - (s - 2)\frac{\alpha^2}{c^2}\right) = 0,$$

$$\left(\frac{\nu^1\alpha^1}{c^1} - \frac{\nu^2\alpha^2}{c^2}\right)\frac{\partial c^1}{\partial\gamma} + \left(\frac{\nu'\alpha'}{\gamma} - (s - 2)\frac{\nu^2\alpha^2}{c^2}\right) = 0.$$

Eliminating $(\partial c^1/\partial\gamma)$ gives

$$(\nu^2 - \nu')(c^1/\alpha^1) + (\nu^1 - \nu')(c^2/\alpha^2) + (S - 2)(\nu^1 - \nu^2)(\gamma/\alpha') = 0 \qquad (21)$$

and this relation can only be consistent with (9) if it is satisfied identically, *i.e.* $\nu^1 = \nu^2 = \nu'$. Repetition of the argument shows that

$$\nu^s = \nu \quad s = 1, 2, \ldots, s$$

so that the relations (12) and (13) are necessarily true.

4. REACTION MECHANISMS AND EXACT SEQUENCES

One of the less satisfactory features of P was its tentative treatment of reaction mechanism. It was confined to the case of a single overall reaction and did not make clear the nature of the degeneracy in the differential equations for the concentrations.

Consider the module $\mathfrak{E}$ consisting of all lists* of the form $(\varepsilon_1 \mathscr{B}_1, \varepsilon_2 \mathscr{B}_2, \ldots, \varepsilon_T \mathscr{B}_T)$ where the ε_t are integers and $\mathscr{B}_1, \ldots, \mathscr{B}_T$ is a set of chemical elements sufficient for the universe of discourse. This may consist of atomic species, indecomposable molecules or radicals. We note in passing that this is the direct sum of the T submodules for which

$$\varepsilon_1 = \varepsilon_2 = \cdots = \varepsilon_{t-1} = \varepsilon_{t+1} = \cdots = \varepsilon_T = 0, \quad t = 1, 2, \ldots, T.$$

This will be called the *module of elements.* The molecular species (without representation of structure) are defined as linear combinations of these elements (P, Def. 1) and we shall consider mixtures of s distinguishable molecular species. The *mixture module* $\mathfrak{M}$ consists of all lists of s species, $(\mathscr{A}_1, \mathscr{A}_2, \ldots, \mathscr{A}_s)$, the only restriction being that they be formed from the fixed set of elements, $\mathscr{B}_1, \ldots, \mathscr{B}_T$. Let $\boldsymbol{\beta} = [\beta_s^t]$ be a matrix of full rank T (T $\leqq$ s), which may be called the *atomic matrix* of the mixture β. We lose no generality by

* The word 'list' has been adopted, following BIRKHOFF & MACLANE (Algebra. New York. Macmillan 1967, p. viii), in place of the ugly "word" *n*-tuple. It will usually be quite clear how many elements are involved; in case of doubt we shall refer to a list of *n* elements.

insisting on its being of full rank T since if it were of lesser rank we could express the mixture in terms of fewer than T elements. Similarly it is no restriction to have at least as many molecular species as we have elements, $S \geqq T$.

The atomic matrix β induces a mapping from the module $\mathfrak{E}$ into the module $\mathfrak{M}$ by mapping the element $(\varepsilon_1 \mathscr{B}_1, \ldots, \varepsilon_T \mathscr{B}_T)$ of $\mathfrak{E}$ into the element $(\beta_1^t \varepsilon_t \mathscr{B}_t, \ldots, \beta_s^t \varepsilon_t \mathscr{B}_t)$ of $\mathfrak{M}$ (summation on t from 1 to T). In fact the image of $\mathfrak{E}$ under the morphism β is a submodule of $\mathfrak{M}$, which, in view of the purpose it will serve, may be called the *reaction mixture module.* Now let $\boldsymbol{\alpha} = [\alpha_r^s]$ be matrix of rank R (R < S). It is called the *stoicheiometric matrix* for the reactions $\mathscr{R}_r \equiv \alpha_r^s \mathscr{A}_s$ over the mixture of S species. The set of all lists of R elements $(\mathscr{R}_1, \ldots, \mathscr{R}_R)$ is a submodule of $\mathfrak{M}$ but we shall call its elements as *sets of reactions* and the module itself as the reaction module $\mathfrak{R}$. The stoicheiometric matrix $\boldsymbol{\alpha}$ defines a morphism α of $\mathfrak{M}$ onto $\mathfrak{R}$. The stoicheiometric matrix $\boldsymbol{\alpha}$ will be called *proper* for the atomic matrix $\boldsymbol{\beta}$ if all the reactions defined by α over mixtures in the reaction mixture module defined by β are proper reactions (*i.e.* they are balanced in the normal chemical sense; *cf. P. Def. 2, Rem. 4.).* Thus the morphism $\alpha: \mathfrak{M} \to \mathfrak{R}$ will map the submodule of the reaction mixture, namely the image of $\beta: \mathfrak{E} \to \mathfrak{M}$, into the null elements $\emptyset$. In other words the kernel of α is the image of β or the pair of morphisms (β, α) is exact at $\mathfrak{M}$ if it corresponds to a real chemical situation. Put in yet other terms we can say that the modules $\mathfrak{E}$, $\mathfrak{M}$ and $\mathfrak{R}$ and the morphisms β and α correspond to a real chemical situation when

$$0 \to \mathfrak{E} \xrightarrow{\beta} \mathfrak{M} \xrightarrow{\alpha} \mathfrak{R} \to 0$$

is a short exact sequence.

This formulation discloses the algebraic structure of what the chemist calls the 'mechanism' of a reaction or set of reactions. To explain the reaction system the chemist will introduce further 'elements', such as catalysts, active centres or even the wall of the reaction vessel. He will introduce further species into the reaction mixture such as adsorbed or activated molecules, atoms or free radicals*, and he will consider a set of proper reactions in this augmented mixture. Let $\bar{T}$ be the number of extra elements introduced, $\bar{S}$ the number of new species and $\bar{R}$ the number of new reactions. Then the $(T + \bar{T})$ elements form an augmented module $\mathfrak{E}^+$,the $(S + \bar{S})$ species an $\mathfrak{M}^+$ and the $(R + \bar{R})$ reactions an $\mathfrak{R}^+$. Since the set of reactions considered in the mechanism must be proper for the augmented reaction mixture there are morphisms $\beta^+: \mathfrak{E}^+ \to \mathfrak{M}^+$ and $\alpha^+: \mathfrak{M}^+ \to \mathfrak{R}^+$ such that

$$0 \to \mathfrak{E}^+ \xrightarrow{\beta^+} \mathfrak{M}^+ \xrightarrow{\alpha^+} \mathfrak{R}^+ \to 0$$

is a short exact sequence. Moreover there is a natural projection, p, of $\mathfrak{E}^+$ onto $\mathfrak{E}$, whose kernel is simply the module consisting of the submodule of lists involving only the $\bar{T}$ Bodenstein elements; in particular $p: \mathfrak{E}^+ \to \mathfrak{E}$ implies

$$(\varepsilon_1 \mathscr{B}_1, \ldots, \varepsilon_T \mathscr{B}_T, \varepsilon_{T+1} \mathscr{B}_{T+1}, \ldots, \varepsilon_{T+\bar{T}} \mathscr{B}_{T+\bar{T}}) \mapsto (\varepsilon_1 \mathscr{B}_1, \ldots, \varepsilon_T \mathscr{B}_T).$$

* FRANK-KAMENETSKY has called these Bodenstein-products (*cf.* P. Rem. 17) and it will be convenient to retain this usage and extend it to the $\bar{T}$ elements and $\bar{R}$ reactions.

Similarly there is a natural projection, q, of $\mathfrak{M}^+$ onto $\mathfrak{M}$ defined by

$$(\mathscr{A}_1^+, \ldots, \mathscr{A}_{\mathrm{S}+\bar{\mathrm{S}}}^+) \mapsto (\mathscr{A}_1, \ldots, \mathscr{A}_{\mathrm{S}}) \quad \text{and} \quad \mathscr{A}_s = \beta_s^t \mathscr{B}_t \quad \text{when} \quad \mathscr{A}_s^+ \mathscr{B}_t + \beta_s^\tau \mathscr{B}_\tau$$

(t summed from 1 to T, τ from T + 1 to T + $\bar{\mathrm{T}}$).

The relation between $\mathfrak{R}^+$ and $\mathfrak{R}$ is given by a matrix of stoicheiometric numbers $\nu = [\nu_r^\rho]$ such that $\mathfrak{R}_r = \nu_r^\rho \mathfrak{R}_\rho^+$ (ρ summed from 1 to R + $\bar{\mathrm{R}}$), and the morphisms of the following diagram commute:

$$\begin{array}{ccccccccc} 0 \rightarrow & \mathfrak{E}^+ & \xrightarrow{\beta^+} & \mathfrak{M}^+ & \xrightarrow{\alpha^+} & \mathfrak{R}^+ & \rightarrow 0 \\ & \downarrow p & & \downarrow q & & \downarrow \nu & \\ 0 \rightarrow & \mathfrak{E} & \xrightarrow{\beta} & \mathfrak{M} & \xrightarrow{\alpha} & \mathfrak{R} & \rightarrow 0. \end{array}$$

We can add a further sequence to this diagram by considering modules $\bar{\mathfrak{E}}$ and $\bar{\mathfrak{M}}$ formed from the Bodenstein elements and products alone. Then there are the natural injections i: $\bar{\mathfrak{E}} \rightarrow \mathfrak{E}^+$ and j: $\bar{\mathfrak{M}} \rightarrow \mathfrak{M}^+$. If $\bar{\beta}$ is the atomic matrix for the Bodenstein products and $\bar{\alpha}$ the stoicheiometric matrix for the proper reactions between them then the Bodenstein sequence

$$0 \rightarrow \bar{\mathfrak{E}} \xrightarrow{\bar{\beta}} \bar{\mathfrak{M}} \xrightarrow{\bar{\alpha}} \bar{\mathfrak{R}} \rightarrow 0$$

is exact. Let μ be the morphism from $\bar{\mathfrak{R}}$ to $\mathfrak{R}^+$ such that $\mu\,\bar{\alpha} = \alpha^+ j$, then the diagram

$$\begin{array}{ccccccccc} & 0 & & 0 & & 0 & \\ & \downarrow & & \downarrow & & \downarrow & \\ 0 \rightarrow & \bar{\mathfrak{E}} & \xrightarrow{\bar{\beta}} & \bar{\mathfrak{M}} & \xrightarrow{\bar{\alpha}} & \bar{\mathfrak{R}} & \rightarrow 0 \\ & \downarrow i & & \downarrow j & & \downarrow \mu & \\ 0 \rightarrow & \mathfrak{E}^+ & \xrightarrow{\beta^+} & \mathfrak{M}^+ & \xrightarrow{\alpha^+} & \mathfrak{R}^+ & \rightarrow 0 \\ & \downarrow p & & \downarrow q & & \downarrow \nu & \\ 0 \rightarrow & \mathfrak{E} & \xrightarrow{\beta} & \mathfrak{M} & \xrightarrow{\alpha} & \mathfrak{R} & \rightarrow 0 \\ & \downarrow & & \downarrow & & \downarrow & \\ & 0 & & 0 & & 0 & \end{array}$$

has exact rows and two exact columns, the first and second. It follows from the 3 × 3 Lemma (*cf.* [11] p. 49) that the last column must be exact and hence that Im μ = Ker ν. It follows that any reactions taking place between the Bodenstein products map into trivial reactions in the original reaction module $\mathfrak{R}$ under the morphism ν that represents the mechanism. If R^+ denotes the total number of independent reactions in $\mathfrak{R}^+$ and $\bar{\mathrm{R}}$ the number in $\mathfrak{R}$, then $\mathrm{R} = \mathrm{R}^+ - \bar{\mathrm{R}}$. But R$'$ is the number of independent stationary conditions and hence $\mathrm{R}^+ - \bar{\mathrm{R}} = \mathrm{R} = \mathrm{S} - \mathrm{T}$. This is SELLERS' extension ([7] p. 16) of the well-known result concerning the number of independent reactions (*cf.* P. Thm. 4).

As well as the abstract account of stoicheiometry there is the metrical aspect. The mean composition in any region may be represented by a point in the positive orthant of composition space. The complete space of $s + \bar{s}$ dimensions may be partitioned into that of s ostensible species and the $\bar{s}$ Bodenstein products. The kinetics of the mechanism determine the motion of the representative point in the complete space, but for the mechanism to be acceptable the projection of this motion into the subspace of the ostensible species must conform with observation. It was not sufficiently clearly brought out in the previous communication (P. § 6) that the relation of ostensible kinetics to the full kinetics of the mechanism is that of the limiting case in the singular perturbation of ordinary differential equations. In particular the work of ACRIVOS and others [12], who first clearly showed this, was overlooked. The case earlier used as an illustration (P. p. 92) has been considered in detail from this point of view [13]. It is interesting that a single parameter, such as the relative smallness of the total catalyst concentration, will often suffice to achieve the required projection as it tends to zero.

5. OF CHEMICAL KINETICS IN GENERAL

The determination of chemical kinetics bears some analogy with the problem of finding the constitutive relations in a continuum. The last few years have shown remarkable progress in the understanding of the latter problem [14–16], the key to which has been the classification of fluids and flows in such a way that an appropriate viscometric flow fully determines the constitutive relations. This change of viewpoint reveals the structure of the subject in a way of which the older techniques of fitting constants in some empirical law were quite incapable. Yet in the analogy it is this older approach that characterizes chemical kinetics; certain rate equations are hypothesized and the constants determined by the best fit of certain data. Even with sophisticated statistical safeguards—and some excellent work has been done on the statistical aspects [17, 18]—this is not really getting at the rational structure. What is needed is the study of classes of reactions of sufficient penetration as to reveal their essential characteristics. This has been done for first order reactions by WEI & PRATER [19], but these share the linearity of Newtonian fluids and progress with non-linear kinetics has been slight [20, 21].

The interaction between the kinetics, the initial composition and the measurements would be taken into account in a full analysis. This is important for a complete aliasing of kinetics can arise under special circumstances. For example, in a continuum of reactions [22] all of which were first order and irreversible, we might denote the fraction of material with rate constant in the interval $(k, k + dk)$ at time t by $x(t, k)$; $x_0(k) = x(0, k)$. Then for parallel first order reaction we would have

$$\frac{\partial}{\partial t} x(t, k) = -k\, x(t, k)$$

and

$$x(t, k) = x_0(k)e^{-kt}.$$

Then $X(t)$ the total fraction of the original remaining would be

$$X(t) = \int_0^\infty x_0(k)e^{-kt}\,dk,$$

and for a suitable initial distribution $x_0(k)$ this might appear to be a single substance disappearing by an entirely different order of reaction. For example, with

$$x_0(k) = k^q K^{-q-1} e^{-(k/K)}/\Gamma(q+1)$$

we would have

$$X(t) = \{1 + Kt\}^{-q-1}$$

or

$$\frac{dX}{dt} = -(q+1)K X^p.$$

This X appears to be the remaining fraction of a substance disappearing by a pth order reaction where $q = (2-p)/(p-1)$.

At first sight it would seem that there is nothing in chemical kinetics that demands more than a thorough understanding of autonomous systems of nonlinear ordinary differential equations, for it could be argued that a continuum of reactions such as we have introduced above is an artificial way of describing a discrete, even if fine-grained, situation. However this may not be so, for from a theoretical point of view one would certainly want to investigate the integrodifferential equations that arise in the limit of an infinite number of species and from a practical stand point this approach has already been proved to be of value. Again at first blush, there would appear to be nothing in chemical kinetics to correspond with the principle of fading memory that has such an interesting role to play in continuum mechanics [15, 16, 23]. The first possible counter example that comes to mind is that of a decaying catalyst. But this would either lead to a case of non-autonomous equations, or, if the decay were due to side reactions, could be accommodated within a larger framework. However, there do appear to be cases in which the rate constants will be functionals of the past temperature history rather than functions of the present temperature. CHASE [24] in his studies of the enzyme Cypridina luciferase found that recovery from the denaturing effect of high temperaures was a comparatively slow process. In the sense in which the temperature must be included in a description of the state and the kinetics are constitutive relations between the state and the forces bringing about a change of state, we have here an instance of fading memory [25]. There are other cases in which the history of concentration, rather than that of the temperature, is important. These lead to some interesting equations currently being studied.

It would seem therefore that the field of chemical kinetics is ripe for the sort of rational analysis that has proved so fruitful in mechanics.

I am indebted to the Graduate School of the University of Minnesota for a Faculty Summer Research Fellowship that enabled the completion of this paper.

REFERENCES

1. Aris, R., Arch. Rational Mech. Anal. **19,** 81 (1965).
2. Kestin, J., Math. Rev. **30,** 4491 (Nov. 1965).
3. Jouguet, E., J. de l'Ecole Poly. II Ser. No. 21, 61, 181 (1921).
4. Shapiro, N. Z., & L. S. Shapley, S.I.A.M. Jour. **13,** 353 (1965).
5. Sellers, P. H., SIAM Jour. App. Math. **15,** 637 (1967).
6. Sellers, P. H., Proc. Nat. Acad. Sci. **55,** 693 (1966).
7. Sellers, P. H., SIAM Jour. App. Math. **15,** 13 (1967).
8. Amundson, N. R., & D. Luss, The Tubular Reactor (in press).
9. Aris, R., Introduction to the Analysis of Chemical Reactors, Englewood Cliffs: Prentice Hall 1965.
10. Blum, E. H., & R. Luus, Chem. Engng. Sci. **19,** 322 (1964).
11. MacLane, S., Homology. Berlin-Göttingen-Heidelberg: Springer 1963.
12. Acrivos, A., J. R. Bowen, & A. K. Oppenheim, Chem. Engng. Sci. **18,** 177 (1963).
13. Heineken, F. G., H. M. Tsuchiya, & R. Aris, Math Biosciences **1,** 95 (1967).
14. Truesdell, C., & R. Toupin, The Classical Field Theories, Hand. der Physik, III/1, p. 700 *et seq.* Berlin-Göttingen-Heidelberg: Springer 1960.
15. Truesdell, C., & W. Noll, The Non-Linear Field Theories of Mechanics, Hand. der Physik. III/3. Berlin-Heidelberg-New York: Springer 1965.
16. Coleman, B. D., W. Noll, & H. Markowitz, Viscometric Flows of Non-Newtonian Fluids, Springer Tracts in Nat. Phil. **5.** Berlin-Heidelberg-New York: Springer 1966.
17. Hunter, W. G., & R. Mezaki, AICHE Journal **10,** 315 (1964).
18. Mezaki, R., & J. R. Kittrell, Ind. Eng. Chem. **29,** 63 (May 1967).
19. Wei, J., & C. D. Prater. Adv. in Catalysis **13,** 203 (1962).
20. Aris, R., Ind. Eng. Chem. Fundamentals **3,** 28 (1964).
21. Wei, J., Ind. Eng. Chem. Fundamentals **4,** 161 (1965).
22. Gavalas, G., & R. Aris, Phil. Trans. Roy. Soc. A **260,** 351 (1966).
23. Coleman, B. D., & V. J. Mizel, Arch. Rational Mech. Anal. **23,** 87 (1966).
24. Chase, A. M., J. Gen. Physiol. **33,** 535 (1950).
25. Wheeler, J. M., private communication.

Department of Chemical Engineering
University of Minnesota

F.*

MODELLING CUBIC AUTOCATALYSIS BY SUCCESSIVE BIMOLECULAR STEPS

R. ARIS†
P. GRAY‡
S. K. SCOTT‡

† *Department of Chemical Engineering and Materials Science, University of Minnesota, Minneapolis, MN 55455, U.S.A.*
‡ *Department of Physical Chemistry, University of Leeds, Leeds LS2 9JT, U.K.*

(*Received* 21 *November* 1986; *accepted* 10 *April* 1987)

Abstract—*The simplest, chemically and thermodynamically satisfactory model of oscillatory behaviour involves only two variables but requires a third-order step:*

$$A + B \rightarrow 3B \qquad (1)$$

$$B \rightarrow C \qquad (2)$$

If the third-order step is replaced by consecutive second-order reactions

$$A + B \rightleftharpoons X \qquad (3, -3)$$

$$X + B \rightarrow 3B \qquad (4)$$

then a 3-variable scheme is produced.

Oscillatory behaviour and exotic stationary-state patterns of extent of reaction versus flow rate in the simplest of open systems survive.

When k_3a_0/k_{-3} *and* (k_4a_0/k_{-3}) *are both small the 3-variable scheme takes a simple analytical form containing both the 2-variable, cubic case and explicit forms for the leading-order corrections. In the limit* $\kappa = \varepsilon \rightarrow 0$ *the cubic case is left.*

1. INTRODUCTION

Until recently it was widely accepted that isothermal oscillatory reactions demanded rather complicated interpretations at all levels. It is now known that this is not the case. A simple representation of autocatalysis that has had great impact in recent years on a wide variety of problems, beginning with the CSTR (Gray and Scott, 1983, 1984) but now extending to systems with diffusive control (Gray and Scott, 1986; Scott, 1987) and spatial non-unifor-

* Reprinted with the permission of Pergamon Press from *Chemical Engineering Science,* volume **43,** number 2, pp. 207–211, 1988.

mity, employes the step

$$\mathrm{A} + 2\mathrm{B} \rightarrow 3\mathrm{B} \quad \text{rate} = k_1 ab^2 \tag{1}$$

coupled with catalyst decay

$$\mathrm{B} \rightarrow \mathrm{C} \quad \text{rate} = k_2 b. \tag{2}$$

For any model to deserve regard as representing "real chemistry" it should be capable of satisfying certain fundamental chemical and physical laws. These include being self-consistent stoichiometrically, as the above system clearly is, conserving mass (which not all models of oscillatory reactions can do) and obeying the requirements, such as detailed balancing, of thermodynamics (Gray, 1970). Recent papers show how the non-linear behaviour exhibited above survives the imposition of these restrictions (Gray *et al.*, 1984; Gray and Scott, 1985; Aris, 1986), in some cases increasing the variety of possible responses (Kay *et al.*, 1987).

There remains one objection—of a less precise kind but felt by many chemists. It is that third-order kinetics as embodied in the representation of step (1) are intrinsically objectionable. If the equations had to be interpreted as representing elementary steps, this would be a weightier consideration, but it has also been asserted that the oscillatory properties of certain other model schemes collapse completely (King, 1983; Gray and Morley-Buchanan, 1985) if the third-order steps therein are replaced. Accordingly it is most desirable to establish whether oscillations and other exotic behaviour arising from a "cubic" rate-law of the form k_1ab^2 can also arise from a series of successive second-order or 'bimolecular' steps. Similar interests have been expressed previously by Tyson (1973) and Tyson and Light (1973).

2. KINETIC SCHEMES AND MASS-BALANCE EQUATIONS

The model we consider here is

$$\mathrm{A} + \mathrm{B} \underset{k_{-3}}{\overset{k_3}{\rightleftharpoons}} X \tag{3}$$

$$\mathrm{B} + \mathrm{X} \xrightarrow{k_4} 3\mathrm{B} \tag{4}$$

again coupled with catalyst decay

$$\mathrm{B} \xrightarrow{k_2} \mathrm{C}. \tag{2}$$

For such a reaction scheme in a CSTR the governing, mass-balance equations for the concentrations a, b and x are

$$\begin{aligned}
\mathrm{d}a/\mathrm{d}t &= k_\mathrm{f}(a_0 - a) - k_3ab + k_{-3}x \\
\mathrm{d}b/\mathrm{d}t &= k_\mathrm{f}(b_0 - b) - k_3ab + k_{-3}x + 2k_4bx - k_3b \\
\mathrm{d}x/\mathrm{d}t &= -k_\mathrm{f}x + k_3ab - k_{-3}x - k_4bx
\end{aligned}$$

where k_f is the inverse of the mean residence-time and it has been assumed that there is no inflow of the complex X. In order to allow easy comparison with the earlier papers, we choose the following dimensionless measures of the concentrations

$$\alpha = a/a_0, \beta = b/a_0, \beta_0 = b_0/a_0 \quad \text{and} \quad \xi = x/a_0 \tag{5}$$

where a_0 is the concentration of reactant A in the feed. We also introduce the groups

$$\kappa = k_3 a_0/k_{-3} = K_3 a_0 \tag{6a}$$

$$\varepsilon = k_4 a_0/k_{-3} \tag{6b}$$

and choose the chemical timescale $t_{ch} = k_1 a_0^2$ by defining k_1 as

$$k_1 = \frac{k_3 k_4}{k_{-3}} \quad \text{so that} \quad k_1 a_0^2 = k_{-3}\kappa\varepsilon. \tag{7}$$

Using the dimensionless times, we write

$$\tau = k_1 a_0^2 t,\ \tau_{res} = k_1 a_0^2/k_f \quad \text{and} \quad \tau_2 = k_1 a_0^2/k_2, \tag{8}$$

and the mass-balance equations finally become

$$\frac{d\alpha}{d\tau} = \frac{1-\alpha}{\tau_{res}} - \frac{1}{\varepsilon}\left(\alpha\beta - \frac{\xi}{\kappa}\right) \tag{9a}$$

$$\frac{d\beta}{d\tau} = \frac{\beta_0 - \beta}{\tau_{res}} - \frac{1}{\varepsilon}\left(\alpha\beta - \frac{\xi}{\kappa}\right) + \frac{2\beta\xi}{\kappa} - \frac{\beta}{\tau_2} \tag{9b}$$

$$\frac{d\xi}{d\tau} = -\frac{\xi}{\tau_{res}} + \frac{1}{\varepsilon}\left(\alpha\beta - \frac{\xi}{\kappa}\right) - \frac{\beta\xi}{\kappa}. \tag{9c}$$

We now investigate analytically the connection between the present, 3-variable model (eq. 9) and its simpler predecessor (eqs. 1 and 2) when ε and κ are relatively small quantities (ε, $\kappa \ll 1$).

From eq. (9c) we obtain

$$\kappa\alpha\beta = \xi\left[1 + \varepsilon\kappa\left(\frac{1}{\tau_{res}} + \frac{\beta}{\kappa} + \frac{\dot{\xi}}{\xi}\right)\right] \tag{10}$$

where $\dot{\xi} = d\varepsilon/d\tau$. Re-arranging (13) gives, for small ε and κ,

$$\xi = \kappa\alpha\beta\left\{1 - \varepsilon\kappa\left(\frac{1}{\tau_{res}} + \frac{\beta}{\kappa} + \frac{\dot{\xi}}{\xi}\right) + \varepsilon^2\kappa^2\left(\frac{1}{\tau_{res}} + \frac{\beta}{\kappa} + \frac{\dot{\xi}}{\xi}\right)^2 - \ldots\right\}. \tag{11}$$

Differentiating (11) with respect to time, we have to the leading order

$$\dot{\xi} = \kappa(\alpha\dot{\beta} + \beta\dot{\alpha}) - \ldots. \tag{12}$$

Since α and β are both less than 1, rates of change of ξ are less by a factor κ than those in α and β.

Combining equations (11) and (12) gives

$$\xi = \kappa\alpha\beta\left\{1 - \varepsilon\beta - \varepsilon\kappa\left[\frac{1}{\tau_{\text{res}}} + \frac{(\alpha\dot{\beta} + \beta\dot{\alpha})}{\alpha\beta}\right] + \varepsilon^2\beta^2 - 0(\varepsilon^2\kappa)\right\}. \tag{13}$$

Substituting for ξ in eqs (9a) and (9b), we obtain for $\mathrm{d}\alpha/\mathrm{d}\tau$ and $\mathrm{d}\beta/\mathrm{d}\tau$

$$\frac{\mathrm{d}\alpha}{\mathrm{d}\tau} = \frac{(1-\alpha)}{\tau_{\text{res}}} - \alpha\beta^2 - \kappa\left\{\frac{\alpha\beta_0 + \beta_0 - \alpha\beta}{\tau_{\text{res}}} - \frac{\alpha\beta}{\tau_2} + \alpha\beta^2(\alpha - \beta)\right\} + \varepsilon\alpha\beta^3 \ldots \tag{14}$$

$$\frac{\mathrm{d}\beta}{\mathrm{d}\tau} = \frac{(\beta_0 - \beta)}{\tau_{\text{res}}} + \alpha\beta^2 - \frac{\beta}{\tau_2}\kappa\left\{\frac{\alpha\beta_0 + \beta_0 - \alpha\beta}{\tau_{\text{res}}} - \frac{\alpha\beta}{\tau_2} + \alpha\beta^2(\alpha - \beta)\right\} - \varepsilon\alpha\beta^3 \ldots \tag{15}$$

These equations clearly reduce to the original model with the termolecular step in the limit $\varepsilon \to \kappa \to 0$.

We now compare the behaviour of (14) and (15) for small, non-zero ε and κ with the zero-order form i.e. the autocatalator. We will also compare these first-order equations with the full, three-variable scheme (9a)–(9c).

3. BEHAVIOUR OF FIRST-ORDER CORRECTION TO AUTOCATALATOR: STATIONARY-STATES AND LIMIT CYCLES

The first-order, stationary-state equations can be derived by setting $\dot{\alpha} = \dot{\beta} = 0$ (16) to give

$$\xi_{\text{ss}} = \kappa\alpha_{\text{ss}}\beta_{\text{ss}}\{1 - \varepsilon\beta_{\text{ss}} + 0(\varepsilon^2, \varepsilon\kappa, \text{etc.})\}. \tag{16}$$

Substitution of this into (9a) and (9b) leads, after some manipulation, to a quartic expression for α_{ss}.

$$(1-\alpha_{\text{ss}})\left(1 + 2\kappa\alpha_{\text{ss}} + \frac{\tau_{\text{res}}}{\tau_2}\right)^3 - \tau_{\text{res}}\alpha_{\text{ss}}(1 + \beta_0 - \alpha_{\text{ss}})^2\left(1 + 2\kappa\alpha_{\text{ss}} + \frac{\tau_{\text{res}}}{\tau_2}\right) - \kappa\alpha_{\text{ss}}(1 + \beta_0 - \alpha_{\text{ss}})\left(1 + 2\kappa\alpha_{\text{ss}} + \frac{\tau_{\text{res}}}{\tau_2}\right)^2 - \varepsilon\tau_{\text{res}}\alpha_{\text{ss}}(1 + \beta_0 - \alpha_{\text{ss}})^3 = 0. \tag{17}$$

This form of the stationary-state condition emphasizes that in the special case $\beta_0 = 0$, one root is

$$\alpha_{\text{ss}} = 1, \quad \beta_{\text{ss}}\xi_{\text{ss}} = 0. \tag{18}$$

The remaining solutions are given by the roots of a cubic: one of them is negative; the other two may either be complex or real and positive. These non-zero, stationary states lie on an isola in the $1 - \alpha_{\text{ss}}$ versus τ_{res} bifurcation diagram. Compared with the zero-order states (α_0), the correction terms serve to reduce the size of the isola as ε and κ increase.

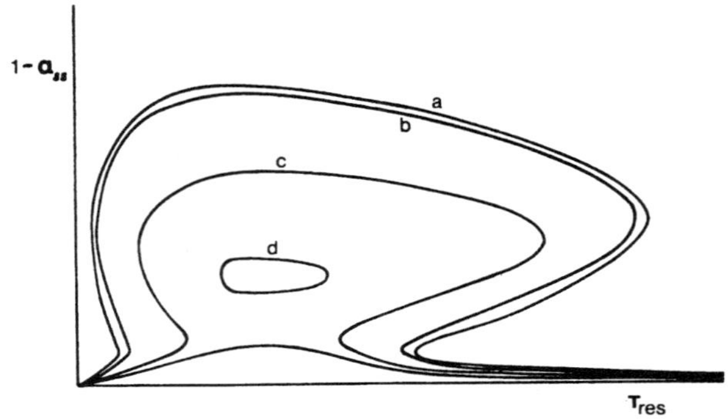

FIGURE 1 Effect of increasing ε or κ on mushroom (a) obtained for cubic autocatalator. As ε or κ increase the mushroom decreases in size, (b) and (c), before becoming an isola (d). Further increases lead to a disappearance of the isola.

For non-zero β_0, eq. (17) is most easily solved as a cubic in τ_{res}. One of the roots is negative (in the limit $\varepsilon = \kappa = 0$, this root is exactly $-\tau_2$): for some values of α_{ss} etc., there are two other real and positive solutions. The stationary-state loci form either unique dependences of $1 - \alpha_{ss}$ on τ_{res}, or isolas or mushrooms. Figure 1 shows how increasing ε and κ serves to decrease the size of a mushroom. The transition from mushroom to isola follows, and ultimately the isola shrinks to a point. The values of ε and κ at which these qualitative changes in the bifurcation diagram arise are not necessarily small compared with unity. Table 1 gives some specific examples.

With $\varepsilon = \kappa = 0$, the autocatalator is capable of displaying Hopf bifurcations and sustained oscillatory solutions. The situation is not much changed for small, non-zero ε and κ. Figure 2 shows the appearance and growth of limit cycles along the upper branch of a mushroom for a typical system with small ε and κ. In general the Hopf bifurcation point moves to lower residence times as ε or κ increase: at a fixed τ_{res}, the amplitude increases and the period lengthens as ε or κ increase. If ε or κ become too large, however, oscillatory behaviour disappears. Again this can require values of these parameters of

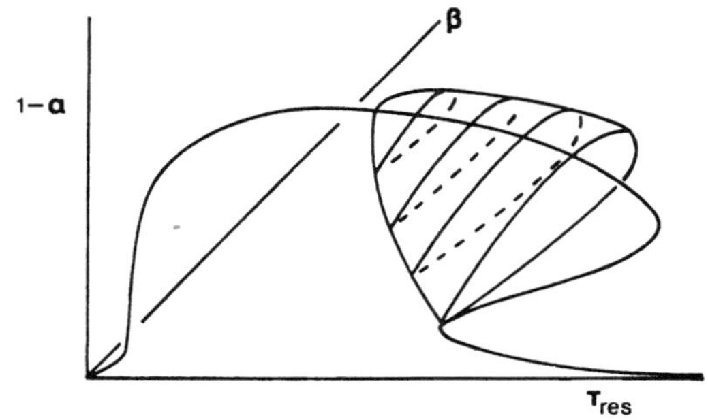

FIGURE 2 The birth and growth of limit cycle oscillations in the $1 - \alpha$, β, τ_{res} space for a system with non-zero ε and κ displaying a mushroom stationary-state pattern. Oscillatory behaviour originates from a supercritical Hopf bifurcation along the upper branch and terminates via homoclinic orbit formation.

TABLE 1 Comparison of 1st order correction and full equations for a system with $\beta_0 = 1/15$ and $\tau_2 = 40$

		1st order correction			Full equations			
Stationary-state solutions at $\tau_{res} = 225$			I			II		
a	$\varepsilon = \kappa = 0$	$\alpha = 0.210$	$\beta = 0.129$	$\xi = 0$	$\alpha = 0.210$	$\beta = 0.129$	$\xi = 0$	
b	$\varepsilon = \kappa = 0.01$	$\alpha = 0.211$	$\beta = 0.129$	$\xi = 2.7 \times 10^{-4}$	$\alpha = 0.210$	$\beta = 0.129$	$\xi = 2.7 \times 10^{-4}$	
c	$\varepsilon = \kappa = 0.1$	$\alpha = 0.216$	$\beta = 0.127$	$\xi = 2.7 \times 10^{-3}$	$\alpha = 0.216$	$\beta = 0.126$	$\xi = 2.7 \times 10^{-3}$	
d	$\varepsilon = \kappa = 1$	(i) $\alpha = 0.292$	$\beta = 0.107$	$\xi = 2.8 \times 10^{-2}$	(i) $\alpha = 0.282$	$\beta = 0.110$	$\xi = 2.8 \times 10^{-2}$	
		(ii) $\alpha = 0.839$	$\beta = 2.74 \times 10^{-2}$	$\xi = 2.24 \times 10^{-2}$	(ii) $\alpha = 0.843$	$\beta = 2.71 \times 10^{-2}$	$\xi = 2.22 \times 10^{-2}$	
		(iii) $\alpha = 0.939$	$\beta = 1.50 \times 10^{-2}$	$\xi = 1.39 \times 10^{-2}$	(iii) $\alpha = 0.939$	$\beta = 1.51 \times 10^{-2}$	$\xi = 1.39 \times 10^{-2}$	
Amplitudes and period of oscillations at $\tau_{res} = 225$								
e	$\varepsilon = \kappa = 0$	$A_\alpha = 0.668$	$A_\beta = 0.492$	$\tau_p = 576.1$	$A_\alpha = 0.668$	$A_\beta = 0.492$		$\tau_p = 576.1$
f	$\varepsilon = \kappa = 0.01$	$A_\alpha = 0.681$	$A_\beta = 0.502$	$\tau_p = 594.4$	$A_\alpha = 0.672$	$A_\beta = 0.493$	$A_\xi = 1.05 \times 10^{-2}$	$\tau_p = 584$
g	$\varepsilon = \kappa = 0.1$	$A_\alpha = 0.855$	$A_\beta = 0.630$	$\tau_p = 1024.3$	$A_\alpha = 0.787$	$A_\beta = 0.578$	$A_\xi = 1.23 \times 10^{-2}$	$\tau_p = 776$
h	$\varepsilon = \kappa = 1$		no oscillation			no oscillation		
Eigenvalues of stationary-state at $\tau_{res} = 225$								
i	$\varepsilon = \kappa = 0$	$2.088 \times 10^{-3} \pm 1.932 \times 10^{-2}i$			$2.088 \times 10^{-3} \pm 1.932 \times 10^{-2}i, -\infty$			
j	$\varepsilon = \kappa = 0.01$	$1.995 \times 10^{-3} \pm 1.926 \times 10^{-2}i$			$1.937 \times 10^{-3} \pm 1.931 \times 10^{-2}i, -1.0 \times 10^4$			
k	$\varepsilon = \kappa = 0.1$	$2.627 \times 10^{-3} \pm 1.790 \times 10^{-2}i$			$3.13 \times 10^{-3} \pm 1.79 \times 10^{-2}i, -104.7$			
l	$\varepsilon = \kappa = 1$	(i) $3.06 \times 10^{-3} \pm 0.76 \times 10^{-2}i$			(i) $3.17 \times 10^{-3} \pm 1.21 \times 10^{-2}i, -1.49$			
		(ii) $2.65 \times 10^{-3}, -5.90 \times 10^{-3}$			(ii) $4.60 \times 10^{-3}, -3.37 \times 10^{-3}, -1.89$			
		(iii) $1.85 \times 10^{-4}, -5.20 \times 10^{-3}$			(iii) $-3.80 \times 10^{-3} \pm 1.58 \times 10^{-3}i, -1.97$			
Changes in bifurcation diagram								
m	mushroom → isola				$\varepsilon = \kappa = 1.47$			
n	disappearance of isola	$\varepsilon = \kappa = 1.43$			$\varepsilon = \kappa = 2.69$			
o	loss of Hopf bifurcation	$\varepsilon = \kappa = 2.21$			$\varepsilon = \kappa = 2.69$			
		$\varepsilon = \kappa = 1.55$						

order unity, requiring the possible inclusion of higher-order terms. To this end we, therefore, confirm that the above patterns are also found for the full three-variable model, eq. (9).

4. COMPARISON OF FIRST-ORDER EQUATIONS WITH FULL, THREE-VARIABLE MODEL

The magnitude of the correction to the simple (zero-order) autocatalator provided by the first-order equations, (14) and (15) for the time dependent behaviour and (17) for the stationary-states, can be seen quantitatively for a typical system from the first column of numerical results in Table 1. The corrections in α_{ss} and β_{ss} appear always to be an order of magnitude less than ε and κ, even when these parameters are of order unity. The second column, which gives the solutions of the full set of three equations (9a)–(9c), shows that the first-order equations actually overestimate the deviation from the $\varepsilon = \kappa = 0$ autocatalator. Returning to column I, we see that the *stationary-state* concentration of X is adequately given by $\xi_{ss} = \kappa\alpha_{ss}$, even for $\varepsilon = \kappa = 0.1$, and is very accurately given by $\xi_{ss} = \kappa\alpha_{ss}\beta_{ss}(1 - \varepsilon\beta_{ss})$ even for $\varepsilon = \kappa = 1$. Notice that for row (d) there are three stationary-state solutions at $\tau_{res} = 225$.

The first-order equations give good, but less spectacular, agreement for the non-stationary behaviour. Rows (e)–(h) show how the oscillatory amplitudes about the unique, unstable stationary-state at $\tau_{res} = 225$ grow as ε and κ increase. A comparison of columns I and II (the exact, three-variable solutions), reveals that eqs (14) and (15) again overestimate the variation from the original autocatalator ($\varepsilon = \kappa = 0$) behaviour, the error in the amplitudes being of order ε and κ. The differences in the oscillatory period between entries in columns I and II increase more quickly. For $\varepsilon = \kappa = 1$, neither model has oscillatory solutions under these conditions. The full eqs (9a)–(9c) also give the amplitude of the pulses in ξ, the dimensionless concentration of X. Typical oscillations for the full scheme are shown in Fig. 3.

Another aspect of the time-dependent behaviour is the behaviour of the eigenvalues representing the local stability of the stationary-states to small perturbations. For ε and κ less than or equal to 0.1, the unique solution has a pair of complex eigenvalues with positive real parts, indicating an unstable focus. The closeness of agreement between the first-order and exact models is again of order ε or κ: the exact equations have a third eigenvalue associated with the third variable; this has a large negative value, representing a rapid decay of order ε^{-2} or κ^{-2}. With $\varepsilon = \kappa = 1$ there are three stationary-states. The agreement between the eigenvalues for the solutions of eqs (16) and (17) and those of the full equation is now not good: the Jacobian formed from eqs (14) and (15) has roots of opposite sign for all three states, concluding that none would be stable, and contrary to computational results. The "fault" can be rectified by looking at the solutions $d\alpha/d\tau = d\beta/d\tau = 0$ from (14) and (15), which have more complex forms that those of (16) and (17). The results given in (i)–(iii) of column I now show resemblance to the exact eigenvalues and indicate that the state of lowest conversion (iii) is a stable solution and it is to this state that all trajectories in the phase-plane approach in both models.

The value of ε and κ at which the mushroom bifurcation diagram becomes

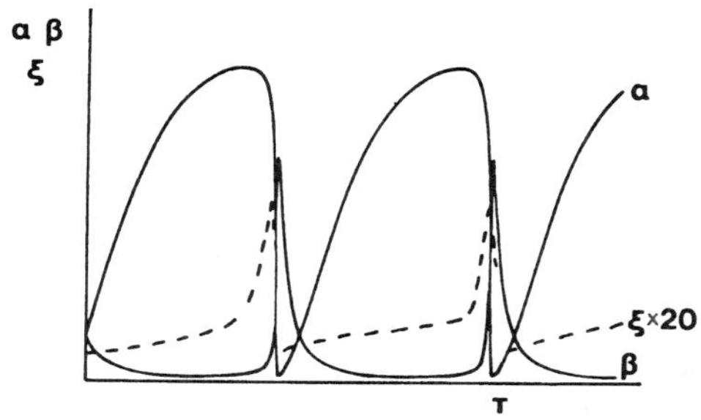

FIGURE 3 Sustaining oscillations in the concentrations α, β and ξ for the full, three-variable model. The concentration ξ is shown $\times 20$.

an isola vary little between the first-order and exact equations. The isola disappears slightly earlier in the former model ($\varepsilon = \kappa = 2.21$) than for eqs (9) ($\varepsilon = \kappa = 2.69$). The biggest difference, reflecting a qualitative as well as a quantitative change of behaviour, is found at the loss of Hopf bifurcations. With the full, three-variable model, Hopf bifurcation along the upper branch of stationary-states survives until the isola shrinks to a point at $\varepsilon = \kappa = 2.69$. For the first-order correction form, however, a second Hopf bifurcation appears on this branch, and the two bifurcation points merge well before the disappearance of the isola.

Additional computations to be reported elsewhere, confirm that the parameter $\kappa = a_0 K_3$ plays a more influential role than $\varepsilon = (k_4/k_3)a_0 K_3$, the behaviour of the system being relatively insensitive to small variations in the latter.

CONCLUSIONS

(1) The cubic autocatalator rate equations can be rigorously derived as a limiting form from a kinetic model involving only bimolecular steps.
(2) Moving away from the above limit the system may still be represented by a pair of mass-balance equations. First-order correction terms involving two small parameters ε and κ can readily be obtained. The multistability and sustained oscillatory behaviour of the autocatalator remains for non-zero ε and κ, and is approached smoothly in the limit $\varepsilon \to 0$, $\kappa \to 0$.
(3) For ε and κ of order unity, the full three-variable model may still display multiple stationary-states, Hopf bifurcations and sustained oscillations.

ACKNOWLEDGMENTS

We are grateful to Professor G. B. Cook for helpful discussions and to the SERC and the EEC for research grants.

REFERENCES

Aris, R., 1986, The mathematical background of chemical reactor analysis II. The stirred tank. *Lectures in Applied Mathematics* **24,** 75–107.

Gray, B. F., 1970, Oscillatory approach to equilibrium in closed chemical systems. *Trans. Faraday Soc.* **66,** 363–371.

Gray, B. F., Scott, S. K. and Gray, P., 1984, Multiplicity for isothermal autocatalytic reactions in open systems: the influence of reversibility and detailed balance. *J. Chem. Soc. Faraday Trans.* 1 **80,** 3409.

Gray, B. F. and Morley-Buchanan, T., 1985, Some criticism concerning the Brusselator model of an oscillating chemical reaction. *J. Chem. Soc. Faraday Trans.* 2 **81,** 77.

Gray, B. F. and Scott S. K., 1985, Multistability and sustained oscillations in isothermal, open systems: cubic autocatalysis and the influence of competitive reactions. *J. Chem. Soc. Faraday Trans.* 1 **81,** 1563.

Gray, P. and Scott, S. K., 1983, Autocatalytic reactions in the isothermal continuous, stirred-tank reactor: isolas and other forms of multistability. *Chem. Engng Sci.* **38,** 29.

Gray, P. and Scott, S. K., 1984, Autocatalytic reactions in the isothermal continuous, stirred-tank reactor: oscillations and instabilities in the system A + 2B → 3B; B → C. *Chem. Engng Sci.* **39,** 1087.

Gray, P. and Scott, S. K., 1986, Chemical reactions in isothermal systems: oscillations and instabilities. In *Nonlinear Phenomena and Chaos* (Edited by S. Sarkar), pp. 69–94: Adam Hilger.

Kay, S. R., Scott, S. K. and Lignola, P.-G., 1987, The application of singularity theory to isothermal autocatalytic reactions: the influence of uncatalyzed reactions. *Proc. R. Soc. Lond. A* **409,** 449.

King, G. A. M., 1983, Reactions for chemical systems far from equilibrium. *J. Chem. Soc. Faraday Trans.* 1 **79,** 75–80.

Scott, S. K., 1987, Isolas, mushrooms and oscillations in isothermal, autocatalytic reaction–diffusion equations. *Chem. Engng Sci.* **42,** 307–316.

Tyson, J. J., 1973, Some further studies of nonlinear oscillations in chemical systems *J. Chem. Phys.* **58,** 3919–3930.

Tyson, J. J. and Light, J. C., 1973, Properties of two-component bimolecular and trimolecular chemical reaction systems. *J Chem. Phys.* **59,** 4164–4173.

G.*

REACTIONS IN CONTINUOUS MIXTURES

RUTHERFORD ARIS

Department of Chemical Engineering and Materials Science University of Minnesota Minneapolis, MN 55455

A continuous mixture is one which is so complex that it is no longer worthwhile to distinguish individual chemical species; instead, an index, such as the simulated boiling point, is chosen and v_i, *the concentration of the species* A_i, *is replaced by* c(x)dx, *the concentration of material with index in the interval* (x, x + dx). *It has been long known that the total concentration of a suitably distributed mixture, each of whose components disappears by a first-order reaction with constant* k(x), *will appear to disappear according to a higher order of reaction. The generalization of this to a mixture that requires two indices for its description is worth considering for three reasons: First, there may well be materials that are so complex as to require this. Second, the second index may be considered to distribute reaction time. Third, this approach seems to answer the mathematical question of how to generalize from the continuum of first-order reactions to one of parallel* N*th-order reactions.*

INTRODUCTION

The notion of a continuous mixture has proved its value in many ways since it was first introduced by de Donder (1931). It arises when the number of distinct species is very large and the difference between adjacent species is relatively slight. Under such circumstances a continuous variable, such as boiling point or chromatographic retention time, is used to describe the composition. This notion has been applied to distillation (Amundson and Acrivos, 1955), to the thermodynamics of the distribution of isomers (Alberty and Oppenheimer, 1984) and other thermodynamical questions (Briano and Glandt, 1984), to polymerization (Zeman and Amundson, 1965), froth flotation (Loveday, 1966), the heterogeneity of catalytic sites (Malhotra and Sadana, 1987), and to reactions in continuous mixtures in general (Aris and Gavalas, 1965; Aris, 1968; Luss and Hutchinson, 1971; Golikeri and Luss, 1971; Kemp and Wojciechowski, 1974; Weekman, 1979; Krambeck, 1984; Astarita and Ocone, 1988; Chou and Ho, 1989; Prasad *et al.*, 1986a, b).

To the best of my knowledge, the work of Prasad *et al.* contains the first mention of the use of two indices, for, in their description of coal liquefaction

* Reprinted with the permission of the American Institute of Chemical Engineers from *AIChE Journal,* volume **35,** number 4, pp. 539–548, 1989.

they take the number of carbon atoms and the number of oxygen atoms to be their two index parameters. They review various models in the context of coal liquefaction and are successful in fitting kinetic kernels (Prasad, 1986). There may also be other materials of sufficient complexity to merit description with two indices, and reactions in such mixtures are worth considering in a more general light. One authority, with much experience in the oil industry, has said that "in any real case, sufficiently complex for a continuous description, I think it is highly unlikely that one could ever determine the initial distribution of reaction rates and that the finest analysis technique will only resolve the mixture into components that still have a continuum of reaction rates" (Krambeck, personal communication, 1984).

It is in the cracking of crude oil that the continuum of parallel reactions in a continuous mixture has found its principal application (Weekman, 1979). It explains how so essentially first-order a process as cracking can give data best correlated by a second-order reaction, for this is just the kinetics of the disappearance of the total concentration of a mixture with exponentially distributed first-order rate constants. Indeed, Krambeck (1984) has shown that, whenever the distribution does not go to zero at the irrefrangible end of the spectrum, the disappearance of the "lump" is as if by second-order reaction. The argument is simple and convincing when the basic reaction is of the first order, for such a reaction is essentially a spontaneous individual process, as when a bond breaks thermally, independent of other species that may be present. It matters not whether the species be discrete or a slice of a continuum; each "does its own thing." It is less easy to see how a continuum of parallel reactions of the second order can be justified. Second-order reaction implies a cooperation between molecules and although this cooperation might reasonably be confined to each of several species when the species are discrete, so that a number of parallel second-order reactions result, this is less plausible when the species are continuous and can scarcely avoid interacting with their neighbors. It is the recognition of this that led Astarita and Ocone (1988) to look at kinetics represented by kinetic laws of the type

$$\frac{d}{dt}c(x,t) = k(x)c(x,t)\boldsymbol{F}\left[\int K(x,z)c(z,t)\,dz\right]$$

Here $c(x, t)dx$ is the concentration of material with index in the slice $(x, x + dx)$ whose rate constant is $k(x)$; $K(x, z)$ describes the interaction of the species. The authors obtain some striking results for uniform systems, as they call those for which K is independent of x (Astarita and Ocone, 1988; Astarita, 1989). Their second-order reaction would imply that each slice reacted with every other, K being a stoichiometric coefficient function. Only if $\boldsymbol{K} = \delta(z - x)$ would we have a continuum of independent parallel second-order reactions. In spite of the physical objections, the mathematical challenge of setting this up properly remains. Ho and Aris (1987) have shown how not to do it. Astarita and Ocone have shown how to do something a little different and probably more sensible physically. We shall see that it can be done quite generally by having a double-indexed mixture with parallel first-order reactions. The first-order kinetics ensures the individuality of the reactions and the distribution in one index simulates a reaction of the Nth order that is distributed by the second index. If we relax the requirement of first-order kinetics and permit uniform Astarita kinetics, we find that, by using his latest results on apparent reaction rate, we can justify the lumping of reactions with quite general nonlinear kinetics. Various special cases will be examined in detail.

Since the dimensionless time for a first-order reaction is the product of the reaction time t and a first-order rate constant k, there is no reason why $k(x)t$ should not be interpreted as $k(x)t(x)$, that is, the reaction time may be distributed over the index space as well as the rate constant. Alternatively, with two indices k might be distributed over one and t over the other as $k(x)t(y)$. We can thus consider a continuum of reactions in a reactor with specified residence time distribution and this is entirely equivalent to the single reaction with the apparent kinetics of the continuum under the segregation hypothesis of residence time distribution theory, a topic that is in the elementary texts. Three indices would be required to distribute the reaction time with a doubly-distributed continuous mixture.

To these three *a priori* reasons for considering the generalization to two indices a fourth may be added *a posteriori*. We shall encounter, in important specific cases, one or two of the rarer special functions associated with the confluent hypergeometric function.

GENERAL FORMULATION FOR A SINGLE INDEX

It is better to define the continuous mixture in terms of an index variable (Aris and Gavalas, 1966; Astarita and Ocone, 1988) rather than to use the dimensional rate constant k. Let x be an index variable, which without loss of generality we can take to be in the interval $[0, \infty)$. The initial concentration of material characterized by indices in the interval $(x, x + dx)$ is

$$c(x, 0)dx = c_0 f(x)dx \tag{1}$$

where c_0 is a concentration. Thus $f(x) \geq 0$.

$$\int_0^\infty f(x)\, dx = 1 \tag{2}$$

and without loss of generality we can scale x so that we also have

$$\int_0^\infty xf(x)\, dx = 1 \tag{3}$$

The distribution $f(x)$ will have one or more shaping parameters and we shall have occasion to use the gamma distribution

$$f(x) = \frac{\alpha^\alpha x^{\alpha-1}}{\Gamma(\alpha)} e^{-\alpha x} = \phi_\alpha(x), \quad \alpha > 0 \tag{4}$$

which, as $\alpha \rightarrow \infty$, approachs $\delta(x - 1)$. We suppose that this initial distribution decays in such a way that

$$\frac{\partial}{\partial t} c(x, t) = -k(x)c(x, t) \tag{5}$$

where $k(x)$ is a monotonic increasing function on $[0, \infty)$ with $k(0) = 0$ and $k(x) \to \infty$ as $x \to \infty$. Thus

$$\begin{aligned} c(x, t) &= c_0 u(x, t) \\ &= c_0 f(x) e^{-k(x)t} \end{aligned} \tag{6}$$

and

$$U(t) = \int_0^\infty c(x, t)\, dx / c_0 = \int_0^\infty f(x) e^{-k(x)t}\, dx \tag{7}$$

If t can be eliminated between $U(t)$ and

$$\dot{U}(t) = -\int_0^\infty k(x) f(x) e^{-k(x)t}\, dx \tag{8}$$

to give a kinetic law for the "lump"

$$\dot{U} = -F(U), U(0) = U_0 = 1 \tag{9}$$

we say that this kinetic law is an alias of the underlying first-order kinetics generated by the distributions $f(x)$, $k(x)$. Since $k(x)$ is monotonic, we can regard $f(x)$ as $f^*(k)$ with $f^*(k)dk = f(x)k'(x)dx$ when necessary. This transformation is the essential feature of Ho's treatment (Chou and Ho, 1988). Let

$$\hat{k} = \int_0^\infty k(x) f(x)\, dx \tag{10}$$

so that $\hat{k}$ has the dimensions of reciprocal time and

$$\hat{k}t = \tau \tag{11}$$

is the dimensionless time. If

$$k(x) = \hat{k}h(x) \tag{12}$$

where $h(x)$ is monotonic and $\int_0^\infty h(x)f(x)dx = 1$, then

$$U(\tau) = \int_0^\infty f(x) \exp - [\tau h(x)]\, dx \tag{13}$$

In most cases it is possible to take $h(x) = x$ and

$$U(\tau) = \int_0^\infty f(x) e^{-x\tau}\, dx \tag{14}$$

and $U(\tau)$ is therefore the Laplace transform of $f(x)$ with x playing the role of the traditional t variable and τ that of s or p, Laplace transform variable.

If we approach the problem from the other direction and ask what distribution will produce a given alias, we have first to solve Eq. 9 and then find the distribution f of which it is the Laplace transform. Now (formally)

$$\begin{aligned} U(\tau) &= \sum_{j=0}^{\infty} (-)^j \frac{\tau^j}{j!} \int_0^\infty x^j f(x)\, dx \\ &= 1 - \tau + \int_{i=2}^{\infty} m_i \frac{(-\tau)^i}{i!} \end{aligned} \tag{15}$$

by the normalizations, Eqs. 2 and 3, and m_i is the ith moment of $f(x)$ about the origin. There is thus an implicit restriction on the function F of Eq. 9, namely that

$$F(1) = 1 \tag{16}$$

We also see immediately that there are certain kinetic laws that cannot be generated. For example the so-called zero-order reaction that just takes place at a constant rate so long as there is any reactant present. Thus F is a constant and by Eq. 1 must be 1, giving

$$U(\tau) = 1 - \tau, \quad 0 \leq \tau \leq 1 \tag{17}$$

But there is no $f(x) \geq 0$ with $m_0 = m_1 = 1$ and $m_i = 0$, $i \geq 1$. For an nth-order reaction with $n \geq 1$ however

$$F(U) = U^n \tag{18}$$

and

$$U(\tau) = \{1 + \tau(n-1)\}^{-1/(n-1)} \tag{19}$$

Inversion of this transform gives the gamma distribution

$$f(x) = \phi_\alpha(x), \quad \alpha = 1/(n-1) \tag{20}$$

which is the well known result, $n = (\alpha + 1)/\alpha$. $U(\tau)$ is shown in Figure 1 for several values of n.

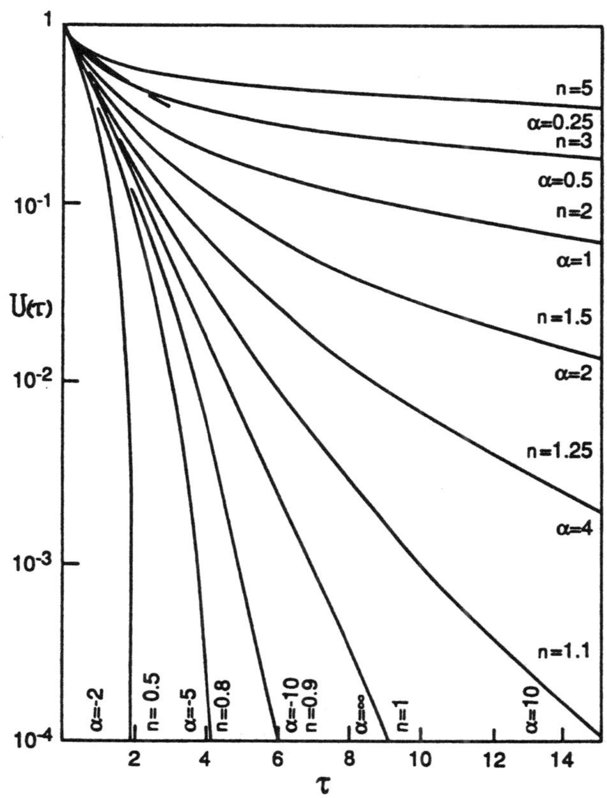

FIGURE 1 Fractional concentration as a function of time for reactions of various orders.

The apparent nth-order rate constant can be obtained from Eq. 19 by writing it in the dimensional form

$$c_0U = c_0\{1 + (n - 1)\hat{k}t\}^{-1/(n-1)} \tag{21}$$

and comparing this with the solution of

$$\frac{dc}{dt} = -k_n c^n, \quad c(0) = c_0 \tag{22}$$

which is

$$c(t) = c_0\{1 + (n - 1)k_k c_0^{n-1} t\}^{-1/(n-1)} \tag{23}$$

Thus,

$$k_n = \hat{k}c_0^{1-n} \tag{24}$$

This points up the desirability of using the dimensionless index variable x and of having the liberty of choosing $h(x)$ in Eq. 12. If $h(x) = x$ then k, the first-order constant, is being linearly distributed. If it is desirable to distribution k_n linearly then $h(x) = x[f(x)]^{n-1}$.

PARALLEL REACTION IN A DOUBLY DISTRIBUTED CONTINUUM

Let $H(x, y)$ and $h(x, y)$ be nonnegative functions that satisfy

$$\int_0^\infty \int_0^\infty H(x, y)\, dx\, dy = 1 \tag{25}$$

$h(0, y) = h(x, 0) = 0$, $\partial h/\partial x > 0$, $\partial h/\partial y > 0$ and

$$\int_0^\infty \int_0^\infty h(x, y)H(x, y)\, dx\, dy = 1 \tag{26}$$

Then, if $c_0u(x, y, t)\, dxdy$ is the concentration of material in the index interval $(x, x + dx) \times (y, y + dy)$,

$$u(x, y, 0) = H(x, y) \tag{27}$$

we have

$$U(\tau) = \int_0^\infty \int_0^\infty H(x, y)e^{-\tau h(x,y)}\, dx\, dy \tag{28}$$

In particular, if $H(x, y) = f(x)g(y)$, where f and g satisfy the normalizations, Eqs. 2 and 3, and $h(x, y) = xy$, $U(\tau)$ becomes

$$\begin{aligned} U(\tau) &= \int_0^\infty f(x) \left[\int_0^\infty g(y)e^{-\tau xy}\, dy \right] dx \\ &= \int_0^\infty f(x)V(\tau x)\, dx \end{aligned} \tag{29}$$

where $V(\tau)$ is the course of the reaction

$$\dot{V} = -G(V) \tag{30}$$

the kinetic alias generated by the distribution g.

Ho and Aris (1987) argued that any formulation of reaction in continuous mixtures must satisfy the single-component identity (SCI), namely that it should reduce to the kinetics of a single component when the mixture is pure. This is true of Eq. 29, for with $f(x) = \delta(x - x_0)$, $U(\tau) = V(x_0\tau)$. The corresponding $H(x, y) = \delta(x - x_0)g(y)$ is not "pure" in the second index, so to speak. The requirement should be enlarged to a discrete component each satisfying the kinetic law given by G. We see that this is

$$H(x, y) = \Sigma u_i \delta(x - x_i) g(y)$$

giving

$$U(\tau) = \Sigma u_i V(\tau x_i)$$

Astarita kinetics satisfy the SCI, but not necessarily the discrete-component identity (DCI). For example, putting $g(x, 0) = f(x) = \Sigma u_i \delta(x - x_i)$ in the uniform kinetic law

$$-\frac{\partial}{\partial t} g(x, t) = k(x) g(x, t) \left\{ \int_0^\infty K(y) g(y, t)\, dy \right\}^{n-1}$$

gives

$$-\dot{g}_i(t) = k_i g_i \{\Sigma u_j k_j g_j\}^{n-1}, \quad g_i(0) = u_i$$

When $n = 2$ this represents the second-order reaction scheme $\alpha_{ik} A_i + A_k \rightarrow$ products with certain constraints on the stoichiometry and kinetic constants. When there is a single component this reduces to a second-order reaction. This cooperative element in the Astarita kinetics is, of course, no defect—indeed it may be its strength.

EXAMPLES

We can have parallel nth-order reactions ($n > 1$) from the y continuum by taking

$$g(y) = \frac{\beta^\beta y^{\beta - 1} e^{-\beta y}}{\Gamma(\beta)}, \quad \beta = 1/(n - 1) \tag{31}$$

to give

$$V(\tau x) = \{1 + \tau x/\beta\}^{-\beta} \tag{32}$$

Thus,

$$\begin{aligned} U(\tau) &= \int_0^\infty f(x)\{1 + \tau x/\beta\}^{-\beta}\, dx = \left(\frac{\beta}{\tau}\right)^\beta \int_0^\infty f(x) \left\{\frac{\beta}{\tau} + x\right\}^{-\beta} dx \\ &= \left(\frac{\beta}{\tau}\right)^{\beta - 1} \int_1^\infty z^{-\beta} f\left[\frac{\beta}{\tau}(z - 1)\right] dz \\ &= \left(\frac{\beta}{\tau}\right)^\beta G_\beta \left\{ f(x); \frac{\beta}{\tau} \right\} \end{aligned} \tag{33}$$

where G_β is the generalized Stieltjes transform of order β (Erdelyi *et al.*, 1954; Widder, 1941).

If for $f(x)$ we use the gamma distribution $\phi_\alpha(x)$ and put $\sigma = \tau/\alpha\beta$, we have

$$U(\tau) = e^{1/2\sigma}\sigma^{\kappa}W_{\kappa,\mu}(1/\sigma) \tag{34}$$

where

$$\kappa = \frac{1}{2}(1-\alpha-\beta), \quad \mu = \frac{1}{2}(\alpha-\beta) \tag{35}$$

and $W_{\kappa,\mu}$ is Whittaker's form of the confluent hypergeometric function. An even more aposit form is the Tricomi function (Spanier and Oldham, 1987):

$$\begin{aligned}\psi(a;c;z) = {} & \frac{\Gamma(1-c)}{\Gamma(1+a-c)}\,{}_1F_1(a;c;z)\\ & + \frac{\Gamma(c-1)}{\Gamma(a)z^{c-1}}\,{}_1F_1(1+a-c;2-c;z)\end{aligned} \tag{36}$$

in terms of which

$$\begin{aligned}U(\sigma) &= \sigma^{-\alpha}\psi(\alpha;1+\alpha-\beta;1/\sigma)\\ &= \sigma^{-\beta}\psi(\beta;1+\beta-\alpha;1/\sigma)\end{aligned} \tag{37}$$

a form that has the advantage showing the invariance under the interchange of α and β. The integral forms of U are

$$\begin{aligned}U(\tau) &= \frac{\alpha^\alpha}{\Gamma(\alpha)}\int_0^\infty \frac{x^{\alpha-1}e^{-\alpha x}}{(1+\tau x/\beta)^\beta}\,dx\\ &= \left(\frac{\alpha\beta}{\tau}\right)^\beta e^{(\alpha\beta/\tau)}\frac{1}{\Gamma(\beta)}\int_1^\infty z^{-\alpha}(z-1)^{\beta-1}e^{-\alpha\beta z/\tau}\,dz\end{aligned} \tag{38}$$

and equivalent forms obtained by interchanging α and β. When α or β is an integer, say $\beta = b$, we have a finite expansion

$$\begin{aligned}U(\tau,\alpha,b) &= \left(\frac{\alpha b}{\tau}\right)^b \frac{e^{\alpha b/\tau}}{(b-1)!}\int_1^\infty z^{-\alpha}(z-1)^{b-1}e^{-(\alpha b/\tau)z}\,dz\\ &= \left(\frac{\alpha b}{\tau}\right)^b e^{\alpha b/\tau}\sum_{j=0}^{b-1}\frac{(-1)^j}{j!(b-1-j)!}\int_1^\infty z^{-\alpha+j}e^{-(\alpha b/\tau)z}\,dz\end{aligned} \tag{39}$$

If $\alpha = a$, an integer, the last integral is $E_{a-j}(ab/\tau)$, E being the exponential integral

$$E_p(s) = \int_1^\infty t^{-p}e^{-st}\,dt; \quad p \geq 1 \tag{40}$$

The formula will produce negative p, so we understand

$$\begin{aligned}E_0(s) &= \alpha_0(s) = (1-e^{-s})/s,\\ E_{-p}(s) &= \alpha_p(s) = \sum_{k=1}^{p}\frac{p!}{k!}\frac{e^{-s}}{s^{p-k+1}} + \frac{p!}{s^p}\alpha_0(s)\end{aligned} \tag{41}$$

Thus

$$U(\tau; a, b) = \left(\frac{ab}{\tau}\right) e^{ab/\tau} \sum_{j=0}^{b-1} \frac{(-1)^j}{j!(b-1-j)!} E_{a-j}(ab/\tau) \tag{42}$$

and in particular

$$U(\tau; 1, 1) = \frac{1}{\tau} e^{1/\tau} E_1(1/\tau) \tag{43}$$

$$\begin{aligned} U(\tau; 2, 1) &= \frac{2}{\tau} e^{2/\tau} E_2(2/\tau) \\ &= \frac{2}{\tau} e^{2/\tau} \{E_0(2/\tau) - E_1(2/\tau)\} = U(\tau; 1, 2) \end{aligned} \tag{44}$$

In general the second form in Eq. 38 can be expanded as

$$\begin{aligned} U(\tau; \alpha, \beta) = \left(\frac{\alpha\beta}{\tau}\right)^{\beta} e^{(\alpha\beta/\tau)} \sum_{j=0}^{\infty} \\ \cdot \frac{(-1)^j}{\Gamma(j+1)\Gamma(\beta-j-1)} E_{\alpha-\beta+1-j}\left(\frac{\alpha\beta}{\tau}\right) \end{aligned} \tag{45}$$

where E for a nonintegral index is the incomplete gamma function

$$E_{\gamma}(s) = s^{\gamma-1}\Gamma(1-\gamma; s) = \int_1^{\infty} t^{-\gamma} e^{-st}\, dt \tag{46}$$

Other special cases are:

$$\alpha = \beta, \quad U(\tau; \alpha, \alpha) = \Gamma(1-\alpha) e^{\alpha^2/2\tau} k_{-2\alpha}(\alpha^2/2\tau) \tag{47}$$

where k is Bateman's function

$$k_{\mu}(x) = \frac{2}{\pi} \int_0^{\pi/2} \cos(x \tan t - \mu t)\, dt \tag{48}$$

$\alpha = \beta + 1/2$,

$$\begin{aligned} U(\tau; \alpha; \alpha - 1/2) = \left[\frac{\alpha(2\alpha-1)}{\tau}\right]^{\alpha-1/2} \\ \cdot e^{\alpha(2\alpha-1)/4\tau} D_{1-2\alpha}\left[\frac{\sqrt{\alpha(2\alpha-1)}}{\sqrt{\tau}}\right] \end{aligned} \tag{49}$$

where $D_{\tau}(x)$ is the parabolic cylinder function; $\alpha = 1$, $\beta = 1/2$,

$$U(\tau; 1; 1/2) = \left(\frac{\pi}{2\tau}\right)^{1/2} e^{1/2\tau} \operatorname{erfc}\left(\frac{1}{2\tau}\right)^{1/2} \tag{50}$$

From the asymptotic expansion of the Tricomi function we get

$$\begin{aligned} U(\tau) = 1 - \tau + \frac{(\alpha+1)(\beta+1)}{\alpha\beta} \frac{\tau^2}{2!} \\ - \cdots = \sum_{j=0}^{\infty} (-)^j \frac{(\alpha)_j(\beta)_j}{(\alpha\beta)^j} \frac{\tau^j}{j!} \end{aligned} \tag{51}$$

which passes to the limit of $e^{-\tau}$ as a α and $\beta \to \infty$. On the other hand the power series expansions that give the behavior for large τ depend on the relative magnitudes of α and β. If

$$\begin{aligned}
\beta > \alpha + 1,\ U(\tau) &\sim \frac{\Gamma(\beta-\alpha-1)}{\Gamma(\beta)}\left(\frac{\alpha\beta}{\tau}\right)^{\alpha}\left(\beta-\alpha-1-\alpha\frac{\alpha\beta}{\tau}\cdots\right)\\
\beta = \alpha + 1,\ U(\tau) &\sim \left(\frac{\alpha\beta}{\tau}\right)^{\alpha}\left\{\frac{1}{\Gamma(1+\alpha)} - \frac{1}{\Gamma(-\alpha)}\frac{\alpha\beta}{\tau}\ln\frac{\alpha\beta}{\tau}\right\}\cdots\\
\alpha < \beta < \alpha + 1,\ U(\tau) &\sim \frac{\Gamma(\beta-\alpha)}{\Gamma(\beta)}\left(\frac{\alpha\beta}{\tau}\right)^{\alpha} + \frac{\Gamma(\alpha-\beta)}{\Gamma(\alpha)}\left(\frac{\alpha\beta}{\tau}\right)^{\beta}\cdots\\
\alpha = \beta,\ U(\tau) &\sim \left(\frac{\alpha^2}{\tau}\right)^{\alpha}\left\{\frac{1}{\Gamma(\alpha)}\ln\tau - \frac{2\gamma-\psi(\alpha)}{\Gamma(\alpha)}\right\}\cdots
\end{aligned} \tag{52}$$

When $\beta < \alpha$ we can reverse the roles of α and β. It is clear that, except when $\alpha = \beta$, $U(\tau)$ is asymptotically $0(\tau^{-\alpha})$ if $\alpha < \beta$ and $0(\tau^{-\beta})$ if $\beta < \alpha$. Thus, if $m = (\alpha + 1)/\alpha$, $n = (\beta + 1)/\beta$, the total concentration ultimately disappears as if by a qth-order reaction where q is the greater of m and n. If $\alpha = \beta$, the asymptotic kinetic law cannot be written down explicitly. $F(U)$ is obtained by eliminating τ between $F = A\tau^{-\alpha-1}\{\alpha \ln \tau - 1\}$ and $U = A\tau^{-\alpha} \ln \tau$ where $A = \alpha^{2\alpha}/\Gamma(\alpha)$. If the instantaneous order is defined as $\mu = d \ln F/d \ln U$, then

$$\mu = m\left\{1 - \frac{1}{\ln \tau(\alpha \ln \tau - 1)}\cdots\right\} \sim m \tag{53}$$

as we would expect. Figure 2 shows $U(\tau;\ \alpha,\ 1)$ for several values of α. The curve for $\alpha = \infty$ is the same as that for the second-order reaction ($n = 2$) in

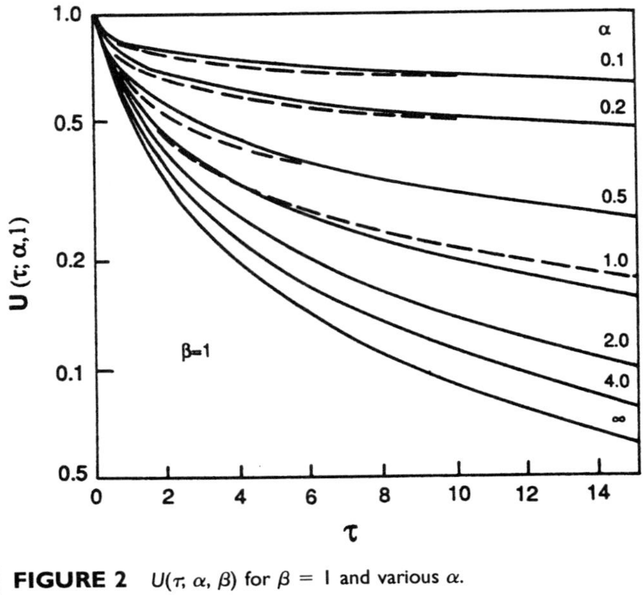

FIGURE 2 $U(\tau,\ \alpha,\ \beta)$ for $\beta = 1$ and various α.

Figure 1. The broken lines in Figure 2 are for nth-order reaction $n = (\alpha + 2)/\alpha$ which show a certain degree of approximation.

GENERALIZED BACKGROUND KINETICS

Astarita and Ocone's uniform kinetics (1988) may be used for the background kinetics (with distribution parameter y). It will be recalled that they showed that kinetics of the form

$$-\dot{u}(y, \tau) = yu(y, \tau)R\left[\int_0^\infty K(z)u(z, \tau)\, dz\right] \tag{54}$$

were patient of the the solution

$$u(y, \tau) = u(y, 0)e^{-y\omega(\tau)} \tag{55}$$

where

$$\dot{\omega}(\tau) = R\left(\int_0^\infty K(z)u(z, 0)e^{-z\omega(\tau)}\, dz\right) \tag{56}$$

For any initial distribution R is a function of ω, and $\omega(\tau)$, the eigenzeit, can be found implicitly by quadrature,

$$\tau = \int_0^{\omega(\tau)} d\omega/\tilde{R}(\omega) \tag{57}$$

where $\tilde{R}$ is the righthand side of Eq. 56 expressed as a function of ω. In particular, Astarita's uniform kinetics with what might be called the α, β gamma distributions

$$\begin{aligned} u(y, 0) &= \frac{\alpha^\alpha y^{\alpha-1}}{\Gamma(\alpha)} e^{-\alpha y}, \\ K(z) &= \frac{\Gamma(\alpha)}{\Gamma(\alpha + \beta)} (\alpha z)^\beta \quad \alpha, \alpha + \beta > 0 \end{aligned} \tag{58}$$

have the beautiful property of being able to imitate any overall reaction kinetics. We have

$$\begin{aligned} U(\tau) &= \int_0^\infty u(y, 0)e^{-y\omega}\, dy \\ &= \frac{\alpha^\alpha}{\Gamma(\alpha)} \int y^{\alpha-1}e^{-y(\alpha+\omega)}\, dy = \left\{1 + \frac{\omega(\tau)}{\alpha}\right\}^{-\alpha} \end{aligned} \tag{59}$$

and

$$\begin{aligned} \int_0^\infty K(z)u(z, \tau)\, dz &= \frac{\alpha^{(\alpha+\beta)}}{\Gamma(\alpha + \beta)} \int_0^\infty z^{\alpha+\beta+1}e^{-z(\alpha+\omega)}\, dz \\ &= \left\{1 + \frac{\omega(\tau)}{\alpha}\right\}^{-(\alpha+\beta)} = U^{(\alpha+\beta)/\alpha} \end{aligned} \tag{60}$$

Now

$$
\begin{aligned}
-\dot{U} &= \left\{\int_0^\infty yu(y,\tau)\,dy\right\} R\left[\int K(z)u(z,\tau)\,dz\right] \\
&= \left\{1+\frac{\omega(\tau)}{\alpha}\right\}^{-(\alpha+1)} R[U^{(\alpha+\beta)/\alpha}] \qquad (61)\\
&= U^{(\alpha+1)/\alpha} R[U^{(\alpha+\beta)/\alpha}] = G(U)
\end{aligned}
$$

so that the kinetics G is imitated if $R(W)$ is taken to be $W^{-(\alpha+1)/(\alpha+\beta)}$ $G[W^{-\alpha/(\alpha+\beta)}]$. The eigenzeit itself is given implicitly by

$$
\frac{\tau}{\alpha} = \int_1^{1+\omega(\tau)/\alpha} dw/R[w^{-(\alpha+\beta)}] = \int_{[1+\omega(\tau)/\alpha]^{-\alpha}}^{1} du/G(u) \qquad (62)
$$

The incorporation of Astarita's uniform kinetics into the separable form of a doubly-distributed reaction is now possible. The specification of $g(y)$ in $u(x, y, 0) = f(x)g(y)$, together with K and R, defines the kinetics G that will then be distributed by x, g, k, and R also define $\omega(\tau)$ and

$$
\begin{aligned}
U(z) &= \int_0^\infty f(x) \int_0^\infty g(y) e^{-y\omega(x\tau)}\,dy\,dx \qquad (63)\\
&= \int_0^\infty f(x) V(x\tau)\,dx \quad \text{where} - \dot{V} = G(V)
\end{aligned}
$$

Since $\tau = kt$ we can regard $x\tau$ as either $(xk)t$ or $k(xt)$. In the first case $f(x)$ distributes the rate constant about the mean k (since $\int_0^\infty xf(x)\,dx = 1$); in the second it distributes the reaction time about the mean t. The gamma distribution again comes to the fore, since a popular configuration is the chain of N stirred tanks for which

$$
f(x) = \frac{N^N x^{N-1}}{(N-1)!} e^{-xN} \qquad (64)
$$

We can in this way distribute reactions of order less than 1 using Astarita's striking result (1989) for $R(w) = w^{g-1}$. By Eq. 61 this will imitate an nth-order reaction where

$$
n = \{\alpha + 1 + (\gamma - 1)(\alpha + \beta)\}/\alpha \qquad (65)
$$

and, since, α, β, and γ are restricted only by $\alpha > 0$, $\beta > -\alpha$, $\gamma > 0$, n can be less than 1 and even negative. Thus

$$
U(\tau) = \int_0^{\xi(\tau)} f(x)\{1 - (1-n)x\tau\}^{1/(1-n)} \qquad (66)
$$

with

$$
\xi(\tau) = 1/(1-n)\tau
$$

is the extension to orders less than 1. This formula is the same as Eq. 38 when, as there, $\beta = 1/(1-n)$ and allowance is made for the concentrations reaching zero in a finite time. For computational purposes a function can conveniently be made zero when negative by averaging it with its absolute value.

DISCRETE DISTRIBUTIONS

Before going on to consider other distributions of the rate constant $k(x)$, it is worth drawing attention to a class of discrete distributions that asymptotically mimic nth-order reactions. Let a discrete mixture of nth-order reactants be described by the vector of concentrations $c_1(t), \ldots, c_p(t), \ldots, c_p, \ldots$ governed by

$$\frac{dc_p}{dt} = -k_p c_p^n, \; c_p(0) = c_{po} \tag{67}$$

Thus

$$c_p(t) = \{[c_{po}]^{1-n} + (n-1)k_p t\}^{-\beta}, \quad \beta = 1/\beta = 1/(n-1) \tag{68}$$

and the total concentration

$$C(t) = \sum_{p=0}^{P} c_p(t) \tag{69}$$

The irrefrangible component, $k_o = 0$, corresponds to $p = 0$ and the fraction of the total reactive material is

$$U = [C(t) - c_o]/C(0) = \sum_{p=1}^{P} c_p(t) \Big/ \sum_{p=0}^{P} c_{po} \tag{70}$$

Without loss of generality the k_p can be assumed to be rational and so made integral multiples of some increment Δk; that is, $k_p = j_p \Delta k$. We may replace the sums over p by sums over j from 0 to ∞ by setting to zero the c_{jo} for any j not equal to a j_p and recover everything thus far, but we may also have a denumerable infinity of components provided the summations coverage.

Consider the initial distribution

$$c_{jo} = c_o\{1 + j(n-1)\Delta k/\hat{k}_n\}^{-\beta}, \quad k_j = j\Delta k \tag{71}$$

Setting

$$u_o = \hat{k}/(n-1)\Delta k \tag{72}$$

we have

$$C(0) = c_o u_o^\beta \zeta(\beta; u_o) \tag{73}$$

where

$$\zeta(\beta; u) = \sum_{j=0}^{\infty} (u + j)^{-\beta} \tag{74}$$

is the so-called Hurwitz function. Substituting this in the formula for $c(t)$ gives

$$U(\tau) = u^\beta \zeta(\beta; u + 1)/u_o^\beta \, \zeta(\beta; u_o) \tag{75}$$

where

$$u = u_o/(1 + \tau) \quad \text{and} \quad \tau = \hat{k}_n c_o^{n-1} t \tag{76}$$

We may pass to the continuous case by letting $\Delta k \to 0$. In this case u and $u_o \to \infty$ and $u^\beta\, \zeta(\beta; u) \sim u/(\beta - 1) + 1/2 + \cdots$ and

$$U(\tau) \sim \frac{1}{1+\tau} \quad \text{as} \quad \Delta k \to 0 \tag{77}$$

Again the apparent reaction order is 2. Of course if Δk is kept fixed, the apparent reaction is n, for, with only $u \to \infty$,

$$U(\tau) \sim \frac{1}{(1+\tau)^\beta} \frac{\zeta(\beta)}{\zeta(\beta; u_o)} \tag{78}$$

DISTRIBUTIONS OF $k(x)$

If instead of taking $\hat{k}h(x, y) = \hat{k}xy$ we were to take it as $k(x)y$, then $V(\tau x)$ in Eq. 29 would be

$$V = \int_0^\infty g(y)[\exp - k(x)ty]\, dy \tag{79}$$

If we again take $g(y) = \phi_\beta(y)$, $\beta = 1/(n-1)$

$$V = \{1 + k(x)t/\beta\}^{-\beta} \tag{80}$$

and

$$U(\tau) = \int_0^\infty f(x)\{1 + k(x)t/\beta\}^{-\beta}\, dx \tag{81}$$

Now, if $k_n - \hat{k}_n x$ then, by Eq. 24,

$$k(x)t = \hat{k}_n c_o^{n-1} t[f(x)]^{n-1} x = x\,[f(x)]^{1/\beta}\tau \tag{82}$$

where $\tau = \hat{k}_n c_o^{n-1} t$ is the dimensionless time. Thus

$$U(\tau) = \int_0^\infty \left\{[f(x)]^{-1/\beta} + \frac{\tau x}{\beta}\right\}^{-\beta} dx \tag{83}$$

We will drop the requirement on $f(x)$ expressed by Eq. 3 and insist only on Eq. 2 being satisfied. Then, if

$$f(x) = \frac{\beta - 1}{\beta}\left(1 + \frac{x}{\beta}\right)^{-\beta}, \beta > 1 \tag{84}$$

we have

$$U(\tau) = \frac{\beta - 1}{\beta} \int_0^\infty \left\{1 + \frac{x}{\beta} + \frac{\tau x}{\beta}\left(\frac{\beta - 1}{\beta}\right)^{1/\beta}\right\}^{-\beta} = \frac{1}{1 + \tau'} \tag{85}$$

where

$$\tau' = \tau\left(\frac{\beta - 1}{\beta}\right)^{1/\beta} \tag{86}$$

This is an exact second-order process in the modified time τ'. Now $c(x, 0) = c_o f(x)$ and $k(x)t = (\hat{k}_n c_o^{n-1} t) f(x)^{1/\beta} x$. But $c_{oo} = f(0) = (\beta - 1)/\beta$, so that $\tau' = \hat{k}_n c_{oo}^{n-1} t$ and the second-order alias depends on the concentration at the irrefrangible limit. The distribution of nth-order reactants, $1 < n < 2$, given by Eq. 71 can appear to be obeying a second-order rate law exactly. We notice that as $n \to 1$, $\beta \to \infty$, and $f(x) \sim e^{-x}$, which agrees with the best known case.

Similarly, a distribution of nth-order reactants with $n < 1$ can simulate a second-order reaction precisely. For again taking $\beta = 1/|n - 1| = 1/(1 - n)$, as in Eq. 46, we have

$$U(\tau) = \int_0^{\xi(\tau)} f(x) \left\{ 1 - \frac{\tau x}{\beta} [f(x)]^{-1/\beta} \right\}^{\beta} dx \tag{87}$$

where $\xi(\tau)$ satisfies

$$\tau \xi [f(\xi)]^{-1/\beta} = \beta \tag{88}$$

Then

$$f(x) = \frac{\beta + 1}{\beta} \left(1 - \frac{x}{\beta} \right)^{\beta} \tag{89}$$

gives

$$U(\tau) = \frac{\beta + 1}{\beta} \int_0^{\beta/(1+\tau')} \left\{ 1 - \frac{x(1 + \tau')}{\beta} \right\}^{\beta} dx \tag{90}$$

where

$$\tau' = \left(\frac{\beta + 1}{\beta} \right)^{-\beta} \tau = \hat{k}_n c_{oo}^{n-1} t \tag{91}$$

is again the dimensionless time.

For a general distribution $f(x)$, and with $k(x)t = x[f(x)]^{1/\beta}\tau$, we have

$$U(\tau) = \int_0^{\infty} f(x) \left\{ 1 + \frac{\tau x}{\beta} [f(x)]^{1/\beta} \right\}^{-\beta}, \quad \text{if } n > 1, \beta = 1/(n - 1) \tag{92}$$

and

$$U(\tau) = \int_0^{\xi(\tau)} f(x) \left\{ 1 - \frac{\tau x}{\beta} [f(x)]^{-1/\beta} \right\}^{\beta}, \quad \text{if } n < 1, \beta = 1/(1 - n) \tag{93}$$

where $\xi(\tau)$ makes the expression in the bracket vanish. The values $n = 2$ and $n = 0$, which both make $\beta = 1/|n - 1| = 1$, are of special simplicity. In the case $n = 2$

$$U(\tau) = \int_0^{\infty} f(x)\{1 + \tau x f(x)\}^{-1} dx \tag{94}$$

and for $n = 0$

$$U(\tau) = \int_0^{\xi(\tau)} [f(x) - \tau x]\, dx \tag{95}$$

where $\xi(\tau)$ is the inverse function of

$$\tau = f(x)/x \tag{96}$$

The meaning of the single-component identity is of interest, for if we take

$$f(x) = \frac{\alpha^\alpha x^{\alpha-1}}{\Gamma(\alpha)} e^{-\alpha x} \tag{97}$$

and let $\alpha \rightarrow \infty$, the integral will be small unless x is near 1. But then $U(\tau) = 1/[1 + \tau f(1)]$ and this can be interpreted as a second-order reaction with time extended by a factor of $f(1)$. Then $\tau' = \hat{k}_2 c_o f(1)t = k_2(c_1)c_1 t$ where $k_2(c_1)$ is this rate constant associated with $x = 1$ and c_1 is the component concentration. This argument can be extended to a general β.

ASYMPTOTIC BEHAVIOR

It should be possible to say something about the asymptotic behavior of integrals like Eqs. 92 and 93 for a general $f(x)$ constrained by Eqs. 2 and 3, but the only case that has so far yielded to my attack is Eq. 94 with $f(x) = e^{-x}$. Letting $x = y/\tau$,

$$\begin{aligned}\tau U(\tau) &= \int_0^\infty e^{-y/\tau}\{1 + y - y(1 - e^{-y/\tau})\}^{-1}\, dy \\ &= \int_0^\infty \frac{e^{-y/\tau}}{1+y} \cdot \sum_{r=0}^{\infty} \left(\frac{y}{1+y}\right)^r (1 - e^{-y/\tau})^r\, dy\end{aligned} \tag{98}$$

and the fact that $y(1 - e^{-y/\tau})/(1 + y) < 1$ allows integration term by term. Also $1 - e^{-y/\tau} = y/\tau\ (1 - 1/2!\ y/\tau + 1/3!\ y^2/\tau^2 \ldots)$ so the rth term will be $O((\tau^{-r})$. Consider terms up to those of order τ^{-3}

$$\begin{aligned}\tau U &= \int_0^\infty \frac{e^{-y/\tau}}{1+y} dy + \frac{1}{\tau}\int_0^\infty \left(\frac{y}{1+y}\right)^2 e^{-y/\tau}\, dy \\ &+ \frac{1}{2\tau^2}\int_0^\infty \frac{y^3(y-1)}{(1+y)^3} e^{-y/\tau}\, dy\end{aligned} \tag{99}$$

Let $y = z - 1$; then the first integral is

$$e^{1/\tau}\int_1^\infty \frac{dz}{z} e^{-z/\tau} = e^{1/\tau}E_1(1/\tau) \sim e^{1/\tau}\left\{\ln \tau - \gamma + 1/\tau + 1/4\tau^2 \ldots\right\}$$

The second is

$$e^{1/\tau}\int_1^\infty \left(\frac{1}{z^2} - \frac{2}{z} + 1\right) e^{-x/\tau}\, dz = e^{1/\tau}$$

$$\cdot \left\{E_2\left(\frac{1}{\tau}\right) - 2E_1\left(\frac{1}{\tau}\right) + - 2E_1(1/\tau) + \tau(1 - c^{-1/\tau})\right.$$

The third is

$$e^{1/\tau}\int_1^\infty \left(z - 5 + \frac{9}{z} - \frac{7}{z^2} + \frac{2}{z^3}\right) e^{z/\tau}\, dz = e^{1/\tau}\left\{2E_3\left(\frac{1}{\tau}\right) - 7E_2\left(\frac{1}{\tau}\right)\right.$$
$$\left. + 9E_1\left(\frac{1}{\tau}\right) - 5\tau(1 - e^{-1/\tau}) + \tau e^{-1/\tau} + \tau^2(1 - e^{-1/\tau})\right.$$

The leading terms in the three cases can be combined to give

$$(\ln \tau)\left(1 - \frac{1}{\tau} + \frac{2}{\tau^2}\cdots\right) - \gamma + \frac{7 + 2\gamma}{2\tau} - \frac{51 + 26\gamma}{4\tau^2}$$

This leads to the expansion:

$$U(\tau) \sim \frac{\ln \tau - \gamma}{\tau} + \frac{2\ln \tau - 7 - 2\gamma}{2\tau^2} + 0(\tau^{-3}\ln \tau)$$
$$= \frac{\ln \tau - 0.57721}{\tau} + \frac{\ln \tau - 4.07722}{\tau^2} \qquad (100)$$

SEQUENTIAL PARALLEL REACTIONS

Let $A(x) \rightarrow B(x) \rightarrow C(x)$ be a parallel continuum of sequential reactions each of the first order. Then if $u(x, t)\, dx$, $v(x, t)\, dx$, and $w(x, t)\, dx$ are the three concentrations of the material in the index slice $(x, x + dx)$,

$$\begin{aligned} \dot{u}(x,t) &= -k(x)u(x,t) \\ \dot{v}(x,t) &= k(x)u(x,t) - k'(x)v(x,t) \\ \dot{w}(x,t) &= k'(x)v(x,t) \end{aligned} \qquad (101)$$

and $u(x, 0)$, $v(x, 0)$, $w(x, 0)$ can be specified. Thus

$$\begin{aligned} u(x,t) &= u(x,0)e^{-k(x)t} \\ v(x,t) &= u(x,0)\{\kappa(x) - 1\}^{-1}\{e^{-k(x)t} - e^{-k'(x)t}\} \\ &\quad + v(x,0)e^{-k'(x)t} \\ w(x,t) &= u(x,0) + v(x,0) + w(x,0) \\ &\quad - u(x,t) - v(x,t) \end{aligned} \qquad (102)$$

where

$$\kappa(x) = k'(x)/k(x) \qquad (103)$$

Thus

$$\begin{aligned} U(t) &= \int_0^\infty u(x,0)e^{-k(x)t}\, dx \\ V(t) &= \int_0^\infty u(x,0)\{\kappa(x) - 1\}^{-1}\{e^{-k(x)t} - e^{-k'(x)t}\}\, dx \\ &\quad + \int v(x,0)e^{-k'(x)t}\, dx \\ W(t) &= U(0) + V(0) + W(0) - U(t) - V(t) \end{aligned} \qquad (104)$$

For example if we use the Γ distribution we can without loss of generality take $k(x) = kx$ and set $\tau = kt$. In the first instance let κ be a constant and $v(x, 0) = w(x, 0) = 0$ with

$$u(x,0) = \frac{\alpha^\alpha x^{\alpha-1}}{\Gamma(\alpha)} e^{-\alpha x}, \quad \alpha > 0$$

Then

$$U(\tau) = (1 + \tau/\alpha)^{-\alpha} \tag{105}$$

and

$$V(\tau) = \{(1 + \tau/\alpha)^{-\alpha} - (1 + \kappa\tau/\alpha)^{-\alpha}\}/(\kappa - 1) \tag{106}$$

The relation between U and V can also be written

$$V = \{1 - [\kappa - (\kappa - 1)U^{1/\alpha}]^{-\alpha}\}U/(\kappa - 1) \tag{107}$$

$V(\tau)$ always goes through a maximum if $v(x, 0) = 0$ as we are assuming here; it is at

$$U_{\max} = \left[\frac{\kappa - \kappa^{1/(\alpha+1)}}{\kappa - 1}\right]^\alpha \tag{108}$$

and has a value

$$V_{\max} = U_{\max}^{(\alpha+1)/\alpha}/\kappa \tag{109}$$

The limiting cases require separate forms and we have

$$\kappa = 1, \quad V = \alpha(1 - U^{1/\alpha})U, \quad U_{\max} = \left(\frac{\alpha}{\alpha+1}\right)^\alpha, V_{\max} = \left(\frac{\alpha}{\alpha+1}\right)^{\alpha+1}$$

$$\alpha \leftarrow \infty, \quad V = (U - U^\kappa)/(\kappa - 1), \quad U_{\max} = \kappa^{-1/(\kappa-1)}, V_{\max} = \kappa^{-\kappa/(\kappa-1)}$$

$$\kappa = 1, \alpha \rightarrow \infty \quad V = U\ln(1/U) \; U_{\max} = e^{-1} = V_{\max}$$

Figure 3 shows plots of $V(U)$ over a range of α for $\kappa = 1$ and Figure 4 for $\alpha = 1$ with κ varying.

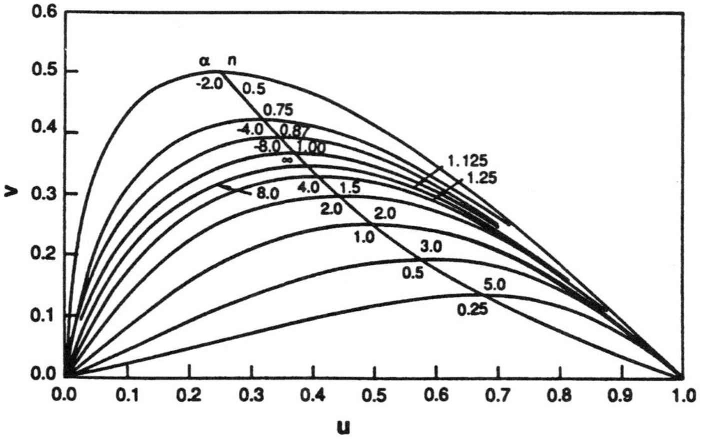

FIGURE 3 $V(U)$ for $\kappa = 1$ and various α.

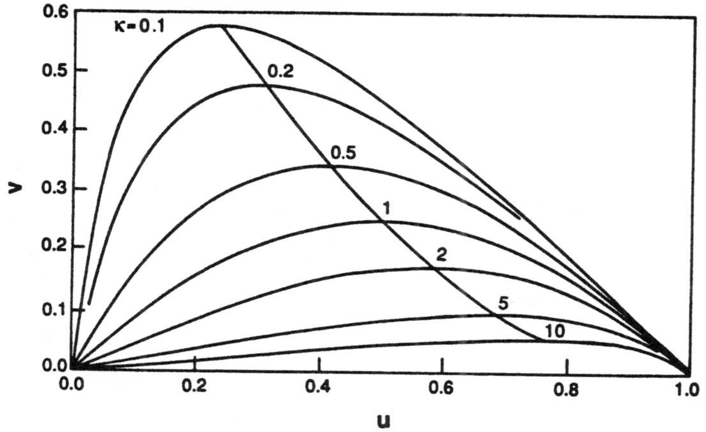

FIGURE 4 $V(U)$ for $\alpha = 1$ and various κ.

The differential equation obeyed by U is, of course, that of a reaction of order $(\alpha + 1)/\alpha$; that governing V can be put in many forms by eliminating τ between V and $\dot{V}$. One such is:

$$\dot{V} = U^{\alpha+1/\alpha} - \frac{\kappa}{\kappa - 1}\{U^{\alpha+1/\alpha} - [U - (\kappa - 1)V]^{\alpha+1/\alpha}\} \tag{110}$$

Since V is not a monotonic function of U we cannot solve Eq. 107 to obtain $\dot{V}$ in the form $U^{(\alpha+1)/\alpha} - f(V)$. In the initial stages of reaction $f(V) = \kappa\alpha + 1/\alpha\ V$, and as we would expect this second reaction starts out as first order. As $U \to 0$ and $V \sim \mu U$, $\mu = (1 - \kappa^{-\alpha})/(\kappa - 1)$, then $f(V) = k_0 V^{\alpha-1/\alpha}$ where

$$k_0 = \kappa(\kappa - 1)^{1/\alpha}(\kappa^{\alpha} - 1)^{-(\alpha+1)/\alpha}(\kappa^{\alpha+1} - 1).$$

We have only considered the simplest case of constant $\kappa(x)$, but there are no reasons to limit the sequential reaction to this form save the wantonness of idle proliferation and the awkwardness of the resulting integrals. Thus if the second step were at a rate independent of x we would have $\kappa(x) = \kappa/x$ and with the usual distribution x,

$$V(\tau) = \int_0^{\infty} \frac{(\alpha x)^{\alpha}}{\Gamma(\alpha)} e^{-\alpha x} \frac{e^{-xz} - e^{-\kappa\tau}}{\kappa - x}\, dx$$

an integral that can be resolved into a series of incomplete gamma functions.

MECHANISMS

Besides having the capital property of being able to imitate any kinetics—see Eq. 61—Astarita's uniform kinetics are often patient of a mechanistic derivation. Thus, for what Astarita and Ocone call uniform generalized Langmuir isotherm kinetic mechanism (UGLIKM), we might take $\theta(x)\, dx$ to be the fraction of catalytic surface occupied by the species $A(x)$ with index in

$(x, x + dx)$. Then if the system is at pseudosteady state

$$r(x) = k_a(x)c(x, t)\left\{1 - \int_0^\infty \theta(x')\,dx'\right\} - k_d(x)\theta(x) = k_r(x)\theta(x)$$

where $k_a(x)$, $k_d(x)$, and $k_r(x)$ are the adsorption, desorption, ad reaction rate constants. Thus

$$\theta(x) = \left\{1 - \int_0^\infty \theta(x')\,dx'\right\} \cdot K(x)c(x, t),$$
$$K(x) = k_a(x)/\{k_d(x) + k_r(x)\}$$

whence

$$\Theta = \int_0^\infty \theta(x)\,dx = K(x)c(x, t)\Big/\left\{1 + \int_0^\infty K(z)c(z, t)\,dz\right\}$$

This leads straight to be UGLIKM kinetics and provides an interpretation for the kernel $K(x)$.

Progress on the same lines is possible for a nonuniform surface but it leads, as might be expected, to a nonuniform GLIKM. Thus if $c(x, t)\,dx$ is again the concentration of material with index in $(x, x + dx)$, $\Theta(y)\,dy$ is the fraction of surface sites characterized by y in $(y, y + dy)$ and $\theta(x, y)\,dx\,dy$ the fraction of y sites occupied by x material. Then

$$r(x, y)\,dx\,dy = k_a(x, y)c(x, t)\,dx\left\{\Theta(y)\,dy - \int \theta(x'y)\,dx'\,dy\right\} - k_d(x, y)\theta(x, y)\,dx\,dy = k_r(x, y)\theta(x, y)\,dx\,dy$$

If the p.s.s.h. holds

$$\theta(x, y) = K(x, y)c(x, t)\left\{\Theta(y) - \int K(x', y)\,dx'\right\}$$

giving

$$\int \theta(x', y)\,dx' = \Theta(y)\int K(x', y)c(x', t)\,dx\Big/\left\{1 + \int K(x', y)c(x', t)\,dx'\right\}$$

Thus

$$r(x, y) = k_r(x, y)K(x, y)\Theta(y)c(x, t)\Big/\left\{1 + \int K(x', y)c(x', t)\right\}$$

and

$$r(x) = \int \frac{k_r(x, y')K(x, y')\Theta(y')}{1 + \int K(x', y')c(x', t)}\,dy'c(x, t)$$

Only if $k_r(x, y)K(x, y)\Theta(y)$ is simply separable—that is, the product of a function of x with a function of y—will the use of the eigenzeit lead to a solution.

ACKNOWLEDGMENT

This paper has benefited greatly from conversations and correspondence with G. Astarita, T. C. Ho, and F. J. Krambeck and from the catalytic interest of this journal's editor in catalysis.

LITERATURE CITED

Alberty, R. A., and I. Oppenheimer, "A Continuous Thermodynamics Approach to Chemical Equilibrium Within an Isomer Group," *J. Chem. Phys.*, **81,** 4603 (1984).

Amundson, N. R., and A. Acrivos, "On the Steady State Fractionation of Multicomponent and Complex Mixtures in an Ideal Cascade." I–VI, *Chem. Eng. Sci.*, **4,** 29; 68; 141; 159; 206; 249 (1955).

Aris, R., "Prolegomena to the Rational Analysis of Systems of Chemical Reactions. II: Some Addenda," *Arch. Rat. Mech. Anal.* **27,** 356 (1968).

Aris, R., and G. Gavalas, "On the Theory of Reactions in Continuous Mixtures," *Phil. Trans. Roy. Soc.*, **A260,** 351 (1966).

Astarita, G., "Lumping Nonlinear Kinetics: Apparent Overall Order of Reaction, *AIChE J.* **35,** 529 (1989).

Astarita, G., and R. Ocone, "Lumping Nonlinear Kinetics," *AIChE J.*, **34,** 1299 (1988).

Briano, J. G., and E. C. Glandt, "Statistical Thermodynamics of Polydisperse Fluids," *J. Chem. Phys.*, **80,** 3336 (1984).

Chou, M. Y., and T. C. Ho, "Continuum Theory for Lumping Nonlinear Reaction Mixtures," *AIChE J.*, **34,** 1519 (1988).

———, "Lumping Coupled Nonlinear Reactions in Continuous Mixtures," *AIChE J.* **35,** 533 (1989).

de Donder, Th., *L'Affinite,* 2[e] pt., Gauthier-Villars, Paris (1931).

Erdelyi, A., ed., *Tables of Integral Transforms, McGraw-Hill, New York (1954).*

Golikeri, S. V., and D. Luss, "Diffusional Effects in Reacting Mixtures," *Chem. Eng. Sci.*, **26,** 237 (1971).

Ho, T. C., and R. Aris, "On Apparent Second-Order Kinetics," *AIChE J.*, **33,** 1050 (1987). The major part of this paper inadvertently repeated a result due to Krambeck (1984) and is only included here because it is more accessible. The authors should have been aware of the result and apologize for the duplication.

Kemp, R. D. D., and B. W. Wojciechowski, "The Kinetics of Mixed Feed Reactions," *Ind. Eng. Chem. Fundam.*, **13,** 332 (1974).

Krambeck, F. J., "Computers and Modern Analysis in Reactor Design," *Proc. ISCRE 8; Ind. Chem. Eng. Symp. Ser.*, **A260,** 351 (1984).

Loveday, B. K., "Analysis of Froth Flotation Kinetics," *Inst. Min. Metall. Trans.*, **C75,** 219 (1966).

Luss, D. G., and P Hutchinson, "Lumping of Mixtures with Many Parallel *N*th-order Reactions," *Chem. Eng. J.*, **2,** 172 (1971).

Malhotra, A., and A. Sadana, "Effect of Activation Energy Microheterogeneity on First-order Enzyme Deactivation," *Biotech. Bioeng.*, **30,** 108 (1987).

Prasad, G. N., "Kinetic Modeling of Coal Liquefaction Reactions," Ph.D. Thesis, Monash Univ., Australia (1986).

Prasad, G. N., J. B. Agnew, and T. Sridhar, "Continuous Reaction Mixture Model for Coal Liquefaction. Theory," *AIChE J.*, **32,** 1277 (1986a).

———, "Continuous Reaction Mixture Model for Coal Liquefaction. II: Comparison with Experiments on Catalyzed and Uncatalyzed Liquifaction of Coals of Different Rank," *AIChE J.*, **32,** 1288 (1986b).

Spanier, J., and K. B. Oldham, *An Atlas of Functions,* Hemisphere, New York (1987).
Weekman, V. W., "Lumps, Models and Kinetics in Practice," *Chem. Eng. Prog. Monogr. Ser.* **75**(11), 3(1979).
Widder, D. V., *The Laplace Transform, Princeton Univ. Press, Princeton (1941).*
Zeman, R. J., and N. R. Amundson, "Continuous Polymerization Models," *Chem. Eng. Sci.,* **20,** 331;637 (1965).

H.*

REACTION OF A CONTINUOUS MIXTURE IN A BUBBLING FLUIDIZED BED

N. R. AMUNDSON† AND R. ARIS††

†Department of Chemical Engineering, University of Houston, Houston, USA ††Department of Chemical Engineering and Materials Science, University of Minnesota, Minneapolis, USA

The simplest model of a bubbling fluidized bed, with uniform bubbles exchanging matter with a dense phase of catalytic particles which promote a continuum of parallel first order reactions is considered. It is shown that the system behaves like a stirred tank with two feeds; the one, direct at the inlet; the other, distributed from the bubble train. The basic results can be extended to cases of catalyst replacement for a single reactant and to Astarita's uniform kinetics for the continuous mixture.

INTRODUCTION

Every subject has its canonical examples. They may be case-studies, models or seminal ideas; they may be expressed with terseness of a theorem or with the amplitude of a monograph, but they are so taut and compelling that they bring instant conviction of their importance and centrality. Take Proposition 11, Problem 6 of Newton's *Principia*, for example, in which he establishes the connection between the ellipticity of the orbit and the inverse square law for the centripetal force tending to the focus[1]. Or, take the work of another Trinity man, Richard Bentley, whom Brink calls 'the Newton of European philological and literary studies'[2], with his youthful *Epistola ad cl(arissimum) v(irum) Iohnnem Millium, S.T.P*[3]. Or his controversial and acerbic *Dissertation on the Epistles of Phalaris*[4] or, more importantly, his *Horace*, of which the contemporary Horatian authority has said that 'his procedure amounts to a fundamentally new and true approach to any text that has come down to us from a remote original through a complex manuscript tradition'[5]. Or, to jump three centuries into the Trinity College of John Davidson himself, take the papers of Sir Geoffrey Taylor, whose name is associated with so many seminal problems[6]—Taylor vortices, the fingering of the front of one liquid pushing another, diffusion by continuous movements, Taylor diffusion, the swimming of snakes and spermatazoa and more.

If chemical engineering needed such a *locus classicus*, it could well be the fluidized bed. Used for heat transfer and mass transfer, separations and

* Reprinted with the permission of the Institution of Chemical Engineers from *Trans* IChemE, volume **71**, part A, pp. 611–617, 1993.

reactions, it is essentially a chemical engineering creature. Of the many chemical engineers who have worked on fluidized beds, not the least is John Davidson, and, of his many publications, his 1963 monograph with David Harrison[7], still stands out as the very model of the adroit combination of physical and mathematical reasoning, of scientific theory and engineering experiment. Though, like an early tall building in a growing city of skyscrapers, it has been overshadowed by the subsequent literature[8], it can still be read for profit as well as for pleasure, and has the terseness of style and the compelling quality of argument that remind one of the earlier expressions of learning that have emanated from Henry the Eighth's foundation.

Catalytic cracking was one of the first uses of the fluid bed as a reactor and there is an exposition of the two basic models—bubble flow through a dense phase which is either completely mixed or in plug flow—in chapter 6 of the Davidson-Harrison monograph. The kinetic law is taken to be first order and this is reasonable, both on aesthetic grounds (linear systems are basic to any situation and, though their predictability deprives them of the excitement of nonlinear systems, they are relatively easy to solve and give easily interpretable results) and from practical considerations (for it does agree with certain experimental results). Curiously, however, the oil industry had for years found the relationship of the fraction of remaining crude to the 'severity' of the reaction, as they call it, fitted the hyperbola of the second order reaction better than the exponential of the first. This observation gave birth to a theory of reactions in continuous mixtures[9] of which this aliasing of reaction order is an elementary result. Let $c(x, t)\mathrm{d}x$ be the concentration of material in the interval $(x, x + \mathrm{d}x)$ that cracks with a first order rate constant $k(x)$, x being an index variable defined on $0 \leq x < \infty$. Let $C(t) = \int c(x, t)\mathrm{d}x$ be the total concentration of material of any index (except where specified all integrals in this paper will run from 0 to ∞), then

$$C(t) = \int c(x, 0) \exp(-k(x)t)\mathrm{d}x \tag{1}$$

and we have only to put $c(x, 0) = C_0 \exp(-x)$, $k(x) = k_2 C_0 x$ to obtain

$$C(t) = C_0/(1 + k_2 C_0 t), \tag{2}$$

which satisfies the quadratic differential equation for the second order disappearance of the 'lumped' species,

$$\mathrm{d}C/\mathrm{d}t = -k_2 C^2, \quad C = C_0 \quad \text{at } t = 0 \tag{3}$$

Three observations may be made at this point. Firstly, that if $k(x)$ is monotonic, we lose no generality by making it proportional to x, for a monotonic $y = k(x)$ has an inverse $x = K(y)$ and

$$\begin{aligned} C(t) &= \int c(x, 0) \exp(-k(x)t)\mathrm{d}x \\ &= \int c(K(y), 0)k'(y) \exp(-yt)\mathrm{d}y. \end{aligned} \tag{4}$$

Secondly, we can stretch the linear scale and write $c(x, 0) = C_0 f(x)$, where the zeroth and first moments $\int f(x)\mathrm{d}x = \int x f(x)\mathrm{d}x = 1$; then $k(x) = \langle k \rangle x$, where $\langle k \rangle = \int k(x)f(x)\mathrm{d}x$. All the integrals are, of course, taken from zero to infinity, since, for any interval $a \leq x \leq b$ not represented in the mixture, $f(x) = 0$. Thirdly, Krambeck[10] has shown that any distribution gives apparent second order kinetics for the lump, as $t \rightarrow \infty$, provided only that $c(0, 0) \neq 0$.

It is of the first importance to remember that, though x may be regarded as dimensionless, $c(x, t)$ is a concentration density (in the same meaning of the word as in 'probability density'), not a concentration. Only integrals of c with respect to x have the dimensions of a concentration.

GAMMA DISTRIBUTIONS

Of all the normalized distributions $f(x)$, the one that has proved to be the most useful and versatile is the gamma distribution

$$g_n(x) = n^n x^{n-1} e^{-nx}/(n-1)! \tag{5}$$

defined for n a positive integer. It can be generalized to any real $v > 0$ by replacing n by v and $(n - 1)!$ by $\Gamma(v)$. As n(or v) tends to infinity, $g_n(x)$ becomes the Dirac delta function $\delta(x - 1)$, but its approach is slow as the standard deviation is $n^{-1/2}$.

To be slightly more general for a moment, the gamma distribution with mean a is obtained by replacing x by x/a (not forgetting the x in the dx that goes along with the density function). To avoid the inherent inaccuracy of multiplying very large and very small numbers it is best calculated by

$$g_n(x, a) = (n^{1/2}/x)\exp\{-nM(x/a) - L(n)\} \tag{6}$$

where

$$M(x/a) = (x/a) - 1 - \ln(x/a)$$

and

$$L(n) = \ln(n-1)! - (n - 1/2)\ln(n) + n. \tag{7}$$

A useful asymptotic formula for an integral $\int h(x)g_n(x)\mathrm{d}x$ is

$$\int h(x)g_n(x)\mathrm{d}x = \sum \mu_{n,p} h^{(p)}(a)a^p/p! \tag{8}$$

where $\mu_{n,p}$ is the dimensionless p^{th} moment about the mean, 1, of $g_n(x)$;

$$\begin{aligned} &\mu_{n,0} = 1, \mu_{n,1} = 0, \mu_{n,2} = 1/n, \mu_{n,3} = 2/n^2, \\ &\mu_{n,4} = (3/n^2) + (6/n^3). \end{aligned} \tag{9}$$

The Gamma distribution has to discretized for some calculations. If M is any integer and $G_n(x)$ the indefinite integral of $g_n(x)$, i.e. $G_n'(x) = g_n(x)$, $G_n(0) = 0$, let $x_{m,n}$ be defined by $G_n(x_{m,n}) = m/M$ and $y_{m,n}$ by $M \int x g_n(x)\mathrm{d}x$ where the integration is from $x_{m-1,n}$ to $x_{m,n}$. This divides the region under the curve into M parts of equal area and associates the centroid of each part with it. Then any integral of the form $\int h(x)g_n(x)\mathrm{d}x$ is replaced by $\Sigma\, h(y_{m,n})/M$.

A formula of great value in connection with the use of the Gamma distribution is

$$\int_0^\infty g_n(x)\mathrm{d}x/(x+y) = n\mathrm{e}^{ny}\mathrm{E}_n(ny) \tag{10}$$

where $\mathrm{E}_n(z)$ is the exponential integral

$$\mathrm{E}_n(z) = \int_1^\infty t^{-n}\mathrm{e}^{-zt}\,\mathrm{d}t. \tag{11}$$

$\mathrm{e}^x\mathrm{E}_n(x)$ is bounded by two hyperbolas that translate into a powerful inequality for the integral in equation (10), namely

$$1/(1+y) < \int_0^\infty g_n(x)\mathrm{d}x/(x+y) \leq 1/(1+y-n^{-1}) \tag{12}$$

APPLICATION OF THE GAMMA DISTRIBUTION

The introductory example may be reworked using the Gamma distribution, since the special case given there is $n = 1$. Let $c(x, 0) = C_0 g_n(x)$ *where* C_0 is the total initial concentration. Let the first order rate constant be $k(x) = kx$ and make time dimensionless as kt. This reaction time or intensity of reaction—severity of reaction as the oil people have it—is really the Damköhler number, Da, for the reactor, with t the time of reaction if it is a batch reactor or the residence time if a PFTR. Thus

$$\gamma(x, Da) = c(x, t)/C_0 = g_n(x)\exp - xDa \tag{13}$$

and

$$\Gamma(Da) = C(t)/C_0 = \int \gamma(x, Da)\mathrm{d}x = (1 + Da/n)^{-n} \tag{14}$$

It follows that the lumped kinetics is of apparent order

$$N = (n+1)/n \tag{15}$$

since $\mathrm{d}\Gamma/\mathrm{d}Da = -(1 + Da/n)^{-n-1} = -\Gamma^N$. All the integrals are well defined if n takes non-integral values v, provided $v > 0$. Hence N can take any value greater than 1. The limit of equation (11) as $n \to \infty$ is clearly exp-Da, the first order reaction result, which is at it should be since the initial distribution is a delta function and the continuous species is really discrete.

Astarita and Ocone[11] generalized the component kinetics from the first order to what they called 'uniform' kinetics

$$\mathrm{d}c(x,t)/\mathrm{d}t = -k\,c(x,t)\,F\left[\int K(y)c(y,t)\mathrm{d}y\right] \tag{16}$$

The uniformity lies in the fact that K is not a function of x and this allows the problem to be reduced to linearity by suitably warping time. Astarita[12] shows that in this way any order of reaction may be achieved for the lump. Indeed Astarita and Aris[13] showed that any kinetics could be imitated by choosing F and K adroitly. We shall return to 'uniform kinetics' later, but, for the moment, will stick with the linear case.

A GENERAL THEOREM FOR SIMPLE, LINEAR REACTOR MODELS

By a reactor model, we mean a system of equations (algebraic, ordinary, or partial differential, functional or integral) which purports to represent a chemical reactor in whole or in part. (The adequacy of such a representation is not at issue here.) It will be called *linear* if all its equations are linear and *simple* if its input and output can be characterized by single, concentration-like variables, u_0 and u. The relation of input and output will also depend on a set of parameters (such as Damköhler number. Thiele modulus, etc.) which may be denoted by $\mathbf{p}$. Let $\Lambda(\mathbf{p})$ be the value of u when $u_0 = 1$. Then, if the input is a continuous mixture with distribution $g(x)$ over an index variable x on which some or all of the parameters may depend, the output is distributed as $\gamma(x) = g(x)\Lambda(\mathbf{p}(x))$ and the lumped output is

$$\Gamma(\mathbf{p}^*) = \int g(x)\Lambda(\mathbf{p}(x))\mathrm{d}x. \tag{17}$$

Here $\mathbf{p}^*$ is the set of characteristic values of the parameters; i.e. $\mathbf{p}(x) = \mathbf{p}^*\omega(x)$ where $\omega(x)$ has values centered on 1. Often we can set $\mathbf{p}^* = \int \mathbf{p}(x)\, g(x)\, \mathrm{d}x$. The proof is really a statement of what linearity means, for if $g(x)\mathrm{d}x$ is the input concentration, $g(x)\mathrm{d}x.\Lambda(\mathbf{p}(x))$ is the output when the parameter values are $\mathbf{p}(x)$. Here x serves merely as an identifying mark, being truly an index variable and the integration in equation (14) follows from the superposition principle for linear systems. The same principle allows an obvious extension to multiple input, multiple output linear systems. Λ becomes a matrix whose elements are the response at one of the out-ports to a unit input at one of the in-ports, the input at all the others being zero. The detail of this case will not be elaborated here, but rather shall we pass to an application.

APPLICATION TO A MODEL OF THE BUBBLING FLUIDIZED BED

Two models of the bubbling bed are referred to by Davidson and Harrison in chapter 6 of their monograph. In both models the bubbles (of average volume V and in number per unit reactor volume N) rise through the particulate, or dense, phase with an absolute velocity of U_A and exchange reactant with it at an equivalent flow rate Q. The difference between the models lies in the assumptions made about the dense phase; in the one this is held to be uniform throughout the reactor, while in the other the dense phase is in plug flow. We shall consider only the first of these; the second can be treated similarly as far as the application of the theorem is concerned, but has ramifications which are still under investigation.

Reaction only takes place in the dense phase since that is where the catalyst particles are. Since the exchange is with a uniform environment where the concentration is c_p, we can see that by the time the bubble has reached the top of the bed, the concentration of reactant in it is $c_p + (c_0 - c_p).\exp - Tr$, where c_0 is the entering concentration, H the height of the bed and $Tr = QH/U_AV$ is a dimensionless transfer number. By doing a mass balance on the dense phase as a whole[14] we obtain a linear equation for c_p in terms of the inlet concentration c_0. The total fractional concentration of reactant that is left in the emerging gas stream is

$$\gamma = \beta e^{-Tr} + (1 - \beta e^{-Tr})^2/(Da + 1 - \beta e^{-Tr}) \tag{18}$$

where $\beta = 1 - U_0/U$ and the Damköhler number is $Da = k'H_0/U$, k' being the rate constant, is a measure of the intensity of reaction. H_0 is the height of the bed at incipient fluidization; $H - H_0 = NVH$.

Equation (18) with a square in the numerator of the second term looks a little odd at first, but, in fact it unfolds itself with unusual clarity. We first observe that even when the reaction is infinitely fast, $Da \to \infty$, a fraction β e^{-Tr} of the inlet concentration will remain in the bubbles when they reach the top of the bed. Thus γ can never be less that β e^{-Tr}; so if we subtract this, to allow $\gamma - \beta e^{Tr}$ to go to zero, and rescale by dividing by $1 - \beta e^{-Tr}$, the effective amount that is available to the dense phase, we have:

$$\gamma^* = (\gamma - \beta e^{-Tr})/1 - \beta e^{-Tr}) = 1/(1 + Da^*) \tag{19}$$

where

$$Da^* = Da/(1 - \beta e^{-Tr}) \tag{20}$$

is the modified Damköhler number. It is enhanced in compensation for the fraction that is not available under any condition, for if we recall the definition and the fact that the available reactant is $qc_0(1 - \beta e^{-Tr})$, we see that $Da^* = k'H_0c_0qc_0(1 - \beta e^{-Tr})$ = maximum reactive flux/available feed flux. In terms of this modified Damköhler number the well mixed dense phase behaves exactly as it should—like a stirred tank. It may be shown that if $f(t)$ is the residence time distribution of the bubble phase [in our plug flow model it is $\delta(t - 1)$] the exponential e^{-Tr} in equations (19) and (20) need only be replaced by the Laplace transform of $f(t)$ with Tr playing the role of the transform variable.

Mathematicians have their names attached to theorems, lemmas and even conjectures; naturalists have species and genera called after them; physicists have their principles and chemists their reactions and reagents. Engineers can claim the dimensionless parameters, for they, who, in their function as designers, have often to work with units, appreciate more than anyone the conceptual beauty of the dimensionless number, invested as it is with full contextual meaning and magnitude. I submit that it would be more than appropriate to call this transfer parameter the Davidson number, for who has done more to elucidate the mechanism of this transfer process than John Davidson? Damköhler has preempted the initial letters *Da*, but that is no matter, for 'transfer' gives us '*Tr*' and we immediately think of 'Trinity' and its present Vice-Master.

We wish to see what the overall conversion of a continuous mixture will be, but, first, we have to ask which parameters will depend on x, the index variable of the continuous mixture. Clearly k' the rate constant in the Damköhler number will be a function of x, and, if monotonic, can be put equal to $Da.x$. The parameter β is clearly hydrodynamic and so, for the most part, are the terms in the Davidson number. The only term in the equation 6.21 of Davidson and Harrison that might depend on x is the gas phase diffusivity, and this appears under a square root sign in the second of two terms. *Tr* was found to be virtually constant, with a value close to three, in a series of experiments by Orcutt which Davidson and Harrison analyze. We will therefore assume that only the Damköhler number varies with x, and that this variation is linear. Then for the well mixed model

$$\Gamma_n(Da, Tr, \beta) = \int [\beta e^{-Tr} + (1 - \beta e^{-Tr})^2/ (Da.x + 1 - \beta e^{-Tr}] g_n(x)dx = \beta e^{-Tr} + [(1 - \beta e^{-Tr})/Da^*] \int g_n(x)dx/ (x + 1/Da^*) \quad (21)$$

Again setting

$$\Gamma_n^*(Da, Tr, \beta) = [\Gamma_n(Da, Tr, \beta) - \beta e^{-Tr}]/(1 - \beta e^{-Tr}) \quad (22)$$

and using the relation (10), we have

$$\Gamma_n^*(Da, Tr, \beta) = (n/Da^*)e^{n/Da^*}E_n(n/Da^*) \quad (23)$$

The inequalities for the exponential integral stated above, equation (12), then give

$$1/(1 + Da^*) < \Gamma_n^*(Da, Tr, \beta) \leq 1/(1 + Da^*(n - 1)/n). \quad (24)$$

These bounds indicate that the curves for various n will not lie very far apart and to separate the curves more effectively $\Delta_n = (1 + Da^*)\Gamma_n^*(Da, Tr, \beta)$ is plotted against Da^* in Figure 1. For small values of Da^* the asymptotic

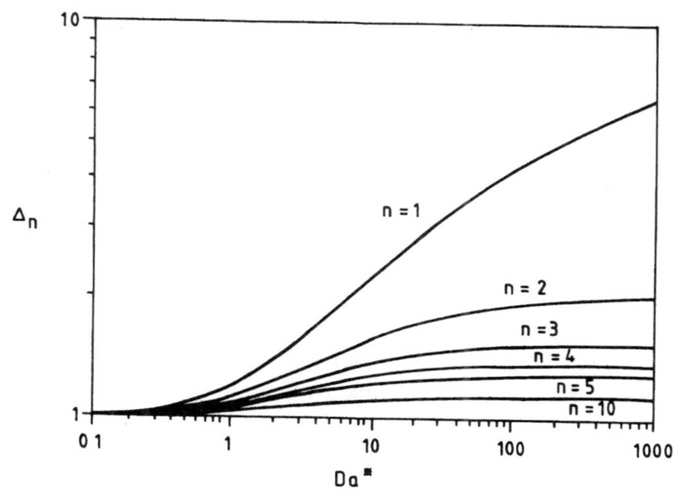

FIGURE 1 Deviation from the output from stirred tank behaviour when the feed is a Γ-distribution with parameter n.

expansion[15] of $E_n(z)$ gives

$$\Delta_n = 1 + Da^{*2}/n - (n+1).2\,Da^{*3}/n^2 + (n+1)(n+2)3\,Da^{*4}/n^3 - \cdots \tag{25}$$

The series expansion for small z gives the expansion for large Da^*. It is messy, but for the record we give here the terms up to order $o(1/Da^*)$;

$$n = 1, \quad \Delta_1 = \ln(Da^*) - \gamma + 2\ln(Da^*)/Da^* + (1 - 2\gamma)/Da^* + \cdots \tag{26}$$

$$n = 2, \quad \Delta_2 = 2 - 4\ln(Da^*/2)/Da^* + (1 + 2\gamma)/Da^* - \cdots \tag{27}$$

$$n = 3, \quad \Delta_3 = 3/2 - 3/Da^* + \cdots \tag{28}$$

$$n > 2, \quad \Delta_n = n/(n-1) - 2n/[(n-1)(n-2)Da^*] + \cdots \tag{29}$$

For large n,

$$\Delta_n \sim 1 + [Da^*/(1 + Da^*)]^2/n + [Da^{*3}(Da^* - 2)/(1 + Da^*)^4]/n^2 \cdots \tag{30}$$

THE DAMKÖHLER NUMBER

We have seen that, if the dense phase is well-mixed and only the rate constant is a function of x, the performance of the fluid bed depends only on the modified Damköhler number, Da^*. If the Davidson number, Tr, depends on x, the linearity may still be exploited but we have to go back to equation (21), which may be rearranged to give

$$\Gamma_n(Da, Tr, \beta) = 1 - \int [1 - \beta e^{-Tr(x)})Da.x/ (Da.x + 1 - \beta e^{-Tr(x)})]g_n(x)dx,$$

where the Tr on the left is the mean value of $Tr(x) = Tr\omega(x)$(say). The integrals no longer come out in terms of known functions, but they are easily calculable.

Diffusion within the catalyst particle can be accounted for by using an effectiveness factor, η, but x should no longer be defined to make $k(x)$ linear in x. Rather, it would be sensible to make ηk linear in x. Of course, straightening out the monotone dependence on x of one parameter, distorts the distribution, as in equation (4), and it may be better to think of the Damköhler number as a function of x. We can alway write it as $Da.\omega(x)$, where $\omega(x)$ has a mean value of 1.

If the catalyst decays as a known function of the length of time it spends in the reactive environment, the catalyst must be drawn off from the dense phase, reactivated and returned, so that the burden is constant. A steady state is quickly built up in which the fraction of catalyst of age in the interval $(a, a + da)$ is $(w/W).\exp - (wa/W)da$, where W is the weight of catalyst in the bed and w the rate of replenishment. If the decay is accounted for by saying that the rate constant should be multiplied by a factor, for example exp

$-\lambda a$, then the Damköhler number must be calculated with a factor of

$$(w/W)\int \exp -(wa/W)\exp -\lambda a\,da = 1/(1 + W\lambda/w). \tag{31}$$

If the decay cannot be expressed solely as a function of age, the linearity is lost and the generalization to continuous mixtures is no longer possible.

If the catalyst were not to decay but for some other reason, perhaps temperature control, the particles were taken out and recycled, then each might be supposed to be in pristine condition on entering the bed. Each particle would then undergo a transition during which the steady state profile of reactant within the particle would be built up. The analysis of Amundson and Aris (1962; this part is not tainted with the error mentioned above in fn. 13) may be used. We assume spherical particles of radius R, and call the profile of concentration at time a, $c(r, a)$. If D is the diffusivity of the reactant and k^* the rate constant per unit volume of catalyst,

$$\partial c/\partial t = (D/r^2)(\partial/\partial r)[r^2(\partial c/\partial r)] - k^*c \tag{32}$$

with initial condition $c(r, 0) = 0$, finiteness at the centre, and with Bi as the Biot number of external mass transfer,

$$(1/Bi)(\partial c/\partial r) + c = c_{\mathrm{p}} \quad \text{at } r = R. \tag{33}$$

In dimensionless form, with

$$\begin{aligned} u &= c/c_{\mathrm{p}}, t = wt'/W, \rho = r/R,\\ \bar{\omega}^2 &= R^2w/DW,\\ \varphi^2 &= k^*R^2/D, \end{aligned} \tag{34}$$

$$\bar{\omega}^2(\partial u/\partial t) = \rho^{-2}(\partial/\partial\rho)[\rho^2(\partial u/\partial\rho)] - \varphi^2 u \tag{35}$$

and

$$(1/Bi)(\partial u/d\rho) + u = 1 \quad \text{at } \rho = 1 \tag{36}$$

This partial differential equation is made ordinary by taking its Laplace transform. It is then solved for the transform of u and the transform of the total reaction rate for the particle calculated. There is no need to invert this transform, for what we want is the average reaction rate in the bed and this is the integral of the product of the reaction rate after a time a and the probability of the particle having age a. Because the ages are exponentially distributed, this is none other than the Laplace transform with the dimensionless transform variable set equal to 1. The end result is that the Damköhler number is

$$Da = [kH_0/\mathrm{U}]/\{1/Bi) + 1/(\Psi \coth \Psi - 1)\} \tag{37}$$

and

$$\Psi^2 = \bar{\omega}^2 + \varphi^2. \tag{38}$$

With as complex an expression as this it is clearly a waste of time to derive an inverse function and the integrals involved can be computed numerically. In fact, much of the numerical work can be done on the desk-top computer using Simpson's rule and watching out for regions of extreme curvature.

THE FLUID BED WITH ASTARITA'S UNIFORM KINETICS

We have seen that this model of the bubbling bed is essentially the same as a stirred tank when the two sources of the feed are recognized. These are the fraction $(1 - \beta)$ that comes with the gas feed at the bottom of the bed and the fraction β in the bubbles which feeds the reactor at all levels and from a diminishing concentration difference. The latter, when referred to the inlet difference $c_0 - c_p$, delivers a fraction $1 - \exp(-Tr)$. Thus the total feed minus outlet is $\{(1 - \beta) + \beta(1 - \exp(-Tr))\}(c_0 - c_p) = \{1 - \beta\exp(-Tr)\}(c_0 - c_p)$ and this is what is equated to the reaction rate, $(kH_0/U)c_p$.

On this view of the reactor it is not surprising to find Astarita and Occone's methods for uniform kinetics work for the bubbling fluid bed. Replacing kc_p by $kc_p(x, t)\,F[\int K(y)c_p(y)dy]$. Thus the balance over the dense phase is

$$\{1 - \beta\exp(-Tr)\}[c_0(x) - c_p(x)] \\ = Da.xc_p(x)F[\int K(y)c_p(y)dy]$$

or

$$c_p(x) = c_0(x)/\{1 + Da^*xF[\int K(y)c_p(y)dy]\} \tag{39}$$

But $F[\int K(y)c_p(y)dy]$ is just a constant, as yet unknown. Call it δ for short, then

$$\delta = F[\int K(y)c_p(y)dy] = F[\int K(y)c_0(y)/\{1 + \delta Da^*y\}dy] \tag{40}$$

and this is an equation for δ. It may be solved by a straightforward calculation that takes δDa^* as parameter and calculates $\delta(\delta Da^*)$ from equation (39) and Da^* as $\delta Da^*/\delta(\delta Da^*)$. When δ is known as a function of Da^* and $c_0(x) = C_0 g(x)$, γ can be written as

$$\gamma(x) = \beta\exp(-Tr) \\ + \{1 - \beta\exp(-Tr)\}g(x)/\{1 + \delta Da^*x\} \tag{41}$$

This leads to the remarkably simple result that the performance of the bed is given be exactly the same function as before, equation (23), save only that Da^* is further modified by the factor δ.

A good example is afforded by Langmuir kinetics of a continuous mixture,

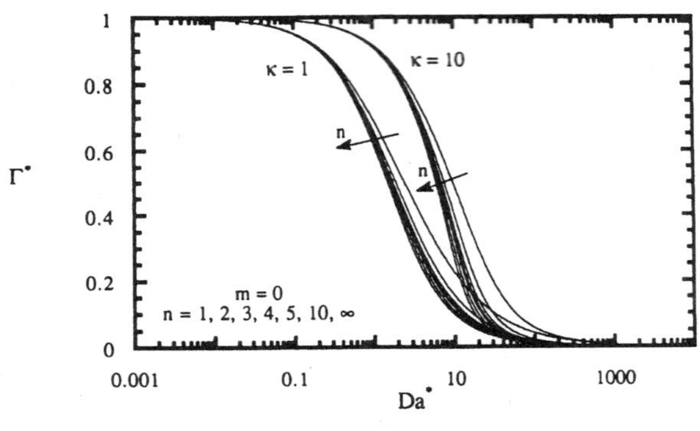

FIGURE 2 Rescaled output as a function of the modified Damköhler number; $n = 1, 2, 3, 4, 5, 10$; $\kappa = 1, 10$; $m = 0$.

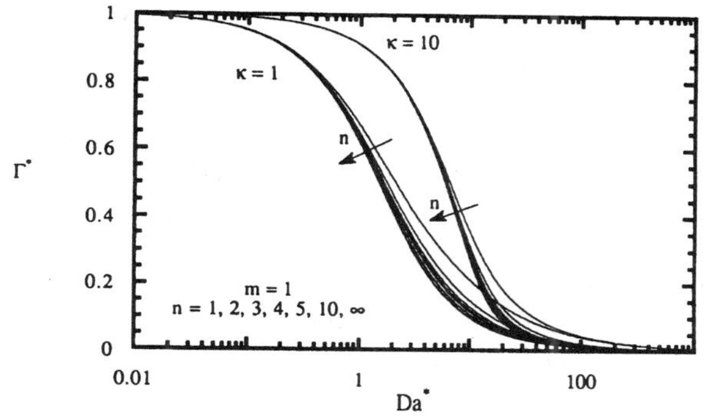

FIGURE 3 Rescaled output as a function of the modified Damköhler number; $n = 1, 2, 3, 4, 5, 10$; $\kappa = 1, 10$; $m = 1$.

a form which can be fully justified by a simple adsorption model[16]. Here $F = 1/\{1 + \int K(y)c(y)dy\}$ and it has been shown that the choice

$$K(y) = K^*\Gamma(n)(ny)^{\mathrm{m}}/\Gamma(m+n), \quad c_0 = C_0 g_{\mathrm{n}}(y),\ \kappa = K^*C_0 \tag{42}$$

is interesting. When lumped in a plug flow reactor, it follows an apparent kinetic expression of $C^{(n+1)/n}/\{1 + K^*C^{(m+n)/n}\}$. Making the same choice for the bubbling bed model gives the equations for δ and Da^* as functions of $q = \delta Da^*$

$$\delta = \{1 + \kappa(n/q)e^{(n/q)}E_{m+n})n/q)\}^{-1} \tag{43}$$

$$Da^* = q\{1 + \kappa(n/q)e^{(n/q)}E_{m+n}(n/q)\} \tag{44}$$

Figure 2 and Figure 3 show Γ^* as a function of Da^* for $\kappa = 1, 10$ and $m = 0, 1$ (resp.). They illustrate how extremely close the results over a wide range of variation of n may be. The variation with κ is more marked and, since an increase in κ reduces the reaction rate, we would expect that a compensating increase in the Damköhler number would be needed to maintain conversion.

ACKNOWLEDGMENT

One of us (RA) is indebted to the PRF of the ACS for continued support of an ongoing investigation of reactions in continuous mixtures (PRF25133-AC7E). The figures and the calculations that lie behind them were done by Paolo Cicarelli.

NOMENCLATURE

(Notation borrowed from Davidson and Harrison (1963) is denoted by DH)

a	age of catalyst
Bi	Biot number for catalyst particle
$C(t)$	total concentration of time t, $\int c(x, t)dx$
C_0	$C(0)$
c	concentration

c_p	concentration of reactant in dense phase. DH
c_0	concentration of reactant in feed
$c(r, a)$	concentration of reactant at radius r of catalyst of age a
$c(x, t)dx$	concentration of material with index in $(x, x + dx)$ at time t
D	diffusivity of reactant within catalyst particle
Da	Damköhler number
Da^*	modified Damköhler number, equation (20)
$E_n(x)$	Exponential Integral. equation (11)
F	nonlinear function in Astarita's kinetics
$g_n(x)$	Gamma distribution, equation (5)
$g_n(x, a)$	Gamma distribution with mean a, equation (7)
$K(y)$	kernel in Astarita's uniform kinetics
$k(x)$	first order rate constant for material of index x
$\langle k \rangle$	average value of k
k'	first order rate constant. DH
k^*	rate constant within catalyst particle
N	apparent order, equation (15); number of bubbles per unit vol. DH
n	parameter of the Gamma distribution
$\mathbf{p}$	set of parameters on which the solution of the system depends
$\mathbf{p}^*$	characteristic values of parameters, $\mathbf{p}(x) = \mathbf{p}^*\omega(x)$
Q	exchange rate between bubble and dense phase. DH
q	parameter for calculation δ and Da^*, equations (42) and (43)
R	radius of catalyst particle
r	radial distance within catalyst particle
Tr	Davidson number, QH/U_AV
U	velocity of fluidizing gas. DH.
U_0	velocity of gas for incipient fluidization. DH.
U_A	absolute velocity of bubble rise. DH.
V	bubble volume. DH.
W	weight of catalyst in bed
w	replacement rate of catalyst
x	index of 'species' in a continuous mixture

Greek letter

β	$1 - U_0/U$. DH.
$\Gamma(Da)$	fraction of total concentration in exit stream, also used with other parameters as arguments, $\int\gamma(x, Da)dx$
Γ^*	$(\Gamma - \beta e^{-Tr})/(1 - \beta e^{-Tr})$
$\gamma(x, Da)$	fraction of inlet concentration remaining in exit stream
γ^*	$(\gamma - \beta e^{-Tr})/1 - \beta e^{-Tr})$
Δ	$(1 + Da^*)\Gamma(Da, Tr, \beta)$
$\Lambda(\mathbf{p})$	ratio of exit to inlet concentrations for linear systems
λ	decay constant for ageing catalyst
$\mu_{n,p}$	p^{th} moment of $g_n(x)$
$\bar{\omega}$	dimensionless replacement rate, $(R^2w/DW)^{1/2}$
φ	Thiele modulus of catalyst particle $R(k^*/D)^{1/2}$
Ψ	$(\varphi^2 + \bar{\omega}^2)^{1/2}$
$\omega(x)$	set of distributed parameters scaled by their characteristic values

REFERENCES

1. Newton, I. 1688. *Philosophiae Naturalis Principia Mathematics,* See *Isaac Newton's Philosophiae Naturalis Principia Mathematics*, 3rd Edition with variant readings, Eds. A. Koyre and I. B. Cohen, 1972, (Cambridge University Press, Cambridge). No less eminent a scientist than S. Chandrasekhar has written of his study of the *Principia* "I first constructed proofs (of some of the well-known propositions) for myself. Then I compared my proofs with those of Newton. The experience was a sobering one. Each time I was left in sheer wonder at the elegance, the careful arrangement, the imperial style, the incredible originality, and above all the astonishing lightness of Newton's proofs; and each time I felt like a schoolboy admonished by his master." From an unpublished ms. "The *Principia*: the intellectual achievement that it is" quoted in *Chandra*, K. C. Wali, 1991, (Chicago U.P. Chicago).
2. Brink, C. O., 1986, *English Historical Scholarship. Historical Reflections on Bentley, Porson and Housman.* (Cambridge. Clarke and New York University Press.) p. 4.
3. Bentley, R., 1691, *Epistola and Cl. V. Johnnem Millium, S.T.P.* published as an appendix to J. Mills' edition of Malelas. Oxford.
4. Bentley, R., 1691, *Dissertation of the Epistles of Phalaris. With an answer to the Objections of the Honourable Charles Boyle, Esquire.* London.
5. Brink, *op. cit.* p. 71.
6. Taylor, G. I., 1958–71, *Scientific Papers.* Ed. G. K. Batchelor. 4 vols. (Cambridge. Cambridge University Press). P. G. Saffman wrote, shortly after Taylor's death in 1975, that he "said when in his seventies that he preferred to work outside the mainstream of scientific activity ... But Taylor had only to be interested in a field and his fundamental discoveries, which followed as a matter of course, would put the topic in the mainstream."
7. Davidson, J. F. and Harrison, D., 1963, *Fluidised Particles.* (Cambridge University Press, Cambridge).
8. This is no place to review the literature, but the later volume *Fluidization* (Academic Press, 1971) which Davidson and Harrison edited deserves mention, as do: chapters 10 and 11 (by Davidson, Harrison, Darton and La Nauze, and Amundson, Bukur and Caram, resp.) in *Chemical Reactor Theory* (Prentice-Hall, Englewood Cliffs. 1977); D. Kunii and O. Levenspeil *Fluidization Engineering* (Wiley, New York. 1969) and P. N. Rowe and J. G. Yates *Fluidized Bed Reactors* (chap. 7 of Chemical Reaction and Reactor Engineering. Ed. J. J. Carberry and A. Varma. Marcel Dekker, New York. 1987).
9. Gavalas, G. and Aris, R., 1966, On the theory of reactions in continuous mixtures. *Phil Trans Roy Soc*, A260: 351. As far as I know this is the first use of continuous mixtures in the context of chemical reaction. They had been used in distillation by Amundson and Acrivos, 1955, *Chem Eng Sci*, 4: 29, in froth flotation by Loveday, 1966, *Inst Min and Metall Trans.* C75: 219 and in thermodynamics by de Donder, 1931, *L'Affinité* (Gauthier-Villars, Paris).
10. Krambeck, F. J., 1984, Computers and modern analysis in reactor design. *Proc ISCRE 8. I Chem E Symp series, No 87*, 733. For a more accessible proof see Ho and Aris, 1987, *AIChE J* 33: 1050.
11. Astarita, G. and Ocone, R., 1988, Lumping nonlinear kinetics. *AIChE J,* 34: 1299.
12. Astarita, G., 1989. Lumping nonlinear kinetics: apparent overall order of reaction. *AIChE J,* 35: 529.
13. Astarita G., and Aris, R., 1989, Continuous lumping of nonlinear chemical kinetics. *Chem Engng and Proc*, 26: 63, and On aliases of differential equations. *Rend Acc Lincei*, LXXXIII; xxx (1989).
14. It is important to do the balance over the whole of the dense phase when this is, by assumption, uniform, failure to do this led to an erroneous result in an earlier paper of ours, Heat transfer in fluidized and moving beds (*Proc. Symp. on Interactions between Fluids and Particles. 1962.* pp 176–182. The error was ultimately corrected in 1990; see Manners makyth Modellers, *Chem Eng Sci,* 46: 1535–44 (1991) and in *Trans IChemE*, 68: 165–174 (1991).
15. Abramowitz, M. and Stegun, I. A., 1964, *Handbook of Mathematical Functions* (Nat. Bur. Standards). Equation 5.1.51, p. 231 gives the asymptotic expansion and 5.1.11 and 12 the series expansion. Equation (29) is obtained from 5.1.52.
16. Aris. R., 1989, Reactions in continuous mixtures. *AIChE J*, 35: 539.

ADDRESS

Correspondence concerning this paper should be addressed to Professor R. Aris, Department of Chemical Engineering, University of Minnesota, MN 55455, USA.

9 STATICS AND DYNAMICS OF CHEMICAL REACTORS

INTRODUCTION

In the 1980s, the same atmosphere of excitement surrounded the work that was being done on forced oscillations as had surrounded the stability and limit-cycle questions 30 years earlier. To have had students of the caliber of Kevrekidis or the profundity of Farr (the others will, I hope, forgive my not forging a semi-alliterative chain of names) was a rare privilege. Lanny Schmidt and I shared a grant that covered the theoreticians as well as part of his very extensive laboratory work. It was difficult to decide which of Kevrekidis' papers [207, 211, 213, 215, 218, 219, 221, 222] to choose, but Reprint I [218] covers a lot of the dynamics of reactors with characteristic verve and panache. Farr's papers, [212, 220, 231], although not so numerous, were also of outstanding depth and delicacy. I have chosen one, Reprint J, written for an issue of *Chemical Engineering Science* and dedicated to N. R. Amundson, in which Farr showed, contrary to all expectations, that the C^* with a nonisothermal $A \rightarrow B \rightarrow C$ is capable of having seven steady-states. It is a contribution to the statics of reactors and a fine example of the use of singularity theory. This line of research was intertwined with the dynamics and the vivacity of work that was exhibited by the graduate students; their interaction with one another and with the mathematics department is one of my most pleasant memories. By training, Farr was and is a mathematician, and Kevrekidis did so much that he picked up a master's degree in mathematics on the side, so that the traditional denominational distinctions were never an issue and natural philosophy would be the best descriptor, did it not sound a shade pompous.

The two *Proceedings of the Royal Society* Papers (Reprints K and L) are a "matched pair," exploring the model reaction that Schmidt and Takoudis had devised [177]; $A + S \leftrightarrow AS$, $B + S \leftrightarrow BS$, $AS + BS + 2S \rightarrow C + 4S$. Here, the autocatalytic element is the vacant site, just as B is in the Gray–Scott reaction and heat is in the non-isothermal exothermic case. The two reprints, although not an absolutely comprehensive treatment of this model, have a satisfying completeness. The tale of students who worked on this class of problem includes Alhumaizi, Cordonier, Farr, Jorgenson, Kevrekidis, McKarnin, and Takoudis; their papers are listed in the Index of Co-Authors.

I.*

SOME COMMON FEATURES OF PERIODICALLY FORCED REACTING SYSTEMS

I. G. KEVREKIDIS, L. D. SCHMIDT and R. ARIS

Department of Chemical Engineering and Materials Science, University of Minnesota, Minneapolis, MN 55455, U.S.A.

(Received 23 January 1985)

Abstract—*Some fundamental qualitative features of the dynamic response of chemical reacting systems to periodic forcing are described. Models considered are a homogeneous autocatalytic reaction (a Brusselator), a bimolecular surface reaction and the nonisothermal CSTR. These systems oscillate spontaneously, and the interaction of cycling an operating variable with the natural frequencies of the systems is examined. Numerical techniques employed include fixed point (shooting) as well as polynomial approximation methods for the computation of periodic trajectories, an algorithm for the computation of invariant tori and pseudo-arc length continuation of solution branches. Phenomena examined are entrainment, frequency locking, bifurcations to invariant tori, breaking of tori and period doubling cascades. Description of these phenomena in forced systems facilitates the understanding of coupling between nonlinear systems in general.*

INTRODUCTION

Chemical reactions and chemical reactors exhibit a wealth of dynamic behaviour patterns that range from steady-state multiplicity and sustained oscillatory behaviour to quasi-periodicity and deterministic spatial and temporal chaos. Local bifurcation theory has been successfully applied to the study of transitions between such patterns in parameter space (Uppal *et al.* 1974, 1976). However, the global characterization of the dynamics of these models in multiparameter space, even with methods like singularity theory (Balakotaiah and Luss, 1982; Jorgensen *et al.* 1984), becomes a very lengthy and expensive task. In attempting to study such systems and to classify and generalize their possible behaviour, one needs to find some fundamental common characteristics in their dynamic responses. We have observed such structures in the modelling of several periodically forced chemical reactions and reactors.

The interest in periodically forced systems extends beyond performance considerations for a single reactor. Stability of structures and control characteristics of chemical plants are determined by their responses to oscillating loads. Epidemics and harvests are governed by the cycle of seasons. Bifurcation and stability analysis of periodically forced systems is especially important in the

* Reprinted with the permission of Elsevier Science from *Chemical Engineering Science,* volume **41,** number 5, pp. 1263–1276, 1986.

current chemical engineering context, where the surge in electronics and computer technology makes the controlled operation of whole chemical processes under periodic and quasi-periodic conditions feasible. The study of forced systems is also of intrinsic interest. Dynamical systems are, and will, continue to be an active research field in a wide spectrum of scientific disciplines, from forced cardiac oscillators (Glass *et al.*, 1984) to the dynamics of Josephson junctions (Levi *et al.*, 1978) and from nonlinear modelling of turbulence (Ruelle and Takens, 1971) to the much-studied B–Z reaction (Hudson *et al.* 1979; Hudson and Mankin, 1981) and the exotic graphics of Julia sets (Mandelbrot, 1983).

The subject of periodic operation of chemical reactors has experienced a revival in the chemical engineering research literature in recent years. The first surge in the late 1960s and early 1970s concentrated on the advantages that periodically operated processes might possess over steady-state processes in terms of mean productivity or selectivity as described in the reviews by Bailey (1973, 1977). Most of that early work was based on drastic assumptions regarding the qualitative response of the system in question, namely the existence of a single periodic attractor for each set of operating conditions. Under such an assumption, one may then define an objective functional of this stable limit cycle (e.g. mean productivity of some species per cycle) which can be optimized (Douglas, 1972).

This approach needs modification as soon as multiple attracting periodic trajectories exist for a particular set of operating parameters. A conceptually different modification will be necessary to account for attractors which are not simply periodic. Quasi-periodic solutions, characterized by multiple frequencies, are the first type one should expect; these are by no means "exotic" but occur generally in several periodically forced systems. Deterministic chaotic situations, arising from the system nonlinearities (and not the stochastic responses due to random noise) need not be discarded as intractable (Wolf *et al.*, 1986; Shaw, 1981).

A particularly appealing characteristic of periodically forced systems is that one of the response frequencies is a tunable parameter which is not changed by system nonlinearities. This allows us to study the interaction of frequencies more clearly than in the general case of autonomous coupled nonlinear oscillators. Forcing, being one-directional coupling, provides a starting point for the study of coupling and synchrony in general. The "clock" that produces the forcing frequency may be considered as an oscillator with infinite inertia or energy compared to the oscillator that it drives (Minorsky, 1962; Hayashi, 1964; Nayfeh and Mook, 1979; Stoker, 1950).

Sincic and Bailey (1977) relaxed the assumption of only one stable attractor for a given set of operating conditions and showed examples of some possible exotic responses in a CSTR with periodically forced coolant temperature. They also probed the way in which multiple steady states or sustained oscillations in the dynamics of the unforced system affect its response to periodic forcing. Several theoretical and experimental papers have since extended these ideas (Hamer and Cormack, 1978; Cutlip, 1979; Stephanopoulos *et al.*, 1979; Hegedus *et al.*, 1980; Abdul-Kareem *et al.*, 1980; Bennett, 1982; Goodman *et al.*, 1981, 1982; Cutlip *et al.*, 1983; Taylor and Geiseler, 1986; Mankin and Hudson, 1984; Kevrekidis *et al.*, 1984).

In the present paper we study common features of the responses of chemical reactor models to periodic forcing, and we consider accurate methods that can be used in this task. In particular, we describe an algorithm for the numerical computation and stability analysis of invariant tori. We shall consider phenomena that appear in a broad class of forced systems and illustrate them through several chemical reactor models, with emphasis on the forcing of spontaneously oscillating systems.

NUMERICAL METHODS

Computation of Periodic Trajectories

In many cases ordinary differential equations (ODEs) provide adequate models of chemical reactors. When partial differential equations become necessary, their discretization will again lead to large systems of ODEs. Numerical methods for the location, continuation and stability analysis of periodic and quasi-periodic trajectories of systems of coupled nonlinear ODEs (both autonomous and nonautonomous) are extensively used in this work. We are not concerned with the numerical description of deterministic chaotic trajectories; where they occur, we have merely inferred them from bifurcation sequences known to lead to deterministic chaos. Extensive literature, as well as a wide choice of algorithms, is available for the numerical analysis of periodic trajectories (Keller, 1976, 1977; Curry, 1979; Doedel, 1981; Seydel, 1981; Schwartz, 1983; Kubicek and Hlavacek, 1983; Aluko and Chang, 1984).

Given the system of ODEs

$$\dot{\mathbf{x}} = \mathbf{f}(\mathbf{x}, t), \mathbf{x} \in \mathbb{R}^n \tag{1}$$

a periodic trajectory is found by solving the two-point free boundary problem

$$\begin{aligned} \dot{\mathbf{x}} &= \mathbf{f}(\mathbf{x}, t) \\ \mathbf{x}(t = 0) &= \mathbf{x}(t = T) \end{aligned} \tag{2}$$

where T is the period of the limit cycle. If eq. (1) is autonomous, T is unknown and varies with the system parameters. One additional scalar equation

$$N(\mathbf{x}, T) = 0 \tag{2a}$$

(an "anchor" equation that eliminates free translation in time) is appended to eq. (2), and this system is solved for the periodic trajectory and its period T (Doedel, 1981).

We obtain initial information regarding the fundamental (or "natural") unperturbed oscillatory states of our systems through AUTO, a FORTRAN code for the bifurcation analysis of systems of autonomous ODEs developed by Doedel (1981). AUTO solves the above system through polynomial interpolation (collocations) and continues the solution branches in parameter space using Keller's pseudo-arc length method (Keller, 1977). For the same purpose and for greater accuracy in the computation of the period of the unperturbed limit cycle—an important quantity in the location of forced subharmonics—we have used a double precision shooting algorithm, which we subsequently modified to locate periodic trajectories of the forced systems.

Integrating eq. (1) with initial conditions $\mathbf{x}(t = 0) = \mathbf{x}_0$ for time τ will give $\mathbf{x}(t = \tau) = \mathbf{F}(\mathbf{x}_0, \tau)$. Solution of the boundary value problem eq. (2) is equivalent to the solution of the system of equations

$$\mathbf{x} - \mathbf{F}(\mathbf{x}, T) = \mathbf{0} \tag{3}$$

demanding that a point on the limit cycle and its image after time T coincide. However, since all points on the limit cycle will coincide with their images after one period T, the anchor equation is necessary to single out one point on the limit cycle and to be solved simultaneously with eq. (3), eliminating its one-parameter infinity of solutions. We see that this shooting algorithm is equivalent to the location of the point lying at the intersection of the limit cycle with the surface defined by the anchor equation: this point is the fixed point of the Poincare map associated with this intersection. Provided that the intersection is transversal (Guckenheimer and Holmes, 1983), we can solve eqs. (3) and (2a) with a Newton method whose Jacobian will have the form

$$\underline{\mathbf{J}} = \left[\begin{array}{c|c} \underline{\mathbf{I}} - \dfrac{\partial \mathbf{F}}{\partial \mathbf{x}} & -\dfrac{\partial \mathbf{F}}{\partial T} \\ \hline \dfrac{\partial N}{\partial \mathbf{x}} & \dfrac{\partial N}{\partial T} \end{array}\right] \tag{4}$$

where $\underline{\mathbf{I}}$ is the unit matrix. During a Newton step, the residuals in eq. (3) will be computed through integration of the system equations (1). The derivatives $\partial \mathbf{F}/\partial \mathbf{x}$ will be computed from the simultaneous integration of the variational equations

$$\dot{\underline{\mathbf{A}}} = \left[\frac{\partial f}{\partial \mathbf{x}}(t)\right] \underline{\mathbf{A}}$$

$$\underline{\mathbf{A}}(t = 0) = \underline{I}$$

to give

$$\underline{\mathbf{A}}(t = T) = \frac{\partial \mathbf{F}(\mathbf{x}, T)}{\partial \mathbf{x}}$$

from which $\partial N/\partial \mathbf{x}$ will also be computed. $\partial \mathbf{F}/\partial T$ (and $\partial N/\partial T$) will be computed from the differential eq. (1) at $\mathbf{F}$.

Upon convergence, the eigenvalues of $\partial \mathbf{F}/\partial \mathbf{x}$ (the characteristic or Floquet multipliers FM_i) are independent of the particular point on the limit cycle (i.e. the particular Poincare section or anchor equation used). One of them, FM_n, is constrained to be unity (Iooss and Joseph, 1980) and this may be used as a numerical check of the computed periodic trajectory; the remaining FMs determine the stability of the periodic orbit, which is stable if and only if they lie in the unit circle in the complex plane ($|FM_i| < 1, 1 \leqslant i \leqslant n - 1$). The multiplier with the largest absolute value is usually called the principal FM (PFM). When (as a parameter varies) the PFM crosses the unit circle, the periodic orbit loses stability and a bifurcation occurs.

A periodically forced system

$$\dot{\mathbf{x}} = \mathbf{f}(\mathbf{x}, p), \quad p(t + 2\pi) = p(t) \tag{5}$$

can be rewritten as autonomous if we agument it with the equations that produce the forcing. In the case of sinusoidal forcing, we may write

$$\begin{aligned} \dot{\mathbf{x}} &= \mathbf{f}(\mathbf{x}, z_1) \\ \dot{z}_1 &= \omega z_2 \\ \dot{z}_2 &= -\omega z_1 \end{aligned} \tag{6}$$

and analyse the system in this autonomous version. It is best, however, to take advantage of the fact that the period of any forced periodic trajectory has to be an integer multiple of the forcing period. We can thus define a "natural" Poincare section, the so-called stroboscopic section, obtained by recording the periodic trajectory not continuously but at every period of the forcing term (at the same phase of the forcing term, hence the term isophase section). In our shooting algorithm we can then solve eq. (3) without the addition of an "anchor" equation; the natural anchor "forcing phase = constant" is automatically satisfied since the period of the trajectory is known.

Beyond its applicability in the simplification of the computations, the stroboscopic representation greatly simplifies the recognition of patterns in the transient response of periodically forced systems. A sustained oscillation appears as a finite number of repeated points, while a quasi-periodic response appears as an invariant circle (see Figs. 3, 4, 6 and 9).

There exists a one-parameter infinity of stroboscopic maps (one associated with each phase of the forcing term). Their fixed points constitute the periodic trajectory. The eigenvalues of $\partial \mathbf{F}/\partial \mathbf{x}$ will, however, be the same for each one of them; the multiplier at unity will correspond to the "additional" equations that produce the forcing. In general, addition of a periodic forcing is equivalent to increasing the dimension of the system by one. Thus, a one-dimensional forced system may exhibit multiple oscillatory states, while a two-dimensional forced system may exhibit everything from bifurcations to two-tori to period doubling cascades to chaos.

The Single Species Forced Reaction

A single autonomous nonlinear ODE may exhibit steady-state multiplicity. If periodically forced, it can consequently exhibit multiple oscillatory states. Let us consider the heterogeneous catalytic reaction A $\rightarrow$ products, occurring in an isothermal reactor with perfectly mixed gas phase. Gaseous A flows in the reactor, adsorbs with Langmuir kinetics on the catalytic surface and reacts to give products that desorb immediately. We assume that the desorption of A is negligible and that the reaction rate is first order in adsorbed A with coverage-dependent activation energy. If the reactor has a differential conversion we may assume the gas phase partial pressure of A to be an operating parameter. After nondimensionalization, the equation of change for the surface coverage of A becomes

$$\dot{\theta} = P(1 - \theta) - \theta \exp(-\alpha\theta) \tag{7}$$

where P is the dimensionless A pressure and α is a measure of the coverage dependence of the reaction activation energy. The bifurcation diagram of eq. (7) with respect to P shows two turning points P_*, P^* for $\alpha < 4$ as shown in Fig. 1(a). For P in $[P_*, P^*]$ the system will have three steady states, one of them unstable.

We now vary the gas phase pressure of A periodically in the form $P = P_0 + a \sin(\omega t)$. Three parameters are necessary to describe the sinusoidal forcing: its midpoint P_0, its amplitude a and the period $T = 2\pi/\tau$. Figure 1(b) shows the result of integrating eq. (7) for every possible initial condition and for one period of the forcing term, its phase being zero at the beginning of each cycle. The final coverage θ^1 is plotted as a function of the initial coverage θ^0. This is the stroboscopic map for the one-dimensional forced system: the points at which the graph intersects the diagonal are fixed points of the map which lie on periodic trajectories that have the period of the forcing. Iterating the map will cause the points to move towards their corresponding attractors, thus sharpening the graph of the map to almost a step, as shown in Fig. 1(c). When (for fast forcing frequencies) the first iterate of the map lies confusingly close to the diagonal, plotting higher iterates helps to visualize the behaviour of the system, as shown in Fig. 1(d). At the limit of small forcing amplitude (FA), the periodic trajectories of the forced system lie close to the steady states of the unperturbed system from which they are born. Furthermore, their stability is the same as the stability of these steady states [a standard result of the method of averaging (Krylov and Bogoliubov, 1947; Hale, 1969)] provided these original steady states are hyperbolic (no linearized eigenvalues

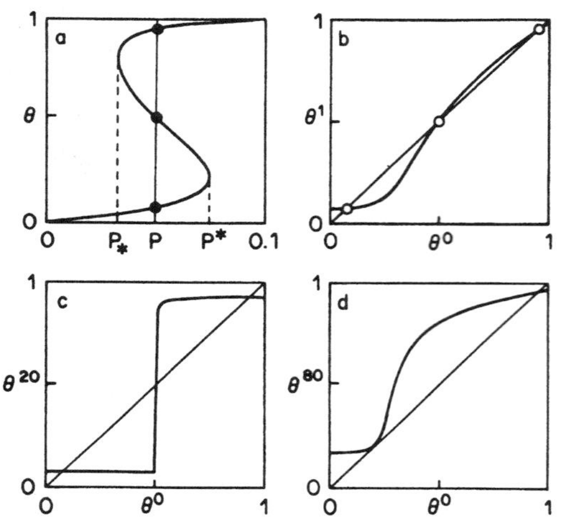

FIGURE 1 The single species forced reaction. (a) The autonomous bifurcation diagram for $\alpha = 6$. For $P = P_0 = 0.05$, three steady states exist and the middle one is unstable. (b) Stroboscopic map for the forced system when $a = 0.001$ and $\omega = 1$. The intersections of the graph with the diagonal lie on limit cycles that have the same stability characteristics as the original autonomous steady states. (c) After 20 iterates of the stroboscopic map all transients have died out. (d) A turning point on a branch of periodic solutions appears as a "pinch" of the graph of any iterate of the map (80th iterate, $P_0 = 0.0755$, $a = 0.05$, $\omega = 20$).

on the imaginary axis). This is equivalent to the demand that the midpoint of the forcing not be a bifurcation or turning point of the unforced bifurcation diagram with respect to the forcing variable—in which case higher-order analysis is required. These forced periodic trajectories will have one *FM* at unity corresponding to the forcing term; it can be shown that at the small *FA* limit, the other multiplier will be $FM_1 = \exp(\lambda T)$, where T is the forcing period and λ is the eigenvalue of the linearization of the unforced system around its steady state. If P_0 is outside the closed interval $[P_*, P^*]$ [Fig. 1(a)] the response to small *FA* forcing will be a single oscillation of the same period, lying close to the autonomous steady state at P_0. The diagram corresponding to Fig. 1(b) will then have only one intersection with the diagonal.

When the forcing amplitude is very small and the midpoint of the forcing oscillation scans the autonomous bifurcation diagram, the qualitative response of the forced system for all frequencies can be deduced from the autonomous system characteristics. As the amplitude of the forcing becomes larger, one cannot predict *a priori* what will occur for a particular system. For this example, the most complicated phenomenon possible is a turning point bifurcation on a branch of periodic solutions where two limit cycles, one stable and one unstable, collide and disappear. This will appear as a "pinch" on the graph of the map [Fig. 1(d)].

Since we are solving a deterministic problem, this graph has to be monotonic because the map is invertible. No bell-shaped graphs [with negative Schwartzian derivatives, of the type of the logistic map (Feigenbaum, 1978; Collet and Eckmann, 1980)] leading to period doubling bifurcations to chaos are possible for one-dimensional forced ODE reactor models.

Spontaneously Oscillating Models

This work is centred around the study of the response to periodic forcing of systems that, when autonomous, had a stable limit cycle surrounding an unstable steady state in their phase plane. For the sake of simplicity—and since many of the fundamental phenomena are the same—we studied two-dimensional systems. We chose two examples of isothermal reactor models; the first is an autocatalytic homogeneous Brusselator (Glansdorff and Prigogine, 1971):

$$\begin{aligned} \dot{x} &= x^2y - bx - x + [\underline{A + a\cos\omega t}] \\ \dot{y} &= bx - x^2y \end{aligned} \tag{8}$$

which was extensively explored by Tomita and coworkers (Tomita *et al.*, 1977; Kai and Tomita, 1979; Tomita, 1982). The forcing variable is the concentration of reactant A.

The second model results from a bimolecular surface reaction, A + B → products, with competitive Langmuir–Hinshelwood kinetics, which occurs in a heterogeneous differential reactor with perfectly mixed gas phase. The reaction is first order in both adsorbed A and B, and two vacant sites are required in the reaction mechanism. If the reaction products desorb immediately, the equations of change for the surface coverages of A and B become

$$\begin{aligned}\dot{\theta}_A &= \alpha_1(1-\theta_A-\theta_B) - \gamma_1\theta_A - \theta_A\theta_B(1-\theta_A-\theta_B)^2\\ \dot{\theta}_B &= [\underline{\alpha_2 + a\cos\omega t}\,](1-\theta_A-\theta_B) - \gamma_2\theta_B - \theta_A\theta_B(1-\theta_A-\theta_B)^2\end{aligned} \tag{9}$$

where the forcing variable is the (nondimensionalized) gas-phase pressure of reactant B. This model was studied by Takoudis *et al.* (Takoudis *et al.*, 1981; Takoudis, 1981).

We also studied the forced nonisothermal CSTR with the $A \rightarrow B$ reaction [Uppal *et al.*, 1974]:

$$\begin{aligned}\dot{x}_1 &= -x_1 + Da_0\tau(1-x_1)\exp(x_2)\\ \dot{x}_2 &= -x_2 + BDa_0\tau(1-x_1)\exp(x_2) - \beta_0\tau(x_2 - [\underline{x_{2c} + a\cos\omega t}]).\end{aligned} \tag{10}$$

The forcing variable is the coolant temperature x_{2c} as in Sincic and Bailey (1977) and more recently in Mankin and Hudson (1984). In eq. (10) x_1 is a dimensionless reactant concentration while x_2 is a dimensionless reactor temperature. These equations hold at the limit of infinite reaction activation energy. All models were thus chosen so that extensive simulation results existed in the literature, and they cover a wide range of lumped reactor types.

We use a parameter of the autonomous system as the forcing variable. In the previous section we showed that there exists a connection between the response of a periodically forced system and the bifurcation diagram of its autonomous version with respect to the parameter that will be used as the forcing variable. The autonomous bifurcation diagram is important to the system response as this parameter is periodically varied. Figures 2(a)–2(c) show sections of the bifurcation diagrams of our model autonomous systems for their respective forcing variables. Details regarding these diagrams are given in the caption. In all cases we have chosen a fixed midpoint for the forcing and studied aspects of the system responses as the forcing amplitude (*FA*) and the forcing frequency (ω/ω_0) were varied. At the small *FA* limit, these responses will be characterized by both the forcing and the system natural frequencies, so that quasi-periodic behaviour is expected.

Computation of Quasi-periodic Trajectories

A quasi-periodic solution to a system of ODEs is characterized by at least two frequencies that are incommensurate (their ratio is an irrational number) (Bohr, 1947; Besicovitch, 1954). Several such frequencies may be present on high-order tori, but for the two-dimensional forced systems we examine, we may have no more than two distinct frequencies (a two-torus, T^2). A quasi-periodic solution is typically born when a pair of complex conjugate *FM*s of a periodic trajectory leave the unit circle at some angle ϕ, where $\phi/2\pi$ is irrational. Such a solution is also expected when we periodically perturb an autonomously oscillating system with a frequency incommensurate to its natural frequency.

It is worth considering the visualization of the response of a two-dimensional system to a forcing oscillation (Fig. 3). The state variables at any instant

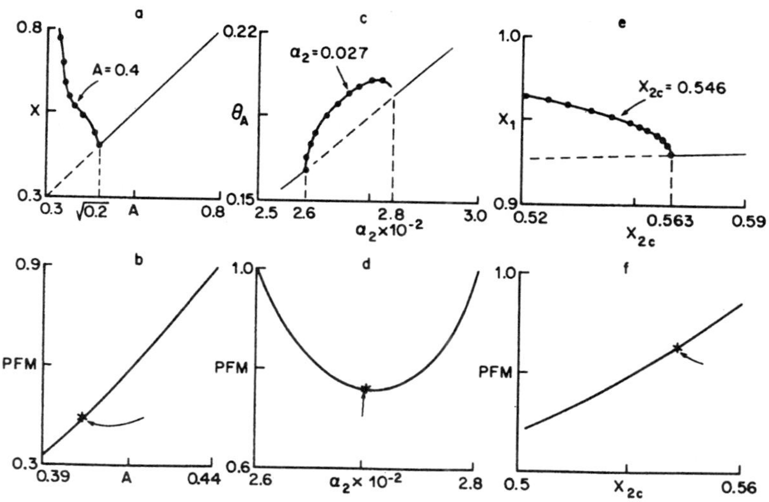

FIGURE 2 Autonomous bifurcation diagrams of the two-dimensional models vs. the parameter that will be used as the forcing variable. (a) The Brusselator [eq. (8)] for $B = 1.2$. Hopf bifurcation at $A = 0.2$. Chosen midpoint for the forcing (indicated by the arrow) $A = 0.4$ when the autonomous limit cycle has $\omega_0 = 0.3750375$. (b) Variation of the principal Floquet multiplier (*PFM*) on the periodic branch vs. *A*. The midpoint PFM is again indicated by an arrow. (c) The surface model [eq. (9)] for $\alpha_1 = 0.016$, $\gamma_1 = 0.001$ and $\gamma_2 = 0.002$. Hopf bifurcations at $\alpha_2 = 0.028$ [subcritical (Takoudis *et al.*, 1981)] and 0.02602. The midpoint for the forcing has been chosen at $\alpha_2 = 0.027$ when $\omega_0 = 0.0081920846$ and the autonomous limit cycle is at its stablest [PFM is at a minimum as shown in (d)]. (c) The forced CSTR [eq. (10)] for $Da_0 = 0.113$, $B = 20$, $\beta_0 = 4$ and $\tau = 0.75$. Hopf bifurcation at $x_{2c} = 0.5630$. Further bifurcations exist for lower values of x_{2c} (two turning points at $x_{2c} = -0.2501$ and 0.1649 and one more Hopf bifurcation at $x_{2c} = 0.1644$). The midpoint for the forcing has been chosen at $x_{2c} = 0.546$ when $\omega_0 = 6.12697752$ and $PFM = 0.77251132$—see (f). The stability of the limit cycle changes monotonically with the bifurcation parameter as in (b). Steady–state branches: ——, stable; – – –, unstable. Periodic branches: ●—●.

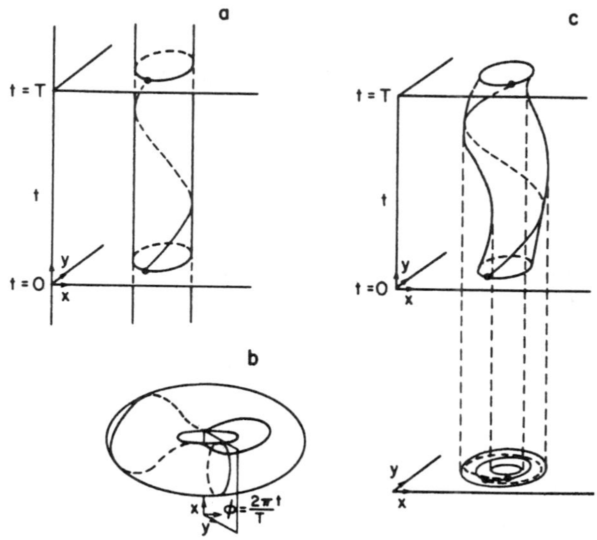

FIGURE 3 The torus picture for a forced system at small forcing amplitude. (a) A view of the unperturbed system; the trajectories eventually wind around a vertical cylinder. (b) Cutting and pasting that cylinder gives us a doughnut surface. (c) As we start forcing, the cylinder (and the resulting doughnut) gets distorted. The projection of a quasi-periodic trajectory on the phase plane gives a space-filling curve.

can be represented by two coordinates in a plane. Taking this to be the horizontal x–y plane, time can be the vertical coordinate z. Suppose that the unperturbed system has an attracting limit cycle surrounding an unstable steady state in its phase plane. Transient trajectories will then eventually wind towards this stable limit cycle. A trajectory that starts on the limit cycle at time $t = 0$ will move on a vertical cylinder whose section at any time t is the limit cycle [Fig. 3(a)]. After time T (where T is the period of the limit cycle), the phase point returns to its original location on the phase plane, and what happened between $t = 0$ and $t = T$ is repeated between $t = T$ and $t = 2T$. We take advantage of this periodicity by cutting the part of this infinite cylinder that lies between the $t = 0$ and $t = T$ planes, bending it around and pasting the two edges together [Fig. 3(b)]. We have thus constructed a doughnut surface from our tube on which the trajectories lie. We are now working on $R^2 \times S^1$ (where S^1 is the unit circle and $\phi \in S^1$, where $\phi = 2\pi t/T$).

We now turn on the forcing with a period approximately the same as that of the original limit cycle. The picture will remain basically the same: the vertical "tube" starts to become distorted but it still exists; trajectories that start away from it will eventually approach the tube surface; trajectories that start on the tube surface will remain on it (the tube surface is said to be invariant under the system equations). The phase point will not [Fig. 3(c)] in general come back to its original position after time T (where T is now the period of the forcing term); the entire closed curve (circle), however, at the intersection of the tube with the $t = 0$ plane will match exactly with the closed curve at $t = T$. This means that we can still use the periodicity to cut, bend and paste the tube into a doughnut surface. The whole doughnut surface (the so-called two-torus or T^2) is invariant under the system equations. If we project the tube on the x–y plane we get an annulus; the projection on the phase plane of a trajectory that winds around the doughnut will eventually fill up this annulus [space filling curve, see Figs. 4(b) and 4(c)].

A quasi-periodic trajectory is winding around the doughnut in the same fashion a conducting wire is wound around an iron core to construct an inductor. It will eventually pass arbitrarily close to any given point on the torus (and will keep returning) but it will never return to precisely the same point twice (or else it would be a periodic trajectory). We say that the quasi-periodic trajectory becomes dense on the torus surface. The doughnut picture is helpful in visualizing the quasi-periodic trajectory (see Fig. 4). It will consist of a main oscillatory pattern (e.g. rotation around the parallel) whose peak amplitudes and peak-to-peak intervals will be modulated through another frequency [e.g. rotation around the meridian, see Fig. 4(a)]. This is often called a "beating" or "breathing" pattern. If we cut the torus with a transverse plane (or by taking some general Poincare map) we obtain an invariant circle (a closed curve) (Iooss *et al.*, 1980; Marsden and McCracken, 1976) in exactly the same fashion that a Poincare section of a limit cycle gave us a fixed point [Fig. 4(d)] (Poincare, 1899). This invariant circle is drawn by the stroboscopic point that hops discontinuously on the curve [Fig. 4(e)] as the quasi-periodic trajectory winding around the torus meets the intersecting plane again and again. This procedure of sectioning can be continued in higher dimensions. For example, the Poincare section of a two-torus in an n-dimensional problem

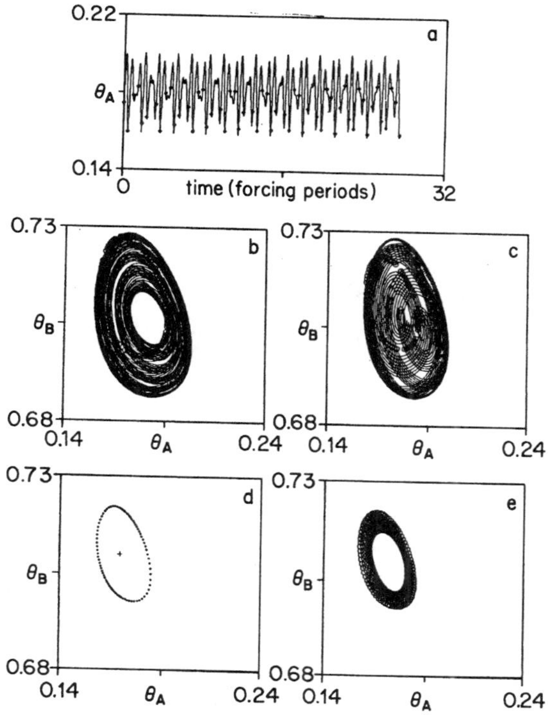

FIGURE 4 Views of a quasi-periodic attractor. (a) Variation of one component of a quasi-periodic trajectory with time. The stroboscopic points are denoted by a ◇. Notice the slight variation every time the basic oscillatory pattern is repeated. Surface model, $\omega_0 = 1.4$, $a/a_0 = 0.5$ ($a_0 = 0.001$). (b, c) Projection of a quasi-periodic attractor on the phase plane: the hole in the middle of the doughnut may [(b), surface model, $\omega_0 = 1.1$, $a/a_0 = 0.1$] or may not [(c), surface model, $\omega_0 = 1.4$, $a/a_0 = 0.5$] be visible. The stroboscopic points marked by ○ are scarcely discernible as the quasi-periodic trajectory fills the space within the projection of the doughnut. (d) The invariant circle of the stroboscopic map that results from cutting the torus in (c) with the cos (ωt) = 1 plane. This circle has been computed with 80 equally spaced discretization points through a polar parameterization whose centre point is marked by (+). (e) shows how the circle is generated by discontinuous hops of the phase point on it; every point and its image are connected with a straight line, and the "caustic" curve in the middle is tangent to all these rays. Such caustics arise in maps of the circle resulting from billiard problems.

will be an invariant curve in n-space. On the other hand, the Poincare section of higher tori will be simpler tori: the section of a T^3 (a torus with three mutually incommensurate frequencies) will be a two-torus and so on.

We are now concerned with the location of two-tori in our systems. We have formulated for that purpose a "shooting on collocations" iterative algorithm that computes an invariant curve of the stroboscopic map for a forced system. Part of this algorithm and its application to maps in general are described elsewhere (Kevrekidis *et al.*, 1985). We briefly describe here its application to a two-dimensional forced system.

(1) Choose a parameterization for the invariant closed curve. In benign cases when the curve is not highly irregular this can be accomplished by

choosing a centre point and then using simple polar coordinates:

$$r = r(\theta), \quad -\pi \leqslant \theta \leqslant \pi, \quad r(\pi) = r(-\pi). \tag{11}$$

Choosing the radial line from which we start measuring the polar angle counterclockwise ($\theta = 0$) is equivalent to "anchoring" a limit cycle.

(2) Make a discretized guess of the closed curve.

$$r_i = r(\theta_i), \quad i = 2, \ldots, N, \quad -\pi \leqslant \theta_i \leqslant \pi. \tag{12}$$

(3) Obtain the coordinates of the images of these points after one forcing period, as well as the derivatives of the image coordinates with respect to the initial point coordinates. These are computed through integration of the system equations and the variational equations for the appropriate number of forcing periods. Notice that again we have as many maps (as many invariant circles) as phases of the forcing term. They all lie on the same invariant doughnut surface.

(4) Since the θ-coordinates of the image points do not in general coincide with the θ-coordinates of the original discretization points r_i, some form of interpolation is necessary to obtain the discretized estimate of the image circle in the form $\rho(\theta_i), i = 1, 2, \ldots, N$. The derivatives $\partial \rho_j / \partial r_i$ will be obtained through the interpolation formula and the use of the periodic boundary condition $r(\theta + 2\pi) = r(\theta)$.

(5) Solve the set of coupled nonlinear collocation equations

$$r_i - \rho_i = 0, \quad i = 1, 2, \cdots, N \tag{13}$$

with an iterative method (in our case, a Newton–Raphson) until a specified external tolerance specification has been met.

It is interesting to consider the Jacobian of this Newton scheme upon convergence. It is a sparse matrix which consists (Figs. 6 and 10) of a diagonal of ones and an off-diagonal band of nonzero entries. The width of this band depends on the interpolation scheme. Its location is an approximation to what is termed "angular function"; it determines approximately the rotation angle between a point on the closed curve and its image point after one iteration of the (stroboscopic) map. Further assumptions regarding the continuity of higher derivatives of $r(\theta)$ with respect to the parameter θ are required in order to prove that the object to which the code converges is indeed a discretization of an invariant circle [see extensive discussion in Kevrekidis *et al.* (1985)].

We now consider the phenomenon of entrainment (the development of resonances) on the torus (Meyer, 1983). When (and if) the off-diagonal band in Fig. 6 crosses the diagonal [Figs. 6(c) and 6(f)], there exist points whose images fall on themselves; they are fixed points of the map we study. These points lie on periodic trajectories that are locked on the torus. Such trajectories appear in pairs in saddle–node bifurcations and are usually termed "subharmonics". When this occurs there is no quasi-periodic attractor winding around the torus surface, but the basic structure of the torus persists: the invariant circle is patched up from the unstable manifolds of the periodic saddle-points with the addition of the node-periodic point (Arnol'd, 1973, 1982). As we continue changing some system parameter the periodic points may come to die in another saddle-node bifurcation (see Fig. 5). Periodic trajectories thus

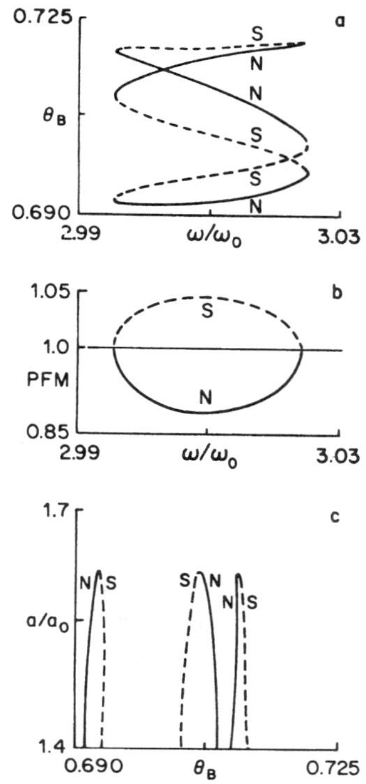

FIGURE 5 Subharmonic saddle–node bifurcations. (a). The subharmonic period 3 isola for the surface model ($a/a_0 = 1.4$, $a_0 = 0.001$). One coordinate of the fixed points of the third iterate of the stroboscopic map is plotted vs. the varying frequency ratio ω/ω_0. Six such points (*S*, *N*) exist simultaneously, three of them (*N*) lying on the stable node period 3 and three (*S*) on the saddle period 3. Notice the two triple turning point bifurcations at $\omega/\omega_0 = 2.9965$ and 3.0286. In (b) the *PFM* of these trajectories on the isola is also plotted. (c) shows another saddle–node bifurcation occurring (for $\omega_0 = 3$) as this time the forcing amplitude is increased to $a/a_0 = 1.655$.

often appear as isolated branches in one-parameter bifurcation diagrams of periodically forced systems. These isolas are dispersed in an "ocean" of quasi-periodic behaviour.

There exists a quantity (the rotation number) which is independent of the particular stroboscopic section we take and characterizes the entire torus. For a radial parameterization of the invariant circle the rotation number is defined—and may be computed—as the limit

$$\tau = \lim_{n\to\infty} \sum_{i=1}^{n} \frac{1}{2\pi n} f^{(i)}(\theta_0)$$

where $f^{(n)}(\theta)$ is the θ-coordinate of the nth return image of the point $r(\theta_0)$. That this limit exists and is independent of θ_0 is a fundamental result (Aronson *et al.*, 1982). When resonances (periodic trajectories) exist on the torus the above ratio will evidently take rational values. We then say that the forcing has entrained the system, because the periods of these trajectories are integer multiples of the forcing period. In general, the rotation number will be a Cantor function of the bifurcation parameter in a one-parameter diagram (Munkres, 1975). It will take rational values for closed intervals in the parameter space; the edges of these intervals are the saddle-node bifurcations we described. These intervals may be extremely narrow and hence difficult to compute.

FORCED DYNAMIC PHENOMENA

Small Amplitude Forcing

We now add a periodic perturbation term [eqs (8)–(10)] to the autonomous oscillating models. We examine the common qualitative traits of the response of all three cases as a function of two variables: the forcing amplitude FA and the ratio of the forcing to the natural frequency (briefly frequency ratio ω/ω_0). The natural frequency of course has meaning only at the limit $FA \to 0$. At that limit all systems—and for all forcing frequencies—have a torus in their phase plane. This torus results from forcing the autonomous limit cycle (Fenichel, 1971). There will exist regions in our two-parameter diagram where a quasi-periodic attractor will wind around this torus. There will also exist regions of entrainment (sometimes referred to as resonance horns or Arnold tongues) where frequency locking will occur and periodic trajectories (subharmonic nodes and saddles) will exist on the torus. These regions typically look like cusps [see Fig. 6(a)]; their boundaries are the loci of the saddle-node bifurcations.

The tip of the typical resonance horn touches the axis $FA = 0$ at points $\omega/\omega_0 = p/q$, where p and q are relatively prime integers. The periodic trajectories within such a horn have a period pT, where T is the forcing period, and we refer to them as p-periodic trajectories or simply period ps. The larger the relatively prime numbers p and q, the narrower the resonance horn and the sharper its tip.

As $FA \to 0$, the projection of the torus on the phase plane [see Fig. 3(c)] is a very thin annulus surrounding the unperturbed limit cycle. Within this annulus there exists a small unstable periodic trajectory (a period 1) resulting from the forcing of the unperturbed unstable steady state within the limit cycle. This unstable period 1 will play a crucial role in the breaking of the torus as the forcing amplitude will become larger.

It is interesting to consider the shapes of the subharmonic trajectories that lock on the torus in the various entrainment regions of order p/q. The subharmonic period 4 at the 4/3 resonance horn is, for example, a three-peaked oscillation in time [Fig. 7(a)] and has three closed loops in its phase-plane projection [Fig. 7(b)], while the subharmonic period 4 at the 4/1 resonance is a single-peaked, single-loop oscillation [Figs. 7(d) and 7(e)]. A subharmonic period 2 at the 2/3 resonance is also included in Figs. 7(g) and 7(h). Multipeaked oscillations observed in chemical systems (Scheintuch and Schmitz, 1977; Flytzani-Stephanopoulos *et al.*, 1980) may thus result from the interaction of frequencies of local oscillators. Such trajectories are the nonlin-

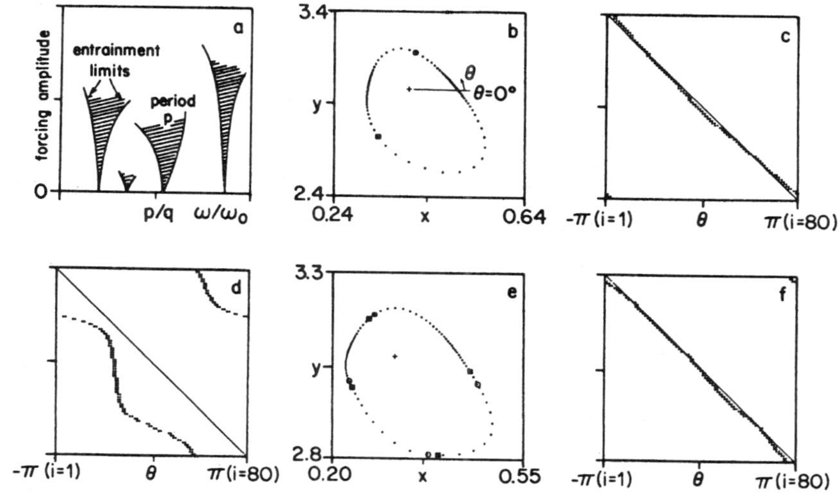

FIGURE 6 Resonances on the torus. (a) Schematic diagram of the typical low-amplitude portion of the two-parameter (FA, ω/ω_0) forced bifurcation diagram. Resonance horns within which an entrained period p locks on the torus touch the $FA = 0$ axis at $\omega/\omega_0 = p/q$, where p and q are relatively prime integers. The boundaries of these horns mark the transition between periodicity and quasi-periodicity. (b) Stroboscopic section of the torus with a node (○) and a saddle (■) period 1 locked on it (Brusselator, $a = 0.0005$, $\omega/\omega_0 = 1$). The unstable period 1 within the doughnut is indicated by (+) and has been used as the centre point for the 80-point even polar mesh computation. (c) shows the Jacobian of this computation. Nonzero elements are denoted by (+) and crossings of this band with the diagonal indicate the location of the resonances. The upper crossing corresponds to the saddle and the lower to the node of (b). (e) shows a torus section with subharmonic periods 4 on it (Brusselator, $\omega/\omega_0 = 4/3$, $a = 0.0072$, same mesh as above). (d) and (f) show the Jacobian of this computation when an invariant circle of the first (d) or the fourth (f) iterate of the map is located. Crossings with the diagonal occur in the latter case.

ear analogue of what is called Lissajoux pictures (Resnick and Halliday, 1966) that so entertain the novice oscilloscope user.

Small-order resonance horns (p, q small) and particularly those with $1 \leqslant p, q \leqslant 4$ are comparatively wide and easier to locate computationally through algorithms that will locate the periodic entrained trajectories. These algorithms, however, will be inadequate for a complete analysis of our systems since (at least as $FA \rightarrow 0$) periodic trajectories appear in disconnected isolas. The motivation behind the construction of our torus-computing algorithm is to provide a means of study of this two-parameter bifurcation diagram that can continue smoothly both within the resonance horns and in the region of quasi-periodicity that separates or—from another point of view—unites them.

Stronger Forcing

As the amplitude of the forcing term becomes larger, the resonance horns will begin to interact. The instantaneous values of the forcing variable will eventually visit intervals in which the autonomous systems do not have a stable limit cycle at all. In such cases the uniformity of behaviour of all systems

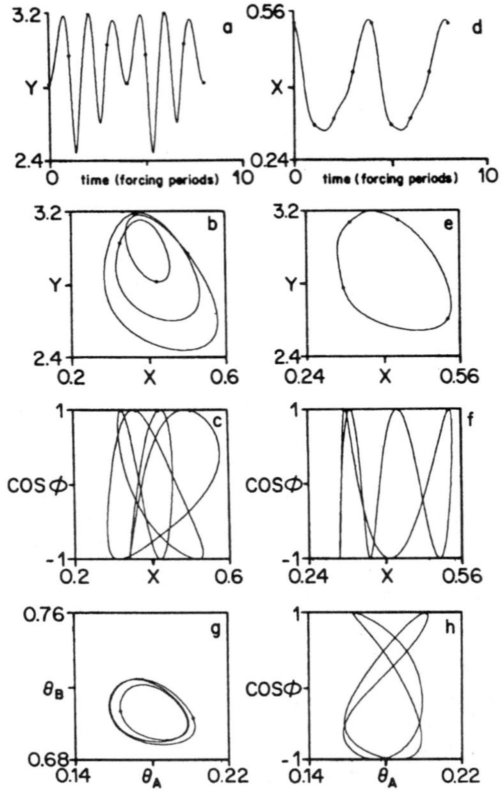

FIGURE 7 Typical shapes of subharmonic trajectories. A subharmonic period 4 within the 4/3 resonance horn is a three-peaked oscillation in time (a), has three loops in its phase plane projection (b), and four loops in its x-cos ϕ projection (c) (Brusselator, $a = 0.0072$, $\omega/\omega_0 = 4/3$). The subharmonic period 4 within the 4/1 resonance horn has one loop in its phase plane projection (e), four loops in the x-cos ϕ projection (f) and is a one-peaked oscillation in time (d). Stroboscopic points are denoted by ◇. Try to imagine them winding around the doughnut in three-dimensional space! An interesting shape shows up at the period 2 resonance in the 2/3 resonance horn (surface model $\omega/\omega_0 = 2/3$, $a/a_0 = 0.1$, $a_0 = 0.001$) (g, h). These shapes are comparatively simple because of the shape of the unperturbed limit cycle which for all cases was a simple closed curve.

(as at small FA) is lost, and what happens will depend on the particular system. The fundamental 1/1 resonance horn in the case of the Brusselator, for example, will persist all the way up to the highest meaningful values of the forcing ($FA = 0.4$). On the other hand, a period doubling bifurcation is observed (at $FA = 0.31499$ for $\omega/\omega_0 = 1$) in the 1/1 resonance horn of the forced CSTR. We would like to examine the phenomena that are common to all systems and prevail the longest as the forcing amplitude increases, before the nonlinearities of each particular system take charge of the dynamics.

The first interesting phenomenon we observe is the breaking of the smooth invariant circle picture when the node-periodic points on the circle become foci [Figs. 8(a) and 8(b)]. The basic structure of the trajectories remains the same, but the "object" we are now studying is no longer a well-defined circle;

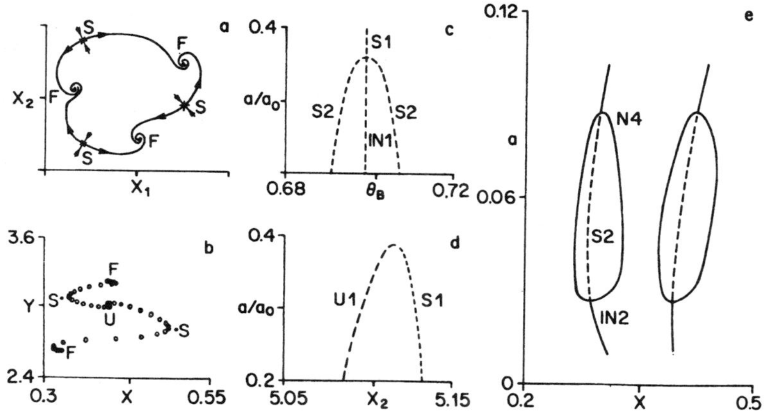

FIGURE 8 Breaking of tori. (a) Schematic diagram of the shape that succeeds the smooth invariant circle when a node–periodic point on it becomes a focus *F*. (b) This situation is illustrated through a transient (Brusselator, $\omega/\omega_0 = 2$, $a = 0.005$). The phase point spirals out from the unstable focus period 1 (U) within the torus. The curving of the two branches is indicative of the location of the saddle-subharmonics, denoted by S. The transient finally winds around the stable focus F subharmonic period 2. (d) Breaking of the torus within the 1/1 resonance horn. The unstable "source" period 1 (U1) within the doughnut hits the subharmonic saddle period 1 (S1) in a saddle–node bifurcation as the forcing amplitude increases (forced CSTR, $\omega/\omega_0 = 1$, $a_0 = 0.0169983$). (c) Breaking of the torus within the 2/1 resonance horn. This time the subharmonic saddle period 2 (S2) is hit in an unstable period doubling bifurcation as the forcing amplitude increases (surface model, $\omega/\omega_0 = 2$, $a_0 = 0.001$). (e) An incomplete cascade in the 2/1 resonance horn of the Brusselator after the torus has broken. The stable period 2 (an inverse stable node IN2) period doubles, changing to a saddle period 2 (S2), but the resulting period 4 (N4) does not continue the cascade; instead, it closes back on the original period 2 (Brusselator, $\omega/\omega_0 = 2$.)

a modification of our algorithm is necessary in order to study it. Such a breaking is the first step towards the breaking of the torus. In some cases, however (Kevrekidis *et al.*, 1984), the foci may re-become nodes in a one-parameter continuation and frequency unlocking will again give quasi-periodic attractors.

The most important characteristic in our test cases, however, is that within the 1/1 and the 2/1 resonance horns the torus will break as *FA* increases. In all models this happens when the unstable "source" period 1 that existed within the torus hits the saddle-periodic trajectories that lie on the torus. This occurs through a saddle-node bifurcation in the 1/1 resonance horn [Fig. 8(d)], and through an unstable period doubling in the 2/1 resonance [Fig. 8(c)]. After these bifurcations the basic structure of the torus has collapsed, and we are left only with the stable entrained periodic trajectories.

As we now change ω/ω_0 the stable periodic trajectories cannot lose stability through a saddle-node bifurcation, since the saddles no longer exist; rather they lose stability through a Hopf bifurcation of the stroboscopic map to a torus (Marsden and McCracken, 1976). This phenomenon, as well as the torus resulting from it, is considerably different from the frequency unlocking case. One of the main differences is that the entire quasi-periodic attractor that bifurcates from a periodic trajectory lies close to it [see Figs. 9(c) and 9(d)], whilst the one that results from unlocking will visit regions in the phase space "far away" from the periodic trajectory that just lost stability. This can be used in an experimental study to identify the type of bifurcation.

Let us examine more closely what occurs on the right-hand boundary of the 1/1 resonance horn [Fig. 9(a)]. In a sequence of one-parameter bifurcation diagrams with respect to ω/ω_0, each taken at a successively higher forcing amplitude FA, we observe that as FA increases, the bifurcation point to a torus changes. The point of exit of the Floquet multipliers of the periodic

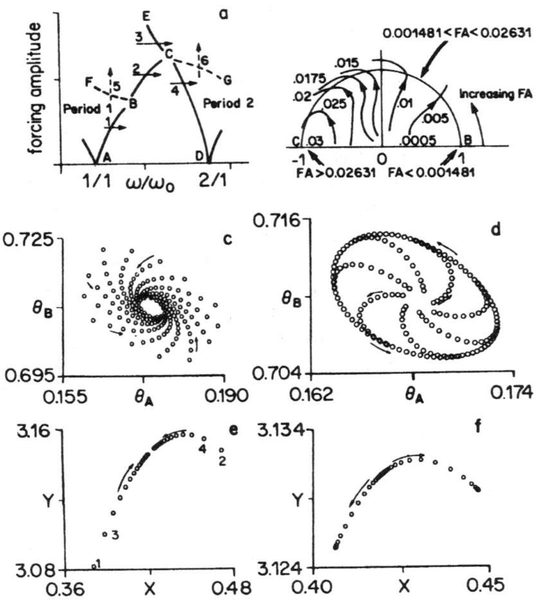

FIGURE 9 Some important "intermediate" forcing amplitude phenomena. (a) Tentative schematic diagram of bifurcations related with the resonance horns at 1/1 and 2/1. The lower right boundary of the 1/1 resonance horn AB is characterized by a saddle–node bifurcation (1) on the torus [the node-period 1 *PFM* meets the saddle–period 1 *PFM* at unity, see (b)]. This changes at the first codimension two bifurcation (at $FA = 0.001481$) and crossing of the BC boundary (2) is a Hopf bifurcation of a period 1 to a torus ($0.001481 < FA < 0.02631$). The *PFM* crossing angle rotates around the unit circle [see (b)] with increasing *FA*. At the subsequent codimension two bifurcation (point C, $FA = 0.02631$) Hopf bifurcation turns into period doubling (3) as the two resonance horns come together. The *PFM* is then crossing the unit circle through -1 [see (b)]. Long before this has happened, the basic torus structure has broken through an unstable saddle–node (5) in the 1/1 horn [Fig. 7(d)] and through an unstable period doubling (6) in the 2/1 horn [Fig. 7(c)]. (4) indicates phase locking of a period 2 on the torus. (c) and (d) illustrate a typical change of the stroboscopic phase plane when a Hopf bifurcation of a periodic point to an invariant circle—a periodic trajectory to a quasi-periodic one—occurs. (c) shows the transient spiralling inwards towards a stable focus (surface model, $a/a_0 = 0.5$, $\omega/\omega_0 = 1.3$) that changes to a spiralling outward and then around the section of the torus ($\omega/\omega_0 = 1.4$). (e) and (f) illustrate the typical change occurring in a period doubling bifurcation. Transient approach to an inverse node period 1, with the phase point alternating between branches is shown in (e) (Brusselator, $\omega/\omega_0 = 1.05$, $a = 0.035$). As ω/ω_0 changes, the period 1 loses stability and becomes an inverse saddle period 1, whose unstable manifold is a Mobius strip (Takoudis *et al.*, 1981). The phase point hops on opposite branches of the stroboscopic section of this manifold away from the saddle and towards the new-born stable period 2 in (f) for $\omega/\omega_0 = 1.1$.

trajectory from the unit circle rotates [see Fig. 9(b)] until, in the end, the bifurcation changes type. The Floquet multipliers become real and one of them exists through -1 giving birth to a period 2 [a typical period doubling bifurcation, Figs. 9(e) and 9(f)]. Thus the resonance horns of 1/1 and 2/1 come together in a codimension two bifurcation [Fig. 9(a), point C], where both Floquet multipliers of a periodic trajectory are located at -1. The other codimension two bifurcation common to all systems [Fig. 9(a), point B] occurs at the point where the saddle–node boundary of the 1/1 resonance turns into a Hopf bifurcation to a torus; at that point both free Floquet multipliers of the periodic point lie at 1.

At interesting phenomenon occurs in the case of other resonance horns; we have studied it for the case of the 3/1 resonance. The torus pattern breaks when the subharmonic periodic trajectories locked on it for small *FA* decollate from the torus as *FA* increases. We are left then with two attractors: a stable period 3 and a stable quasi-periodic trajectory. This is a spectacular case of multistability (co-existence of periodic and quasi-periodic oscillations). The initial conditions will determine the attractor to which the system will eventually converge. This decollation of the subharmonics from the torus was predicted by Greenspan and Holmes (1984). They also predicted chaotic trajectories close to the parameter values where the subharmonic decollation occurs.

The detailed study of these bifurcations (which we shall describe in subsequent publications) is important for several reasons. It will be helpful in understanding some of the mechanisms that synchronize chemical oscillators, by breaking the pattern of higher-dimensional tori that exist when they are weakly coupled. It also provides a very convenient setting for studying on practically any oscillating system phenomena that are, by comparison, less frequently observed in autonomous systems.

We have limited our considerations to common characteristics of the response of these systems. Of course, each of the systems individually exhibits a variety of dynamic phenomena in its parameter space (e.g. both the Brusselator and the CSTR when forced may exhibit a period doubling cascade to chaos at large forcing amplitudes). We present here only one example of the so-called "incomplete cascade" [Fig. 8(e)]. Within the 2/1 resonance horn of the Brusselator, the boundaries of successive period doublings appear as a series of "shells", each contained in the previous one in the two-parameter plane. Taking a one-parameter cut, one might miss the inner shells and only find a few branchings that subsequently close again. Such one-parameter behaviour will then suggest the necessity of a two-parameter study to outline the whole portrait; this study often helps in understanding and linking together phase portrait traits that have no apparent connection in the one-parameter pictures (Aronson *et al.*, 1982).

Other Models and Different Types of Forcing

We have examined some fundamental qualitative features of the dynamic response of several chemical models to periodic forcing. Most of the phenomena we observe are not restricted to two-dimensional systems or to sinusoidal forcing. They may—and will—be found in the modelling of any physical process through coupled nonlinear ODEs, at least at the small forcing amplitude limit, irrespective of the type of the forcing and the forcing variable. Sinusoidal forcing was chosen to avoid the discontinuities in the time derivatives of the bang-bang forcing (Sincic and Bailey, 1977). Although any periodic forcing can be decomposed to its Fourier components, the responses to each component cannot be superposed in nonlinear systems; at the small amplitude limit, however, the qualitative response should remain the same irrespective of the type of the forcing (bang-bang, sawtooth, etc.).

Comments on the Computational Methods

The shooting algorithms coupled with Newton iteration and pseudo-arc length continuation proved efficient for the computation of forced periodic branches and their bifurcations. The main computational cost lies in the sometimes stiff integration of a large system of simultaneous equations: the equations of change coupled with the variational equations and the sensitivity equations with respect to the parameters necessary for the continuation of solution branches. This cost may be lowered if we take advantage of the observation that the entries in the Jacobian of the variational and sensitivity equations have already been computed in the Jacobian of the equations of change at the current moment in time.(See also the occupancy in Fig. 10.)

Quasi-Newton methods may be used instead of our full Newton iteration. We have used the fast (quadratic) convergence rate of our Newton algorithm as a numerical check to discriminate between periodic and very slowly changing quasi-periodic trajectories; the accurate computed elements of the Jacobian in a Newton iteration can be used in stability computations for the located periodic trajectories. There are deficiencies in the use of a full Newton algorithm, such as its sometimes small radius of convergence (Schwartz, 1983). Several other possibilities for continuation methods also exist (Doedel, 1986; Seydel and Hlavacek, 1986). The pseudo-arc length continuation was sufficient for our calculations.

In our computation of invariant circles of maps the main cost lies in the

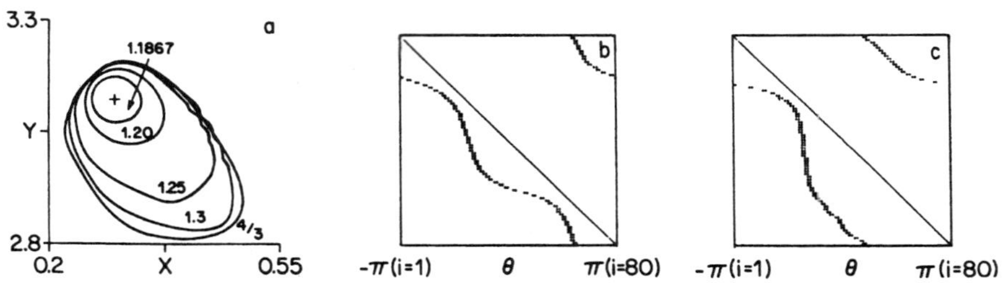

FIGURE 10 Change of the torus and the angular function with ω/ω_0. (a) A succession of computed sections of invariant tori for various values of ω/ω_0 (Brusselator, a = 0.0072). The centre point is indicated by (+). (b, c) The occupancy of the converged Jacobian for ω/ω_0 = 1.186667 and 1.3, respectively. The "bumps" on some of the circles are artifacts of the mesh and are associated with the almost vertical parts of the nonzero band. They can be eliminated by mesh adaptation.

efficient solution of a large system of sparse linear equations repeatedly (e.g. through the YALE Sparsepack). In the computation of sections of invariant tori, however, constructing the entries in the sparse matrix and the residuals is by far the most expensive task. We have to solve the same system of ODEs for a large set of initial conditions. In this task the parallel processing capabilities of a supercomputer like the U.C.C. Cray-1 at Minnesota can be of particular help; we are currently constructing a vectorized version of our integrator for that purpose.

An alternative formulation for the torus-computing algorithm is to solve for an invariant circle along with a nonlinear change of coordinates that makes the action of the stroboscopic map conjugate to a rigid rotation on the circle. This is equivalent to the parameterization

$$\mathbf{F}(\mathbf{x}(s)) = \mathbf{x}(s + \tau), \quad 0 \leqslant s \leqslant 1, \quad \text{periodic in } s.$$

It is based on Denjoy's theorem, and τ is the rotation number. This algorithm, implemented by Chan (1983) computes invariant circles with irrational rotation numbers. We may, of course, discretize and solve for the whole invariant surface and not just for a section of it. Instead of having to integrate the system equations, we will then be solving for a much larger number of unknowns resulting from the additional dimension we had suppressed in the shooting approach we used.

The main complication in the application of the algorithm (Kevrekidis *et al.*, 1985) lies in the difficulty of locating a simple parameterization of the invariant circle valid for many different shapes. A per cent total arc length parameterization is applicable in most cases, but it lacks the computationally appealing characteristic of producing a sparse Jacobian. Indeed, since the total arc length depends on the coordinates of all discretization points, the Jacobian will be a full matrix. With appropriate modifications the algorithm can, however, be used to locate invariant tori (both stable and unstable) of both autonomous and periodically forced systems as well as in the location of invariant manifolds in general. Stability information on the located torus (the tangent bundle) is contained in the converged Jacobian of the problem and may be used to locate further bifurcations of the tori. A special application of the algorithm is that it may be used—with essential boundary conditions—to locate both homoclinic and heteroclinic orbits.

Control Applications

A periodically forced system may be considered as an open-loop control system. The intermediate and high amplitude forced responses can be used in model discrimination procedures (Bennett, 1981; Cutlip *et al.*, 1983). Alternate choices of the forcing variable and observations of the relations and lags between various oscillating components of the response will yield information regarding intermediate steps in a reaction mechanism. Even some unstable phase plane components of the unforced system will become apparent through their role in observable effects (such as the codimension two bifurcations described above where they collide and annihilate stable, observable responses).

The natural parameterization of a forced periodic trajectory through the forcing term phase provides a very convenient way of closing the control loop and stabilizing unstable periodic trajectories. Consider such a trajectory $\gamma(\phi)$, $0 < \phi < 2\pi$ and the proportional control system

$$\dot{\mathbf{x}} = \mathbf{f}(\mathbf{x}, \phi) + \underline{\mathbf{A}} \cdot (\mathbf{x} - \gamma(\phi))$$

obtained from the forced system with the addition of the control term. Through the tunable parameters in A we can relocate the *FM*s of the periodic trajectory $\gamma(\phi)$ without changing it. Such a procedure can be used to stabilize a desirable—under some performance criterion—limit cycle, as has been used in the past to stabilize steady states. Bailey and co-workers have recently used such a procedure to drive a system to bifurcation in order to elucidate its dynamical characteristics (Lyberatos *et al.*, 1984).

CONCLUSIONS

Chemical reacting systems will exhibit a rich variety of responses to periodic forcing. For small and intermediate forcing amplitudes these responses are organized around several common qualitative patterns. These patterns include the interplay between entrainment and quasi-periodicity for small forcing amplitude, when the identity of both the forcing and the oscillating system are preserved. We have also observed several routes through which this low-amplitude toroidal pattern breaks as the forcing increases. A study of these patterns assists the analysis of the individual system. More importantly, it helps elucidate the interaction between individual frequencies and the coupling mechanisms between local oscillators. Dynamical phenomena such as the interaction of entrainment regions are more amenable to computation and understanding than in autonomous systems.

We believe that the qualitative phenomena we presented, as well as the type of analysis that led to them, are much more informative than a mere set of bifurcation diagrams. The interaction of oscillators is a vital problem in all levels of chemical engineering research (Neu, 1979; Cohen and Neu, 1979; Schmitz and Tsotsis, 1979; Jensen and Ray, 1980; Scheintuch and Pismen, 1981; Chang, 1983), from the coupling of microscopic catalyst sites to the interaction of large-scale units in a plant. Fundamental theoretical and experimental research, as well as an abundance of new tools (supercomputing, computing algebra, fast and affordable on-line data acquisition and processing), is continuously improving our ability to deal with these phenomena. It is important to recognize and understand the dynamical structure that underlies them, the problems they pose, and the perspectives they open to our design capabilities.

ACKNOWLEDGMENTS

This work partially supported by NSF under grant No. CPE 8313497.

REFERENCES

Abdul Kareem, H. K., Silveston, P. L. and Hudgins, P. R., 1980, Forced cycling of the catalytic oxidation of CO over a V_2O_5 catalyst. I—Concentration cycling. *Chem. Engng Sci.* **35,** 2077–2084; II—Temperature cycling. *Chem. Engng Sci.* **35,** 2085–2088.

Abraham, R. and Shaw, C., 1983, *Dynamics: The Geometry of Behavior*, Vols. 1 and 2. Aerial Press, Santa Cruz.

Aluko, M. and Chang, H.-C., 1984, PEFLOQ: an algorithm for the bifurcational analysis of periodic solutions of autonomous systems. *Comp. Chem. Engng* **8,** 355–365.

Arnol'd, V. I., 1973, *Ordinary Differential Equations.* M.I.T. Press, Boston, MA.

Arnol'd, V. I., 1982, *Geometrical Methods in the Theory of Ordinary Differential Equations.* Springer, New York.

Aronson, D. G., Chory, M. A., Hall, G. R. and McGehee, R. P., 1982, Bifurcations from an invariant circle for two-parameter families of maps of the plane: a computer assisted study. *Comm. Math. Phys.* **83,** 303–354.

Bailey, J. E., 1973, Periodic operation of chemical reactors: a review. *Chem. Engng Commun.* **1,** 111–124.

Bailey, J. E., 1977, Periodic phenomena in chemical reactor theory. In *Chemical Reactor Theory* (Edited by Lapidus, L. and Amundson, N. R.). Prentice-Hall, Englewood Cliffs, NJ.

Balakotaiah, V. and Luss, D., 1982, Structure of the steady state solutions of lumped parameter chemical reacting systems. *Chem. Engng Sci.* **37,** 1611–1623.

Bennett, C. O., 1982, Understanding heterogeneous catalysis through the transient method. *ACS Symp. Ser.* **178,** 1–32.

Besicovitch, A. E., 1954, *Almost Periodic Functions.* Dover, New York.

Bohr, H., 1947, *Almost Periodic Functions.* Chelsea, New York.

Chan, T. N., 1983, Numerical bifurcation analysis of simple dynamical systems. M. Comp. Sci. Thesis, Concordia University, Montreal.

Chang, H. C., 1983, The domain model in heterogeneous catalysis. *Chem. Engng Sci.* **38,** 535–546.

Cohen, D. S. and Neu, J. C., 1979, Interacting oscillatory chemical reactors. In *Bifurcation Theory and Applications in Scientific Disciplines.* N.Y. Acad. Sci., New York.

Collet, P. and Eckmann, J.-P., 1980, *Iterated Maps on the Interval as Dynamical Systems*, Progr. Phys. Vol. 1. Birkhäuser, Boston.

Curry, J. H., 1979, An algorithm for finding closed orbits. In *Global Theory of Dynamical Systems* (Edited by Nitecki, Z. and Robinson, C.), Lecture Notes in Math., Vol. 819, Springer, New York.

Cutlip, M. B., 1979, Concentration forcing of catalytic surface rate processes. *A.I.Ch.E. J.* **25,** 502–508.

Cutlip, M. B., Hawkins, C. J., Mukesh, D., Morton, W. and Kenney, C. N., 1983, Modelling of forced periodic oscillations of carbon monoxide over platinum catalyst. *Chem. Eng. Commun.* **22,** 329–344.

Doedel, E. J., 1981, AUTO: a program for the automatic bifurcation analysis of autonomous systems. *Cong. Num.* **30,** 265–284.

Doedel, E. J., 1984–1985, Private communications.

Doedel, E. J., 1986, Continuation techniques in the study of chemical reaction schemes. In *Proc. Special Year in Energy Math., University of Wyoming* (Edited by Gross, K. I.). SIAM (in press).

Douglas, J. M., 1972, *Process Dynamics and Control*, Vol. 2, Chap. 10. Prentice-Hall, Englewood Cliffs, NJ.

Feigenbaum, M. J., 1978, Quantitative university for a class of nonlinear transformations. *J. Stat. Phys.* **19,** 25–52.

Fenichel, N., 1971, Persistence and smoothness of invariant manifolds for flows. *Ind. Univ. Math. J.* **21,** 193–226.

Flytzani-Stephanopoulos, M., Schmidt, L. D. and Caretta, R., 1980, Steady state and transient oscillations in NH_3 oxidation on Pt. *J. Catal.* **64,** 346–355.

Glansdorff, P. and Prigogine, I., 1971, *Thermodynamic Theory of Structure, Stability and Fluctuations*, p. 233. Wiley Interscience, New York.

Glass, L., Guevara, M., Belair, J. and Shrier, A., 1984, Global bifurcations of a periodically forced biological oscillator. *Phys. Rev. A* **29,** 1348–1357.

Goodman, M. G., Kenney, C. N., Morton, W. and Cutlip, M. B., 1981, in *Proceedings of the 2nd World Congress of Chemical Engineering*, Canada.

Goodman, M. G., Kenney, C. N., Morton, W., Cutlip, M. B. and Mukesh, D., 1982, Transient studies of carbon monoxide oxidation over platinum catalyst. *Surf. Sci.* **120,** L453–460.

Greenspan, B. and Holmes, P., 1984, Repeated resonance and homoclinic bifurcations in a periodically forced family of oscillators. *SIAM J. Math. Anal.* **15,** 69–97.

Guckenheimer, J. and Holmes, P., 1983, *Nonlinear Oscillations, Dynamical Systems and Bifurcations of Vector Fields.* Springer, New York.

Hale, J. K., 1969, *Ordinary Differential Equations.* Wiley, New York.

Hamer, J. W. and Cormack, D. E., 1978, Influence of oscillating external pressure on gas-phase reactions in porous catalysts. *Chem. Engng Sci.* **33,** 935–944.

Hayashi, C., 1964, *Nonlinear Oscillations in Physical Systems.* McGraw-Hill, New York.

Hegedus, L. L., Chang, C. C., McEwen, D. J. and Sloan, E. M., 1980, Response of catalyst surface concentrations to forced concentration oscillations in the gas phase: the NO, CO, O_2 system over a-alumina. *Ind. Engng Chem. Fundam.* **19,** 367–373.

Hudson, J. L. and Mankin, J. C. 1981, Chaos in the Belousov–Zhabotinski reaction. *J. Chem. Phys.* **74,** 6171–6177.

Hudson, J. L., Hart, M. and Marinko, D., 1979, An experimental study of multiple peak periodic and nonperiodic oscillations in the Belousov–Zhabotinski reaction. *J. Chem. Phys.* **71,** 1601–1606.

Iooss, G. and Joseph, D., 1980, *Elementary Stability and Bifurcation Theory.* Springer, New York.

Iooss, G., Arneodo, A., Coullet, P. and Tresser, C., 1980, Simple computations of bifurcating invariant circles for mappings. *Lect. Notes Math.* **898,** 192–211.

Jensen, K. F. and Ray, W. H., 1980, A microscopic model for catalytic surfaces—I. Catalytic wires and gauzes. *Chem. Engng Sci.* **35,** 2439–2457.

Jorgensen, D. V., Farr, W. W. and Aris, R., 1984, More on the dynamics of the stirred tank with conservative reactions. *Chem. Engng Sci.* **39,** 1741–1752.

Kai, T. and Tomita, K., 1979, Stroboscopic phase portrait of a forced nonlinear oscillator. *Progr. Theor. Phys.* **61,** 54–73.

Keller, H. B., 1976, *Numerical Solution of Two-Point Boundary Value Problems.* Regional Conf. Series in Appl. Math., SIAM.

Keller, H. B., 1977, Numerical solution of bifurcation and nonlinear eigenvalue problems. In *Applications of Bifurcation Theory* (Edited by Rabinowitz P. H.). Academic Press, New York.

Kevrekidis, I. G., Schmidt, L. D. and Aris, R., 1984, On the dynamics of periodically forced chemical reactors. *Chem. Engng Commun.* **30,** 323–330.

Kevrekidis, I. G., Aris, R., Schmidt, L. D. and Pelikan, S., 1985, Numerical computation of invariant circles of maps. *Physica* **16D,** 243–251.

Krylov, N. M. and Bogoliubov, N. N., 1947, *Introduction to Nonlinear Mechanics.* Princeton University Press, Princeton, NJ.

Kubicek, M. and Hlavacek, V., 1983, *Numerical Solution of Nonlinear Boundary Value Problems with Applications.* Prentice-Hall, Englewood Cliffs, NJ.

Levi, M., Hoppensteadt, F. and Miranker, W., 1978, Dynamics of the Josephson junction. *Quart. Appl. Math.* **35,** 167–198.

Lyberatos, G., Kuszta, B. and Bailey, J. E., 1984, Discrimination and identification of dynamic catalytic reaction models via introduction of feedback. *Chem. Engng Sci.* **39,** 739–750.

Lyberatos, G., Kuszta, B. and Bailey, J. E., 1985, Normal forms for chemical reaction systems via the affine transformations. *Chem. Engng Sci.* **40,** 199–208.

Mandelbrot, B. B., 1983, On the quadratic mapping $z \rightarrow z^2 - m$ for complex m and z: the fractal structure of its M set and scaling. *Physica* **7D,** 224–239.

Mankin, J. C. and Hudson, J. L., 1984, Oscillatory and chaotic behavior of a forced exothermic chemical reaction. *Chem. Engng Sci.* **39,** 1807–1814.

Marsden, J. E. and McCracken, M., 1976, *The Hopf Bifurcation and its Applications,* Appl. Math. Sci. Vol. 19. Springer, New York.

Meyer, K., 1983, Tori in resonance. IMA preprint No. 13, University of Minnesota.

Minorsky, N., 1962, *Nonlinear Oscillations.* Van Nostrand, New York.

Munkres, J. R., 1975, *Topology, a First Course.* Prentice-Hall, Englewood Cliffs, NJ.
Nayfeh, A. H. and Mook, D. T., 1979, *Nonlinear Oscillations.* J. Wiley, New York.
Neu, J. C., 1979, Coupled chemical oscillators. *SIAM J. Appl. Math.* **37,** 307–315.
Poincare, H., 1899, *Les Methodes Nouvelles de la Mecanique Celeste.* Gauthier-Villars, Paris.
Resnick, R. and Halliday, D., 1966, *Physics,* Part 1. Wiley International.
Ruelle, D. and Takens, F., 1971, On the nature of turbulence. *Comm. Math. Phys.* **20,** 167–192; **23,** 343–344.
Scheintuch, M. and Pismen, L. M., 1981, Inhomogeneities and surface structures in oscillatory catalytic kinetics. *Chem. Engng Sci.* **36,** 489–497.
Scheintuch, M. and Schmitz, R. A., 1977, Oscillations in catalytic reactions. *Catal. Rev. Sci. Engng* **15,** 107–172.
Schmitz, R. A. and Tsotsis, T. T., 1979, On the possibility of spatially patterned states in systems of interacting catalyst particles. A.I.Ch.E. 72nd Annual meeting, San Francisco, Paper No. 109d.
Schwartz, I. B., 1983, Estimating regions of existence of unstable periodic orbits using computer-based techniques. *SIAM J. Num. Anal.* **20,** 106–120.
Seydel, R., 1981, Numerical computation of periodic orbits that bifurcate from stationary solutions of ordinary differential equations. *Appl. Math. Comput.* **9,** 257–271.
Seydel, R. and Hlavacek, V., 1986, Role of continuation in engineering analysis. *Chem. Engng Sci.* (submitted).
Shaw, R., 1981, Strange attractors, chaotic behavior and information flow. *Z. Naturforsch.* **36A,** 80–85.
Sincic, D. and Bailey, J. E., 1977, Pathological dynamic behavior of forced periodic chemical processes. *Chem. Engng Sci.* **32,** 281–286.
Stephanopoulos, G., Fredrickson, A. G. and Aris, R., 1979, The growth of competing microbial populations in a CSTR with periodically varying inputs. *A.I.Ch.E. J.* **25,** 863–872.
Stoker, J. J., 1950, *Nonlinear Vibrations.* J. Wiley, New York.
Takoudis, C. G., 1981, Ph.D. Thesis, University of Minnesota.
Takoudis, C. G., Schmidt, L. D. and Aris, R., 1981, Isothermal sustained oscillations in a very simple surface reaction. *Surf. Sci.* **105,** 325–333.
Taylor, T. W. and Geiseler, W., 1986, Periodic operation of a stirred flow reaction with limit cycle oscillations. *Ber. Bunsenges Phys. Chem.* (submitted).
Thoulouze-Pratt, E., 1983, Numerical analysis of the behaviour of an almost periodic solution to a periodic differential equation, an example of successive bifurcations of invariant tori. In *Rhythms in Biology and Other Fields of Application,* Lect. Notes in Biomath, Vol. 49, pp. 265–271.
Tomita, K., 1982, Chaotic response of nonlinear oscillations. *Phys. Rep.* **86,** 113–167.
Tomita, K., Kai, T. and Hikami, F., 1977, Entrainment of a limit cycle by a periodic external excitation. *Progr. Theor. Phys.* **57,** 1159–1177.
Uppal, A., Ray, W. H. and Poore, A. B., 1974, On the dynamic behavior of continuous stirred tank reactors. *Chem. Engng Sci.* **29,** 967–985.
Uppal, A., Ray, W. H. and Poore, A. B., 1976, The classification of the dynamic behaviour of continuous stirred tank reactors—influence of the reactor residence time. *Chem. Engng Sci.* **31,** 205–214.
Wolf, A., Swift, J. B., Swinney, H. L. and Vastano, J. A., 1986, Determining Lyapounov exponents from a time series. *Physica D* (submitted).

J.*

"YET WHO WOULD HAVE THOUGHT THE OLD MAN TO HAVE HAD SO MUCH BLOOD IN HIM?"—REFLECTIONS ON THE MULTIPLICITY OF STEADY STATES OF THE STIRRED TANK REACTOR

W. W. FARR and R. ARIS

Department of Chemical Engineering and Materials Science, University of Minnesota, Minneapolis, MN 55455, U.S.A.

(Received 13 *December* 1985)

Abstract—*The question of the multiplicity of the steady states of a chemical reactor was one of the concerns in the pioneering work of Bilous and Amundson. Their diagrams showed quite clearly the geometry of the situation, and this kind of analysis sufficed for many years. It remained for Balakotaiah and Luss, using the methods of singularity theory, to give a comprehensive treatment of the question. After a brief survey, we take up the case of consecutive first-order reactions and show that up to seven steady states are possible.*

INTRODUCTION

It goes without saying that the quotation (*Macbeth* V, 1, 42–44) forming the first part of our title is not to be read as any allusion to the colleague to whom these papers are all dedicated, but rather to the system on which his early work shed so much light. This system, the stirred tank or well-mixed reactor, can still evoke amazement—though fortunately with none of the macabre and melancholy overtones of Lady Macbeth's—that so much of the life blood of the subject is to be found in this the simplest of reactors, or, to give it another interpretation, that there can be so many steady states in a comparatively simple reaction system.

It has been commonplace in chemical reaction engineering circles, since Luss first discovered the reference, that Liljenroth (1918) was the first to mention the existence of multiple steady states and the associated stability

* Reprinted with the permission of Elsevier Science from *Chemical Engineering Science,* volume **41,** number 6, pp. 1385–1402, 1986. Copyright 1986 Elsevier Science.

phenomena. He gave an essentially correct explanation from a static point of view by means of an empirical heat generation curve and a calculated removal curve. There was a long pause until the 1940s when some important, but largely overlooked, Russian work was published (Frank-Kamenetski, 1940, 1941). No doubt this was hard to find until some time after the war and neither it, nor later work by Salnikov (1948), seems to have had any real influence in the West. Denbigh considered some of the dynamic aspects (Denbigh, 1947; Denbigh *et al.*, 1948) and the regions of attraction of different steady states, a concept which, he noted, had been mentioned in a physiological context by Burton (1939) who referred to it as "equifinality". Van Heerden's paper on autothermic reactors (1953) contains an argument for stability from the slopes of the heat generation and removal curves which is valid in the context of adiabatic reactors in which his paper is set.

Such was the state of the art when Amundson and Bilous's paper was published in the first volume of the newly founded *A.I.Ch.E. Journal* (Bilous and Amundson, 1955). This for the first time treated the reactor as a dynamical system and, using Lyapounov's method of linearization, gave a pair of algebraic conditions for local stability. One of these corresponded to the slope condition of previous analyses, and there was a brief flurry of attempts to invest the other with a similarly physical explanation. For the global picture they introduced the "phase plane" (another feature of the theory of dynamical systems) and, with consummate skill, Bilous conjured the now classic figures from a Reeves electronic analogue computer. Even in this early paper, they had touched upon the consecutive reaction scheme $A \rightarrow B \rightarrow C$ and had shown that up to five steady states might be expected under some conditions.

This is not the place to attempt a detailed history of the development of the understanding of stirred tank behaviour; it is sufficient to point out that Amundson and Bilous's paper is effectively the source of the two streams that have dominated much of the last 30 years' work on the stirred tank. The statics, so to speak, with its emphasis on the multiplicity question has flowered in a remarkable series of papers by Balakotaiah and Luss (1981, 1982a, b, c, 1983, 1984) who have built upon the methods of singularity theory for systems with a distinguished parameter developed by Golubitsky and Schaeffer (1979) and applied to the single-reaction CSTR by Golubitsky and Keyfitz (1980). The present paper is in that tradition with respect to the consecutive reaction scheme. The dynamical current, emphasizing the question of stability and attempting to fill out the gallery of phase portraits, has also flowed strongly in the subsequent work of Amundson and his colleagues (Amundson and Aris, 1958; Amundson and Schmitz, 1963; Amundson and Goldstein, 1965; etc.) and in work which shows his influence even at one or two removes. Thus Uppal *et al.* (1974, 1976) gave the first comprehensive picture though this has been added to by others (Vaganov *et al.*, 1978; Chang and Calo, 1979; Williams and Calo, 1981; Kwong and Tsotsis, 1983). Gray and his colleagues have shown that the algebraically simpler system of an autocatalytic, isothermal reaction gives many of the features of the non-isothermal, first-order scheme (Gray and Scott, 1984). And if proof of how much blood the old man has yet in him were needed, we have only to look at the richness of the system that Jensen and Planeaux discuss in this very issue.

THE SYSTEM

In this section we will develop the equations needed to apply singularity theory to the problem of two sequential reactions in a CSTR. The well-known equations for c_A, c_B and T, the concentrations of A, B and the temperature in a reactor of volume V through which a homogeneous stream of rate q flows, are:

$$V\frac{\mathrm{d}c_A}{\mathrm{d}t'} = q(c_{Af} - c_A) - Vc_A A_1 \exp(-E_1/RT) \tag{1}$$

$$V\frac{\mathrm{d}c_B}{\mathrm{d}t'} = q(c_{Bf} - c_B) + Vc_A A_1 \exp(-E_1/RT) - Vc_B A_2 \exp(-E_2/RT) \tag{2}$$

$$\rho C_p V\frac{\mathrm{d}T}{\mathrm{d}t'} = q\rho C_p(T_f - T) - UA_c(T - T_c) + (-\Delta H_1)Vc_A A_1 \exp(-E_1/RT) + (-\Delta H_2)Vc_B A_2 \exp(-E_2/RT) \tag{3}$$

(the notation is given fully in the Notation). In dimensionless form we have the variables:

$$t = qt'/V, \quad u = c_A/c_{Af}, \quad v = c_B/c_{Af}, \quad w = (E_1/R\overline{T})(T - \overline{T})/\overline{T}, \quad \overline{T} = (T_f + \kappa T_c)/(1 + \kappa) \tag{4}$$

and the parameters:

$$\alpha = (VA_1/q)\exp(-E_1/R\overline{T}) \tag{5}$$

$$\beta = \{(-\Delta H_1)c_{Af}/(\rho C_p \overline{T})\}(E_1/R\overline{T}) \tag{6}$$

$$\gamma = E_1/R\overline{T} \tag{7}$$

$$\kappa = UA_c/q\rho C_p \tag{8}$$

$$\nu = E_2/E_1, \quad \rho = \Delta H_2/\Delta H_1 \tag{9}$$

$$\sigma = (A_2/A_1)\exp\{(E_1 - E_2)/R\overline{T}\} \tag{10}$$

the simplification:

$$c_{Bf} = 0 \tag{11}$$

the abbreviation:

$$E(w) = \exp\{\gamma w/(\gamma + w)\} \tag{12}$$

and the equations:

$$\dot{u} = 1 - u\{1 + \alpha E(w)\} \tag{13}$$

$$\dot{v} = u\alpha E(w) - v\{1 + \sigma\alpha E^{\nu}(w)\} \tag{14}$$

$$\dot{w} = -(1 + \kappa)w + \beta u\alpha E(w) + \beta\rho\sigma v\alpha E^{\nu}(w). \tag{15}$$

We have simplified the situation slightly by letting $c_{Bf} = 0$, but still have seven parameters. Further approximations are often made, namely that

$\gamma \gg w$ so that $E(w)$ reduces to $\exp(w)$ and that the activation energies of the two reactions are the same. In Jorgensen *et al.* (1984) these hypotheses were called URP and LB after groups (Uppal *et al.*, 1974; Balakotaiah and Luss, 1982b) who used them earlier. The consequences of using these hypotheses have been subjects to controversy for some time (Uppal *et al.*, 1974; Balakotaiah and Luss, 1982a, b, 1983, 1984; Chicone and Retzloff, 1981), but their use is nearly universal owing to the algebraic difficulties encountered without them. The present work was undertaken to study carefully the effects of the LB/URP hypothesis.

Attention is also often restricted to the case when both reactions are exothermic, though Balakotaiah and Luss (1982b, 1984), Kahlert *et al.* (1981) and Jorgensen *et al.* (1984) allowed one of the reactions to be endothermic. We will not make this restriction either, so the parameters ρ and β are allowed to be of either sign. Since we are only concerned with multiplicity, we can define

$$B = \beta/(1 + \kappa) \tag{16}$$

and reduce the number of independent parameters to six. Thus we would expect that the highest-order singularity would be of codimension five. (For our purposes, codimension can be defined as the number of defining conditions minus two.) As we shall see, the notion of the highest-order singularity as the organization centre for all of the qualitatively different bifurcation diagrams does not work for this system and the analysis is correspondingly more difficult, but a clear picture (probably not obtainable without the use of singularity theory) emerges nevertheless.

The species balance equations (13) and (14) can be solved for the steady-state values of u and v, and the resulting expressions substituted in the temperature equation. After using eq. (16) the expression

$$w = B\frac{\alpha E(w)}{1 + \alpha E(w)}\left(1 + \rho\frac{\sigma\alpha E^{\nu}(w)}{1 + \sigma\alpha E^{\nu}(w)}\right) \tag{17}$$

results. This equation is often put over a common denominator; the result is eq. (18) below.

$$\sigma B(1 + \rho)\alpha^2 E^{\nu+1} + B\alpha E - \sigma\alpha E^{\nu} w(1 + \alpha E) - w(1 + \alpha E) = 0. \tag{18}$$

The solutions of this equation have some general properties that are of interest, as they will help to guide our later investigations. We first note that the equation is a quadratic in α. Earlier investigators, including the prescient Bilous and Amundson (1955), noted that if both reactions are exothermic, only one root of the quadratic is meaningful, but if this restriction is not made both roots can be positive and hence meaningful. To see this, we first rewrite this equation in the form.

$$\alpha^2(B(1 + \rho) - w)\sigma E^{\nu+1} + \alpha(B - w \times (1 + \alpha E^{\nu-1}))E - w = 0 \tag{19a}$$

or

$$\alpha_2\alpha^2 + \alpha_1\alpha + \alpha_0 = 0. \tag{19b}$$

Necessary and sufficient conditions for two positive roots to exist are

$$(1)\ a_0 a_2 > 0 \tag{20a}$$

$$(2)\ a_1 a_2 < 0 \tag{20b}$$

$$(3)\ a_1^2 - 4a_0 a_2 > 0 \tag{20c}$$

or

$$(1)\ w(r_1 - w) < 0 \tag{21a}$$

$$(2)\ w(r_2 - w) > 0 \tag{21b}$$

where

$$r_1 = B(1 + \rho), \quad r_2 = B(1 + \sigma E^{\nu-1})^{-1}.$$

Note that if two exothermic reactions are considered, we have $0 < r_2 < r_1$, but if we allow the other three cases, r_1 and r_2 are free to take any values. For these to be two positive roots for some values of B and w, conditions (1) and (2) can be combined in the single condition.

$$(1 + \sigma E^{\nu-1})^{-1} > w/B > 1 + \rho. \tag{22}$$

Note that the left-hand side also depends on w for $\nu \neq 1$, and that this condition can never be satisfied if both reactions are exothermic or endothermic. The interesting cases where the heats of reaction have different signs are summarized in Fig. 1, leaving aside for now the local multiplicity questions best addressed via singularity theory. In a later section we will include these considerations; here we wish to understand the gross (or perhaps asymptotic is a better term) features of our bifurcation diagrams. Precisely because singularity theory is local, asymptotic information of this sort is crucial and plays a major role in integrating the results of the former. To return to Fig. 1, we have treated the case $\nu = 1$ and $B > 0$. Each of the terms in the inequality can be represented by straight lines in the w–$f(w)$ plane. The first and third terms are horizontal lines and the middle term is a straight line of slope $1/B$ passing through the origin. The line $f(w) = 1 + \rho$ will take one of the three positions relative to the line $(1 + \sigma)^{-1}$ and the w axis labelled a, b or c in Fig. 1. With one more bit of information we can determine the gross features of the bifurcation diagram from this figure. Condition (3) [eq. (20c)] can be rewritten in the form

$$B^2((E + \sigma E^{\nu})^2(x - (1 + \sigma E^{\nu-1})^{-1})^2 + 4x(1 + \rho - x)\sigma E^{\nu+1}) > 0 \tag{23}$$

where $x = w/B$. By inspection, this inequality is satisfied when $x = 1 + \rho$ but is not if eq. (22) is satisfied and x is close to $(1 + \sigma E^{\nu-1})^{-1}$; hence for some value of x satisfying eq. (22), the roots of the quadratic become complex. For values of x satisfying eq. (22) less than this special value the quadratic will have two positive roots. Now the diagrams of distinct asymptotic behaviour can readily be obtained from the figure by denumerating the qualitatively different ways in which the three curves can intersect. The three resulting diagrams are presented at the bottom of Fig. 1, identified as a, b or c to correspond with the upper diagram. If $B < 0$, then three more diagrams are

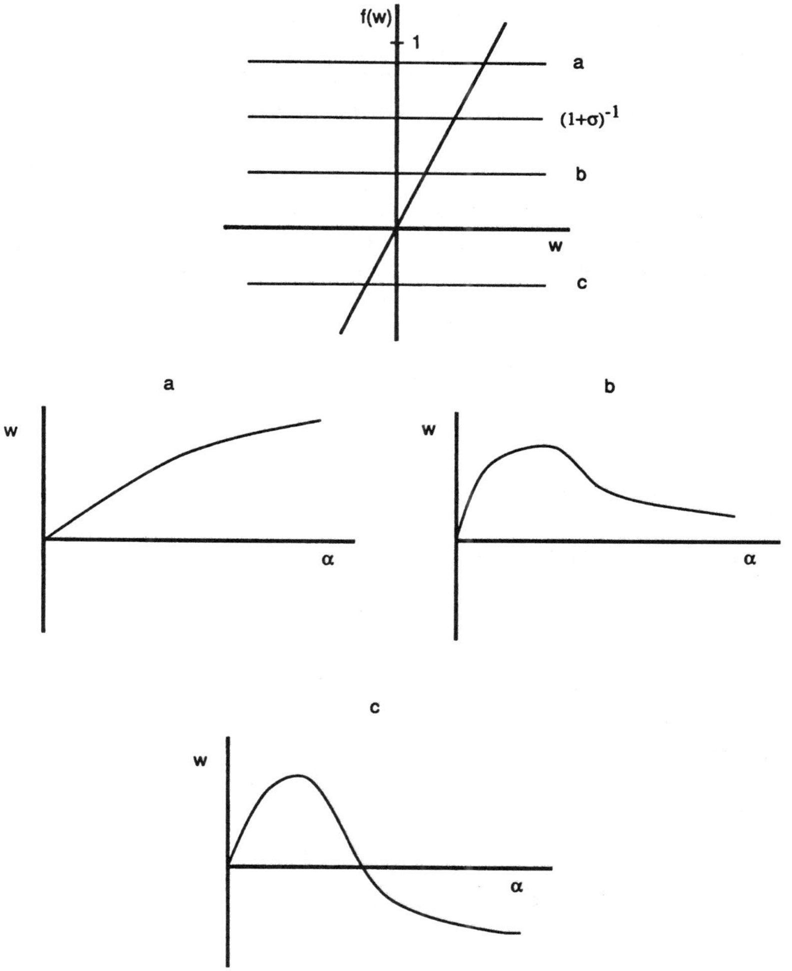

FIGURE 1 Three distinguished (a, b, c) possible values for $\rho + 1$ for $\nu = 1$ and $B > 0$ and the corresponding asymptotic bifurcation diagrams.

possible which can be generated by reflecting the diagrams a, b and c across the α-axis. For $\nu \neq 1$, the first term is no longer a horizontal line and while the situation is slightly more complicated, no new asymptotic diagrams are obtained. It is easy to see that no other asymptotic diagrams are possible, since they can only cross the w-axis at the origin and both roots are negative for $|w|$ large. If we use the nomenclature of Jorgensen *et al.* (1984) where X stands for exothermic and N for endothermic and combinations such as XN denote that the first reaction is exothermic and the second endothermic, we can say that an XX system always has a diagram of type a. That is, for a system of this type the bifurcation diagram can be parametrized globally by w. If the system is of type XN, then any of a, b or c is possible. For systems with the first reaction endothermic, one reflects the diagrams across the α-axis and replaces X with N and vice versa.

We now begin the singularity theory analysis of eq. (18). This equation may be regarded for our purposes as the function

$$G(w, \alpha; B, \rho, \sigma, \nu, \gamma) = 0 \tag{24}$$

in which w is the variable, α is the distinguished parameter, and B, ρ, σ, ν and γ are fixed parameters. In order to apply the machinery of singularity theory to eq. (18) we first simplify the equation symbolically by defining

$$A = \sigma\alpha^{1-\nu}B(1+\rho), \quad \Gamma = \sigma\alpha^{1-\nu}, \quad \varepsilon(\alpha, w) = \alpha E(w), h(w, \varepsilon) = w(1+\varepsilon) \tag{25a}$$

$$g(w, \varepsilon) = \varepsilon^{\nu} w(1+\varepsilon) \quad \text{and } k = \varepsilon^{\nu+1} \tag{25b}$$

and rewriting eq. (18) as

$$Ak + B\varepsilon - \Gamma g - h = 0. \tag{26}$$

This is the equation to which we will apply singularity theory (or at least its mechanics), though when certain delicate mathematical questions encountered in the next section arise we will be forced to return to eq. (18).

Our first task is to find the largest value of n for which the series of functions

$$G = G_w = G_{ww} = \cdots = \partial^n G/\partial w^n = 0 \tag{27}$$

has a simultaneous solution. We know from the work of Balakotaiah and Luss (1982b) that under LB/URP the largest value of n is 4 so we try to solve eq. (27) with $n = 4$ for w, α, σ, ρ and B, keeping ν and γ as parameters. In symbolic form we have

$$Ak + B\varepsilon - \Gamma g - h = 0 \tag{28a}$$

$$Ak_w + B\varepsilon_w - \Gamma g_w - h_w = 0 \tag{28b}$$

$$Ak_{ww} + B\varepsilon_{ww} - \Gamma g_{ww} - h_{ww} = 0 \tag{28c}$$

$$Ak_{www} + B\varepsilon_{www} - \Gamma g_{www} - h_{www} = 0 \tag{28d}$$

$$Ak_{wwww} + B\varepsilon_{wwww} - \Gamma g_{wwww} - h_{wwww} = 0 \tag{28e}$$

We will eliminate A, B, Γ and ε from this system and obtain a single polynomial equation in w, ν and γ. Before doing so, we develop a systematic method for obtaining the w-derivatives of ε^p, where p is any positive real number. Under URP we have the simple relation

$$\partial^i \varepsilon^p / \partial w^i = p^i \varepsilon^p. \tag{29}$$

In general, the right-hand side of eq. (29) will be multiplied by a function depending on p and the derivatives of order less than or equal to i of the natural logarithm of $E(w)$. It is convenient to define a new variable z by

$$z = 1 + w/\gamma \tag{30}$$

and a series of functions $f_i(p, z)$ by

$$\partial^i \varepsilon^p / \partial w^i = \varepsilon^p f_i(p, z) \tag{31}$$

since the derivatives of $\ln(E(w))$ can be expressed simply as powers of γ and z. The first five f_is in the series are given below.

$$f_0 = 1 \tag{32a}$$

$$f_1 = pz^{-2} \tag{32b}$$

$$f_2 = p^2z^{-4} - 2p\gamma^{-1}z^{-3} \tag{32c}$$

$$f_3 = p^3z^{-6} - 6p^2\gamma^{-1}z^{-5} + 6p\gamma^{-2}z^{-4} \tag{32d}$$

$$f_4 = p^4z^{-8} - 12p^3\gamma^{-1}z^{-7} + 36p^2\gamma^{-2}z^{-6} - 24p\gamma^{-3}z^{-5}. \tag{32e}$$

We mention that these functions are easily generated from $f_0 = 1$ and the recursion relation

$$f_{i+1} = pz^{-2}f_i + \gamma^{-1}\partial f_i/\partial z \tag{33}$$

Finally before returning to eq. (28) we make some additional notational definitions:

$$\partial^i k/\partial w^i = kk_i, \quad \partial^i \varepsilon^\nu/\partial w^i = \varepsilon^\nu \mathrm{f}_i, \quad \partial^i \varepsilon/\partial w^i = \varepsilon\varepsilon_i, \quad \partial^i h/\partial w^i = h_i \tag{34a}$$

$$\partial^i g/\partial w^i = \varepsilon^\nu g_i \quad (\text{i.e. } k_i = f_i(\nu + 1, z), \quad \mathrm{f}_i = f_i(\nu, z), \varepsilon_i = f_i(1, z)). \tag{34b}$$

The special linear structure of eq. (28) in A, B and Γ can be utilized to solve eqs. (28a)–(28c) for them via Cramers rule, giving

$$Ak = \frac{\begin{vmatrix} h_0 & 1 & g_0 \\ h_1 & \varepsilon_1 & g_1 \\ h_2 & \varepsilon_2 & g_2 \end{vmatrix}}{\begin{vmatrix} 1 & 1 & g_0 \\ k_1 & \varepsilon_1 & g_1 \\ k_2 & \varepsilon_2 & g_2 \end{vmatrix}} \tag{35a}$$

$$B\varepsilon = \frac{\begin{vmatrix} 1 & h_0 & g_0 \\ k_1 & h_1 & g_1 \\ k_2 & h_2 & g_2 \end{vmatrix}}{\begin{vmatrix} 1 & 1 & g_0 \\ k_1 & \varepsilon_1 & g_1 \\ k_2 & \varepsilon_1 & g_2 \end{vmatrix}} \tag{35b}$$

$$-\Gamma\varepsilon^\nu = \frac{\begin{vmatrix} h_0 & 1 & h_0 \\ h_1 & \varepsilon_1 & h_1 \\ h_2 & \varepsilon_2 & h_2 \end{vmatrix}}{\begin{vmatrix} 1 & 1 & g_0 \\ k_1 & \varepsilon_1 & g_1 \\ k_2 & \varepsilon_2 & g_2 \end{vmatrix}} \tag{35c}$$

We can replace the determinants symbolically by defining, Λ, M, N and Δ by

$$Ak = \Lambda/\Delta, \quad B\varepsilon = M/\Delta, \quad -\Gamma\varepsilon^{\nu} = N/\Delta. \tag{36}$$

As long as $\Delta \neq 0$, we can rewrite eqs. (28d) and (28e) in the new forms

$$\Delta k_3 + M\varepsilon_3 + Ng_3 - \Delta h_3 = 0 \tag{37a}$$

$$\Delta k_4 + M\varepsilon_4 + Ng_4 - \Delta h_4 = 0. \tag{37b}$$

Note that eqs (37a) and (37b) no longer involve exponentials, but are rational equations. After substituting for the symbolic quantities and performing considerable manipulation [details are in Farr (1986)], one can obtain an equivalent system of equations which are quadratics in ε with coefficients depending on ν, γ and z.

With further algebraic effort ε can be eliminated between the quadratics, resulting in a polynomial in z whose coefficients depend only on ν and γ. If ν and γ are fixed, each solution of this polynomial will determine one value of ε and in turn values of the other parameters which will satisfy eqs (28a)–(28e), unless at those values any of the conditions $\Delta \neq 0$, $\varepsilon \neq 0$, $w \neq \gamma$ are violated. Below we treat the exceptional cases where one of these conditions is violated, but before continuing we mention another special case. It is possible for the quadratics to be linearly dependent for some particular values of z, ν and γ, so that for these values there will be two values of ε instead of one. It turns out that this can happen only for $\nu = 1$ and results in the solution

$$w = 3\gamma/(\gamma - 3) \tag{38a}$$

$$\varepsilon = (2\gamma - 6 \pm \sqrt{3(\gamma - 3)^2 + 9})/\gamma. \tag{38b}$$

These solutions constitute a generalization of the results of Balakotaiah and Luss (1982b, 1984) to finite γ. If $\nu \neq 1$, we can proceed to eliminate ε and obtain an equation which is a quintic in z, but can be reduced to a quartic since $z = \gamma/(\gamma - 2)$ is a root for all values of ν. This solution also satisfies $\Delta = 0$ for all ν and so it is not a proper root. If one returns to the system (28a)–(28c), it can be shown (Farr, 1986) that for this root the only solution with bounded values of the parameters corresponds to hysteresis points for the well-studied (Uppal *et al.*, 1974; Balakotaiah and Luss, 1981, 1982a; and others) single reaction system (with $\rho = 0$ and σ arbitrary) and that it never gives rise to butterfly points in the feasible region. Once this degenerate root is eliminated it is convenient to return to w as the variable instead of z, and to factorize the resulting expression partly to obtain

$$\begin{aligned} & w^4\gamma^{-5}(\gamma - 2)(-12\nu^2(\gamma - 2)^2(\gamma - 3) + 4(\gamma - 3)((\gamma - 3)^2 + 3)) \\ & + w^3\gamma^{-3}(\gamma - 2)\begin{pmatrix} -3\nu^4(\gamma - 2)^3 + 2\nu^2\gamma^{-1}(\gamma - 2)(\gamma^3 - 12\gamma^2 \\ + 84\gamma - 144) + \gamma^{-1}(\gamma - 6)\gamma^3 - 36\gamma + 96) \end{pmatrix} \\ & w^2\gamma^{-2}(\gamma - 2)\begin{pmatrix} 18\nu^4(\gamma - 2)^2 - 4\nu^2\gamma^{-1}(4\gamma^3 - 45\gamma^2 + 180\gamma \\ -216) - 2\gamma^{-1}(5\gamma^3 - 18\gamma^2 - 108\gamma + 432) \end{pmatrix} \\ & + w\gamma^{-1}(\gamma - 2)\begin{pmatrix} -36\nu^4(\gamma - 2) + 48\nu^2\gamma^{-1}(\gamma - 3)(\gamma - 4) \\ + 12\gamma^{-1}(3\gamma^2 - 2\gamma - 48) \end{pmatrix} \end{aligned} \tag{39}$$

The roots of eq. (39) with ν and γ considered as parameters determine branches of candidates for butterfly singularity points, which advance to that status if certain conditions are not violated. We have already seen how the conditions $\Delta \neq 0$, $\varepsilon \neq 0$, $w \neq -\gamma$ arise naturally in the derivation of eq. (39) as guarantees that solutions of the latter equation are valid solutions of (28a)–(28e). A separate class of conditions arise from the theory of Golubitsky and Schaeffer, allowing one to relate the quasi-global behaviour of the system of interest to that of simple polynomial functions. For example, the minimum conditions from the theory for a butterfly point are that eqs (28a)–(28e) be satisfied and also

$$\partial G/\partial \alpha \neq 0 \tag{40a}$$

and

$$\partial^5 G/\partial w^5 \neq 0. \tag{40b}$$

(A further condition in the form of a non-zero determinant guarantees that one has a universal unfolding in a neighbourhood of the butterfly point, but this condition is often ignored in practice. The approach of Balakotaiah and Luss (1984) circumvents this issue by computing the transition varieties globally—presumably if a universal unfolding were not present, one would also detect it using their methods.) It must be emphasized that violations of the conditions $\Delta \neq 0$, $\varepsilon \neq 0$, $w \neq -\gamma$ indicate mathematical problems peculiar to the particular system and are outside the scope of singularity theory while if eqs (40a) and (40b) are violated the theory provides a hierarchy of additional conditions to check so that in most cases useful conclusions will be obtained. In the next section we will consider in detail the effects of violations of eq. (40b) and in doing so will obtain some results that could not have been found using the LB/URP hypotheses. Equation (40a) is also violated in this system, but never in the physical region at a point where $\partial G/\partial w = 0$ for this non-dimensionalization.

DISCUSSION I: BUTTERFLY POINTS

The effort required in reducing eqs (28a)–(28e) to eq. (39) plus some non-degeneracy conditions now pays off in the relative ease of analysing a single polynomial function instead of a system of transcendental equations. For example, we conclude immediately from eq. (39) that there are no more than four butterfly points for each fixed ν, γ pair, that we can expect the qualitative nature of the solutions of eq. (36) to change as γ passes through 3 (coefficient of w^4 vanishes) or 2 (all coefficients vanish). These conclusions are certainly not obvious from eqs (28a)–(28e). With more effort, we can obtain qualitative information from special cases of eq. (39) useful in guiding and understanding its numerical solution. If ν is set equal to 1, the resulting equation factors easily giving two double roots. One root is given by eq. (38a) and the other is $w = -\gamma$ which cannot qualify as a butterfly point, but does provide a starting point for numerical work. As ν approaches ∞, three of the roots of eq. (39) approach

$$w = 2\gamma/(\gamma - 2), \quad \varepsilon = 1 - 4/\gamma \tag{41}$$

the degenerate root removed above. The other root can be shown to be asymptotically proportional to $\pm\ \nu^2$, the sign depending on the value of γ.

There are two other simple cases where analytic roots of eq. (39) can be obtained. If we set $w = 0$, the resulting quadratic in ν^2 can be solved to give

$$\nu^2 = 1 - 3/\gamma + \sqrt{3 + 9\gamma^{-2}} \tag{42}$$

and setting $w = \gamma$ we obtain

$$\nu^2 = \frac{\gamma^2 - 12\gamma + 48}{\gamma(\gamma - 4)}. \tag{43}$$

It is also possible to find solutions with $\varepsilon = 0$. These are not solutions of eqs (28a)–(28e) in the strict sense nor are they physically meaningful, but they will be useful in several connections later. One such solution is $w = \gamma$ and ν given by eq. (43); the other two sets are given by

$$\nu = 1 \tag{44a}$$

$$w = -\gamma \tag{44b}$$

and

$$\nu^2 = \frac{2(\gamma - 6)}{3(\gamma - 4)} \tag{45a}$$

$$w = \frac{\gamma}{8}\left(\gamma - 5 + \frac{9}{\gamma - 3} \pm \left(\gamma - 1 - \frac{3}{\gamma - 3}\right)\sqrt{\frac{\gamma - 12}{\gamma - 4}}\right). \tag{45b}$$

The solution (44a), (44b) turns out to be a limit point of eq. (39). While the solution given by eqs (45a) and (45b) is a regular point for eq. (39), it is worth noting that it only makes sense if $\gamma \geqslant 12$ or $\gamma < 4$. A final qualitative result concerns no a solution, but the lack of one. A root of eq. (39) will have a vertical asymptote if the coefficient of w^4 vanishes, which happens (for $\gamma \neq 3$) if

$$\nu^2 = \frac{(\gamma - 3)^2 + 6}{3(\gamma - 2)^2}. \tag{46}$$

It is clear from these last few equations that certain special values of γ will be important in determining qualitative changes in the behaviour of the roots of eq. (39). The values $\gamma = 2$ and $\gamma = 3$ have already been mentioned; we can now add $\gamma = 12$, $\gamma = 6$ and $\gamma = 4$ to this list. It does not seem obvious why they should be integers—just as it was not obvious that the transcendental equations (28a)–(28e) could be reduced to a single polynomial equation. [It appears that if one considered n reactions in series, the same sort of procedure would work if one assumed that all of the activation energies were the same. If, however, the activation energies are all different and one attempts to eliminate variables, as was done to eqs (28a)–(28e) by writing them as a set of linear equations, it is not possible to define enough independent constants (i.e. analogous to A, B and Γ) to correspond to the number of independent powers of ε. The essential problem is that the number of possible constants grows like $3n$ and the number of independent powers of ε grows like 2^n. Chicone and Retzloff (1981) gave a bound on the multiplicity of such a system of $2^m - 1$ (where m is the number of reactions plus 1) when all the reactions are exothermic. As we shall see, it is possible to have seven steady states with two reactions but only when the system is of type XN; hence their upper bound is not sharp. (It is worth noting, however, that they did successfully predict the values of ν where seven steady states might be obtained.) This question is still under investigation, and will be addressed as progress is made.]

In an earlier paper (Jorgensen *et al.*, 1984), we treated this system under the URP hypothesis and a brief review of those results will aid us in understanding the similarities and differences when that hypothesis is removed. The URP hypothesis is often described by saying that γ is infinite. This is not strictly proper since an infinite γ would also affect the values of α, σ and B. What one really wants to claim is that w is small compared to γ, so that $E(w)$ is closely approximated by e^w. In practice, however, we can recover the results obtained in that paper by taking the limit as γ becomes large of eq. (39), and obtain qualitative and quantitative behaviour by using the same procedure on eqs (41)–(46). In Fig. 2 we have reproduced the plots of w and ε vs. ν

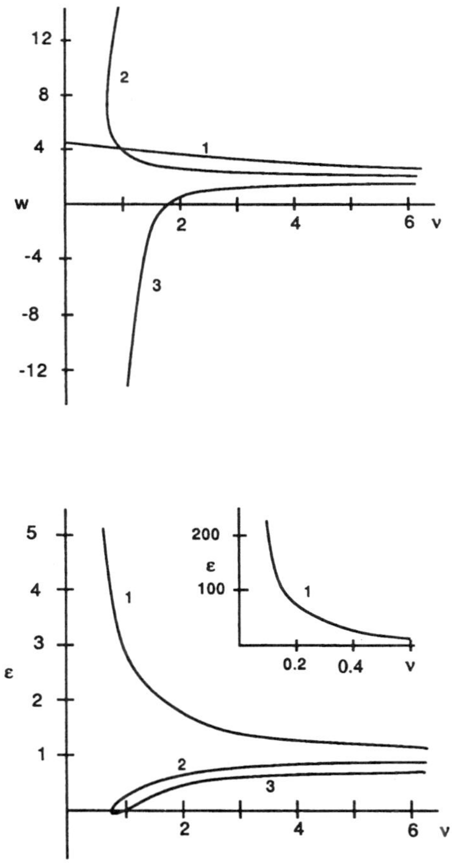

FIGURE 2 Butterfly candidates w and ε vs. ν under URP from Jorgensen *et al.* (1984).

from that paper, and the analogues of eqs (41)–(46) correctly predict the value of ν where $w = 0$, $\varepsilon = 0$, etc. The quartic reduces to a cubic in this case, so there can be no more than three roots. Figure 3 presents the parameters for each of the three roots, identified with a number to correspond to the particular root. Details of asymptotic behaviour are shown in the insets in both figures. The numbering system is based rather arbitrarily upon characteristics of each root at $\nu = 1$. Root 1 is the branch which has $w = 3$ and $\varepsilon = 2 + \sqrt{3}$, root 2 has $\varepsilon = 2 - \sqrt{3}$. From these figures, it is a simple task to determine when the butterfly points lie in the feasible region and, if so, to determine their type. A feasible butterfly point must have α and σ positive, and the type is then determined by the signs of B and ρ. As described in Jorgensen *et al.* (1984), feasible butterfly points of type XX (roots 1 and 3), XN (root 2) and NX (root 3) can all be found in this system. One thing

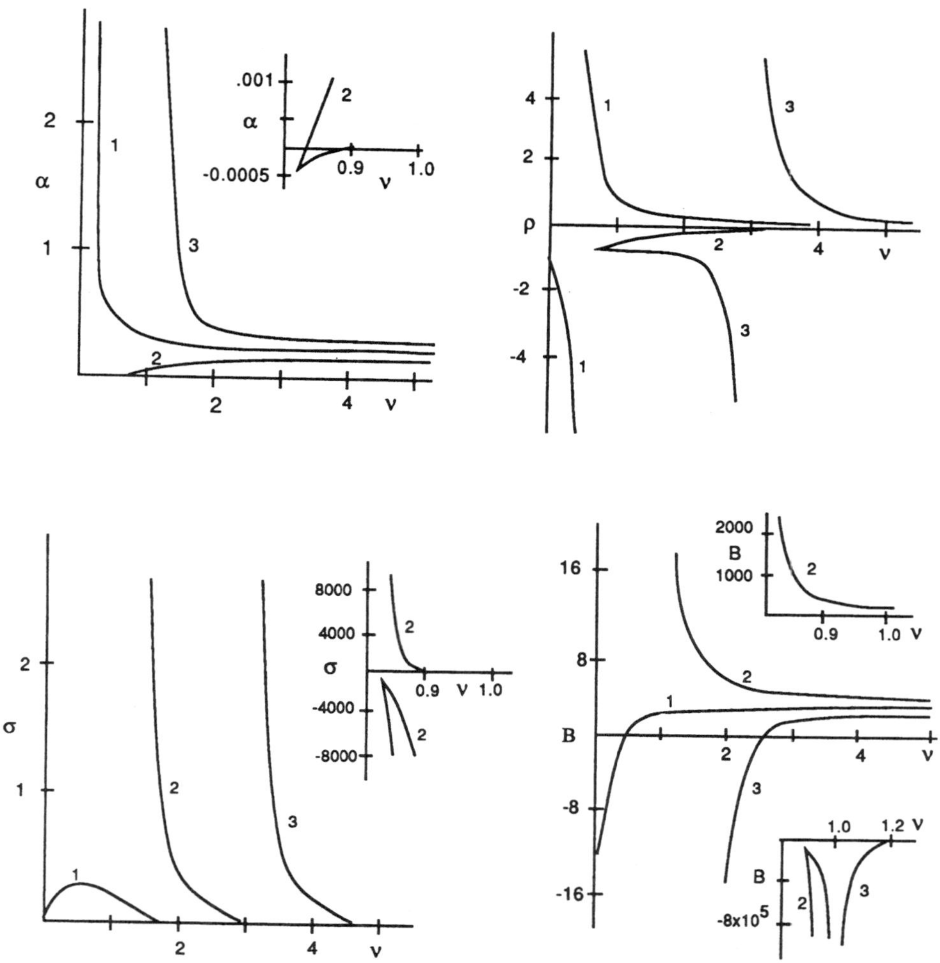

FIGURE 3 Parameters α, ρ, σ, B vs. ν for butterfly candidates under URP.

recognized since that paper was written is that there is a singularity of higher order than the butterfly present, though not in the feasible region. It can be shown (Farr, 1986) that a simple limit point on the w vs. ν curve implies that one more w-derivative of G is zero at that point, but the cusp structures in the parameter plots should have been dead giveaways. This higher-order singularity goes by the name of the wigwam in the catastrophe theory literature (Woodcock and Poston, 1974). Perhaps the most striking feature of Figs. 2 and 3 is the essential monotonicity of the parameter plots, broken by discontinuities rather than local maxima or minima, reflecting the rational nature of eqs (35a), (35b) and (35c).

We turn now to the solutions of eq. (39). Using the methods of Golubitsky and Schaeffer (1979), one can show (Farr, 1986) that the solution of this equation (regarding ν as our distinguished parameter) results in the four qualitatively different diagrams shown in Fig. 4. Actually the methods of the

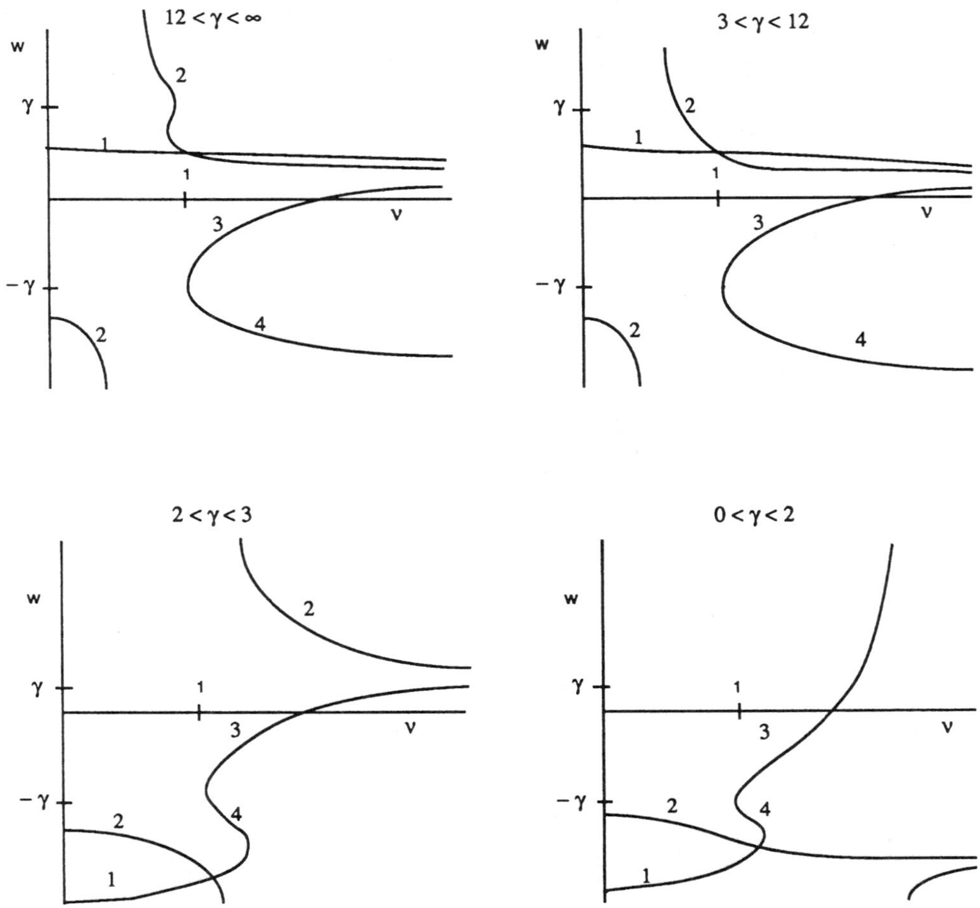

FIGURE 4 Four qualitatively different solution sets of eq. (39), each showing the four branches of butterfly candidates for a different range of γ.

above authors have to be extended somewhat, since eq. (39) is not well behaved at certain values of ν and γ where the coefficient of the quartic term vanishes or the asymptotic behaviour changes. For large γ, three of the roots can be identified with those in Fig. 2 and are numbered accordingly. The new root, labelled 4, is always less than $-\gamma$ for ν close to 1 and is never physically meaningful, and in fact for very small values of γ, up to three of the roots lie in this region and can be dismissed immediately. Figure 5 presents the same plots with these non-physical roots excised. Roots with $w > \gamma$ are not proscribed, but are perhaps unlikely to be achieved. If one considers the top leftmost graph in Fig. 5 and imagines it to be cut off above $w = \gamma$, the similarity to Fig. 2 is transparent. The parameters for $\gamma = 100$ are plotted in Fig. 6 (schematically since the scales are so diverse), identified by number as to which root they correspond. The root 1 and root 3 plots are quite similar to the URP case. If one were to remove the root 2 parameter curves for values of $w > \gamma$, the resulting plots would be quite similar to those in Figs. 2 and 3 except that roots 2 and 3 no longer connect at the origin in the ε and ρ plots since when $w = \gamma$ and $\varepsilon = 0$ for root 2, ν as given by eq. (43) is strictly less than one. The new portion of the ε curve in Fig. 6 loops around to intersect the ν-axis at exactly the same value as the first crossing as given by eq. (45a) before becoming negative again. These three crossings of the ν-axis plus a zero of Δ produce the vertical asymptotes responsible for the violent oscillations of

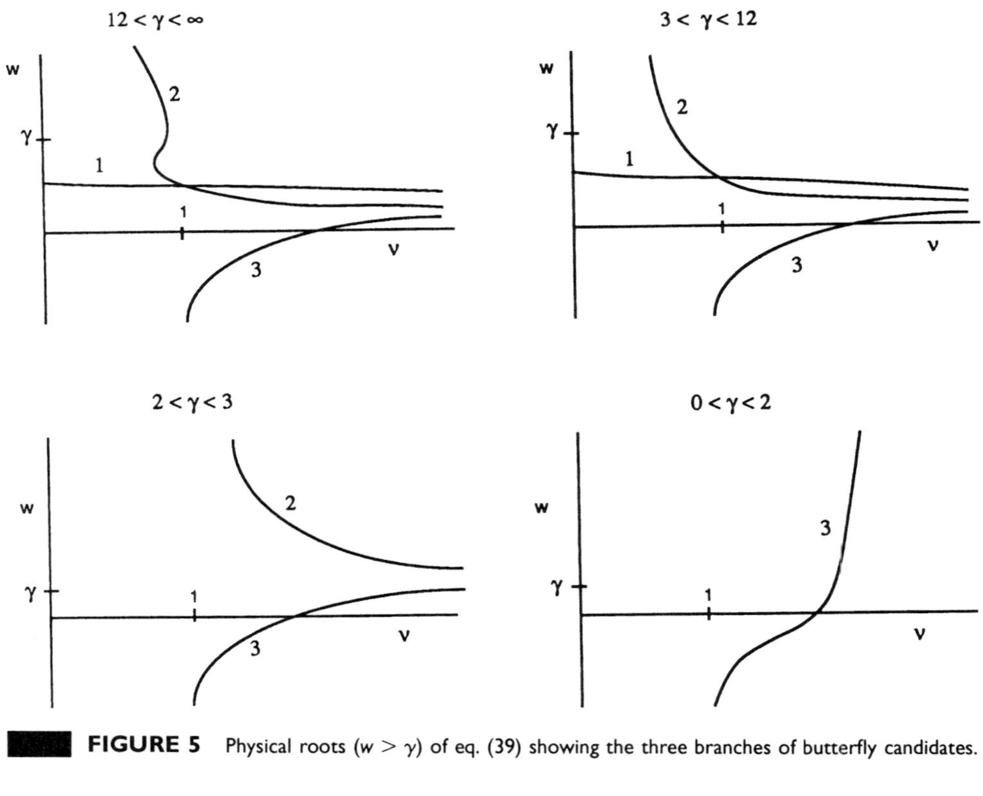

FIGURE 5 Physical roots ($w > \gamma$) of eq. (39) showing the three branches of butterfly candidates.

σ and B. The two cusp structures correspond to the two limit points of the w vs. ν curve and both correspond to wigwam singularities, one of which is (marginally) in the feasible region. These higher-order singularities will be discussed in detail below where bifurcation diagrams not previously reported will be presented. It is worth noting that for a small ν interval, there are three type XN butterfly points in the feasible region. This occurs very close to the upper wigwam point; the upper cusp structure in the σ-plot where this situation obtains would not be visible if it were drawn to scale.

A representative of the class $3 < \gamma < 12$ is furnished by Fig. 7 for $\gamma = 10$. Now differences are apparent in root 1. For this root, the parameter σ crosses the ν-axis at a larger value and the butterfly point moves out of the feasible region; this happens before the transition from XX to NX that happens for larger γ can occur. Root 3 seems unchanged, though the transition from NX to XX has shifted to a slightly smaller ν value. The wigwam singularities of root 2 have coalesced, and the value of ν at which the ε curve for this root crosses the axis is at a minimum. The two vertical asymptotes in the σ and B plots are due to one $\varepsilon = 0$ root [given by eq. (43)] and a zero of Δ, but the cusp structures have vanished with the wigwam points.

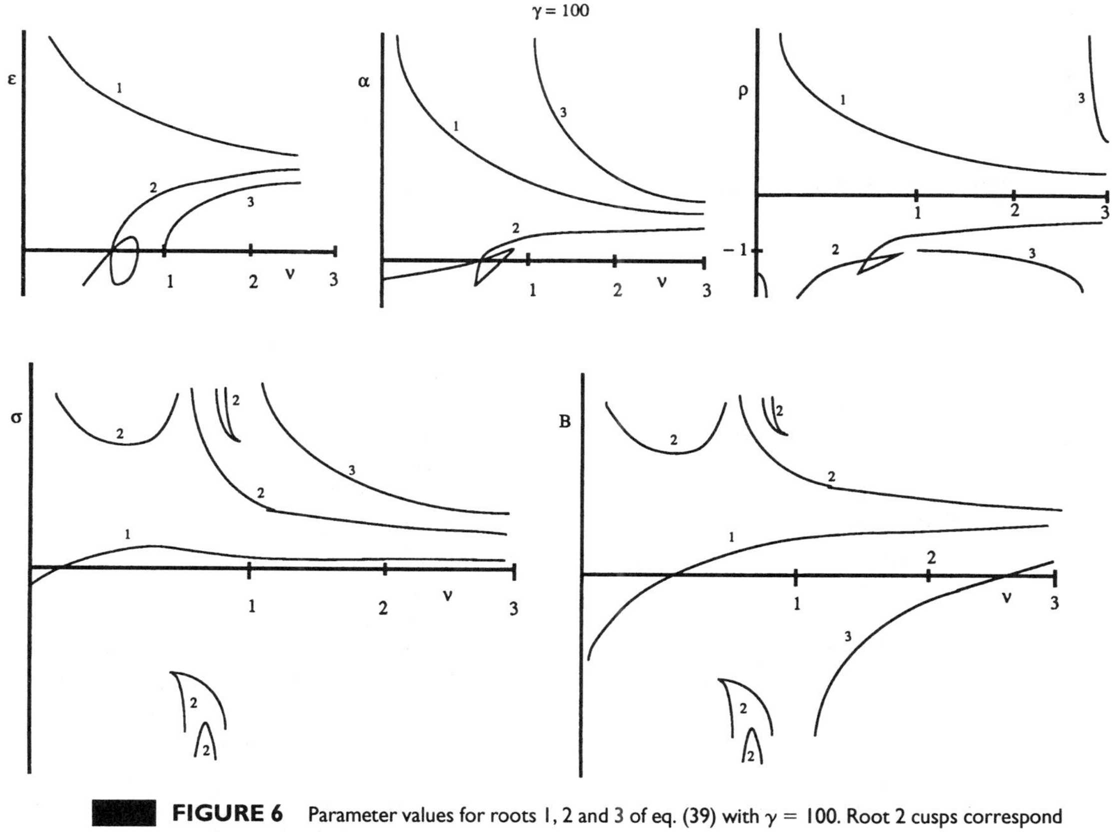

FIGURE 6 Parameter values for roots 1, 2 and 3 of eq. (39) with $\gamma = 100$. Root 2 cusps correspond to wigwam singularities.

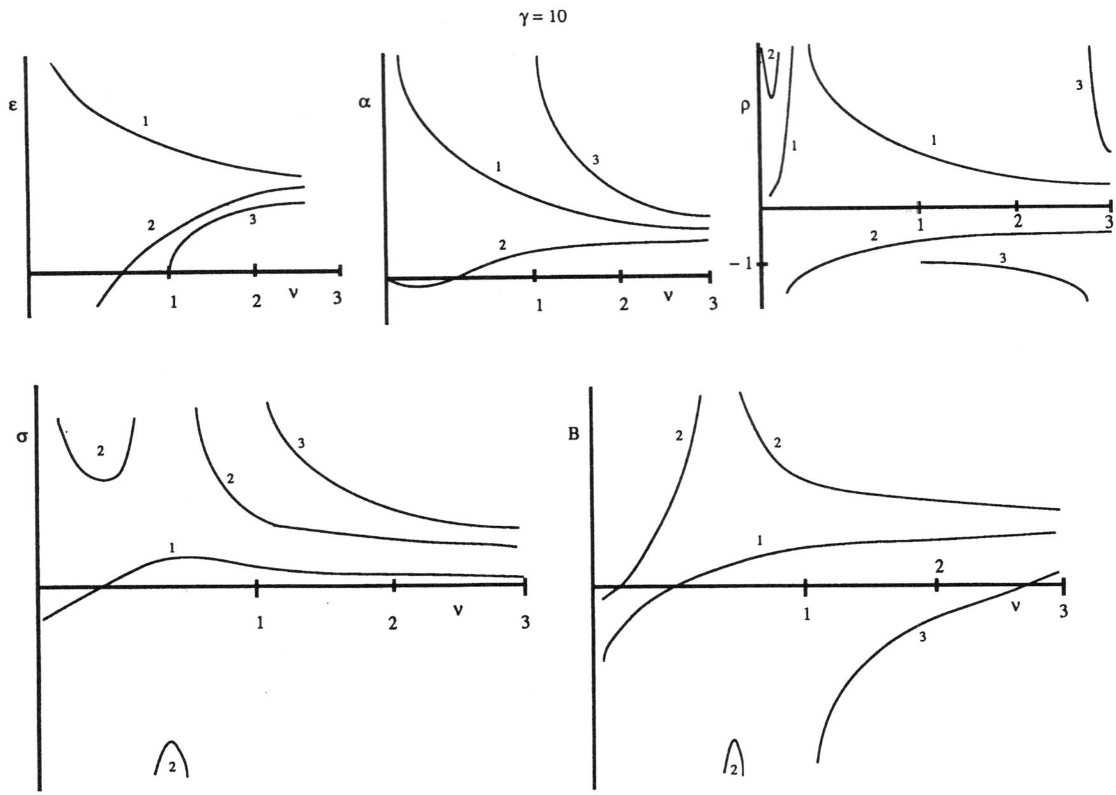

FIGURE 7 Parameter values for roots 1, 2 and 3 of eq. (39) with $\gamma = 10$.

More of these plots could be presented, but we can summarize the results by mapping out in ν–γ space the feasibility boundaries and type of the butterfly points for each of the three roots. This has been done in Fig. 8. From this figure one can determine for each ν, γ pair whether butterfly points exist in the feasible region and their number and type. If one considers the realistic case where γ is fairly large (larger than 15–20) and ν is close to 1 then aside from the narrow strip where there are three type XN butterfly points associated with root 2, then the number and type of butterfly points do not change much in going from "infinite" γ to $\gamma = 20$. This is not to say, however, that nothing new is learned by going to all the trouble to treat the general case. Under LB/URP, for example, root 3 is not even detected. This root is interesting because it leads to bifurcation diagrams similar to b and c in Fig. 1 (except they are reflected across the α-axis) as long as it remains of type NX. Interestingly enough, the butterfly point always appears to the right of the horizontal tangent and so multiple solutions appear in this region also. Numerical calculations indicate that as the interval of multiplicity grows, the region where there

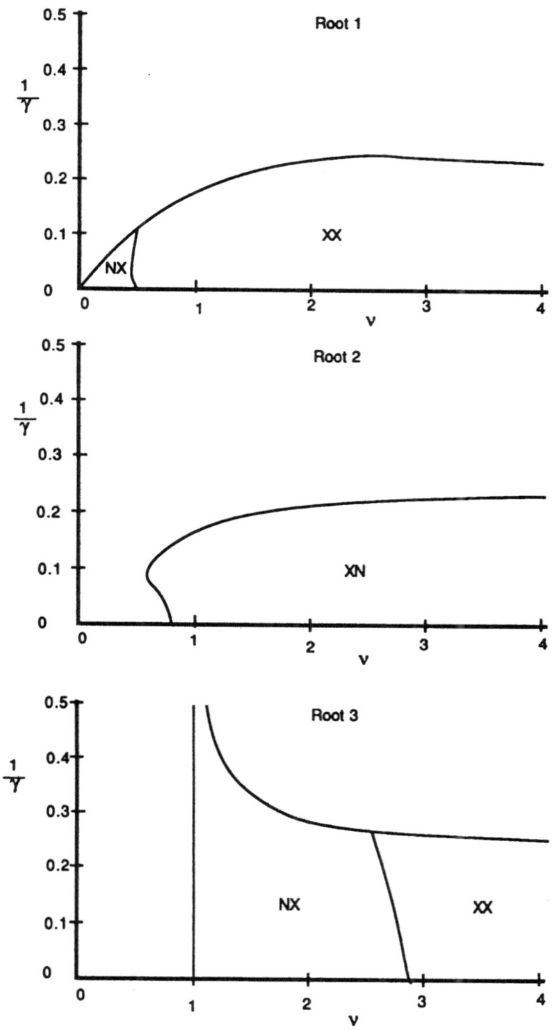

FIGURE 8 Feasibility boundaries for the three physical butterfly candidate points. No butterfly points exist outside of the labelled regions.

are two α-values at the same w-value shrinks correspondingly: the Isola variety does not exist for this nondimensionalization so a limit point and a point of horizontal tangency can never coalesce. Similar diagrams are also obtained for small values of ν when root 1 is of type NX. Root 2 has a very small region where its diagrams are like b and c in Fig. 1, very close to where ε crosses the ν-axis at the smallest value of w (cf. Figs 5–7). Two example diagrams, chosen to illustrate the novel aspects rather than multiplicity, are shown in Fig. 9.

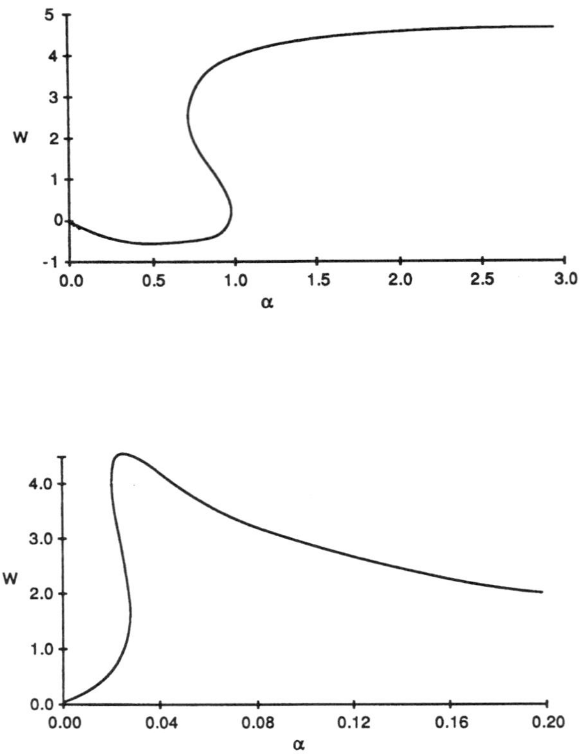

FIGURE 9 Example bifurcation diagrams showing double roots of eq. (18). Top, $B = -5$; bottom, $B = 20$, both $\gamma = 15$, $\nu = 0.75$, $\rho = -2$, $\sigma = 1$.

DISCUSSION II: MAXIMUM MULTIPLICITY

According to the scheme of Golubitsky and Schaeffer (1979), as expanded upon by Balakotaiah and Luss (1984), diagrams with the maximum number of solutions can be found by following a systematic, stepwise procedure beginning with finding the value of n defined by eq. (22). The latter authors do not say so explicitly, but their procedure is based on the assumption that for some choice of distinguished parameter (not necessarily the most natural one physically) one can obtain a full unfolding of the appropriate cuspoid singularity, and the steps in the procedure become pellucid if one simply writes down the singularity and its derivatives. The cuspoid singularity (of order k) is the simplest example encountered in the theory and has a universal unfolding of the form

$$x^k + \alpha_{k-2}x^{k-2} + a_{k-3}x^{k-3} + \ldots + a_1 x - \lambda = 0. \tag{47}$$

From the discussion above, we need only look for multiple roots of eq. (39) to find the maximum value of n for our problem. For $\nu = \sqrt{2}/2$ and $\gamma = 12$, $w = 12$ is a triple root of eq. (39), the most degenerate root obtained for any values of ν and γ. Therefore, we would expect that the value of n is 6 (star

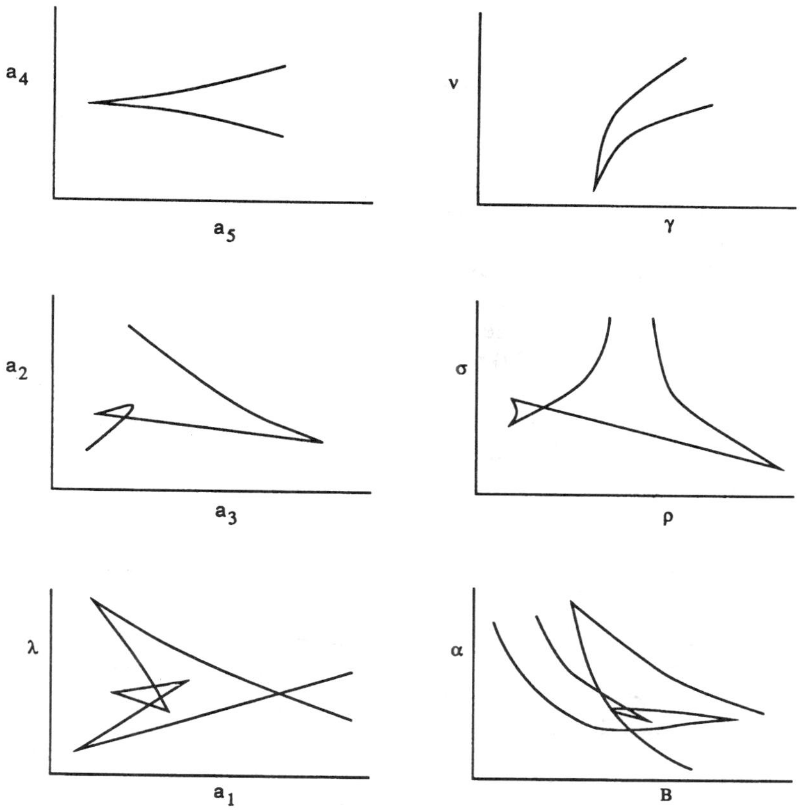

FIGURE 10 Comparison of cross-sections of parameter space for star singularity (left-hand side) and eq. (18) (right-hand side) treated as a star singularity.

singularity, $k = 7$) and the maximum number of solutions is 7. Unfortunately, ε as calculated exactly for the values of ν, γ and w stated above is 0, and hence the values of B and σ are undefined at this point. Since this triple root is obtained when two double roots come together, one can look at the values of B and σ as the triple root is approached and see that B and σ are of one sign on each of the branches, but tend to $\pm\infty$ as the triple root is approached. This breaks up the expected star singularity structure, as shown in Fig. 10 where the three stages of the procedure of Balakotaiah and Luss (1984) are shown for both a star singularity and eq. (18). At the first stage, both exhibit a cusp made up of wigwam singularities. The second stage consists of the swallowtail points with two of the parameters fixed to lie inside the cusp of the first stage. The star singularity exhibits the characteristic catastrophe set of the butterfly singularity, but the swallowtail points of eq. (18), when projected onto the ρ–σ plane, do not. One problem is that one must choose ν and γ very close to the upper branch of wigwam points for all three butterfly points to lie in the feasible region, but even when this is done the projection of the swallowtail points does not match that for the star singularity. At the final stage of limit points of the two functions with four of the parameters fixed, the difference is even more apparent, with the fold points of eq. (18) breaking up into two disconnected components. Diagrams with seven steady states can be found using this method, but results from singularity theory are more difficult to use in guiding the search.

This suggests that one could take a different approach and consider only the highest-order singularity that occurs in the feasible parameter space, i.e. the wigwam singularities that occur for $\gamma > 12$. This procedure works quite well as long as one keeps in mind that the results of singularity theory are strictly local in nature. The 19 different local diagrams that result from a universal unfolding of the wigwam singularity are shown in Fig. 11. As Golubitsky and Schaeffer (1979) remarked, the denumeration of the possible diagrams follows from purely logical considerations. Limit points always appear in pairs, so the diagrams of the wigwam singularity can be classified roughly into groups having either 1, 3 or 5 limit points. Within each group, subclassifications are assigned based on the allowable relative positions of the limit points. This

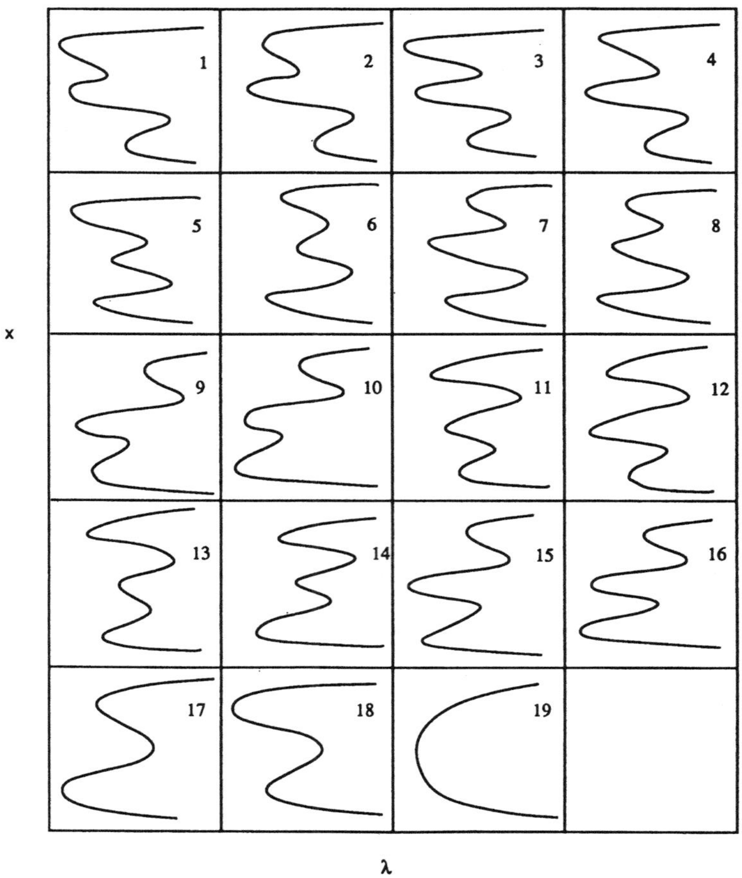

FIGURE 11 Nineteen qualitatively different bifurcation diagrams for wigwam singularity.

can be made explicit by assigning to each limit point a number from 1 to 5 going from the smallest to largest x-value and then giving each diagram a "name" by writing out the string of digits in the order of the λ-values. For example, we would denote diagram 1 in Fig. 11 by 53412 and diagram 17 by 132. Certain rules of order prevail so not all combinations of five and three digits result in allowable diagrams. Proper "names" for diagrams with five limit points must have 1 and 3 appearing to the left of 2, 4 to the right of 3, and 5 to the left of 4. The resulting 16 possible diagrams with five limit points and the two diagrams with three limit points are those appearing in Fig. 11.

Because the solutions of eq. (18) are connected components that pass through the origin, the resulting 19 global diagrams are as shown in Fig. 12. The total number of distinct diagrams that can exist for this system is thus 23; the seven from the butterfly singularity plus the 16 with six limit points close to the wigwam singularity. Note that it is not possible to find all possible diagrams in the neighborhood of an "organizing centre", since the limit point at the smallest value of w does not interact with the other five. Finding regions

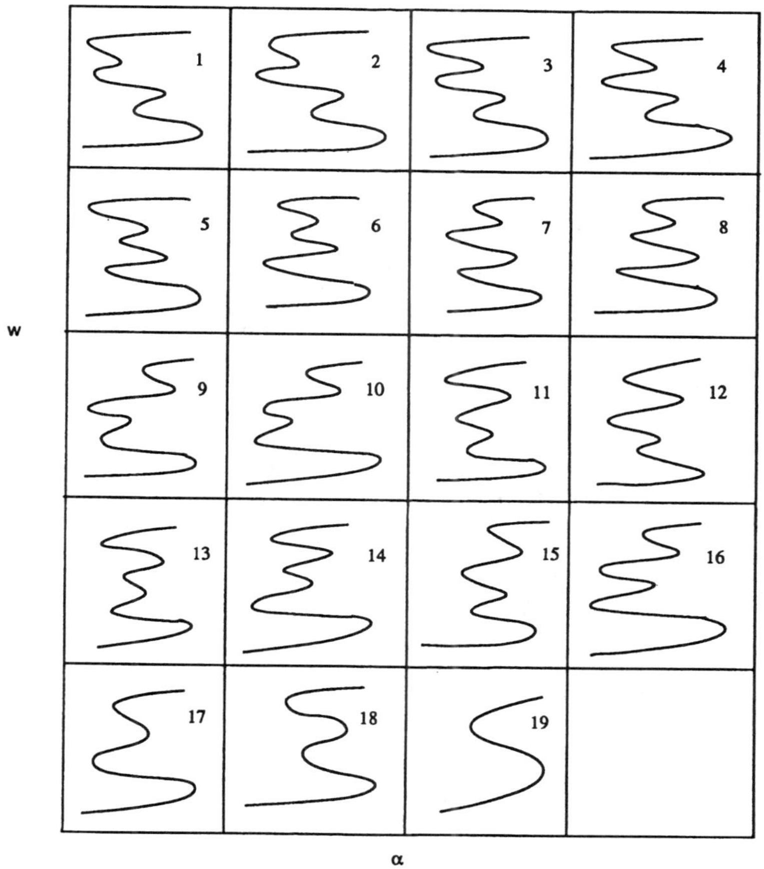

FIGURE 12 Global bifurcation diagrams for eq. (18) near a wigwam singularity.

in parameter space where each diagram in Fig. 12 appears follows from the procedure laid down in Balakotaiah and Luss (1984). For this problem, however, the distinction laid down in that paper between finding diagrams with a specific number of solutions and finding the qualitatively distinct diagrams becomes blurred because the highest-order singularity considered is contact equivalent to a wigwam singularity with α as the distinguished parameter. It is also a singularity of even order, so the procedure of Balakotaiah and Luss (1984) for finding a specific number of solutions includes the computation of the hysteresis variety. The next step suggested by the authors is the computation of the limit points, which now form a set of isolated points. The qualitative nature of the diagram, however, is completely determined by the hysteresis and double limit varieties; hence the computation of the limit points is much less efficient than simply including the double limit variety with the hysteresis variety.

An example will help to clear this point up. Fixing the value of γ at 15, we can compute the cusp of fifth-order (or butterfly) singular points originating from the sixth-order singularity in the feasible region. When these points are plotted in the ν–B plane, the result is the upper graph in Fig. 13. Choosing ν and B to have the values 0.7520 and 930,000, the hysteresis variety projected

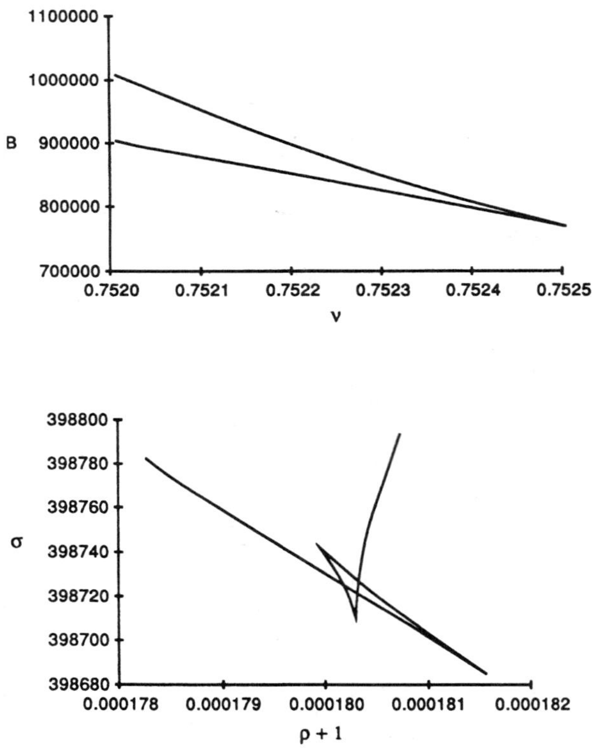

FIGURE 13 Fifth-order singularities (top) and hysteresis variety (bottom) for $\gamma = 15$, near wigwam singularity.

on the ρ–σ plane is shown in the lower graph in the same figure. For clarity, the double limit variety is shown only in Fig. 14 where we have distorted both varieties somewhat and blown up the portion where the various diagrams are concentrated in the lower half of the figure. Each open region is identified with a number corresponding to the identification numbers in Figs 11 and 12. We note that only 16 of the 19 diagrams of Fig. 11 appear. Analysis of the wigwam singularity as well as extensive numerical computations indicate that this is to be expected; the nonpersistence varieties (H and DL) are of codimension three in a four-dimensional space and the classification of all possible two-dimensional cross-sections has not yet been done, though considerably more detail is in Farr (1986). It seems worthwhile, however, to point out that for these simple cuspoid singularities, much about the disposition of the DL variety can be deduced from the H variety. If we use the limit point numbering system described above then each variety can be broken into pieces, each of

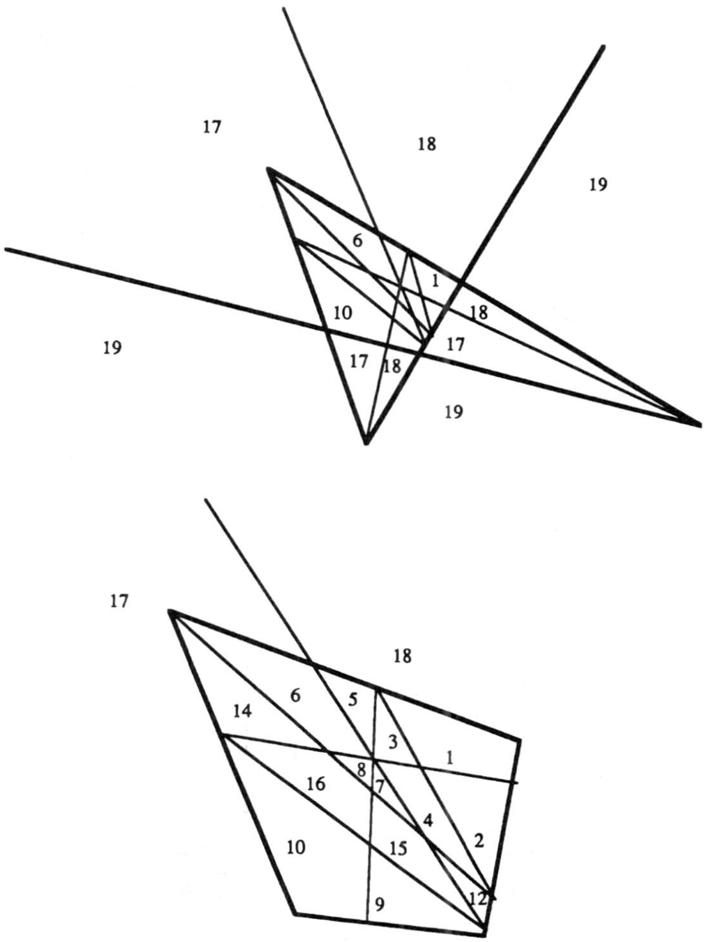

FIGURE 14 Details of hysteresis and double limit varieties for wigwam singularity.

which can be designated by the two digits of the limit points involved. Thus H12 refers to the coalescence of limit points 1 and 2 and DL53 to when limit points 5 and 3 occur at the same λ-value. Topological considerations can give us a lot of information on the intersections of the two varieties, allowing us to build up a good picture without having to compute it explicitly. For example, the DL53 branch must have one end at the swallowtail point where H34 and H45 come together in a cusp and the other end on H23 at the same point

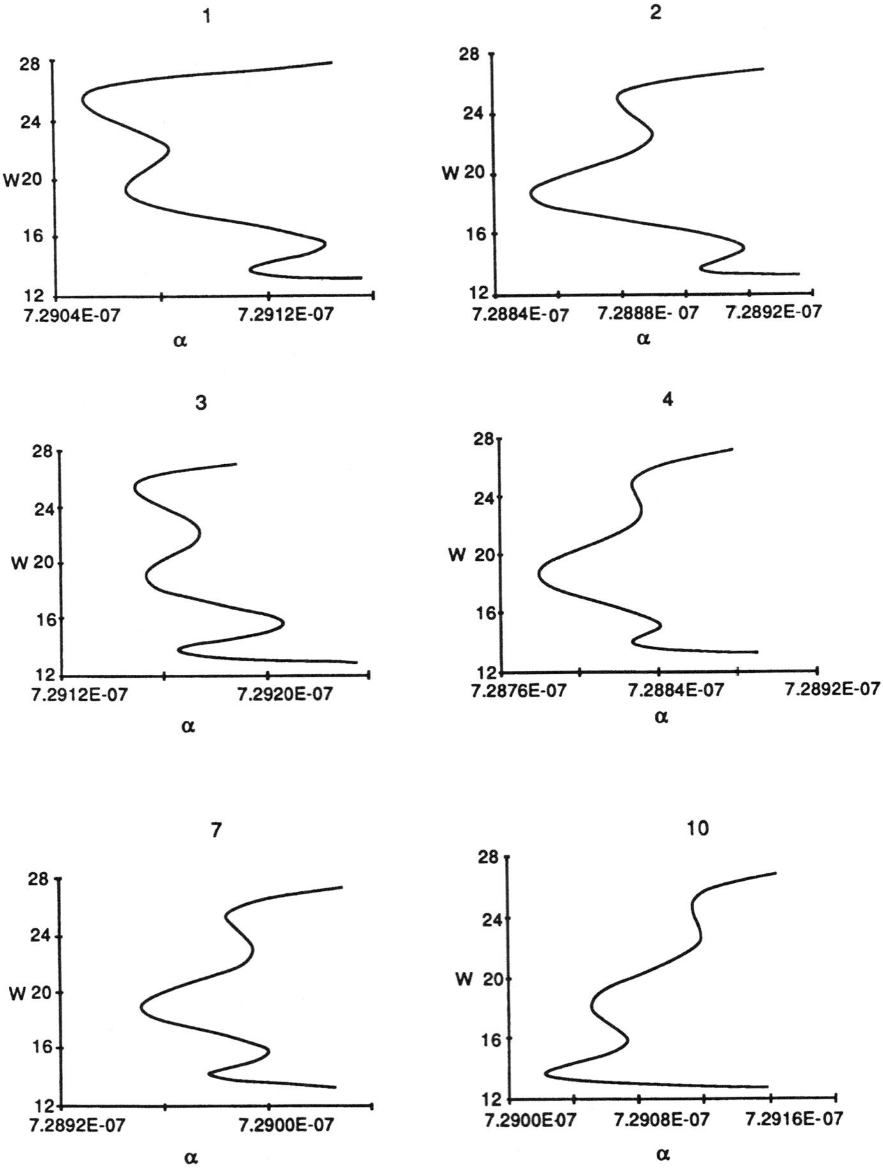

FIGURE 15 Diagrams corresponding to regions 1, 2, 3, 4, 7 and 10 of Fig. 14.

that DL52 intersects H23. This process of logical deduction helps greatly to guide the numerical computations necessary in any particular case.

Examples of several of the diagrams found in this example will be presented below, but first they need to be put into perspective. The hysteresis variety in Fig. 13 is actually only a portion of the whole. There is another swallowtail point at much smaller values of w, related to the limit point that

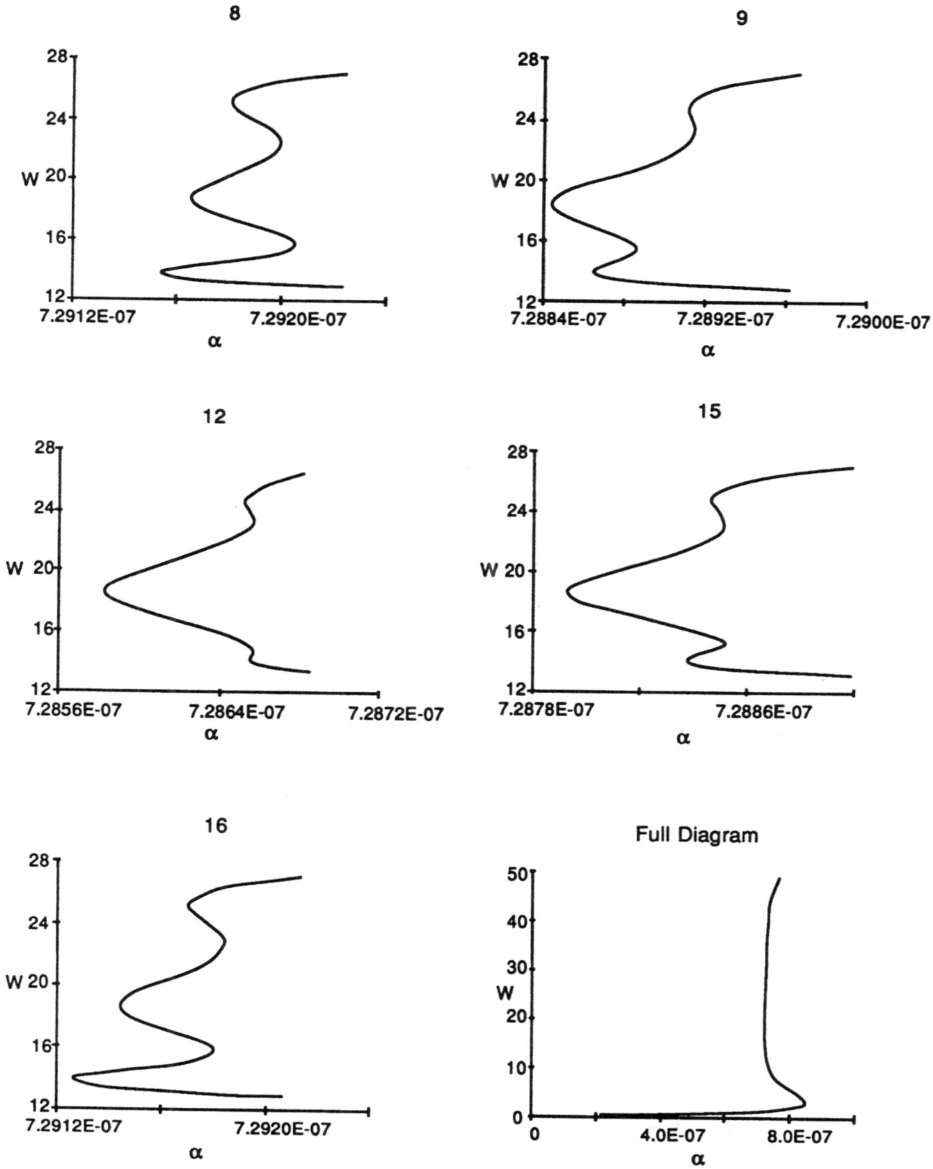

FIGURE 16 Diagrams corresponding to regions 8, 9, 12, 15 and 16 of Fig. 14 and full diagram showing scale (lower right).

must be added to the diagrams in Fig. 11 to obtain those in Fig. 12. This swallowtail point does not interact with the three shown in Fig. 13 and if it were included, the details shown in Fig. 13 would hardly be visible. We take this as further evidence that a full unfolding of the star singularity is not present in this problem and also conclude that the fine structure from the unfolding of the wigwam singularity will be very difficult to see. This is certainly true as shown in Fig. 15 and 16 where we have plotted several example diagrams. The first limit point as well as the lowest branch of solutions is not shown since the diagrams had to be cut to show the fine structure. The diagram in the lower left-hand corner of Fig. 16 includes the lower portion of the diagram and demonstrates this point. (The near verticality of the intermediate portion of this diagram is explained by the proximity to a wigwam singularity.) The parameter values for each diagram in Figs 15 and 16 are given in Table 1. Note that the values are so closely spaced that it seems most unlikely that they would have been found without the aid of singularity theory.

CONCLUSIONS

In this paper we have described the steady-state behaviour of the system $A \rightarrow B \rightarrow C$ in a CSTR with particular attention to the effects of the often-used LB/URP hypotheses. We have determined the steady-state multiplicitly structure for a wide range of activation energies and found diagrams that occur for a small region of parameter space exhibiting seven simultaneous steady states. Singularity theory with a distinguished parameter as developed by Golubitsky and Schaeffer (1979, 1985) and extended by Balakotaiah and Luss (1981, 1982b, c, 1983, 1984) has become one of the most powerful tools for investigating steady-state multiplicity, and is especially useful for unravelling the quite complicated behaviour of the system $A \rightarrow B \rightarrow C$ in a CSTR. The region of seven steady states is exceedingly small and not likely to be encountered experimentally but the case when all of the activation energies are the same is equally unlikely and, as this study shows, is in a very real sense not descriptive of the situation when $\nu > 1$ and the first reaction is endothermic.

This study was inspired partly by Balakotaiah and Luss (1982b), where the question of multiplicity for two consecutive or parallel reactions was treated under very general conditions but without using singularity theory. In that paper, the results were expressed mainly in terms of cross-sections of the bifurcation set (fold points). Of particular interest were cross-sections when both reactions were exothermic with two regions each having five steady states. It was hoped that by using the more powerful methods of singularity theory it would be possible to bring these two regions together and find the higher-order singularity responsible. This could not be done and it is now clear that the two regions come from two independent butterfly points (i.e. roots 1 and 3) which never coalesce. Also interesting were the cross-sections of a type XN system. The authors do not exhibit cross-sections with more than one region of five steady states, but the present study shows that there are cross-sections with two such regions (cf. diagrams 1 and 2 in Fig. 12). The region where such cross-sections exist is extremely small and it is not surprising

TABLE I Values of ρ and σ for the diagrams in Figs 15 and 16. In all cases ν = 0.752 and B = 930000

Fig.	Diagram	$(\rho + 1) \times 10^4$	$\sigma \times 10^{-5}$
15	1	1.7978	3.987260
15	2	1.7980	3.987240
15	3	1.7976	3.987262
15	4	1.79804	3.987225
15	7	1.7978	3.987240
15	10	1.7975	3.98724
16	8	1.7975	3.887258
16	9	1.79785	3.987225
16	12	1.79829	3.98720693
16	15	1.7980	3.987254
16	16	1.7970	3.98728

that they were not found earlier, especially in view of the numerical difficulties discussed below.

This example shows that if one is interested in handling a wide range of activation energies, numerical problems result if one is not willing or able to reduce the series of eqs (22) to a single equation with relatively simple properties. The problem is that parameters go off to infinity and numerical methods which depend on Newton's method applied to a multidimensional system and continuation in a parameter would not be able to cope with this easily. This difficulty could perhaps be circumvented by redefining some of the parameters to be their reciprocals, but the original parameters also pass through zero and one merely exchanges one problem for another. On the other hand, the algebraic difficulties associated with the reduction to a single equation were considerable and it is difficult to see how it could be done for a more complicated systems. The use of symbolic manipulators could permit further results to be obtained, especially for cases of equal activation energies. This study shows, however, that one loses information by using the LB/URP hypotheses.

The non-dimensionalization used in this work is perhaps the simplest, but it suffers from the defect that important physical bifurcation parameters are not isolated. The simple cuspoid diagrams are probably not those that would be obtained from experiments, where the residence time is a convenient parameter. Balakotaiah and Luss (1983) considered such a formulation for two parallel or simultaneous reactions; the diagrams for the case of sequential reactions are similar, at least when the activation energies are equal. The maximum multiplicity question, however, is independent of the formulation and we conjecture that diagrams with seven steady states could be found in a small region of parameter space, though we have not looked for them.

ACKNOWLEDGMENT

W.W.F. was supported in the course of this research by a Kodak fellowship.

NOTATION

A	parameter, eq. (25a)
A_1, A_2	pre-exponential factors in the rate constants
A_c	area of cooling surface
B	combined heat of reaction and cooling parameter, eq. (16)
c_A, c_B	concentrations of A and B
c_{Af}, c_{Bf}	feed concentrations of A and B
C_p	heat capacity
$E(w)$	function defined by eq. (12)
E_1, E_2	activation energies
ΔH_1	heat of reaction for A → B
ΔH_2	heat of reaction for B → C
q	volumetric flow rate
R	gas constant
T	temperature
$\overline{T}$	mean temperature $(T_f + \kappa T_c)/(1 + \kappa)$
T_c	coolant temperature
T_f	feed temperature
t'	time
t	dimensionless time
u	dimensionless concentration of A
V	volume of reactor
v	dimensionless concentration of B
w	dimensionless temperature
α	Damköhler number
β	heat of reaction number, eq. (6)
Γ	parameter, eq. (25a)
γ	Arrhenius number, eq. (7)
ε	$\alpha E(w)$
κ	dimensionless cooling rate, eq. (8)
ν	ratio of activation energies, eq. (9)
ρ	ratio of heats of reaction in eq. (9) and thereafter, fluid density in eqs (3), (6)
σ	selectivity ratio, eq. (10)

REFERENCES

Amundson, N. R. and Aris, R., 1958, An analysis of chemical reactor stability and control—I. *Chem. Engng Sci.* **7,** 121–131.

Amundson, N. R. and Goldstein, R. P., 1965, An analysis of chemical reactor stability and control—Xa. *Chem. Engng Sci.* **20,** 195–236.

Amundson, N. R. and Schmitz, R., 1963, An analysis of chemical reactor stability and control—Va. *Chem. Engng Sci.* **18,** 265–289.

Aris, R., 1977, Num in olla agitata papilio est? or, catastrophes and chemical reactors. In *Catastrophes and Other Important Matters,* University of Minnesota.

Balakotaiah, V. and Luss, D., 1981, Analysis of multiplicity patterns of a CSTR. *Chem. Engng Commun.* **13,** 111–132.

Balakotaiah, V. and Luss, D., 1982a, Analysis of multiplicity patterns of a CSTR (letter). *Chem. Engng Commun.* **19,** 185–189.

Balakotaiah, V. and Luss, D., 1982b, Exact steady-state multiplicity criteria for two consecutive or parallel reactions in lumped-parameter-systems. *Chem. Engng Sci.* **37,** 433–445.

Balakotaiah, V. and Luss, D., 1982c, Structure of the steady-state solutions of lumped-parameter chemically reacting systems. *Chem. Engng Sci.* **37,** 1611–1623.

Balakotaiah, V. and Luss, D., 1983, Dependence of the steady-states of a CSTR on the residence time. *Chem. Engng Sci.* **38,** 1709–1729.

Balakotaiah, V. and Luss, D., 1984, Global analysis of the multiplicity features of multi-reaction lumped-parameter systems. *Chem. Engng Sci.* **39,** 865–881.

Bilous, O. and Amundson, N. R., 1955, Chemical reactor stability and sensitivity. *A.I.Ch.E. J.* **1,** 513–521.

Burton, A. C., 1939, The properties of the steady state compared to those of equilibrium as shown in characteristic biological behavior. *J. cell. comp. Physiol.* **14,** 327–349.

Chang, H.-C. and Calo, J. M., 1979, Exact criteria for uniqueness and multiplicity of an *n*th order chemical reaction via a catastrophe theory approach. *Chem. Engng Sci.* **34,** 285–299.

Chicone, C. and Retzloff, D. G., 1981, Dynamics of the CR equations modelling a constant flow stirred tank reactor, *Nonlinear Anal.* **6,** 983–1000.

Denbigh, K. G., 1944, Velocity and yield in continuous reactions—I. *Trans. Faraday Soc.* **40,** 352–373.

Denbigh, K. G., 1947, Velocity and yield in continuous reactions—II. *Trans. Faraday Soc.* **43,** 648–660.

Denbigh, K. G., Hicks, M. and Page, F. M., 1948, The kinetics of open reaction systems. *Trans. Faraday Soc.* **44,** 479–494.

Farr, W. W., 1986, Thesis, University of Minnesota.

Frank-Kamenetski, D. A., 1940, *Zh. Fiz. Khim.* **14,** 30–35 (in Russian).

Frank-Kamenetski, D. A., 1941, *Usp. Khim.* **10,** 372–415 (in Russian).

Golubitsky, M. and Keyfitz, B., 1980, A qualitative study of the steady-state solutions for a continuous flow stirred tank reactor. *SIAM J. Math. Anal.* **11,** 316.

Golubitsky, M. and Schaeffer, D. G., 1979, A theory for imperfect bifurcation via singularity theory. *Commun. pure appl. Math.* **32,** 21–98.

Golutibsky, M. and Schaeffer, D. G., 1985, *Singularities and Groups in Bifurcation Theory,* Vol. 1. Springer, New York.

Gray, P. and Scott, S. K., 1984, Autocatalytic reactions in the isothermal continuous stirred tank reactor. *Chem. Engng Sci.* **39,** 1087–1097.

Halbe, D. C. and Poore, A. B., 1981, Dynamic of the continuous stirred tank reactor with reactions $A \rightarrow B \rightarrow C$. *Chem. Engng J.* **21,** 241–253.

Hlavacek, V., Kubicek, M. and Visnak, K., 1972, Modelling of chemical reactors—XXVI. *Chem. Engng Sci.* **27,** 719.

Jorgensen, D. V., Farr, W. W. and Aris, R., 1984, More on the dynamics of the stirred tank with consecutive reactions. *Chem. Engng Sci.* **39,** 1741–1752.

Kahlert, C., Rossler, O. E. and Varma, A., 1981, *Modelling of Chemical Reaction Systems.* Springer, New York.

Kwong, V. K. and Tsotsis, T. T., 1983, Fine structure of the CSTR parameter space. *A.I.Ch.E. J.* **29,** 343–347.

Liljenroth, F. G., 1918, *Chem. Metall. Engng* **19,** 287.

Salnikov, I. E., 1948, Thermokinetic model of homogeneous periodic reactions. *Dokl. Akad. Nauk.* **60,** 405.

Uppal, A., Ray, W. H. and Poore, A. B., 1974, On the dynamic behavior of continuous stirred tank reactors. *Chem. Engng Sci.* **29,** 967–985.

Uppal, A., Ray, W. H. and Poore, A. B., 1976, The classification of the dynamic behavior of continous stirred tank reactors—influence of reactor residence time. *Chem. Engng Sci.* **31,** 205–214.

Vaganov, D. A., Samoilenko, N. G. and Abramov, V. G., 1978, Periodic regimes of continuous stirred tank reactors. *Chem. Engng Sci.* **33,** 1133.

Van Heerden, C., 1953, Autothermic processes. *Ind. Engng Chem.* **45,** 1242–1247.

Williams, D. C. and Calo, J. M., 1981, "Fine structure" of the CSTR parameter space. *A.I.Ch. E. J.* **27,** 514–516.

Woodcock, A. E. R. and Poston, T., 1974, *A Geometrical Study of the Elementary Catastrophes,* Lecture Notes in Mathematics 373. Springer, New York.

K.*

AUTONOMOUS BIFURCATIONS OF A SIMPLE BIMOLECULAR SURFACE-REACTION MODEL

BY M. A. MCKARNIN†, R. ARIS, and L. D. SCHMIDT

Department of Chemical Engineering and Materials Science, University of Minnesota, Minnesota, Minneapolis 55455, U.S.A.

(Communicated by P. Gray, F.R.S.—Received 24 *April* 1987*)*

A pair of ordinary differential equations describing the Langmuir–Hinshelwood mechanism of a bimolecular surface reaction is presented as a simple model for a heterogeneously catalyzed chemical reaction. Numerical and analytical techniques are combined to determine the possible dynamics of this model, which, even though it consists only of low-order polynomials, exhibits very complicated behavior. This includes sustained oscillations, which appear or disappear at Hopf bifurcations, turning points on periodic branches, global homoclinic and metacritical Hopf bifurcations and codimension-two bifurcations with double eigenvalues of zero. The emphasis of the presentation is on geometrical representation of the observed dynamics.

1. INTRODUCTION

Simple dynamical systems have proved valuable as models of certain classes of physical systems in many branches of science and engineering. In mechanics and electrical engineering Duffing's and van der Pol's equations have played important roles and in physical chemistry and chemical engineering much has been learned from the study of simple, even artificially simple, systems. In calling them simple we mean to imply that their formulation is as elementary as possible; their behaviour may be far from simple. Models should have the two characteristics of feasibility and actuality. By the first we mean that a favourable case can be made for the proposed reaction, perhaps by some further elaboration of mechanism but within the framework of accepted kinetic principles. Thus irreversible reactions are acceptable provided that they can be obtained as the limit of a consistent reversible set. By actuality we mean that they are set in an actual context, as taking place in a stirred tank, on a catalytic surface or in a porous medium. It is not usually necessary to assume the reaction to take place in a closed system with certain components held constant presumably by being in excess.

Historically, the irreversible, first-order, exothermic reaction $A \rightarrow B$ taking place in a continuous flow stirred tank has held pride of place in the study of

* Reprinted with the permission of the editor of the *Proceedings of the Royal Society,* **A,** volume **415,** pp. 363–387, 1988.

† Present address: Shell Development Co., Westhollow Research Centre, P.O. Box 1380, Houston, Texas 77251–1380, U.S.A.

well mixed reactors as dynamical systems (see, for example, Amundson & Bilous 1955). More recently Gray & Scott (1984) have used the isothermal autocatalytic system, $A + 2B \rightarrow 3B$, $B \rightarrow C$, as a simpler system with qualitatively similar behaviour. In their model, with lower-case letters for the concentrations of A and B, a subscript f for feed values, ks for the rate constant and θ for the residence time, the equations representing the balances on A and B are:

$$\theta(\mathrm{d}a/\mathrm{d}t) = a_{\mathrm{f}} - a + \theta k_1 ab^2, \tag{1}$$

$$\theta(\mathrm{d}b/\mathrm{d}t) = b_{\mathrm{f}} - b + \theta k_1 ab^2 - \theta k_2 b. \tag{2}$$

These are both algebraically simpler and less stiff than the corresponding mass and heat balances for the non-isothermal reaction $A \rightarrow B$:

$$\theta(\mathrm{d}a/\mathrm{d}t) = a_{\mathrm{f}} - a - \theta k(T)a, \tag{3}$$

$$\theta(\mathrm{d}T/\mathrm{d}t) = T_{\mathrm{f}} - T + J\theta k(T)a - h(T - T_{\mathrm{c}}), \tag{4}$$

where T is the temperature, $k(T) = A \exp(-E/RT)$ the rate constant, J the adiabatic temperature rise, h a heat transfer coefficient and T_{c}, the coolant temperature.

In both these cases there is an autocatalytic element, i.e. one which is both the product of the reaction and which tends to increase its rate. This is the substance B in the first case and heat in the second. It is this element of feedback that is the source of the 'interesting' behaviour. The first two terms in both of these equations represent the access to the site of reaction, in this case the stirred tank, in accordance with the criterion of actuality. The feasibility of these simple reaction schemes can be established by showing that they can be embedded in a fully reversible mechanism and the simple system recovered by limiting processes that do not violate the laws of thermodynamics or kinetics (for example, the Wegscheider condition). Yablonskii and his coworkers (Bykov *et al.* 1978, 1979*a*, *b*, *c*; Yablonskii & Bykov 1979; Gol'dshtein *et al.* 1986) have considered a number of simple models from which it is clear that the autocatalytic feature is essential. In the bimolecular surface reaction the autocatalytic role is played by the vacant sites which are indeed the product of the main reaction which releases those previously held by adsorbate and, at the same time, are a positive influence on the rate of reaction.

After describing the reaction we will look at the steady states of the reaction rate and their stability. This may change either by a saddle-node bifurcation, in which one eigenvalue slips through the origin into the right half plane, or by a Hopf bifurcation, in which both eigenvalues pass across the imaginary axis as a pair of complex conjugates. The degenerate Hopf transitions in which both eigenvalues pass through the origin together, and the homoclinic bifurcation are also considered.

2. SURFACE REACTION MODEL

The system we have selected for this bifurcation study is a variation on simple Langmuir-Hinshelwood kinetics for a bimolecular heterogeneously catalyzed reaction which was first proposed by Takoudis *et al.* (1981*a*). The reactants and products are gases which adsorb on a catalytic surface where they react. For simplicity, it will be assumed that the reaction is isothermal and the products desorb immediately after reaction. A mechanism for such a reaction is given by:

$$\left.\begin{aligned} \mathrm{A} + \mathrm{S} &\rightleftharpoons \mathrm{AS}, \\ \mathrm{B} + \mathrm{S} &\rightleftharpoons \mathrm{BS}, \\ \mathrm{AS} + \mathrm{BS} + 2\mathrm{S} &\rightarrow \text{products}, \end{aligned}\right\} \tag{5}$$

where A and B are the gaseous reactants, S represents a vacant adsorption site on the catalyst surface, and AS and BS represent adsorbed reactants. The first two steps are reversible and the third step, in which two additional vacant sites are required, is irreversible. The two vacant sites can be required for several reasons. If the reactants have large groups which interact, then steric affects may necessitate two additional surface sites even though the actual adsorption is on only two sites. The two additional sites may also be required for intermediates or products. The requirement of two vacant sites for reaction is an important factor in the dynamics of this particular system because it provides the strong nonlinearity in the model. If only one vacant site is required in the reaction step, then the autocatalytic generation of vacant sites is not strong enough to give sustained oscillations.

To model the process, a mass balance is made for the adsorbed species, and the partial pressures of the reactants are considered as constant. The surface coverage of each adsorbed species is defined as the fraction of the total number of available adsorption sites occupied by that species and denoted by θ_A or θ_B. The differential equations for this mechanism are:

$$\left.\begin{aligned} \mathrm{d}\theta_A/\mathrm{d}\tau &= k_1 P_A(1-\theta_A-\theta_B) - k_{-1}\theta_A - k_3\theta_A\theta_B(1-\theta_A-\theta_B)^2, \\ \mathrm{d}\theta_B/\mathrm{d}\tau &= k_2 P_B(1-\theta_A-\theta_B) - k_{-2}\theta_B - k_3\theta_A\theta_B(1-\theta_A-\theta_B)^2, \end{aligned}\right\} \tag{6}$$

where P_A and P_B are the reactant partial pressures and the rate constants are functions of temperature. The first term in each equation represents the rate of adsorption of each species. The second term represents desorption and the last term is the rate of disappearance by reaction. By the definition of the surface coverages, the state variables θ_A and θ_B must satisfy the constraints:

$$0 \leqslant \theta_A, \theta_B, \theta_A + \theta_B \leqslant 1. \tag{7}$$

Following the work of Takoudis *et al.* (1981*a*), these equations can be put into dimensionless form by dividing through by k_3. If we define:

$$\begin{aligned} &\alpha_1 = k_1 P_A/k_3, \quad \alpha_2 = k_2 P_B/k_3, \\ &\gamma_1 = k_{-1}/k_3, \quad \gamma_2 = k_{-2}/k_3, \quad t = k_3\tau, \end{aligned} \tag{8}$$

then (6) becomes:

$$\left.\begin{aligned} \mathrm{d}\theta_A/\mathrm{d}t &= \alpha_1(1-\theta_A-\theta_B) - \gamma_1\theta_A - \theta_A\theta_B(1-\theta_A-\theta_B)^2 = f_1, \\ \mathrm{d}\theta_B/\mathrm{d}t &= \alpha_2(1-\theta_A-\theta_B) - \gamma_2\theta_B - \theta_A\theta_B(1-\theta_A-\theta_B)^2 = f_2. \end{aligned}\right\} \tag{9}$$

The steady-state equations are obtained by setting $f_1 = f_2 = 0$ and combining the resulting equations to give:

$$\theta_A = \frac{(\alpha_1 - \alpha_2) - (\alpha_1 - \alpha_2 - \gamma_2)\theta_B}{(\alpha_1 - \alpha_2 + \gamma_1)}, \tag{10}$$

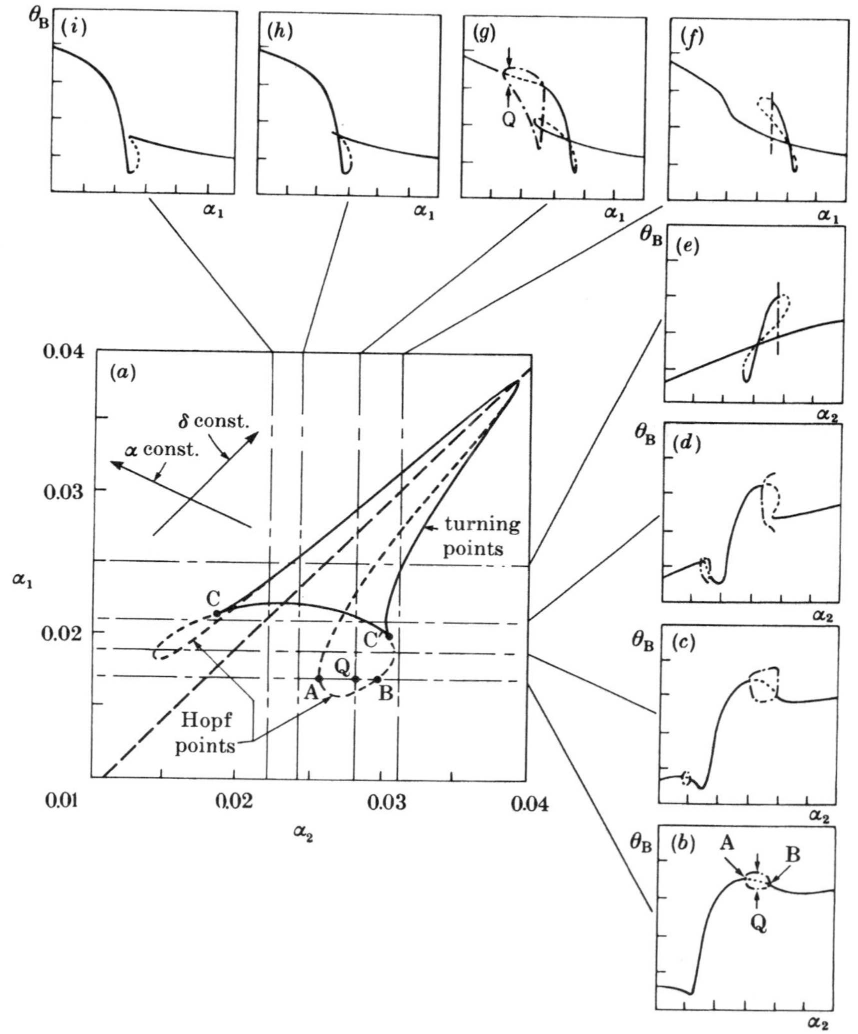

FIGURE 1 Steady-state bifurcation diagrams for variations in the reactant partial pressures. (*a*) Partial two-parameter bifurcation diagram representing the projection of turning points, Hopf bifurcation points, and apparent triple points. (*b*)–(*i*) One parameter sections of the steady-state θ_B surface. The vertical axes are the steady-state θ_B and range from 0 to 1. The horizontal axes correspond to the appropriate axis of the two parameter diagram (*a*). Steady-states are stable or unstable for solid or dashed curves respectively and periodic branches are denoted by pairs of chained curves which represent the minimum and maximum values of θ_B on the limit cycle. The periodic branches all terminate in Hopf bifurcations or, when a saddle is present, homoclinic (infinite period) bifurcations. (*b*)–(*e*) α_1 = 0.017, 0.019, 0.021, 0.025; (*f*)–(*i*) α_2 = 0.031, 0.028, 0.024, 0.022.

$$g(\theta_B) = \lambda(1 - (1 + \nu)\theta_B) - \mu\theta_B - \theta_B[1 - (1 - \kappa)\theta_B][1 - (1 + \nu)\theta_B]^2 = 0, \tag{11}$$

where

$$\lambda = \frac{\alpha_2(\alpha_1 - \alpha_2 + \gamma_1)^2}{\gamma_1(\alpha_1 - \alpha_2)}, \mu = \frac{\gamma_2(\alpha_1 - \alpha_2 + \gamma_1)^3}{\gamma_1^2(\alpha_1 - \alpha_2)}, \nu = \frac{\gamma_2}{\gamma_1}, \kappa = \frac{\gamma_2}{(\alpha_1 - \alpha_2)}. \tag{12}$$

Equation (11) is a quartic equation for the steady-state value of θ_B. Because the only values of θ_B that make physical sense are those between zero and one, Sturm's method (Takoudis *et al.* 1981*b*) can be used to show that there can be either one or three real roots within this interval depending upon the values of the parameters. The region of triplicity is the finger-shaped domain in the α_1, α_2-plane of figure 1. Its boundary is labelled 'turning points' because its equation may be found by eliminating θ_A and θ_B between $f_1 = f_2 = 0$ and $\det[\partial(f_1, f_2)/\partial(\theta_A, \theta_B)] = 0$.

3. BIFURCATION ANALYSIS

The form of the model that will be used most often in this section is given by (9). Although there are four parameters, the two primary parameters are α_1 and α_2, for these are proportional to the partial pressure of the reactants and would be the parameters most easily varied in an experiment.

(a) Model Symmetry

A useful feature of this model is that it has several symmetry properties, because the differential equations are invariant under the simultaneous interchange of the αs, the γs and the θs. This has the important result that the phase portrait of the system is merely reflected about the $\theta_A = \theta_B$ axis when the parameters are interchanged in pairs. The eigenvalues of the jacobian matrix are also invariant under this interchange of parameters and any two-parameter bifurcation diagram using the pair α_1 and α_2, or γ_1 and γ_2 to describe bifurcations that depend on the eigenvalues of the jacobian matrix such as turning points or Hopf points will also be reflected about a 45° axis when the parameter pair not used in the diagram is interchanged. This means that the parameter space for which one must search for bifurcations is effectively reduced by half, and without loss of generality we can take $\gamma_1 \leqslant \gamma_2$.

The domain for the state variables is given by the interior of the unit triangle as defined by the physical constraints in (7). It can easily be shown that the vector field for this system of differential equations is directed inward everywhere on the boundary of the domain (see figure 9, for example). Therefore, a trajectory emanating from any point initially in the triangular domain will remain within this domain.

(b) Steady-State Bifurcations

The simplest bifurcation diagrams are one parameter continuations of one of the steady-state variables as a single parameter is varied. Such diagrams are shown in figure 1 (*b*)–(*i*) where the bifurcation parameter has been chosen to be α_2 for (*b*)–(*e*) and α_1 for (*f*)–(*i*) and the ordinate is the coverage of B. For periodic branches, the maximum and minimum values of θ_B on the limit cycle are traced out using dash-dot curves. These appear in figure 1 (*b*)–(*g*) and will be discussed later. For the moment only the curve of steady states (solid where these are stable, broken where unstable) should be followed. These steady-state curves were obtained by a modified version of the DERPAR algorithm described by Kubicek & Marek (1983). In this method, the bifurcation parameter is added to the state variables and the arc length of the branch is used as the independent variable. This avoids singularity at the turning points at the price of an extra dimension.

The bifurcation diagrams in figure 1(*b*)–(*e*) were constructed for fixed values of α_1, γ_1 and γ_2 and those in figure 1(*f*)–(*i*) for fixed α_2, γ_1 and γ_2. The steady-state curves are cross sections of the surface shown in figure 2(*a*), (*b*), where the cross sections are for constant α_1 and constant α_2 respectively. As α_1 is increased in figure 2*a*, the curve of steady states develops a distinct peak and valley followed by a pinching off in the middle. The remaining cross sections consist of 'figure eights' with a smooth curve passing through the middle. The size of these 'figure eights' decreases until only a smooth, stable steady-state curve remains.

The cross sections of constant α_2 shown in figure 1 (*f*)–(*i*) and figure 2(*b*), (*c*) evolve rather differently. Figure 1 (*f*) shows the same figure-eight cross section, which in the constant-α_1 sections formed one of a sequence of sections diminishing like a pair of touching fingers as α_1 increases (figure 2*a*). The two loops of the figure eight are on different sides of a smooth surface continuous with the smooth one-valued surface for small α_1. The two loops of the figure eight in the constant-α_2 sections start on either side of the surface near the 'finger tip' (A in figure 2*b*) at high values of α_2, but as α_2 diminishes, the two loops move to the same side by the transition shown in figure 2*c* (curve 2). This 'bow-tie' structure unties itself as if someone were pulling on the right-hand end. After the bow-tie is untied, a very sharp turning point remains. This turning point then unfolds in another cusp or fold catastrophe.

An interesting feature of the catastrophe surface of figure 2 is related to the point of intersection of each figure eight. This is a line of triple points since three branches of the surface pass through the part of the line $\alpha_2 = \alpha_1 + \gamma_1$, $\theta_B = \gamma_1/(\gamma_1 + \gamma_2)$ which lies within the finger-shaped region of triplicity (outside this region the line lies in the surface but the value of θ_B is unique). On this line θ_A has three distinct values though the three critical points have common value of θ_B as is clear from such a phase portrait as is shown in figure 9*a* for the parameter values $\alpha_1 = 0.025$, $\alpha_2 = \alpha_1 + \gamma_1 = 0.026$. the triangular region of triplicity, shown by the solid line in figure 1, is determined by eliminating θ_B between $g(\theta_B) = 0$ and $g'(\theta_B) = 0$. This can be done by dialytic elimination but yields algebraic expressions of unmanageable complexity. The two cusp points, C and C′ in figure 1, are the solutions of

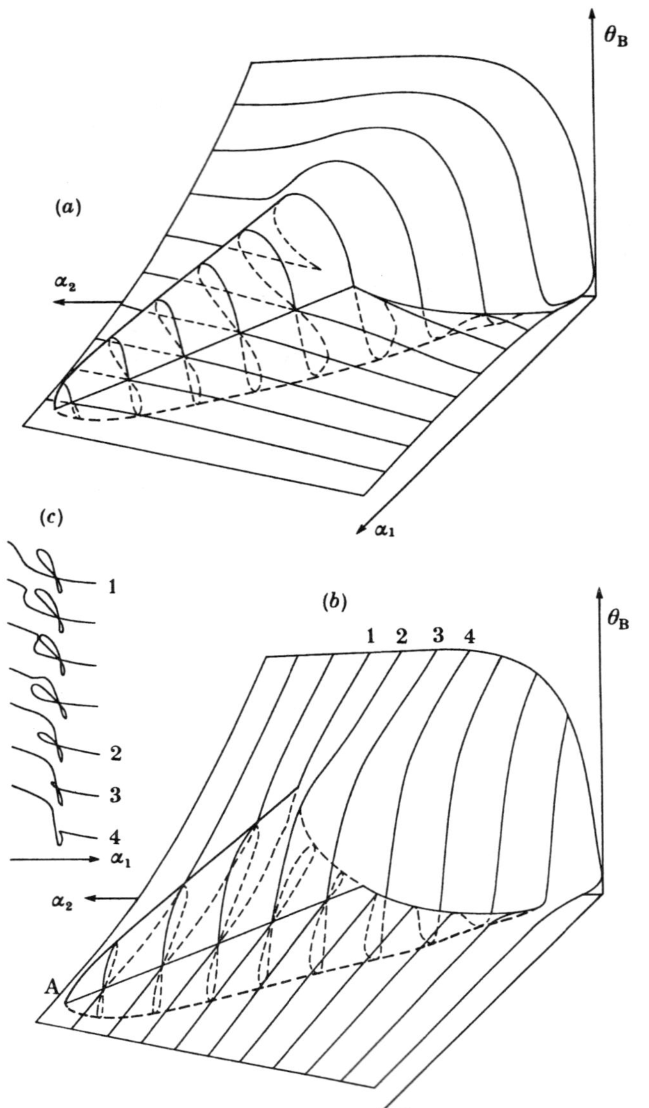

FIGURE 2 Catastrophe surface showing the steady-state coverage of component B as a function of the partial pressures of the reactants (α_1 and α_2) for $\gamma_1 = 0.001$, $\gamma_2 = 0.002$. (*a*) Cross sections of constant α_1, (*b*) cross sections of constant α_2, (*c*) intermediate sections of constant α_2 showing how the shape of the steady-state branches change.

$g = g' = g'' = 0$ and correspond to ($\alpha_1 = 0.021\ 422$, $\alpha_2 = 0.018746$) and ($\alpha_1 = 0.020\ 133$, $\alpha_2 = 0.030\ 296$) respectively.

In the above discussion, we have examined the θ_B steady-state catastrophe surface and a similar picture would have been obtained if we had looked at the surface $\theta_A(\alpha_1, \alpha_2)$. Although the use of one of the state variables as the ordinate is the traditional way to show bifurcation results, we are not restricted to this selection. Because the model is for a chemical reaction, another reasonable choice for the ordinate would be the dimensionless rather of reaction:

$$r = \theta_A \theta_B (1 - \theta_A - \theta_B)^2. \tag{13}$$

To obtain a three-dimensional picture of this catastrophe surface, we could again calculate the points of the surface numerically by solving the quartic equation, $g(\theta_B) = 0$, and substituting the steady-state values of the coverages into the above rate expression. The r_s-catastrophe surface is shown in figure 3 and resembles a shark's fin. This surface is simpler than the θ_B-catastrophe surface. This is because the dependence of r_s upon the steady-state coverages is nonlinear and some of the singularities such as the triple point line have cancelled out. The two surfaces do, however, have the same region of triplicity in the α_1, α_2-plane. If the r_s-catastrophe surface is cut as shown in the insert to figure 3 by any vertical plane in a region where there are multiple steady-states, then there are two types of steady-state bifurcation diagrams. One is the common fold, which here has two pairs of turning points; the other a smooth curve with an isolated, closed curve (isola) above it. The dotted curve in the inset shows that the steady-state loses stability at Hopf points (H, H′) rather than at turning points (T, T′) for this cross section.

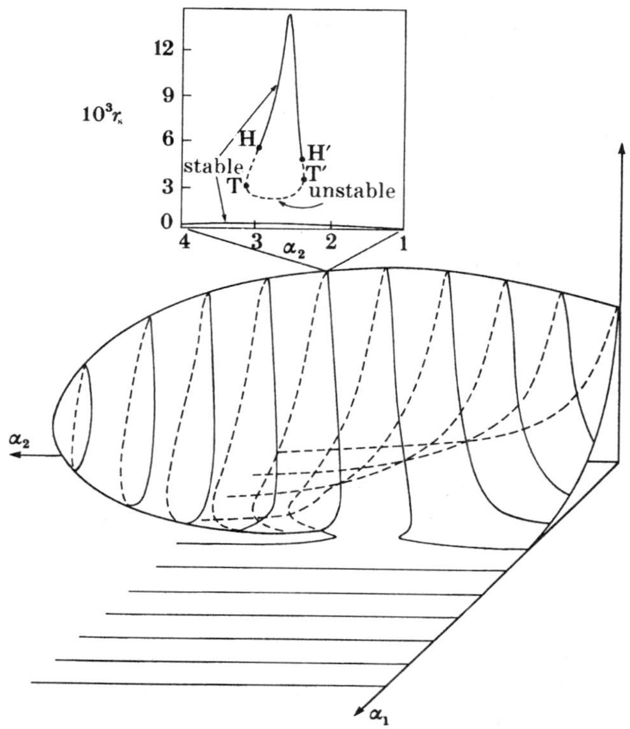

FIGURE 3 Three-dimensional view of the steady-state reaction rate surface when $\gamma_1 = 0.001$, and $\gamma_2 = 0.002$. The inset shows a isola for a section of constant α_1 where H and H′ are Hopf bifurcations and T and T′ are turning points.

Further insight can be obtained by looking at the r_s-catastrophe surface analytically. We begin by defining the new parameter:

$$\delta = \alpha_1 - \alpha_2. \tag{14}$$

At steady state, the rate of reaction can be obtained by equating the f_1 and f_2 of (9) to zero to give:

$$r_s = \alpha_1(1 - \theta_A - \theta_B) - \gamma_1\theta_A = \alpha_2(1 - \theta_A - \theta_B) - \gamma_2\theta_B. \tag{15}$$

Equations (15) can now be considered as linear equations in the two unknown steady-state coverages, the solution of which gives:

$$\theta_A = \frac{\alpha_1\gamma_2 - r_s(\gamma_2 - \delta)}{\alpha_1\gamma_2 + \alpha_2\gamma_1 + \gamma_1\gamma_2}, \; \theta_B = \frac{\alpha_2\gamma_1 - r_s(\gamma_1 + \delta)}{\alpha_1\gamma_2 + \alpha_2\gamma_1 + \gamma_1\gamma_2}. \tag{16}$$

If these equations are then substituted back into the rate expression in (13), one obtains after some rearrangement:

$$r_s(\alpha + 1)^4 = \left[\frac{\delta + \alpha\gamma_2}{\gamma_1 + \gamma_2} + \left(\frac{\delta}{\gamma_1\gamma_2} - \frac{1}{\gamma_1}\right) r_s\right]\left[\frac{\alpha\gamma_1 - \delta}{\gamma_1 + \gamma_2} - \left(\frac{\delta}{\gamma_1\gamma_2} + \frac{1}{\gamma_2}\right) r_s\right] \times \left[1 + \left(\frac{1}{\gamma_1} + \frac{1}{\gamma_2}\right) r_s\right]^2, \tag{17}$$

where the new parameter α has been defined as:

$$\alpha = \alpha_1/\gamma_1 + \alpha_2/\gamma_2. \tag{18}$$

Equation (17) is quadratic in δ and has the roots:

$$\delta = \frac{(\gamma_1 - \gamma_2)(1 + \rho)(\alpha - \rho) \pm \sqrt{D}}{2(1 + \rho)^2}, \tag{19}$$

where the discriminant is given by:

$$D = (\gamma_1 + \gamma_2)^2(1 + \rho)^2(\alpha - \rho)^2 - 4\gamma_1\gamma_2(\gamma_1 + \gamma_2)(\alpha + 1)^4\rho \tag{20}$$

and ρ is defined by:

$$\rho = \frac{\gamma_1 + \gamma_2}{\gamma_1\gamma_2} r_s. \tag{21}$$

The coordinates α_1 and α_2 have been replaced by the new coordinates α and δ. The coordinate transformation given by (14) and (18) has an inverse:

$$\alpha_1 = \frac{\gamma_1(\alpha\gamma_2 + \delta)}{\gamma_1 + \gamma_2}, \quad \alpha_2 = \frac{\gamma_2(\alpha\gamma_1 - \delta)}{\gamma_1 + \gamma_2}. \tag{22}$$

The lines of constant α and δ (see figure 1) show that this transformation provides a way of looking across the region of triplicity from a direction (that of α = constant) such that only two intersections with its boundary are made. Figure 4 shows how this is true for the whole surface $r(\alpha_1, \alpha_2)$, the two roots of the quadratic, given by (19) and (20), being the δ coordinates of the near and far sides of the fin from the 'constant-α viewpoint'. Lines of constant δ

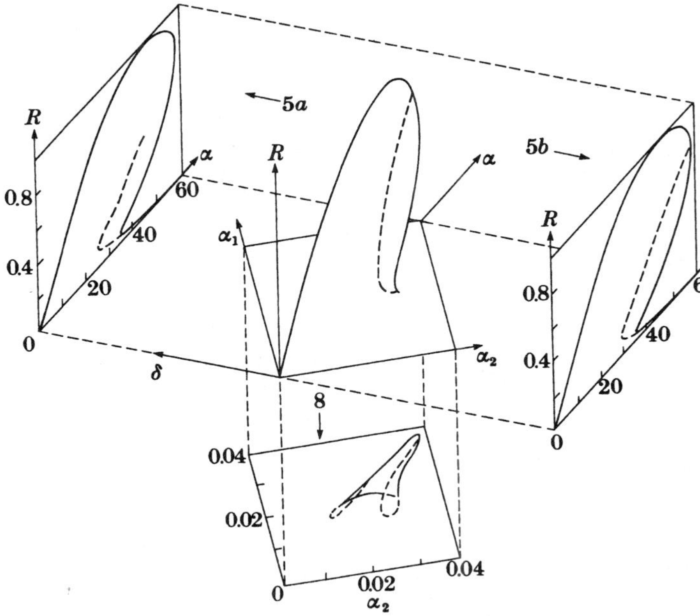

FIGURE 4 Relation between Figures 3, 5 and 8. The steady-state reaction-rate surface has bifurcation curves scribed on its surface which can be projected onto the parameter plane below or along lines of constant α to give side view.

represent variation in the reactant partial pressures such that the relative changes are inversely proportional to the adsorption constants: $\Delta P_A/\Delta P_B = k_2/k_1$. Lines of constant α correspond to variations in reactant pressures such that: $\Delta P_A/\Delta P_B = -K_2/K_1 = -(k_2/K_{-2})/(k_1/k_{-1})$, where K_2 and K_1 are the constants in the equilibrium Langmuir adsorption isotherm.

Equations (19)–(22) offer a method for generating level curves of the r_s-catastrophe surface by solving a quadratic equation, rather than the quartic $g(\theta_B) = 0$. This is done by choosing a fixed rate and then calculating δ for several values of α. The α_1 and α_2 coordinates are then calculated from (22).

When the r_s-catastrophe surface is viewed from the side along lines of constant α, the outline is given by the equality of the two roots of the quadratic (19), i.e. the discriminant must be zero. Setting D in (20) equal to zero gives:

$$4\Gamma\rho = \left(\frac{1+\rho}{1+\alpha}\right)^2\left[1 - \left(\frac{1+\rho}{1+\alpha}\right)\right]^2, \tag{23}$$

where the parameter Γ is defined by:

$$\Gamma = \frac{\gamma_1\gamma_2}{\gamma_1 + \gamma_2}. \tag{24}$$

The outline curve can now be parameterized by defining:

$$z = (1 + \rho)/(1 + \alpha) \tag{25}$$

to obtain from (19) and (23):

$$R = 16z^2(1 - z)^2 \tag{26}$$

and $$\alpha = 1/z + (1/4\Gamma)z(1 - z)^2 - 1,\ \delta = \tfrac{1}{2}(\gamma_1 - \gamma_2)((1/z) - 1), \tag{27}$$

where the new dimensionless rate is defined by:

$$R = 64\Gamma\rho = 64r. \tag{28}$$

It can be easily shown that the feasible ranges for z and R are given by:

$$0 \leqslant z \leqslant 1,\ 0 \leqslant R \leqslant 1. \tag{29}$$

An important consequence of (26) is that the maximum rate of $R = 1$ always occurs for $z = \frac{1}{2}$, or in terms of the original dimensionless reaction rate:

$$r_{\max} = \tfrac{1}{64}. \tag{30}$$

At this maximum the values of the parameters are:

$$\alpha(R_{\max}) = 1 + 1/32\Gamma, \quad \delta = \tfrac{1}{2}(\gamma_1 - \gamma_2), \tag{31}$$

$$\alpha_1 = \tfrac{1}{2}\gamma_1 + \tfrac{1}{32}, \quad \alpha_2 = \tfrac{1}{2}\gamma_2 + \tfrac{1}{32}. \tag{32}$$

Also, (26) can be inverted and one can substitute the resulting value of z into (27). This yields, for a given level of R, the values of α which correspond to the two points on the outline curve:

$$\alpha = \frac{1 \mp \sqrt{(1 - \sqrt{R})} + R/32\Gamma}{1 \pm \sqrt{(1 - \sqrt{R})}}. \tag{33}$$

For a given reaction rate R, the only values of α for which there are realizable steady states are those between the two values given by (33).

An important result that can be obtained from (26) and (27) is a criterion for multiple steady-states. The outline (or side view) of the catastrophe surface is generated by plotting R against α as z is varied from zero to one as shown in figure 5 where the outline has been plotted by using (26) and (27). Also shown on these plots are the regions of characteristic stability which will be discussed in the next section. An appropriate representation of the stability is to scribe the bifurcation curves across which the character of the steady-state changes directly upon the catastrophe surface. Figure 5 actually shows the projection of these curves in the direction of constant α. The relation between figures 1*a*, 3 and 5 is shown in figure 4.

Because a fold in the outline of the surface implies a fold in the surface itself, there will be multiple steady states if the outline curve develops a pair of turning points. The presence of turning points on this curve requires that $d\alpha/dz = 0$ at the two turning points. If one performs the differentiation of (27) and sets the result equal to zero, the following condition is obtained:

$$h(z) = z^2(1 - z)(1 - 3z) = 4\Gamma; \tag{34}$$

$h(z)$ has its maximum at:

$$z_{\max} = \tfrac{1}{6}(3 - \sqrt{3}) = 0.211325 \tag{35}$$

and $$h(z_{\max}) = \sqrt{3}/18 - \tfrac{1}{12} = 0.012892. \tag{36}$$

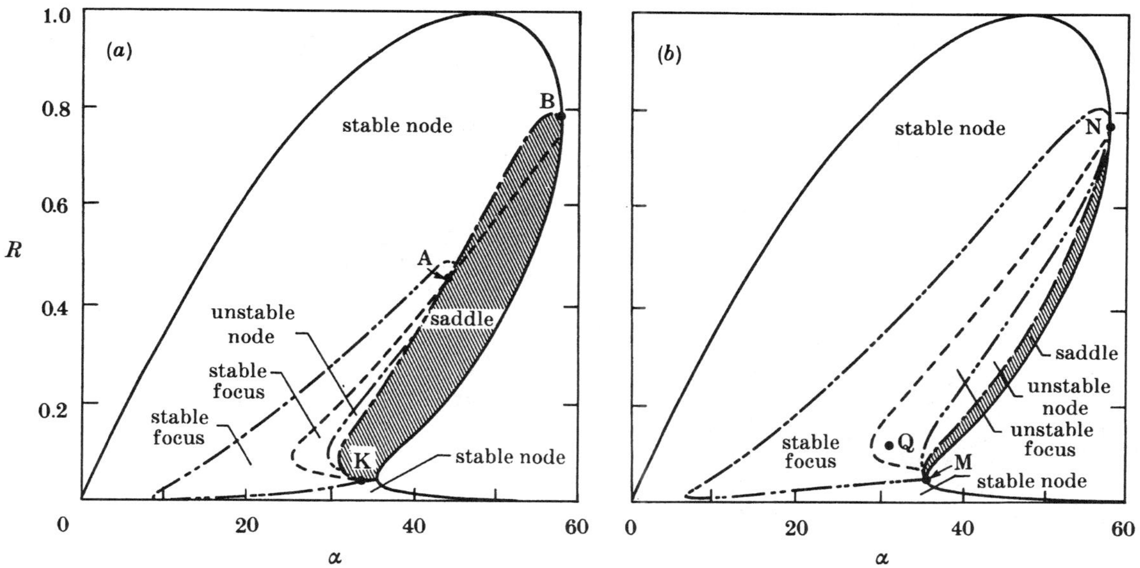

FIGURE 5 Side views of the steady-state reaction rate surface for $\gamma_1 = 0.001$ and $\gamma_2 = 0.002$. (*a*) Left-hand or $\delta > 0$ side of the surface; (*b*) right-hand or $\delta < 0$ side of the surface. Points K, L, M and N correspond to double zero eigenvalues and Q is the projection of the state labelled the same in Figures 1 and 9.

Therefore, by (34), there will only be a fold in the r_s-catastrophe surface if Γ satisfies:

$$\Gamma < \tfrac{1}{4}h(z_{\max}) = \sqrt{3}/72 - \tfrac{1}{48} = 0.0032229. \tag{37}$$

The condition that multiplicity obtains for sufficiently small Γ suggests asking what happens when γ_1 and γ_2 tend to zero. From the basic equations we see that either $\theta_A + \theta_B = 1$ or $\alpha_1 = \alpha_2 = \theta_A\theta_B(1 - \theta_A - \theta_B)$ at steady state. In the first case, $r = 0$, i.e. the flat part of the $r(\alpha_1, \alpha_2)$ surface has degenerated into the base plane. In the second case, $\alpha_1 = \alpha_2$ shows that the fin degenerates into a vertical plane standing on the diagonal. Now at a point $\alpha_1 = \alpha_2 = A$ (say), r can take any value between the greatest and least values of

$$r = \theta_A\theta_B(1 - \theta_A - \theta_B)^2 = A(1 - \theta_A - \theta_B) \tag{38}$$

subject to

$$\theta_A\theta_B(1 - \theta_A - \theta_B) = A.$$

The curves of constant $A (0 \leq A \leq \frac{1}{27})$ are the curvilinear triangles shown in figure 6*a* and it is clear from (38) that the greatest and least values of r must be where $(1 - \theta_A - \theta_B)$ is greatest and least (i.e. at D and E in figure 6*b*). By symmetry this is on the line $\theta_A = \theta_B$, so that by taking their common value (θ say, $0 \leq \theta \leq \frac{1}{2}$) as parameter we can calculate the outline of the degenerate fin for $\gamma_1 = \gamma_2 = 0$ from $R = 64\theta^2(1 - 2\theta)^2$, $A = \theta^2(1 - 2\theta)$. This is a closed loop shown in figure 6*c* leaving the origin as $R = 64A$ and reaching $R = 1$ at $A = \frac{1}{32}$ $(\theta = \frac{1}{4})$, having a turning point at $R = \frac{64}{81}$, $A = \frac{1}{27}$ $(\theta = \frac{1}{3})$

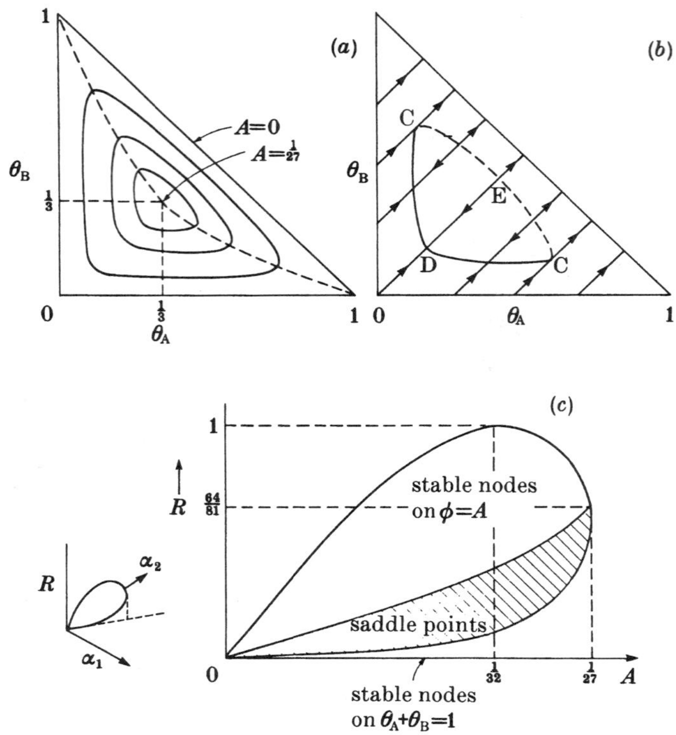

FIGURE 6 Demonstration of the relation between the function ϕ and the shape of the phase plane. (*a*) Phase plane contours of the function $\phi = \theta_A\theta_B(1 - \theta_A - \theta_B) = A$. (*b*) Phase portrait for the special case of negligible desorption and identical absorption rates ($\alpha_1 = \alpha_2 = A$; $\gamma_1 = \gamma_2 = 0$). (*c*) Shape of the non-unique steady-state reaction rates for variations in reactant partial pressures for the case of negligible desorption.

returning to the origin as $R \sim 512A^2$. Any reaction rate between these limits is possible, for every point of the closed curve $\theta_A\theta_B(1 - \theta_A - \theta_B) = A$ is a possible steady state.

4. THE STABILITY OF THE STEADY STATES

It might be thought that the preceding discussion of a trivial special case was uninteresting, but this is not so, for the structure of the system in the degenerate case $\gamma_1 = \gamma_2 = 0$ contains the clue to the structure of the general case. Writing the equations for this case as:

$$\left.\begin{aligned}\dot{\theta}_A &= [\alpha_1 - \theta_A\theta_B(1 - \theta_A - \theta_B)](1 - \theta_A - \theta_B),\\ \dot{\theta}_B &= [\alpha_2 - \theta_A\theta_B(1 - \theta_A - \theta_B)](1 - \theta_A - \theta_B),\end{aligned}\right\} \quad (39)$$

we see that the solutions when $\alpha_1 = \alpha_2 = A$ are lines $\theta_A = \theta_B + B$ where B is a constant, $-1 \leqslant B \leqslant 1$. Those that intersect the curve

$$\phi(\theta_A, \theta_B) \equiv \theta_A\theta_B(1 - \theta_A - \theta_B) = A$$

have three steady states, one at each intersection of the line with $\phi = A$ and one on the line $\theta_A + \theta_B = 1$. The last is always stable and of the two intersections that with the lower value of $\theta_A + \theta_B$ is stable and the other unstable. This can be recorded on the degenerate 'fin' shown in figure 6*c* as the rate at C or C′ where the line $\theta_A = \theta_B + B$ is just tangent to $\phi = A$. The phase plane for $\alpha_1 = \alpha_2 = A$, $\gamma_1 = \gamma_2 = 0$ is shown in figure 6*b* where clearly each point of both $\theta_A + \theta_B = 1$ and of $\phi(\theta_A, \theta_B)$ is a steady state. The stable points on the diagonal and on CDC′ are not asymptotically stable because one of the eigenvalues is zero. When the picture is distorted by making γ non-zero or the αs different, the smallness of the terms in α_1, α_2, and the reaction rate in (9) still characterizes the phase portrait and we shall find the spoor of $\phi(\theta_A, \theta_B)$ everywhere that there are multiple steady states.

If we move away from the line $\alpha_1 = \alpha_2$ (still keeping $\gamma_1 = \gamma_2 = 0$) there is only the line of stable, but dead, steady states $\theta_A + \theta_B = 1$ where the surface is fully saturated. This is because the curves $\phi = \alpha_1$ and $\phi = \alpha_2$, do not intersect. If $\alpha_1 > \alpha_2$ then in the region between $\phi = \alpha_1$ and $\phi = \alpha_2$, $\dot{\theta}_A > 0$ and $\dot{\theta}_B < 0$, though both can be quite small if α_1 and α_2 differ only a little. The trajectories are therefore distorted from their linearity when $\alpha_1 = \alpha_2$ (figure 6*b*) to the picture shown in figure 7*a*. The state moves rapidly to the zone between $\phi = \alpha_1$ and $\phi = \alpha_2$ and then drifts slowly to the southeastern end and then follows a path to saturation. It is remarkable that under these circumstances all initial points in the region OPQSR would lead to the steady-state F and the rest of the line would attract from a smaller basin of steady-states. The picture would be similar but with a northwesterly flow if α_2 were greater than α_1.

If now we return to $\alpha_1 = \alpha_2 = A$, $\gamma_1 = \gamma_2 = \gamma$, there are three steady states on the line $\theta_A = \theta_B$ at points close to the intersections with $\phi = A$ and $\theta_A + \theta_B = 1$. In fact, if $\theta = \theta*(A)$ and $\theta*(A)$ are the roots of $\phi(\theta, \theta) = A$ the three roots for small γ are at $\theta_A = \theta_B = \theta* - \gamma/[2(1 - 2\theta*)(1 - 3\theta*)]$, $\theta* + \gamma/[2(1 - 2\theta*)(3\theta* - 1)]$ and $\frac{1}{2} - \gamma/(4A)$ respectively. Their character remains unchanged for small values of γ. The old line of neutrally stable steady states is distorted very slightly and becomes a line of very slow drift toward the stable node near $\theta_A = \theta_B = \frac{1}{2}$. The unstable part of the locus $\phi = A$ is distorted slightly into the stable manifold of B (i.e. the set of all points that are attracted to the saddle point near $\theta*$; (DBE in figure 7*c*). From a point such as F or G to the left of DBE the trajectory goes rapidly to a point near $\phi = A$ and does a sharp turn to come back slowly along HC or JC to the stable steady state C. Again the imprint of the locus $\phi(\theta_A, \theta_B) = A$ is very clear.

If we destroy the symmetry by introducing unequal γs (i.e. $\nu = \gamma_2/\gamma_1 \neq 1$) we obtain three steady states deriving respectively from intersections of the line $\theta_A = \nu\theta_B$ and the loci $\phi(\theta_A, \theta_B)$ and $\theta_A + \theta_B = 1$, provided $N > \nu > 1/N$, where $N = N(A)$ is the slope of the tangent from the origin to the curve $\phi(\theta_A, \theta_B) = A$. Such a situation is shown in figure 7*d* where the three steady states are again called A, B and C. The stable manifold of B is again marked DBE and, close to D, one again finds a point F that leads to the path FHC with a sharp corner at H. Between D and F is a point K which also leads to C but via a sharp turn at L. The unstable manifold of the saddle

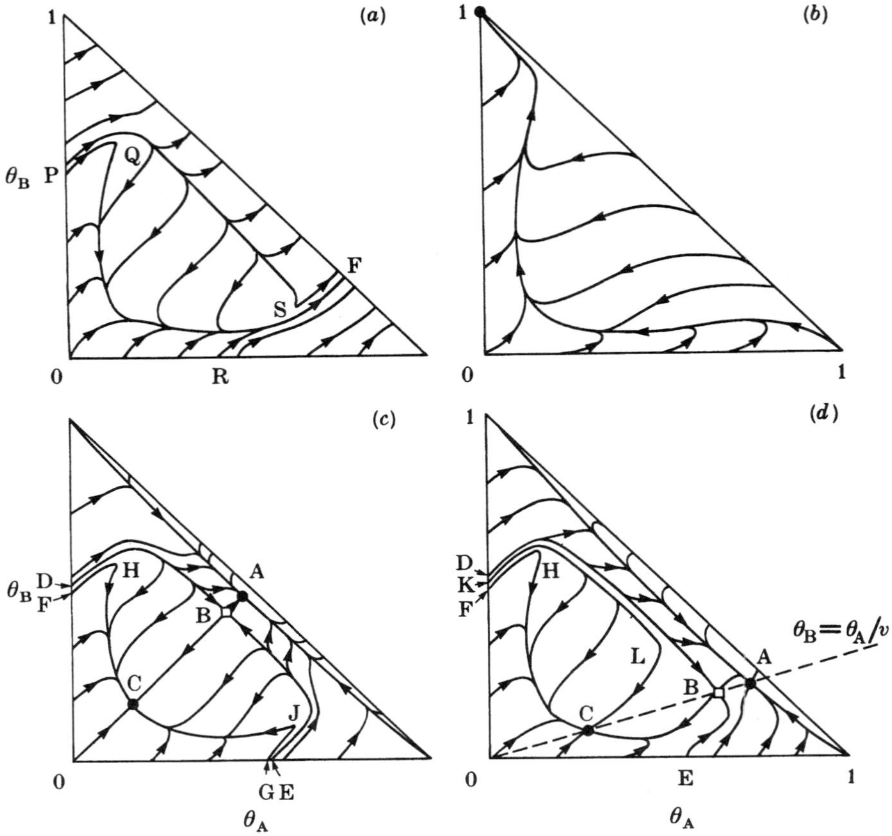

FIGURE 7 Phase portraits for various deviations from the simple case: $\alpha_1 = \alpha_2 = A$, $\gamma_1 = \gamma_2 = 0$. In each case, the remnants of the level curve $\phi = A$ can still be seen and determine the shape of the slow approach to steady state. (*a*) $\alpha_1 > \alpha_2$, $\gamma_1 = \gamma_2 = 0$; (*b*) $\alpha_1 = \alpha_2$, $\gamma_1 > 0$, $\gamma_2 = 0$; (c) $\alpha_1 = \alpha_2 = A$, $\gamma_1 = \gamma_2 = \gamma > 0$; (*d*) $\alpha_1 = \alpha_2 = A$, $\gamma_2 = \nu\gamma_1$.

point B leads in one direction to C and in the other to A. Again the domain of attraction of C is the region to the left of DBE.

These simple departures from the degenerate case allow us to see more readily what is happening in the general case. Referring back to figure 5 we see that more complicated transitions than the simple saddle-stable node bifurcation of figure 6 can occur. The stable nodes are still to be found at the top of the fin and the saddles on the lower side, but only across arc AB of figure 5*a* is the transition from stable node to saddle a direct one. In other cases the stable node becomes a stable focus which becomes unstable and then changes back to being a node before one eigenvalue passes through the origin to give a saddle in the shaded region. Figure 5*b* shows the transitions on the other side of the fin and it convenienty separates the character of the three possible steady states. It must however be read in conjunction with a projection with the α_1, α_2-plane such as is shown in figure 8. Both are drawn

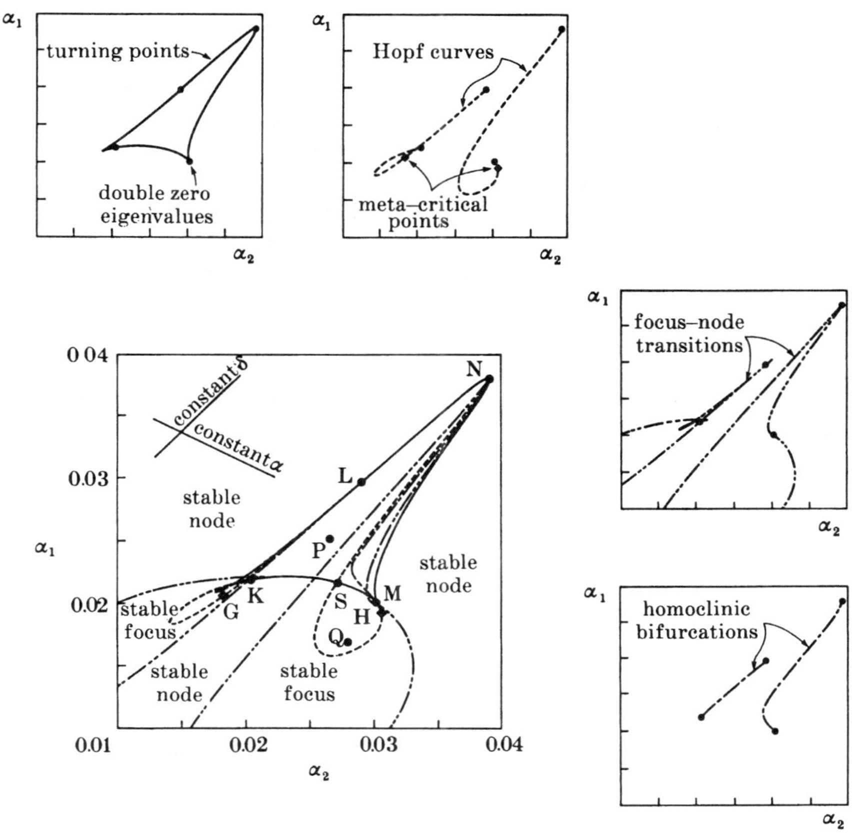

FIGURE 8 Detailed bifurcation diagram for variations in the reactant partial pressures (α_1 and α_2) when $\gamma_1 = 0.001$ and $\gamma_2 = 0.002$. The middle diagram is the superposition of the four surrounding bifurcation curves. The points K, L, M and N correspond to double zero eigenvalues and the points G and H are metacritical Hopf points.

for $\gamma_1 = 0.001$, $\gamma_2 = 0.002$. In this figure, as in figure 1, the interior of the finger-like region with the two cusps is the region of multiple steady-states. This is the projection of the shaded region shown as seen from the two sides in figure 5 for when there are three steady-states the middle one is always a saddle. Likewise the steady state nearest the line $\theta_A + \theta_B = 1$ and with the lowest reaction rate is always a stable node, so that the stability characteristics shown in figure 8 are those of the steady-state of highest reaction rate within the region of triplicity or of the unique steady-state outside this region.

The triple point line is shown in figure 1, but not in figure 8 since it is not of crucial importance. If α_1, α_2 lies on it (i.e. $\alpha_2 = \alpha_1 + \gamma_1$) the value of θ_B at the three steady-states happens to be the same, but the phase portrait is essentially of the same character as we have anticipated in the reasoning behind figure 7*d*. Such a phase plane is shown in figure 9*a* for $\alpha_1 = 0.025$, $\alpha_2 = 0.026$, $\gamma_1 = 0.001$, $\gamma_2 = 0.002$ and corresponds to the point P in figure 8.

5. HOPF BIFURCATIONS

The Hopf bifurcation where a stable focus becomes unstable and sheds or absorbs a periodic solution is an important transition which has received a great deal of attention (for a review see Marsden & McCracken 1976). Clearly the lines over which it can take place are the loci of steady-states whose eigenvalues are purely imaginary. These are shown on the sides of the 'fin' in figure 5. Because this is a two-dimensional system we can write down the condition quite explicitly. Writing the equations:

$$\begin{aligned}\dot{\theta}_A &= [\alpha_1 - \phi(\theta_A, \theta_B)](1 - \theta_A - \theta_B) - \gamma_1\theta_A,\\ \dot{\theta}_B &= [\alpha_2 - \phi(\theta_A, \theta_B)](1 - \theta_A - \theta_B) - \gamma_2\theta_B,\end{aligned} \tag{40}$$

where $\phi(\theta_A, \theta_B) = \theta_A\theta_B(1 - \theta_A - \theta_B)$ as before, we see that the jacobian is:

$$J = \begin{bmatrix} -\alpha_1 - \gamma_1 - \dfrac{\phi(\theta_A, \theta_B)(1 - 3\theta_A - \theta_B)}{\theta_A} & -\alpha_1 - \dfrac{\phi(\theta_A, \theta_B)(1 - \theta_A - 3\theta_B)}{\theta_B} \\ -\alpha_2 - \dfrac{\phi(\theta_A, \theta_B)(1 - 3\theta_A - \theta_B)}{\theta_A} & -\alpha_2 - \gamma_2 - \dfrac{\phi(\theta_A, \theta_B)(1 - \theta_A - 3\theta_B)}{\theta_B} \end{bmatrix}. \tag{41}$$

The condition for the eigenvalues to be purely imaginary is:

$$\operatorname{tr} J = 0, \det J > 0 \tag{42}$$

and the curve in the α_1, α_2-plane along which this condition holds can be calculated and plotted either on the α_1, α_2-plane as in figures 1 and 8 or on the sides of the $r(\alpha_1, \alpha_2)$ surface as in figure 5. The bordering inserts on figure 1 are a useful way of showing what happens between the Hopf bifurcations. In figure 1*a*, for example, the lowest horizontal section of constant $\alpha_1 = 0.017$

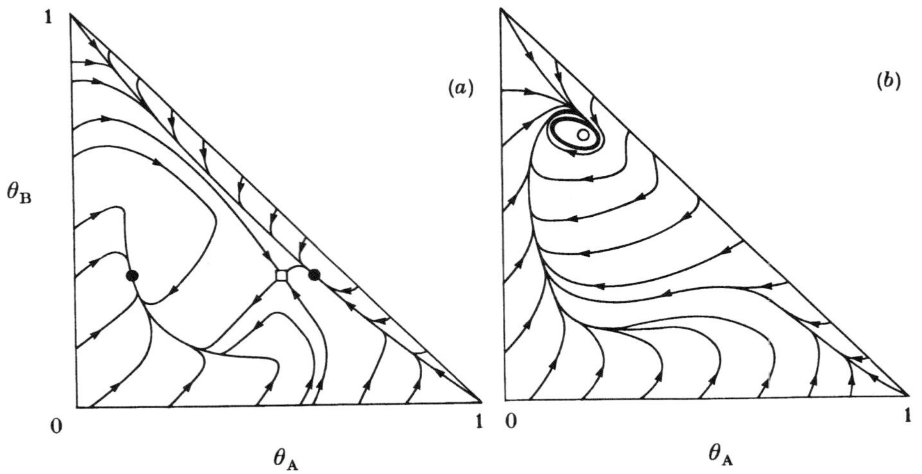

FIGURE 9 Phase portraits for the system when $\gamma_1 = 0.001$ and $\gamma_2 = 0.002$. (*a*) Phase diagram at an apparent triple point (labelled P in Figure 8; $\alpha_1 = 0.025$, $\alpha_2 = 0.026$) where only the value of θ_B is identical for the three steady-states. (*b*) Limit cycle surrounding an unstable focus corresponding to the point labelled Q in Figures 1 and 8 ($\alpha_1 = 0.017$, $\alpha_2 = 0.028$).

has two intersections, A and B, with the Hopf curve. For a value of α_2 between these points (e.g. $\alpha_2 = 0.028$, the point Q on figures 1 and 8) there is a stable limit cycle surrounding an unstable focus and the phase plane is as in figure 9*b*. This is indicated in figure 1*b* by the 'bubble' between the points A and B; the broken line in the middle is the now unstable steady-state value of θ_B and the curves above and below are the greatest and least values of θ_B on the limit cycle. Similarly the sections shown in figures 1*c*–*g* have Hopf bifurcations. The vertical section shown in figure 1*g* also goes through the point Q and these positions are indicated.

Though the most usual Hopf transition is the supercritical case we have just described in which a stable periodic solution arises about an unstable steady-state, there are cases of subcritical Hopf bifurcation. This can occur when an unstable steady-state becomes stable, not by a stable limit cycle shrinking down upon it, but by its shedding and unstable limit cycle which grows until it meets the stable limit cycle. This was in fact the type of Hopf transition uncovered by Amundson & Aris (1958). An illustration is shown in figure 10 where the triangular limit cycle in 10*a* is indeed shrinking as the parameter increases but it is of finite amplitude where the steady state regains stability. In figure 10*b* we see that this stable steady state attracts only points

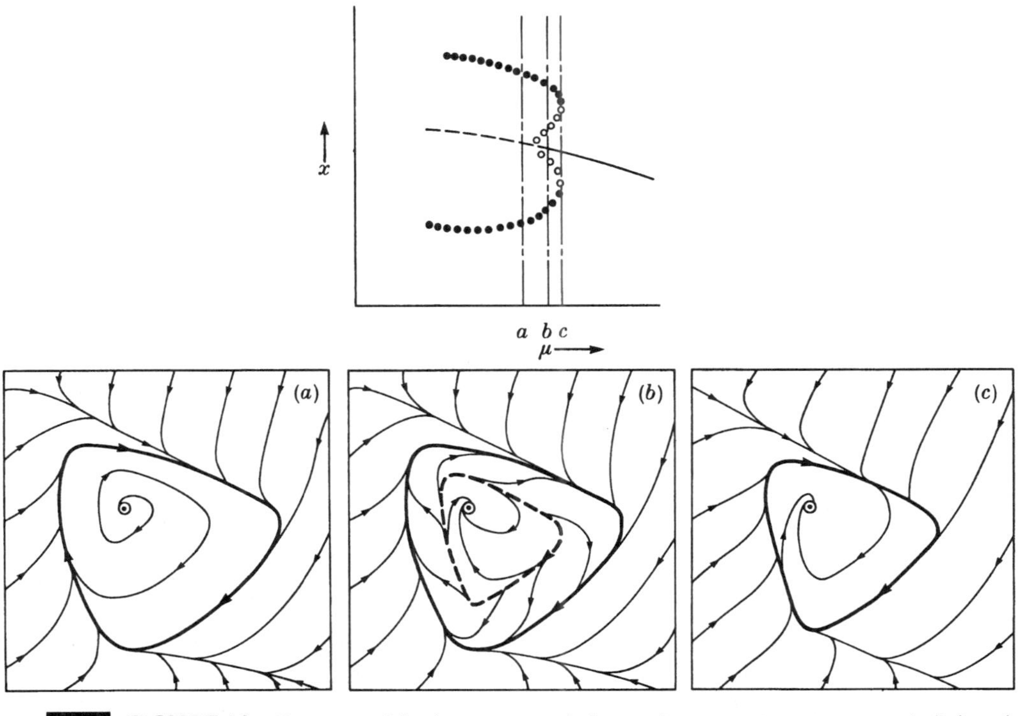

FIGURE 10 Illustration of the disappearance of a limit cycle via a turning point on a periodic branch near a subcritical Hopf bifurcation. (*a*) A stable limit cycle surrounding an unstable focus; (*b*) the unstable focus undergoes a subcritical Hopf bifurcation and leaves an inner unstable limit cycle surrounding a stable focus; (*c*) the two limit cycles combine into a metastable configuration and disappear altogether as the parameter is further increased.

within the inner limit cycle. Because the amplitude of a limit cycle grows like the square root of the difference between the parameter and its critical value at the Hopf bifurcation, it is not surprising that the system of the two limit cycles about an unstable steady-state does not last long and the region of multiple limit cycles is not large. Unstable limit cycles such as this have been successfully calculated two different ways. The easiest is by simulation in which the differential equations are simply integrated backwards in time so that the stability of all limit sets is reversed. The second method (a shooting method) uses Newton-Raphson iteration to find a point in the phase plane which will come back to itself after integrating forward for an unknown time T. One of the coordination of the starting point is held fixed and the unknowns to be solved for are the period T and the remaining coordinate.

Subcritical Hopf transitions are found on the segments HM and GL of the Hopf curve and all other transitions are supercritical. The points H and G in figure 8 are located at ($\alpha_1 = 0.019308$, $\alpha_2 = 0.030686$) and ($\alpha_1 = 0.020668$, $\alpha_2 = 0.018330$) respectively, and might be called metacritical. They are bifurcations of codimension two so that we expect only isolated metacritical points on the Hopf curve. These have to satisfy not only the conditions of (42), but also:

$$\begin{aligned}&[f_{1,12}(f_{1,11} + f_{1,22}) - f_{2,21}(f_{2,11} + f_{2,22}) - f_{1,11}f_{2,11} + f_{1,22}f_{2,22}]\\&\quad + \omega[f_{1,111} + f_{1,122} + f_{2,211} + f_{2,222}] = 0.\end{aligned} \tag{43}$$

Where f_1 and f_2 are defined in (9) and suffix 1 (or 2) after the comma denotes a differentiation with respect to θ_A (or θ_B); ω^2 is the det$J$ of the Hopf condition given in (42).

At the points where the Hopf curve, tr$J$ = 0, meets the curve of turning points, det$J$ = 0, there is another form of degenerate Hopf bifurcation, namely that caused by a double eigenvalue of zero. The double zero eigenvalue points are labelled as K, L, M and N in figure 8 and correspond to ($\alpha_1 = 0.021\,970$, $\alpha_2 = 0.020\,491$), ($\alpha_1 = 0.029\,730$, $\alpha_2 = 0.029\,133$), ($\alpha_1 = 0.020\,146$, $\alpha_2 = 0.030\,287$), ($\alpha_1 = 0.037\,975$, $\alpha_2 = 0.039\,159$) respectively. Although the point M nearly coincides with the cusp point C′ (labelled in figure 2), this is purely accidental. Takens (1974), Bogdanov (1975), and Guckenheimer & Holmes (1983) have discussed the normal form and its universal unfolding about such a point. Terminating at the double zero eigenvalue point is a curve of Hopf bifurcations and a curve of homoclinic bifurcations (discussed in the next section), both of which have quadratic tangencies to the saddle-node curve. Also passing through the double zero eigenvalue point is a curve of focus-node transitions as shown in figure 8. The fact that several codimension-one bifurcation curves converge upon codimension two bifurcation points makes the double zero eigenvalues a centre of interesting behaviour.

The point S of figure 8 at which the Hopf bifurcation curve crosses the boundary of the multiplicity region is not a double zero degeneracy, for the upper steady state (i.e. that with the larger θ_B) is undergoing the Hopf bifurcation at the same time as the lower steady-state undergoes a saddle-node bufurcation, i.e. the conditions tr$J$ = 0 and det$J$ = 0 apply at different points. It does, however, serve to show the four combinations of the two most common bifurcations. It is difficult to get the point S because the movement of the steady states is extremely sensitive but the pair $\alpha_1 = 0.021\,691$, $\alpha_2 = 0.027\,285$ gives a size of limit cycle and a distance between saddle and node each of which is in the third decimal place. If the calculation could be done exactly the neutrally stable focus A, at $\theta_A = 0.165$, $\theta_B = 0.637$, would serve as a global attractor, but with two peculiar features (cf. figure 11S). Firstly, the spiralling down of the trajectory would be much slower than the usual exponential decay since the real part of the eigenvalues is zero and the stability depends on higher-order terms. Secondly, the bifurcating saddle-node (B, C), at $\theta_A = 0.522$, $\theta_B = 0.421$, serves as a resting place of one-sided stability and theoretically any starting point in the region D (B, C) EF would lead to (B, C). However, the least disturbance would send the state of the system on to A. The inserts in the bottom right corner of figure 11 show a close up of the trajectories in the neighbourhood of the saddle-node bifurcation just before and just after the coalescence of B and C. It is clear from this how the flow is pinched and slowed down in this neighbourhood. Figure 11 also shows the phase planes for four values of α_1 and α_2 differing by ± 0.0005 from the critical

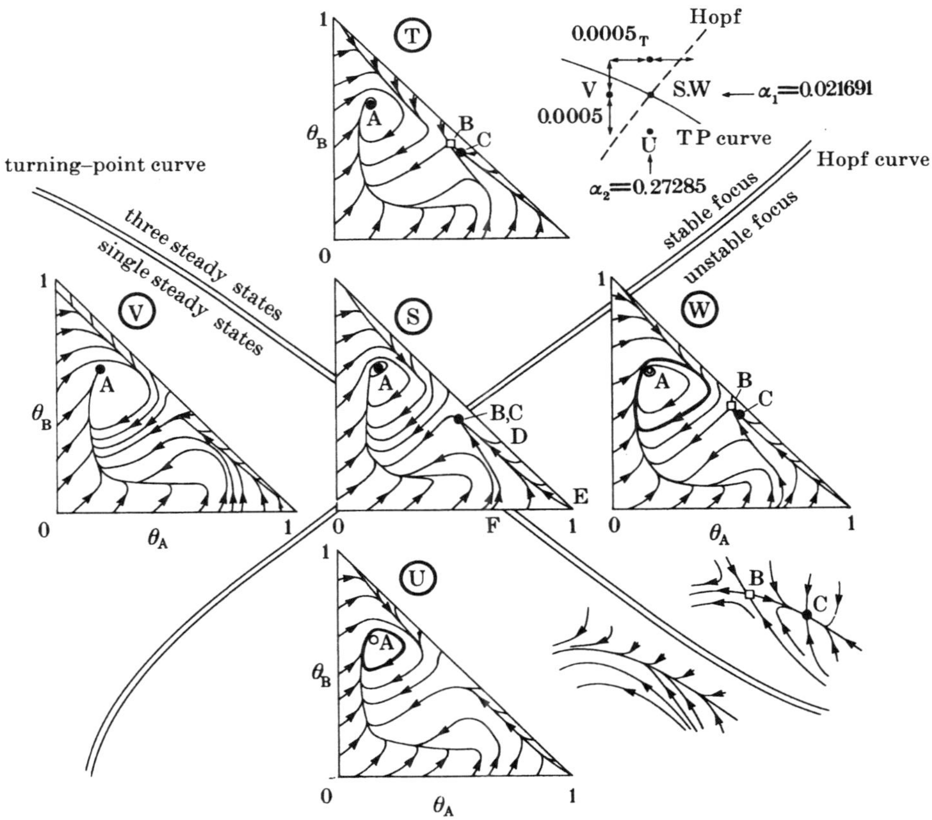

FIGURE 11 Phase portraits for the point labelled S in Figure 8 and its four neighbours T, U, V and W. The curves of turning points and Hopf points correspond to different steady states.

values at S. Increasing either α_1 or α_2 (T or W) gives three steady-states; increasing α_2 (W) or decreasing α_1 (U) gives a limit cycle. The saddle-node combination, though it has disappeared in figure 11 U and V, leaves its mark behind as a region of very slow progress near $\theta_A = 0.522$, $\theta_B = 0.421$. The growth of the limit cycles (shown as a heavier curve in figure 11 U and W) is remarkable for such a small change in parameter and reflects the square root dependence.

6. HOMOCLINIC BIFURCATIONS

Besides disappearing by shrinking to zero or coalescing with an unstable limit cycle, periodic solutions may vanish with a finite amplitude by growing into the angle of a saddle point. This is called the homoclinic bifurcation because the transitional case is obtained when one outgoing eigentrajectory of a saddle point becomes an incoming eigentrajectory of the same saddle point (see figure 12*b*). The state would in principle follow round this trajectory but would come to rest at the saddle point B. This is a semistable situation, as are all of these transitions. For, if the state were perturbed into the interior of the loop (i.e. towards A), the trajectory would spiral outwards and resume its ever lengthening 'periodic' course; whereas, if the state were tipped to the outside, it would quickly go to the stable state C. Figure 12(*a*) and (*c*) shows

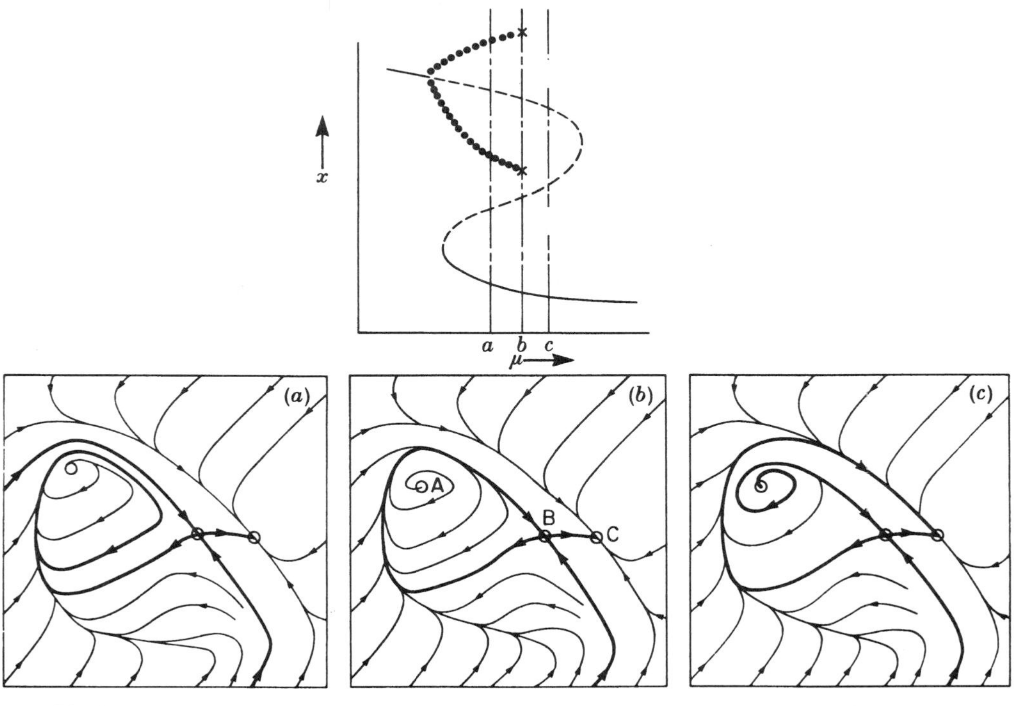

FIGURE 12 Phase portrait illustrations of the homoclinic bifurcation. (*a*) The stable and unstable manifolds of the saddle surrouund the stable limit cycle; (*b*) the saddle manifolds join together with the limit cycle; (*c*) the saddle manifolds continue to cross and the limit cycle disappears.

phase portraits before and after the transition and the figure at the top shows their relation to one another in the bifurcation diagram (x represents any state variable, such as θ_B in figure 1, and μ a parameter such as α_1 or α_2). As the parameter passes the critical value, the limit cycle disappears and the manifolds of the saddle pass through one another so that they appear to be inside-out from their original positions. The status of the unstable focus or node has remained unchanged throughout the bifurcation. In fact, nothing peculiar has happened locally about any of the steady states, and the bifurcation cannot be monitored with local information. This type of bifurcation is called a global bifurcation (as opposed to the local bifurcations that have been presented thus far) and is much more difficult to analyse because no linearizations or eigenvalue criteria can be used. Some analytical progress has been made in the study of this type of bifurcation (for example, the Melnikov theory described by Guckenheimer & Holmes 1983; Chow *et al.* 1980), but any continuation of the bifurcation with respect to other parameters must be done numerically. The numerical continuation of the homoclinic bifurcation was performed by Kevrekidis's method (Kevrekidis 1988). This method involves finding the straight line that bisects the stable and unstable eigenvectors of the saddle and observing by numerical integration the point where the stable and unstable manifolds of the saddle hit this line. The distance $d(\alpha_1, \alpha_2)$ between the two intersection points on the straight line is a measure of the distance of the system from the homoclinic bifurcation and will be zero at the bifurcation. The equation

$$\mathrm{d}(\alpha_1, \alpha_2) = 0 \qquad (44)$$

now defines a curve $\alpha_1 = \alpha_1(\alpha_2)$ which can be found using the arc-length continuation method discussed earlier.

A two-parameter continuation of the homoclinic bifurcation is shown in figure 8 superimposed on the bifurcations discussed previously. Crossing the curve in this parameter space represents the appearance or disappearance of a limit cycle via the homoclinic bifurcation. As expected, the homoclinic bifurcation curve is confined within the closed curve of turning points since the presence of a saddle is required for the bifurcation to occur. As the values of γ_1 and γ_2 are changed, and provided inequality (37) is still satisfied so that a saddle is available, it may be possible for these two curves of homoclinic bifurcations to interact. As mentioned in the previous section, the homoclinic bifurcation curves terminate at the double zero eigenvalue points (labelled K, L, M, and N in figure 8). This is an important feature of these bifurcation points because local (eigenvalue) information can be used in this case to predict the approximate location of a global bifurcation curve.

7. DISCUSSION

The mathematical analysis of this model has shown that this system can have multiple steady states and steady oscillatory states. However, it is not immediately obvious what the underlying physical mechanism for this behaviour is. For this, one must return, not to the model, but to the assumed mechanism that generated the model.

Most of the interesting phenomenon occur because of the two vacant sites required for reaction. This requirement means that the highest rates of reaction occur at intermediate coverages of the reactants; to be precise, r_{max} occurs at $\theta_A = \theta_B = \frac{1}{4}$ which implies that the 'reactants' (adsorbed A and B and vacant sites) are present in stoichiometric proportions. With this in mind, we can now make physical arguments to help understand the causes of multiple steady states and oscillations. If one assumes that initially the partial pressures of both reactants are low and the coverages on the catalyst surface are high, then the rates of adsorption and reaction will be smaller than the rate of desorption. This will cause the coverages to decrease. As this occurs, the rate of adsorption, r_a, increases and the rate of desorption, r_d, decreases. The dominant change, however, is the increase in the rate of reaction, r_R, due to its quadratic dependence on the vacant sites. Therefore, the coverages continue to decline until r_R passes through its maximum and begins to decrease. Only then does the increase in r_a offset r_d and r_R. The system comes to a steady state at low coverages. Because r_R is small in these circumstances, one can consider the system to be in adsorption–desorption equilibrium and Langmuir–Hinshelwood kinetics are applicable. The same procedure can be applied to high partial pressure of the reactants. In this case, the coverages of the reactants will be high and again the rate of reaction will be relatively small leading to the adsorption–desorption equilibrium situation which predicts the compatible situation of high total coverage for high partial pressures of reactants. The relative distribution of species A to species B on the surface at these near equilibrium states will be determined by the ratio of the adsorption and desorption constants α_1, α_2 and γ_1, γ_2.

Intermediate coverages of both reactants will only be found with intermediate partial pressures. It is in this operating régime where the rate of reaction is relatively large and adsorption–desorption equilibrium may not occur. This intermediate state may merely represent the case between the two cases described above, which is obtained as one changes the partial pressures from low to high. However, it is possible for both of the states described above to exist at the same time at intermediate pressure and the state at intermediate coverages represents a third state. Because this third state occurs at intermediate coverages, the rate of reaction r_R will be high. Because $r_a = r_R + r_d$ must be satisfied at steady state, a high rate of reaction implies a low rate of desorption. The relative magnitude of r_R and r_d can be controlled by the surface temperature, T_S. Therefore, we would only expect this intermediate state to exist for certain values of T_S. This is confirmed by the result shown in (37). The parameter Γ depends on the rate constants according to

$$\Gamma = \frac{1}{k_3/k_{-1} + k_3/k_{-2}}. \tag{45}$$

If we assume the usual exponential expressions for the rate constants, we obtain

$$\Gamma = \frac{1}{(\nu_R/\nu_A)\exp[(E_{dA} - E_R)/RT_s] + (\nu_R/\nu_B)\exp[(E_{dB} - E_R)/RT_s]}, \tag{46}$$

where

E_{dA}, E_{dB} are the activation energies for desorption,
E_R is the activation energy for reaction,
ν_R, ν_A, ν_B are the pre-exponential factors for reaction and desorption.

From this expression, one can show that, for reasonable parameter values, there can be no multiple steady states for very large T_S. This will also be the case, independent of T_S, if E_R is larger than both E_{dA} and E_{dB}. However, if E_R is less than either desorption activation energy, then it is possible to reduce Γ below the critical value required for multiple steady states by decreasing the surface temperature. The actual values of the reactant partial pressure and surface temperature for which multiple steady states occur can vary by orders of magnitude due to the sensitivity of (46) to the parameters (and the fact that preexponential factors can vary by orders of magnitude between different systems).

The mechanism by which oscillations occur also involves the vacant site requirement for reaction. For critical values of the partial pressures, the coverage of one of the reactants decreases leading to an increase in the rate of reaction due to the availability of vacant sites. This accelerates the decrease in coverage until the rate of reaction subsides. The large number of vacant sites then increases the rate of adsorption until the surface coverage returns to its previous state to complete the cycle.

Characterization of the different states and bifurcations that a system can exhibit not only gives insight into the possible mechanisms, but also gives valuable information for process control and model discrimination applications (Chang & Chen 1984; Lyberatos *et al.* 1984). For example, it can be seen from figure 3 that if the reaction is started by slowly sending one reactant into the reactor up to its operating partial pressure, followed by the other reactant, then the rate of reaction will be very low corresponding to that part of the reaction rate surface under the protruding 'shark's fin'. If, on the other hand, the reactants are simultaneously brought up to their operating pressure, then the rate of reaction will move along the upper ridge of the 'shark's fin' and the final state of the reactor will be ignited. The final state of the reactor therefore depends on the parameter path used to get there during start up.

This research was sponsored by NSF under grant no. CBT8611542.

REFERENCES

Amundson, N. R. & Aris, R. 1958 *Chem. Engng Sci.* **7,** 212.
Amundson, N. R. & Bilous, O. 1955 *A.I.Ch.E. Jl* **1,** 513.
Bogdanov, R. I. 1975 Versal deformations of a singular point on the plane in the case of zero eigenvalues. *Functional Analysis Appl.* **9,** 144–145.
Bykov, V. I., Ivanova, A. N. & Yablonskii, G. S. 1979 On one class of kinetics models of oscillatory catalytic reactions. *Kinet. Phys. Chem. Oscill.* **2,** 468–476.
Bykov, V. I., Yablonskii, G. S. & Kim, V. F. 1978 *Dokl. Akad. Nauk. SSSR* **242** (3), 637.
Bykov, V. I., Yablonskii, G. S. & Kumbilieva-Buda, K. 1979 *React. Kinet. Catal. Lett., U.S.S.R.* **11** (2), 97.

Bykov, V. I., Yablonskii, G. S. & Kuznetsova, I. V. 1979 *React. Kinet. Catal. Lett., U.S.S.R.* **10** (4), 307.

Chang, H. & Chen, L. 1984. Bifurcation characteristics of nonlinear systems under conventional PID control. *Chem. Engng Sci.* **39,** 1127–1142.

Chow, S. N., Hale, J. K. & Mallet-Paret, J. 1980 An example of bifurcation to homoclinic orbits. *J. diffl Equat.* **37,** 351–373.

Gol'dshtein, V. M., Sobolev, V. A. & Yablonskii, G. S. 1986 Relaxation self-oscillations in chemical kinetics: a model, conditions for realization. *Chem. Engng Sci.* **41** (11), 2761–2766.

Gray, P. & Scott, S. K. 1984 Autocatalytic reactions in the isothermal continuous stirred tank reactor: oscillations and instabilities in the system $A + 2B \rightarrow 3B$; $B \rightarrow C$. *Chem. Engng Sci.* **39** (6), 1087–1097.

Guckenheimer, J. & Holmes, P. 1983 In *Nonlinear oscillations, dynamical systems, and bifurcations of vector fields,* pp. 364–376. New York: Springer-Verlag.

Kevrekidis, I. G. 1988 A numerical study of global bifurcations in chemical dynamics. I. systems with two degrees of freedom. *A.I.Ch.E. Jl* (In the press.)

Kubicek, M. & Marek, M. 1983 In *Computational methods in bifurcation theory and dissipative structures,* pp. 39–47. New York: Springer-Verlag.

Lyberatos, G. Kuszta, B. & Bailey, J. E. 1984 Discrimination and identification of dynamic catalytic reaction models via introduction of feedback. *Chem. Engng Sci.* **39,** 739.

Marsden, J. E. & McCracken, M. 1976 In *The Hopf bifurcation and its applications.* New York: Springer-Verlag.

Takens, F. 1974 Singularities of vector fields. *Publ. Math. IHES* **43,** 47–100.

Takoudis, C. G., Schmidt, L. D. & Aris, R. 1981 Isothermal sustained oscillations in a very simple surface reaction. *Surf. Sci.* **105,** 325.

Takoudis, C. G., Schmidt, L. D. & Aris, R. 1981 Multiple steady states in reaction controlled surface catalyzed reactions. *Chem. Engng. Sci.* **36,** 377.

Yablonskii, G. S. & Bykov, V. I. 1979 *Theor. Exp. Chem., U.S.S.R.* **15** (1), 41.

L.*

FORCED OSCILLATIONS OF A SELF-OSCILLATING BIMOLECULAR SURFACE REACTION MODEL

BY M. A. MCKARNIN†, L. D. SCHMIDT, and R. ARIS

Department of Chemical Engineering and Materials Science, University of Minnesota, Minneapolis, Minnesota, 55455, U.S.A.

(Communicated by P. Gray, F.R.S.—Received 24 *August* 1987 — *Revised 8 December* 1987*)*

The effects of forced oscillations in the partial pressure of a reactant is studied in a simple isothermal, bimolecular surface reaction model in which two vacant sites are required for reaction. The forced oscillations are conducted in a region of parameter space where an autonomous limit cycle is observed, and the response of the system is characterized with the aid of the stroboscopic map where a two-parameter bifurcation diagram for the map is constructed by using the amplitude and frequency of the forcing as bifurcation parameters. The various responses include subharmonic, quasi-periodic, and chaotic solutions. In addition, bistability between one or more of these responses has been observed. Bifurcation features of the stroboscopic map for this system include folds in the sides of some resonance horns, period doubling, Hopf bifurcations including hard resonances, homoclinic tangles, and several different codimension-two bifurcations.

I. INTRODUCTION

Forced oscillation is a well-known technique for the characterization of linear systems and is referred to as a frequency response method in the process control field. By contrast, the response of nonlinear systems to forcing is much more diverse and not yet fully understood. In nonlinear systems, the forced response can be periodic with a period that is some integer multiple of the forcing period (a subharmonic response), or quasi-periodic (characterized by more than one frequency) or even chaotic, when the time series of the response appears to be random. In addition, abrupt transitions or bifurcations can occur between any of these responses as one or more of the parameters is varied and there can be more than one possible response for a given set of parameters depending on the initial conditions or recent history of the system.

* Reprinted with the permission of the editor of the *Proceedings of the Royal Society,* **A,** volume **417,** pp. 363–388, 1988.

† Present address: Shell Development Company, Westhollow Research Center, P.O. Box 1380, Houston, Texas 77251-1380, U.S.A.

The study of forced oscillations in nonlinear systems crosses disciplinary interests and can help in understanding the response of nonlinear electronic devices to AC voltages, of the heart to a pace-maker, the behaviour of RF plasma discharges in low-pressure plasma reactors, the motion of objects subject to mechanical vibrations, and many other important problems. In chemical engineering, there has been considerable interest in the periodic operation of heterogeneously catalysed reactors to increase either their yield or selectivity above what could be achieved by steady-state operation. There have been numerous theoretical studies on the optimization of continuously stirred tank reactors or CSTRS (Codell & Engel 1971; Watanabe *et al.* 1982; Hoffmann & Schadlich 1986) as well as packed bed catalytic reactors (Denis & Kabel 1970; Bailey & Horn 1971, 1972; Lee & Bailey 1974; Cutlip 1979), including chromatographic reactors (Gore 1967). Experimental studies have lagged behind but show great promise. Reaction rate increases as much as 200% or more have been reported for Fischer–Tropsch hydrocarbon synthesis (Silveston *et al.* 1986), acetic acid addition to ethylene on an H_2SO_4-impregnated support (Renken *et al.* 1984), and carbon monoxide oxidation on various supported catalysts (Kareem *et al.* 1980; Cutlip 1979; Zhou *et al.* 1986). Rate increases of 45% are reported for the Haber process of ammonia synthesis (Wilson & Rinker 1981) as well as the oxidation of both butene and cyclohexane on a platinum gauze (Wandrey & Renken 1973). Many other reports of more modest reaction rate enhancements due to concentration cycling exist for oxidation reactions (Unni *et al.* 1973; Renken *et al.* 1976) as well as hydrogenation reactions (Al-Taie & Kershenbaum 1978; Bilimoria & Bailey 1978; Helmrich *et al.* 1974) on a variety of catalytic structures and even includes the enhanced performance of automotive catalytic converters by the high-frequency concentration cycling between CO-rich and O_2-rich feed (Hertz 1982; Metcalfe & Sundaresan 1986).

All of the above studies are concerned with finding a region in parameter space where specific reaction systems can be enhanced. There are very few studies that try to understand the dynamic behaviour or bifurcations of nonlinear chemical reactors to forcing. Some progress has been made for limiting cases such as very low-frequency forcing (quasi-static operation) and very high-frequency forcing (pseudo-steady-state operation) as well as small amplitude perturbations. For a review of these limiting cases and how they relate to reaction rate enhancement, see Bailey (1973, 1977). In addition to better reactor performance, forced oscillations may also find applications in chemical engineering as an analytical laboratory probe for model discrimination, determination of characteristic times and rate limiting steps (Unni *et al.* 1973), and in process control, either for systems with noisy inputs characterized by a primary frequency, or systems downstream from naturally oscillating (or poorly controlled) processes. It may even be desirable to introduce forced oscillations, not to improve a reactor yield or selectivity, but rather to improve its dynamic characteristics. The forced system may be less sensitive to disturbances (Codell & Engel 1971) or may result in a limit cycle with smaller amplitude than the autonomous limit cycle (Fjeld 1972, 1974).

Before such process control strategies can be implemented, either for intentional or unavoidable forced oscillations, one must be able to predict the open-loop response of the chemical reactor to forced oscillations as a function of the forcing parameters. We will examine a simple bimolecular, isothermal surface reaction model represented by a pair of coupled ordinary differential equations in which autonomous oscillations occur. The response of the reaction system, including bifurcations which occur as the forcing parameters are varied, were studied numerically with the aid of the stroboscopic map. This is the third autonomously oscillating model to be studied in this fashion. Kia & Tomita (1979) examined the response of the Brusselator model to forced oscillations and found a region in forcing amplitude and frequency space where concentric ovals of period doublings accumulated upon a relatively large region of deterministic chaos. Later, similar techniques were used by Kevrekidis *et al.* (1986) in the study of forced oscillations in the non-isothermal CSTR for the reaction A → B. In this case the region of parameter space chosen to introduce forced oscillations was much more complicated. Not only was a limit cycle present in the autonomous system, but for near-by parameter values there was a homoclinic bifurcation, a supercritical Hopf bifurcation, a turning point on the periodic branch, and a saddle-node bifurcation. This added degree of complication in the autonomous bifurcation diagram appeared to be carried over into the bifurcation diagram for the stroboscopic map of the forced system even though chaotic solutions were restricted to smaller regions of parameter space than for the Brusselator.

To obtain more information about what kinds of bifurcations are typical for forced oscillators and to further explore any possible relations between the autonomous bifurcation diagrams and the bifurcation diagrams for the stroboscopic map of the forced system, we have chosen to force our model in a region where a limit cycle exists. This not only permits us to compare results with the previous two models, but is a legitimate control concern because autonomous limit cycles have been experimentally observed in so many heterogeneously catalysed systems (see the reviews by Sheintuch & Schmitz (1977) and Slin'ko & Slin'ko (1978)), and in the light of the above studies on reactor enhancement, it is reasonable to ask how such systems would respond to forcing for a set of parameters in which the autonomous system would oscillate naturally. In addition, these studies may provide insight into the problem of coupled oscillators because forced oscillations in a self-oscillating system can be thought of as one-way coupling between two oscillators.

2. SURFACE REACTION MODEL

The model that will be used for forced oscillation studies is one which was first proposed by Takoudis *et al.* (1981) as a simple example of an isothermal surface reaction without coverage dependent parameters in which limit cycles can occur. The bimolecular reaction between species A and B is presumed to occur as a Langmuir–Hinshelwood bimolecular process except that two adjacent vacant sites on the surface are required for the reaction to take place.

The mechanism is represented by

$$\left.\begin{aligned} A + S &\underset{-1}{\overset{1}{\rightleftharpoons}} AS, \\ B + S &\underset{-2}{\overset{2}{\rightleftharpoons}} BS, \\ AS + BS + 2S &\xrightarrow{3} \text{products} + 4S, \end{aligned}\right\} \tag{1}$$

where S is a vacant adsorption site and AS and BS are adsorbed states for the reactants. Because we consider only the fast desorption of products, this vacant-site requirement introduces a reaction step which is autocatalytic in vacant sites and provides the strong feedback required for oscillations to occur. This mechanism yields the following pair of dimensionless ordinary differential equations, where the state variables, θ_A and θ_B, are the surface coverages of the two adsorbed reactants:

$$\left.\begin{aligned} d\theta_A/dt &= \alpha_1(1-\theta_A-\theta_B) - \gamma_1\theta_A - \theta_A\theta_B(1-\theta_A-\theta_B)^2 = f_1, \\ d\theta_B/dt &= \alpha_2(1-\theta_A-\theta_B) - \gamma_2\theta_B - \theta_A\theta_B(1-\theta_A-\theta_B)^2 = f_2. \end{aligned}\right\} \tag{2}$$

The surface coverages are constrained by: $0 \leqslant \theta_A, \theta_B, \theta_A + \theta_B \leqslant 1$, and the dimensionless parameters are

$$\left.\begin{aligned} \alpha_1 &= k_1 P_A/k_3, \quad \alpha_2 = k_2 P_B/k_3, \\ \gamma_1 &= k_{-1}/k_3, \quad \gamma_2 = k_{-2}/k_3, \quad t = k_3\tau, \end{aligned}\right\} \tag{3}$$

in which P_A and P_B are the partial pressures of species A and B respectively. In each equation of (2), the first term represents adsorption of the reactants, the second term is desorption, and the last term represents the reaction step. The parameters α_1 and α_2 are directly proportional to the partial pressure of the reactants above the source and are the primary bifurcation parameters because the other constants are functions only of the surface temperature, which is assumed constant.

The autonomous bifurcation features of this model were examined in detail in a previous paper (McKarnin *et al.* 1988) in which multiple steady states and oscillations were found. This model, although algebraically simple, exhibited a large variety of bifurcations including isola formation, subcritical and supercritical Hopf bifurcation, homoclinic bifurcations, cusp points, and bifurcation points with double eigenvalues of zero. A simplified two-parameter bifurcation diagram for (2) is shown in figure 1*a* for $\gamma_1 = 0.001$ and $\gamma_2 = 0.002$. The curves of Hopf bifurcations extend well below the region of multiple steady states and it is the section of $\alpha_1 = 0.019$ for which the one-parameter bifurcation diagram is shown in figure 1*b*. In that figure, the stable steady state loses stability and a limit cycle is born via supercritical Hopf bifurcations at four points. The parameter that will be forced to oscillate is α_2, whereas α_1 will be held at a value of 0.019.

The centre of the forced oscillation (mean value of α_2) has been chosen as $\overline{\alpha}_2 = 0.028$ which lies in the middle of the second oscillatory region where the autonomous system has a natural period of $T_0 = 911.98$ or frequency of

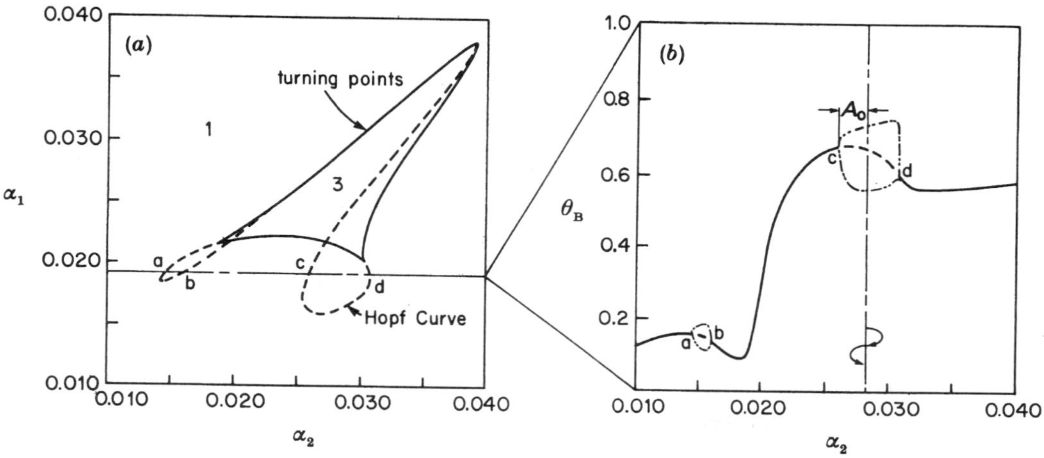

FIGURE 1 Bifurcation diagrams of the autonomous system for $\gamma_1 = 0.001$, $\gamma_2 = 0.002$. (*a*) Multiple-steady states are found inside the finger-shaped region, and limit cycles are born upon crossing the Hopf curves. (*b*) Steady-state and limit-cycle branches for $\alpha_1 = 0.019$. The location of the average value of α_2 used for forced oscillations is 0.028, which is in the oscillatory region and a distance of A_0 from the left Hopf bifurcation point labelled *c*.

$\omega_0 = 2\pi/T_0 = 0.006\,889\,6$. This point is near the midpoint between the two Hopf bifurcation points (labelled c and d in figure 1) located at $\alpha_2 = 0.025\,903$ and $\alpha_2 = 0.030\,728$. The Hopf bifurcation closest to the centre of forcing is the one on the left (labelled c) and, as will be shown later, the distance $A_0 = 0.002\,097$ from the centre of forcing to Hopf point c is a measure of the largest forcing amplitude for which the response of the reaction system can be qualitatively predicted *a priori*. This is not surprising, especially for lower frequencies, because the forcing parameter would begin to visit regions of parameter space where the autonomous system no longer has a limit cycle. The set of non-autonomous or forced differential equations to be studied is then given by

$$\left.\begin{aligned} \mathrm{d}\theta_A/\mathrm{d}t &= \alpha_1(1-\theta_A-\theta_B)-\gamma_1\theta_A-\theta_B(1-\theta_A-\theta_B)^2 = f_1, \\ \mathrm{d}\theta_B/\mathrm{d}t &= (\alpha_2+A\cos(\omega t))(1-\theta_A-\theta_B)-\gamma_2\theta_B-\theta_A\theta_B(1-\theta_A-\theta_B)^2 = f_2. \end{aligned}\right\} \tag{4}$$

3. MATHEMATICAL AND NUMERICAL FRAMEWORK

The mathematical method that will be used to characterize the system response is the stroboscopic map that was used by Kai & Tomita (1979) and Kevrekidis *et al.* (1984, 1986). If a point in the phase plane is used as an initial condition for the integration of the ordinary differential equations (ODES) (4), a trajectory will be traced out. After integrating for one forcing period (from $t = 0$ to $t = T$), the trajectory will arrive at a new point in the phase palne. This new point is defined as the stroboscopic map of the original point and integration for more than one period will generate a series of points or iterates of the map. The result is as if a stroboscope of the same period as the forcing is shined upon the phase plane as a trajectory is integrated. The stroboscopic map is a type of Poincaré map (Guckenheimer & Holmes 1983) for the autonomous three-dimensional system formed from (4) by the introduction of the forcing phase as a third variable, and the Poincaré section is simply taken as the surface of constant phase. If the response of the system to forcing is periodic, then the period must be an integer multiple, pT, of the forcing period. This response will appear as a cycle of p points for the first iterate of the map that will constitute p fixed points for the pth iterate of the map. The stroboscopic map effectively reduces the dimension of attractors in the phase space by one dimension. For example, limit cycles appear in the phase plane as closed trajectories but are reduced to fixed points by the map. Likewise, quasi-periodic trajectories appear in the phase plane as the projection of dense windings on a torus but are reduced to a closed curve by the map. Consequently, the study of the dynamics and bifurcations of the original system of ODES has been reduced to the study of the dynamics and bifurcations of the stroboscopic map.

The conditions for the existence of a pT-periodic solution to the forced system can be written in terms of a fixed point equation for the pth iterate of the stroboscopic map

$$\boldsymbol{F}^p(\boldsymbol{x}, \mu) - \boldsymbol{x} = 0, \tag{5}$$

where $\boldsymbol{F}^p(\boldsymbol{x}, \mu)$ is obtained by integrating (4) from $(\theta_A, \theta_B) = \boldsymbol{x}$ at $t = 0$ to $t = pT$ ($T = 2\pi/\omega$ = forcing period) and μ represents the system parameters $(\alpha_1, \bar{\alpha}_2, \gamma_1, \gamma_2, A, \omega)$. Equation (5) can be solved by using Newton's method where the required jacobian matrix is given by

$$\boldsymbol{J}(\boldsymbol{x}, \mu) = \mathrm{D}\boldsymbol{F}^p(\boldsymbol{x}, \mu) - \boldsymbol{I}, \tag{6}$$

and the matrix $D\boldsymbol{F}^p(\boldsymbol{x}, \mu)$ represents the jacobian of the stroboscopic map is evaluated by integrating the following variational matrix equations

$$\left.\begin{aligned} &\mathrm{d}\boldsymbol{x}/\mathrm{d}t = \boldsymbol{f}(\boldsymbol{x}, \mu, t); \quad \boldsymbol{x}(t=0) = \boldsymbol{x}_0, \\ &\mathrm{d}\boldsymbol{M}/\mathrm{d}t = \frac{\partial \boldsymbol{f}}{\partial \boldsymbol{x}}(\boldsymbol{x}(\boldsymbol{x}_0, \mu, t), \mu, t)\cdot \boldsymbol{M}; \quad \boldsymbol{M}(t=0) = \boldsymbol{I}, \\ &D\boldsymbol{F}^p(\boldsymbol{x}_0, \mu) = \boldsymbol{M}(t = pT). \end{aligned}\right\} \tag{7}$$

The differential equation for $\boldsymbol{M}$ in (7) is non-antonomous and involves evaluation of the jacobian of the forced-model equations at the current value of the trajectory $\boldsymbol{x}(\boldsymbol{x}_0, \mu, t)$ for each time step so that it must be integrated simultaneously with the system equations. Upon convergence on a fixed point, the matrix $\boldsymbol{M}$ becomes the monodromy matrix whose eigenvalues are those of the jacobian of the stroboscopic map evaluated at the fixed point and are called the Floquet multipliers of the periodic solution.

The Floquet multipliers determine the stability and character of the fixed point (or limit cycle) much in the same way as the eigenvalues of the jacobian matrix of a set of ODES determine the stability type of steady-states. A fixed point will be stable if all of the multipliers have magnitude less than unity: $|\lambda_i| < 1$, $i = 1, n$. For a two-dimensional forced system like (4), the fixed points can be classified according to:

$$\begin{aligned}
&\lambda\text{s real and } -1 < \lambda_1, \lambda_2 < 1, \quad \text{stable node,}\\
&\lambda\text{s real and } |\lambda_1|, |\lambda_2| > 1, \quad \text{unstable node,}\\
&\lambda\text{s complex and } |\lambda_1| = |\lambda_2| < 1, \quad \text{stable focus,}\\
&\lambda\text{s complex and } |\lambda_1| = |\lambda_2| > 1, \quad \text{unstable focus,}\\
&\lambda\text{s real and } |\lambda_1| < 1, |\lambda_2| > 1, \quad \text{unstable saddle.}
\end{aligned}$$

If a periodic solution has period equal to pT, then there will be p fixed points for the pth iterate of the stroboscopic map $\boldsymbol{F}^p$ and each fixed point will have precisely the same Floquet multipliers. In addition, the multipliers for the pth iterate of the map are related to those of the first iterate according to: $\lambda_i(\boldsymbol{F}^p) = \lambda_i^p(\boldsymbol{F})$. The analogy between steady-states of ODES and fixed points for maps can be taken yet a set further to include loss of stability in bifurcations. The simple codimensional-one local bifurcations for maps occur when one real multiplier or a pair of complex multipliers cross the unit circle in the complex plane. If a single real multiplier passes through $+1$, the fixed point disappears in a collision between a saddle and a node (saddle-node bifurcation). If a single real multiplier passes through -1, the fixed point changes from a node to a saddle (or vice versa) and spawns a new cycle of periodic points that has the same stability character as the original fixed point before bifurcation (supercritical case) but has double the period. If a pair of complex multipliers cross the unit circle at an angle that is an irrational fraction of 2π a (soft) Hopf bifurcation occurs in which a fixed point (limit cycle) spawns an invariant closed curve (torus) as it changes its own stability; this is directly analogous to the Hopf bifurcation for ODES. If the complex pair of multipliers cross the unit circle at a point where $\lambda^3 = 1$ or $\lambda^4 = 1$, then the bifurcation is more complicated than the standard Hopf and is referred to as a hard resonance and represents a codimension-two bifurcation. If the passage is at $\lambda^q = 1$, $q = 5, 6, \ldots$, then two period q cycles are created on the new invariant circle consisting of saddles and nodes separating one another. This kind of irregular Hopf bifurcation is often called a 'soft' resonance. Other possible local codimension-two bifurcations include cases where both Floquet multipliers leave the unit circle at $+1$ or -1 simultaneously.

Critical fixed points that are undergoing bifurcation can be found by augmenting the set of fixed-point equations (5) with one of these Floquet multiplier conditions:

$$\boldsymbol{F}^p(\boldsymbol{x}, \mu) = 0, \quad G(\boldsymbol{x}, \mu) = 0, \tag{8}$$

where $G(\boldsymbol{x}, \mu)$ is some condition for bifurcation such as $\det(D\boldsymbol{F}^p \pm \boldsymbol{I}) = 0$. This three-dimensional set of equations can then be solved for the bifurcation fixed point and the value of one of the parameters at which the bifurcation occurs. In practice, this will require that the third row of the jacobian matrix for Newton's method be approximated numerically but the first two elements of the third column can be calculated variationally by

$$\left.\begin{aligned}
&\frac{dy}{dt} = \frac{\partial f}{\partial x}(x(x_0, \mu, t), \mu, t) \cdot y + \frac{\partial f}{\partial \mu_i}(x(x_0, \mu, t), \mu, t); \; y(t = 0) = 0, \\
&\frac{\partial F^p}{\partial \mu_i}(x_0, \mu) = y(pT) - \delta(\mu_i)\frac{2\pi p}{\omega^2} f(F^p(x_0, \mu), \mu, pT); \; \delta(\mu_i) \equiv \begin{cases} 0, \mu_i \neq \omega \\ 1, \mu_i = \omega \end{cases},
\end{aligned}\right\} \tag{9}$$

where μ_i is the bifurcation parameter being solved for. These equations can be derived by formally writing the stroboscopic map F^p as the time integral from $t = 0$ to $t = pT$ of the ODEs and then differentiating both sides with respect to the parameter and invoking the chain rule inside the integral. For the case where one of the parameters is the forcing frequency, Leibnitz's rule must be used because the upper limit of integration is $pT = 2\pi p/\omega$. This results in the second contribution to the derivative of F^p in (9) and accounts for the fact that a change in ω not only changes the system of ODEs being integrated, but also changes the length of integration.

4. EXCITATION DIAGRAM

A natural way to characterize the response of a nonlinear system to forcing is to divide the forcing frequency and amplitude parameter plane into the various domains where the system response is qualitatively the same. This is done by making a two-parameter continuation of the various codimension-one bifurcations of the stroboscopic map, so that for each value of the forcing amplitude, (8) can be solved for the bifurcation fixed point as well as the value of the forcing frequency at bifurcation. The resulting two-parameter (forcing amplitude and frequency) bifurcation diagram is called the excitation diagram. To be able to go around turning points while generating these curves in the forcing parameter plane, a modified version of the DERPAR algorithm for true arc-length continuation given by Kubicek & Marek (1983) was used. A fundamental requirement for using arc-length continuation of the solution of (8) is that the jacobian matrix of (8) be augmented by one more column for the derivative of (8) with respect to the continuation parameter. In general, the derivatives of the bifurcation condition G are best computed numerically. The derivatives of F^p with respect to x and both forcing parameters A and ω can be calculated by integrating the variational equations (7) and (9) simultaneously. Note that the jacobian of the model equations $\partial f/\partial x$ is required at each time step for the more accurate integration packages (we used EPISODE), but the variational equations require no extra computations because they are linear with $\partial f/\partial x$ as their coefficient matrix. Although the surface reaction model (4) can be moderately stiff at times, it can be shown that the set of variational equations cannot have eigenvalues spread any further apart than those of (4) so that including the variational equations will not increase the stiffness of the problem.

Because of the enormous amount of integration required, computation costs on conventional computers would have been prohibitive and thus, all of our computations were performed on the Cray 1 and Cray 2 supercomputers at the University of Minnesota. The results are shown in figure 2 where the forcing frequency ω has been scaled by the system's natural frequency ω_0

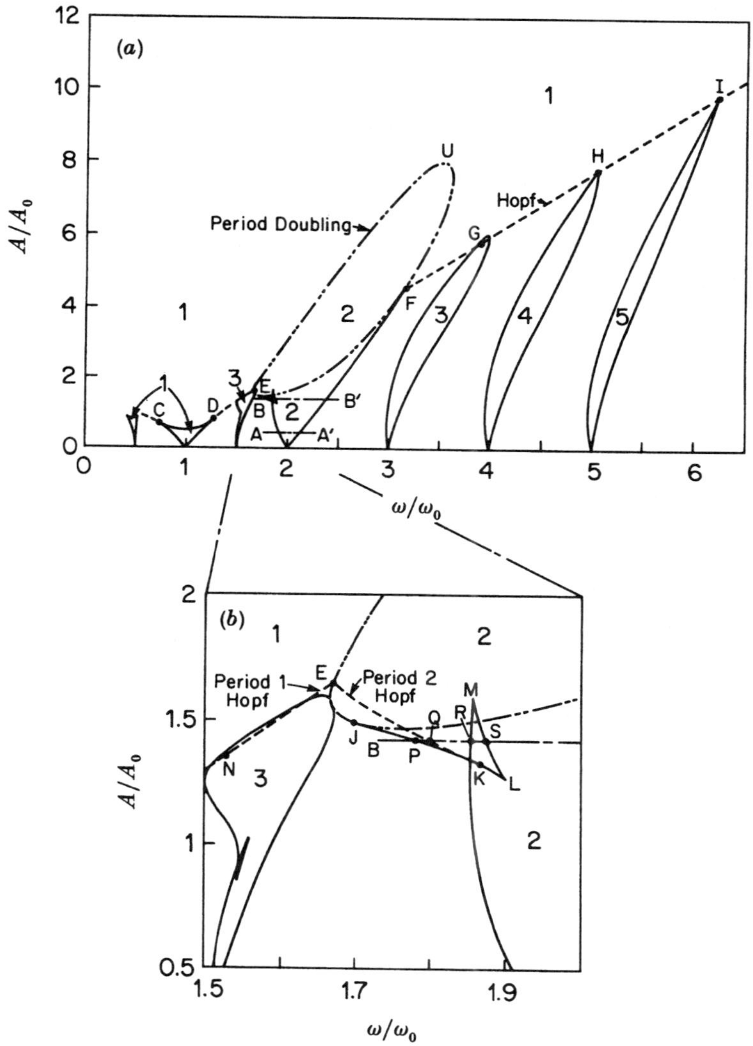

FIGURE 2 The excitation diagram for the model when $\gamma_1 = 0.001$, $\gamma_2 = 0.002$, $\alpha_1 = 0.019$, and α_2 is forced to oscillate sinusoidally with an average of $\alpha_2 = 0.028$, (*a*) The regions numbered *N* indicate the existence of stable periodic responses of period $NT = N2\pi/\omega$. The regions without numbers are composed largely of quasi-periodic responses. (*b*) Enlargement of the region of the excitation diagram in (*a*) near the junction of the 2/1 and 3/2 resonance horns with the period-doubling and Hopf curves.

(frequency of the autonomous limit cycle) and the forcing amplitude A has been scaled by the distance A_0 from $\overline{\alpha}_2$ to the nearest autonomous Hopf bifurcation. In this figure, regions where stable solutions with period p exist are labelled by the appropriate numeral.

(a) Small and Large Forcing Amplitudes

In the limiting case of very large-amplitude forcing, the system is completely dominated by the forcing and the only response possible is a stable limit cycle of period 1 (same period as the forcing). The amplitude at which this becomes the case varies greatly and seems to increase with the forcing frequency.

In the case of low-amplitude forcing, the perturbation methhod of averaging (Hale 1969; Guckenheimer & Holmes 1983) can be used to transform the nonautonomous system into an autonomous averaged system that approximates the dynamics of the stroboscopic map. In general, it is expected that the stable limit cycle surrounding the unstable steady state in the autonomous system will become a stable torus (quasi-periodic trajectories characterized by both the forcing frequency and the natural frequency of the limit cycle) surrounding an unstable limit cycle (corresponding to the unstable autonomous steady state). This is the case in the areas in between the V-shaped regions that occur at low amplitudes in figure 2. As ω/ω_0 approaches some rational number p/q, the system nears a resonance of order p and the system will exhibit a subharmonic solution of period p (p times the forcing period or q times the natural period) and gives fixed points for the pth iterate of the stroboscopic map. This phenomena is also called subharmonic phase locking or entrainment and the region inside the V where it occurs is called a resonance horn. The resonance occurs via period p saddle-node bifurcations where p saddles and p nodes are formed and separate one another on the invariant circle. This means that a stable limit cycle and an unstable limit cycle (with node and saddle stability type, respectively) have developed on the torus. The bifurcation is accompanied by the passing of a Floquet multiplier (eigenvalue of DF^p) through $+1$. The resulting stable oscillation will have q peaks per period in the time plot.

An example is shown in figure 3 for section AA′ near the bottom of the 2/1 resonance horn of figure 2. As the frequency is increased from left to right, the torus becomes phase locked as a pair of period 2 saddle nodes develop on it. The saddle nodes then separate with the saddles alternating with the node and the invariant circle is now composed of the unstable manifolds of the saddles whereas the stable manifolds of the saddles come from the unstable period 1 focus in the middle of the circle and from infinity. As the frequency is increased further, the saddles rotate around the circle and recombine with their neighbouring nodes in another saddle-node bifurcation.

In general, there should be a resonance horn at every rational value of $\omega/\omega_0 = p/q$, but as q increases, the apices become more narrow (Hall 1984) and only the larger ones can be located numerically without great difficulty. The resonance horns shown in figure 2 are for resonances of $p/q = 1/2, 1/1, 3/2, 2/1, 3/1, 4/1$, and $5/1$. This entrainment behaviour has been recently observed experimentally in several forced oscillatory chemical systems (Dolnik

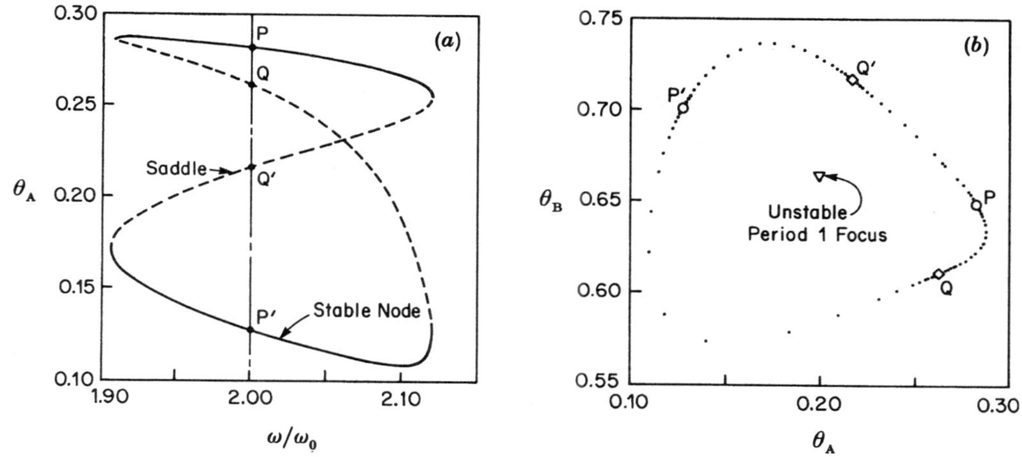

FIGURE 3 (*a*) The low-amplitude section AA′ of the 2/1 resonance horn at $A/A_0 = 0.5$ (see figure 2*a*). (*b*) The stroboscopic phase portrait in the 2/1 resonance horn for $A/A_0 = 0.5$, $\omega/\omega_0 = 2.0$. The period 2 saddles and nodes alternate and the unstable manifolds of the saddles (squares) make up the structure of the phase-locked torus.

et al. 1986; Dulos & DeKepper 1983; Hudson *et al.* 1986; Capsaskis & Kenney 1986; Pugh *et al.* 1986) and has also been studied in connection with cardiac pace-makers (Guevara *et al.* 1981).

(b) Local Codimensional-One Bifurcations

The three standard local codimensional-one bifurcations are the saddle-node, Hopf, and period doubling bifurcations and several have been continued numerically for this model and appear in figure 2. We have chosen not to show the curves of focus-node transitions because they do not represent any changes in stability, only changes in the approach to the steady behaviour. The saddle-node bifurcations that occur during phase locking of the torus at low amplitudes continue upward and either close upon themselves as in the case of the period 3 resonance horns or the terminate in some codimension-two bifurcation.

In the 1/1 entrainment region each side of the resonance horn terminates at points C and D respectively. These points are codimension-two bifurcations and correspond to double +1 multipliers. As the saddle-node curve at the right horn boundary rises from zero amplitude towards point D, one multiplier remains at unity (the criterion for a saddle-node bifurcation) as the other free-multiplier of the saddle-node increases until it is also equal to unity upon arrival at point D. The same thing occurs for the left boundary of the resonance horn. The arc CD is also a saddle-node bifurcation curve but is different from those on the sides of the resonance horn. As arc CD is crossed from below, the period 1 saddle combines not with its companion stable node, but with the unstable node that was in the centre of the phase locked torus. As the pair collides, the invariant circle is lost and only the stable node remains. Exactly the same scenario is observed for the 1/2 resonance horn as well.

Emanating from points C and D are curves of period 1 Hopf bifurcations. The single stable period 1 node that exists above Hopf curve DE becomes a stable focus as it approaches the Hopf curve and then undergoes a supercritical Hopf bifurcation where the focus becomes unstable as the complex pair of multipliers leave the unit circle and spawns a stable torus that grows quadratically near the bifurcation. This is consistent with phase locking because this invariant closed curve surrounding an unstable period 1 fixed point is precisely what exists between the various resonance horns. The bifurcating fixed point lying on curve DE has a complex pair of multipliers that are located on the unit circle at some angle $\pm\theta$. As one moves along curve DE towards endpoint D, θ approaches π where both multipliers become real at the double -1 point E. This double -1 point is yet another codimension-two bifurcation, which will be discussed in detail later. Another period 1 Hopf curve extends from point F through points G and H. F is another double -1 point and, as one moves away from F along the Hopf curve, the angle at which the complex multipliers leave the unit circle decreases from π. The points G and H correspond to angles $\pm\frac{2}{3}\pi$ and $\pm\frac{1}{2}\pi$ respectively and are 'hard' resonances of the Hopf bifurcation because the Floquet multipliers leave the unit circle at third and fourth roots of unity, respectively. Points G and H are both important codimension-two bifurcation points and will be discussed in detail in the next section. The Hopf curves described above are for period 1 fixed points. Subharmonic solutions (fixed points of period greater than one) can also bifurcate to tori via Hopf bifurcations. Such a curve exists for period 2 and extends from point E to K, where it terminates on a period 2 saddle-node curve. The angle at which the complex Floquet multipliers leave the unit circle approaches zero at either point of the curve.

The large oval at the top of the 2/1 resonance horn is a curve of period doublings and figure 4 demonstrates how they occur. Along curve EUF at the top of the oval, a stable period 1 node becomes an unstable saddle while shedding a stable period 2 node. This corresponds to one real multiplier through -1. As one moves along the period doubling curve towards the double -1 point E, the free multiplier also decreases and passes through -1 at E. Therefore, the bifurcation character of curve EJ in figure 4 is different. In this case, the period doublings occur as an unstable period 1 node becomes a saddle and spawns a pair of unstable period 2 nodes. The nodes involved in the bifurcation are now unstable because they are below the period 1 and period 2 Hopf curves, which terminate at E, and the period doubling occurs inside the torus which surrounded the unstable period 1 node before the period doubling. The nature of the period doubling changes once again at point J where a curve of period 2 saddle-node bifurcations hits the period doubling curve. In crossing curve JF from below, a pair of period 2 saddles collides with the unstable period 1 node that lies in the middle of the phase locked torus. A period 1 saddle is left and separates the remaining stable period 2 nodes. Therefore, only the direction of the bifurcating period 2 branch changes at point J (subcritical to supercritical, respectively) whereas the nature of the period 1 fixed point, which doubles, changes at E. Stroboscopic phase portraits are shown in figure 5 for the four regions labelled *a–d* in figure 4. The location of the Floquet multipliers in the complex plane is also shown

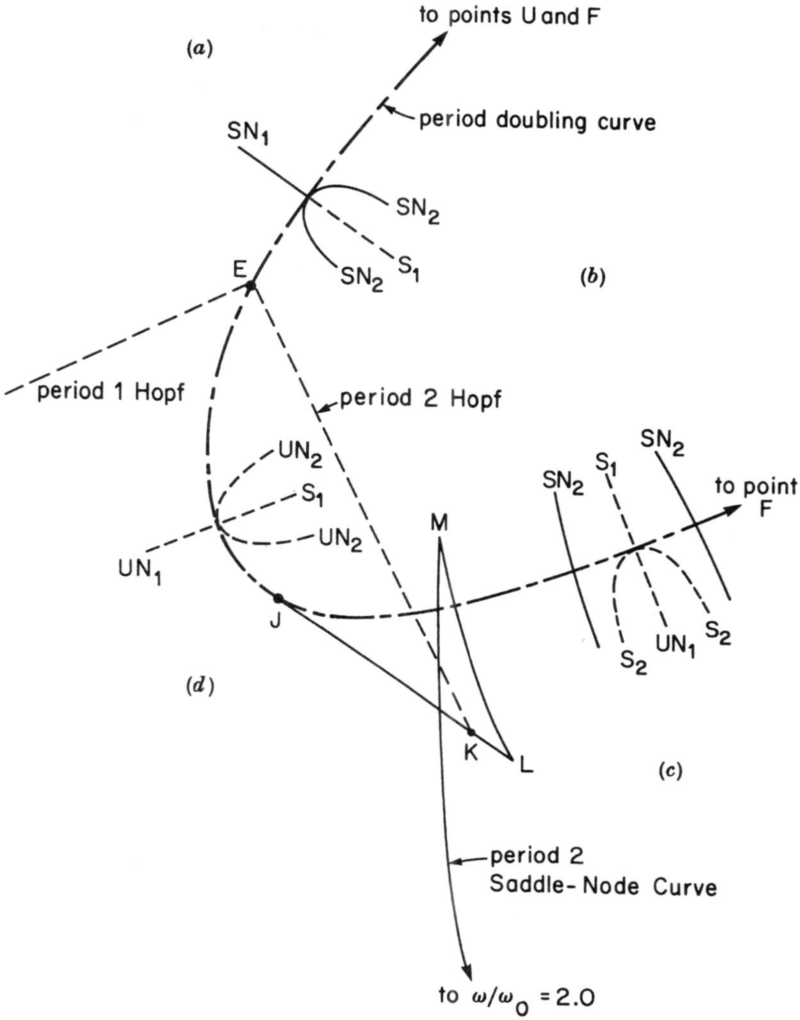

FIGURE 4 Illustration of the three qualitatively different period doublings that occur on the segments FUE, EJ and JF. Point E has two Floquet multipliers at −1 and point J is a metacritical period doubling bifurcation point.

for the period 1 fixed point in each portrait. If one starts at point *a* in figure 4 and goes clockwise through points *b*, *c*, *d* and back to *a* again, the following scenario occurs. The single stable period 1 node becomes a saddle in a period-doubling bifurcation, which spawns two stable period 2 nodes. The period 1 saddle then becomes an unstable node in another period-doubling bifurcation as a pair of period 2 saddles is formed. The period 2 saddles and nodes form an invariant circle surrounding the unstable period 1 node so that the system has a phase locked torus of second-order resonance. Next, the saddles move

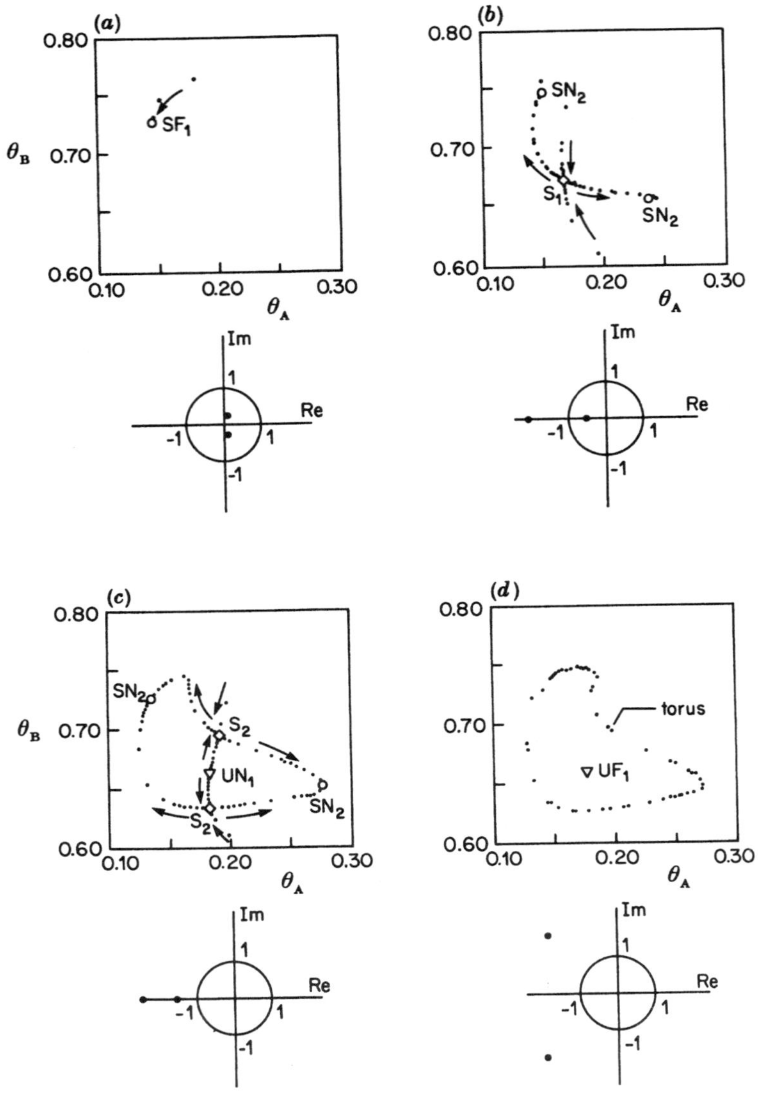

FIGURE 5 Sample phase portraits taken from the regions labelled *a–d* in figure 4. The pair dots around each circle indicate the location of the Floquet multipliers of the period 1 fixed point with respect to the unit circle in the complex plane. (*a*) $\omega/\omega_0 = 1.60$, $A/A_0 = 2.00$; (*b*) $\omega/\omega_0 = 1.90$, $A/A_0 = 2.00$; (*c*) $\omega/\omega_0 = 2.00$, $A/A_0 = 1.45$; (*d*) $\omega/\omega_0 = 1.75$, $A/A_0 = 1.30$.

towards the nodes along the invariant circle and collide together in a period 2 saddle-node bifurcation, which leaves a torus surrounding the unstable period 1 node. The unstable period 1 node then becomes a focus and the torus shrinks around it, collapsing in a Hopf bifurcation. This leaves only a stable period 1 node, which completes the cycle.

(c) Local Bifurcations of Codimension-Two and Three

Several codimension-two bifurcations have already been mentioned. Although they occur in restricted subspaces of parameter space and would therefore be difficult to locate experimentally, their usefulness lies in their role as centres for critical behaviour. Emanating from each local codimension-two point will be two or more of the above codimension-one bifurcation curves. Their usefulness in studying dynamics is akin to that of the triple point in thermodynamic phase equilibria in which boundaries between three different phases come together at a point in a two-parameter diagram. Because some of these codimension-two points have been studied and classified analytically, finding one can provide clues about what other codimension-one bifurcation curves to expect near by and thus aids in the continuation of all of the bifurcation curves in the excitation diagram.

(i) Two Floquet Multiplier at −1

The double −1 points at E and F have been studied by Arnold (1977, 1983) by averaging the forced system in the neighbourhood of the bifurcation. The resulting autonomous system approximates the phase behaviour of the stroboscopic map and predicts two possibilities for the double −1 point. The one that was observed in our model predicts the termination of a period 1 Hopf curve on a period-doubling curve from the side where there is no period 2 solutions. From the other side, a period 2 Hopf curve also terminates at the double − 1 point. Our results for point E in figure 2*b* agree quite well with the above prediction. In addition to the local bifurcation curves mentioned above, the averaged system also predicts two global bifurcation curves extending from the double −1 point on the same side of the period doubling curve as the period 2 Hopf. The first is a homoclinic bifurcation involving both pairs of manifolds of the period 1 saddle crossing and destroying the unstable period 2 torus created by the subcritical period 2 Hopf and leaving an unstable period 1 torus behind. The second curve is a curve of turning points for the period 1 torus branch which has been examined by Chenciner (1985). Bifurcations such as these are very difficult to track because of their global nature (the bifurcations cannot be monitored by the Floquet multiplier of any fixed point) and because of the extreme sensitivity of the structure of the phase portraits. Consequently we could not prove conclusively that all of these predicted global bifurcation curves exist in our model.

(ii) Two Floquet Multipliers at −1 (Bogdanov Points)

The next co-dimension two bifurcation point is the so called Bogdanov point where two real Floquet multipliers are equal to unity such as point K in figure 2*b*. This bifurcation point has been extensively studied by Bogdanov (1975, 1981) as well as Takens (1974). The analytical results predict a Hopf bifurcation curve terminating with quadratic tangency on a saddle-node curve. The node involved in the saddle-node bifurcation is stable on one side of the Hopf curve and unstable on the other. Also terminating at the Bogdanov point is a homoclinic bifurcation curve, which is responsible for the destruction of the torus created by crossing the Hopf curve. This is precisely what is observed for our model at the period 2 Bogdanov point K. The Bogdanov points C and D on the 1/1 resonance horn as well as the analogous ones for the 1/2 resonance horn (figure 2*a*) differ slightly from the one at K. For these points, the curve of Hopf bifurcation terminate at the curve of saddle nodes with quadratic tangency just as before. This time, however, the two parts of the saddle-node bifurcation curve (corresponding to stable and unstable bifurcating nodes respectively) form a cusp instead of a continuous parabola. There is also no curve of homoclinic bifurcations. This is because the saddle-node bifurcations occurring on the right-hand side of the 1/1 resonance horn involve phase locking the torus so that the torus is destroyed in a straightforward way. At the period 2 Bogdanov point K, the saddle-node bifurcations that occur along KL (figure 2*b*) do not create saddle-node pairs that lie on any invariant circle (phase locking of the torus is not involved) and the only way to destroy the period 2 torus created by the Hopf is by a global homoclinic bifurcation.

(iii) Metacritical Period Doubling

Another interesting codimension-two bifurcation is labelled as point J in figure 2*b* where a period 2 saddle-node curve terminates on a period-doubling curve. To our knowledge, no one has studied this bifurcation analytically. However, as was explained in the preceding section, the period-doubling curve is supercritical along EJ, but is subcritical along JF (see figure 4). Therefore, point J corresponds to a metacritical period doubling. Guckenheimer & Holmes (1983) discuss the normal form for the period-doubling bifurcation and the parameter

$$a = \frac{1}{2}\left(\frac{\mathrm{d}^2h}{\mathrm{d}x^2}\right)^2 + \frac{1}{3}\left(\frac{\mathrm{d}^3h}{\mathrm{d}x^3}\right), \tag{10}$$

where $h(x)$ is the one-dimensional map resulting from restricting the stroboscopic map to its centre manifold (the manifold tangent to the eigendirection corresponding to the -1 Floquet multiplier). As the parameter a in (10) changes sign, both the direction and stability of the bifurcating period 2 branch changes. Therefore, the criteria for the metacritical period-doubling point J must be

$$\det\,[\boldsymbol{DF}(\boldsymbol{x}, \mu) + \boldsymbol{I}] = 0, \quad a(\boldsymbol{x}, \mu) = 0. \tag{11}$$

The prediction of the period 2 saddle-node curve terminating at J remains until someone develops a universal unfolding for this co-dimension two bifurcation.

(iv) Saddle-Node Cusp Points

In figure 2*b*, there are clearly folds in the left-hand side of the 3/2 and 2/1 resonance horns. This phenomenon had not (when we observed it) been seen in other forced oscillators such as the Brusselator model (Kai & Tomita 1979) and the non-isothermal CSTR (Kevrekidis *et al.* 1986), although it may have been missed in previous numerical studies that did not use arc-length continuation. It is however also to be found in unpublished work of Marek's group.

The cusp points at M and L are quite different from the 'apparent cusp'

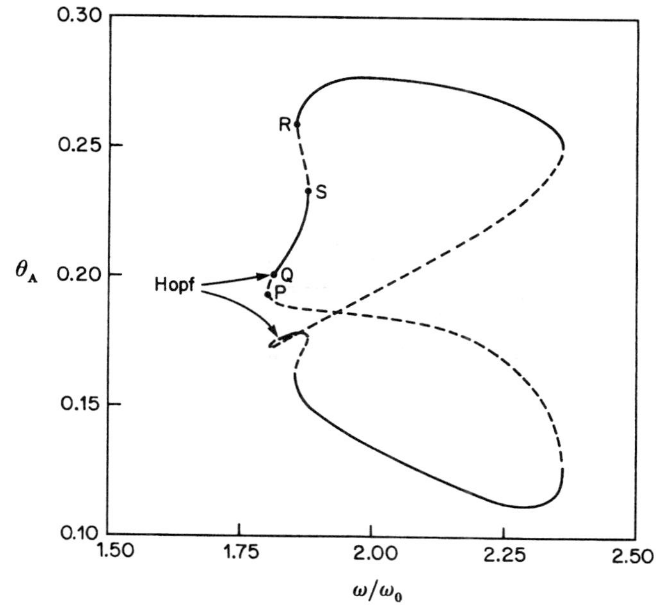

FIGURE 6 The section BB′ of the fold in the left-hand side of the 2/1 resonance horn at A/A_0 = 1.40 (see figure 2). The points P, R, and S are turning points and point Q is Hopf point.

at the Bogdanov points C or D in the 1/1 resonance horn. In the latter case, the two saddle-node curves that meet at a cusp are qualitatively different since the node involved is stable along the right-hand side of the 1/1 resonance horn and unstable along curve CD. Also, the free Floquet multiplier never passes through +1 or −1 at points M and L. A one-parameter cut was made across the fold in the 2/1 resonance horn for $A/A_0 = 1.40$ and is shown in figure 6. It is basically the same as that for low amplitudes (figure 3*a*) except an indentation has developed in the left side of the lobes of the figure-eight and there is now a period 2 Hopf bifurcation labelled *Q*. The three turning points P, R and S correspond to the saddle-node curves JK, M down to the 2/1 point at zero amplitude, and ML, respectively. The folds have the important implications of bistability, because the oscillating reaction can have two different periodic responses, both having a period of $2T$, but one residing on the phase-locked torus (corresponding to the top and bottom branches in figure 6) whereas the other is in the interior of the torus. Figure 7 is an illustration of how the figure-eight cut that exists at low amplitude changes as the amplitude is increased through the fold and the period-doubling region.

(v) Hopf Bifurcations with Hard Resonances

Among all of the points on the period 1 Hopf curve, some will have complex Floquet multipliers λ with a phase angle θ of $\pm(m/n)2\pi$ with $n = 3$ or 4 (i.e. third or fourth roots of unity) and are called hard reasonances. Because these points are fixed points for $\boldsymbol{F}^n$ that have multipliers equal to $\lambda^n = 1$, it is not surprising to find that subharmonic fixed points of period n are involved in addition to the bifurcating period 1 fixed point.

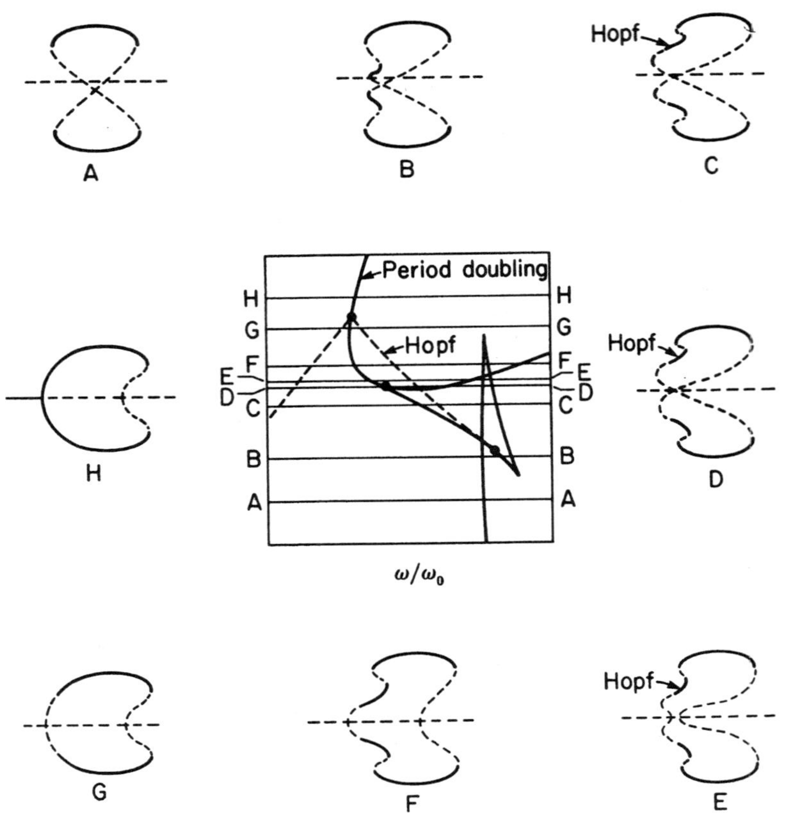

FIGURE 7 Illustration of the metamorphoses of constant amplitude sections of the 2/1 resonance horn as the amplitude is raised from the phase locked region through the fold and period doubling regions. The horizontal line in each of the surrounding figures is a period 1 node whereas the closed curves are of period 2.

The case of $n = 3$ occurs at points N and G in the top of the 3/2 and 3/1 resonance horns of figure 2 and is shown in detail for the tip of the 3/1 resonance horn in figure 8 where the co-dimension two bifurcation point has been labelled F. This bifurcation has been studied by Ioos & Joseph (1983) where it was determined that an unstable subharmonic period 3 solution bifurcates from both sides of the Hopf curve. This means that three period 3 saddles collide with the period 1 fixed point that is undergoing the Hopf bifurcation as shown in section CC′ of figure 8*d*. It was shown by Aronson *et al.* (1986) that this crossing of the saddles is responsible for the change in the saddle-node pairings that occur in one parameter cuts of the period 3 resonance horn, as shown in section AA′ and BB′ of figure 8*b*, *c*. They also made mathematical arguments that led to the conjecture that all period 3 resonance horns should close off smoothly in such a way as to surround the $1\frac{1}{3}$ bifurication point. This has been confirmed for all of the period 3 resonance horns that have been studied in our model as well as the Brusselator and CSTR

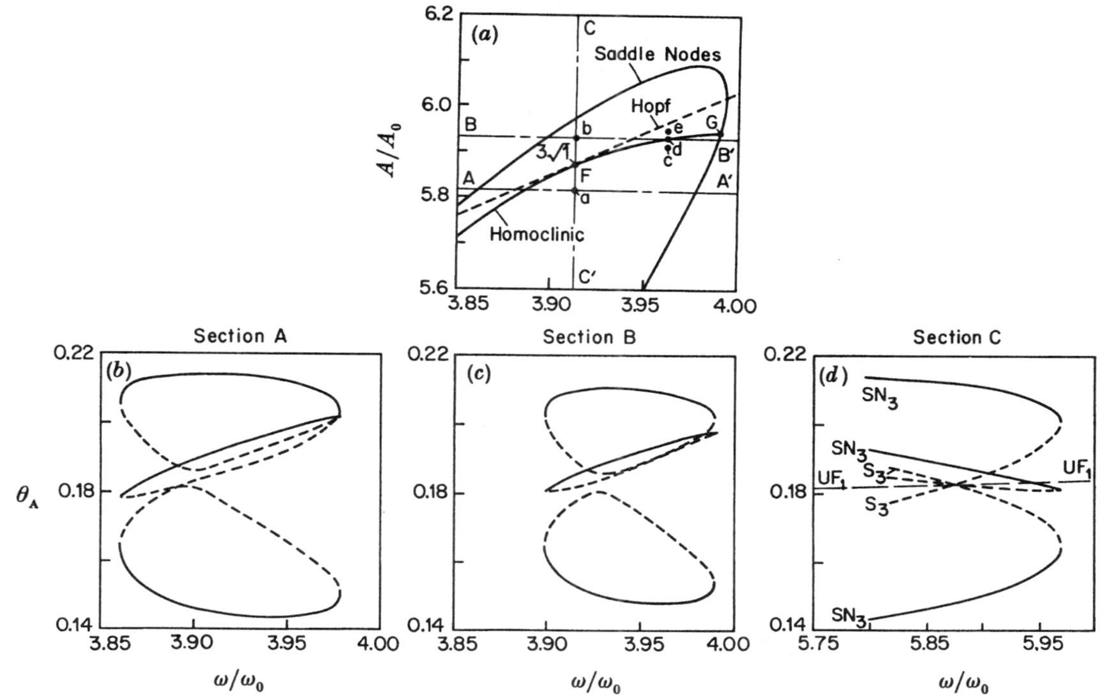

FIGURE 8 (*a*) Detail of the tip of the 3/1 resonance horn illustrating typical way in which period 3 resonance horns close around a point with Floquet multipliers at the third root of unity (point F). (*b*) and (*c*) The saddle-node pairings change from section AA′ to section BB′ and the unstable manifolds of the period 3 saddles no longer make up a phase locked torus. (*d*) A one-parameter vertical cut through the third root of unity point. The three saddles coalesce with the period one focus that is undergoing the Hopf bifurcation.

models. Phase portraits for the system at the points labelled a and b in figure 8*a* are given in figure 9*a*, *b* and show the qualitatively different phase plane structure that exists below and above the third root of unity point, F.

In addition to these local results, Arnold (1974, 1983) has shown that a curve of homoclinic bifurcations for the period 3 saddles touch the Hopf curve at the $1\frac{1}{3}$ point with quadratic tangency as shown in figure 8*a*. It is believed that this curve terminates somewhere on the resonance horn boundaries at points like the one labelled G in figure 8*a*. Therefore, crossing the resonance horn boundary below G will create saddle nodes on the torus thus phase-locking it, whereas crossing the resonance horn above G involves saddle nodes being born off the torus. In this latter case, the stable manifolds of the period 3 saddles divide the phase plane into two regions so that bistability exists between a subharmonic period 3 response and a quasi-periodic response. This curve of homoclinic bifurcations explains how the torus gets from a phase-locked state to a free state so that it can undergo the normal Hopf bifurcation at the period 1 Hopf curve. Phase portraits of the system at the points labelled c, d and e in figure 8*a* are given in figure 9*c*, *d*, *e*. These portraits show how

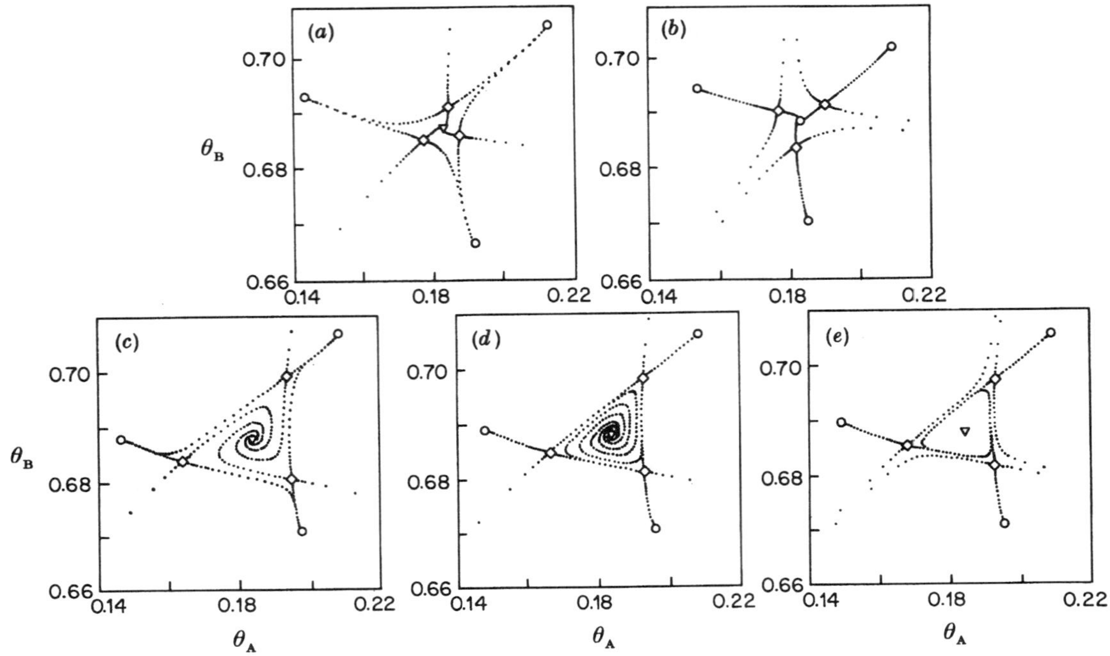

FIGURE 9 Stroboscopic phase portraits for the points on figure 8 labelled (*a*)–(*e*). (*a*) Below the third root of unity point (labelled F in figure 8) the phase portrait is structurally a period three phase locked torus; (*b*) above point F, the period 1 fixed point in the centre is now stable and the phase locked torus has disappeared; (*c*)–(*e*) before, during, and after a period 3 homoclinic bifurcation to the right of point F; $\omega/\omega_0 = 3.97$ for each, and $A/A_0 = 5.90$, 5.93 and 5.95 for (*c*)–(*e*) respectively. The period 3 phase locked torus is transformed to a free torus as the stable manifold of each saddle crosses the unstable manifold of an adjacent saddle.

the system goes from a period 3 phase locked torus to a free torus with the period 3 saddles and nodes on the outside of the torus. The homoclinic process involves the stable manifolds of the saddles crossing the unstable manifolds of an adjacent saddle.

The next hard resonance is for $n = 4$, or the fourth root of unity, and occurs at point H at the top of the 4/1 resonance horn in figure 2. Arnold (1973) and Iooss & Joseph (1980) have also examined this case analytically, but the results are not complete. In general, there can be three possibilities: no period 4 solutions exist on either side of the bifurcation, a set of period 4 saddles bifurcate on both sides of criticality, or period 4 saddles and period 4 nodes (stable or unstable) both bifurcate on one side of the Hopf bifurcation curve. The first and latter cases are what is observed in our model and the $1^{\frac{1}{4}}$ point at H is the mechanism by which the 4/1 resonance horn closes off. If one passes through point H from above along a line that is between the tangents of the tip of the 4/1 resonance horn, then a period four saddle-node bifurcation occurs leaving a phase locked torus. If the line of approach to H is not in between the two tangents, then a simple Hopf bifurcation occurs and no period 4 solutions are involved. The second possibility, of having period

4 saddles on either side of the bifurcation, would correspond to the period 4 resonance horn encircling the $1\frac{1}{4}$ point similar to what was observed for the period 3 horns. Although we know of no cases of period 4 resonance horns closing off smoothly like period 3 horns, it has been discussed as a possibility by Aronson *et al.* (1986).

If the Hopf bifurcation occurs as multipliers leave the unit circle at an angle of $\pm(m/n)2\pi$ for $n \geqslant 5$, then the bifurcation is called a soft resonance. According to Arnold (1977) these bifurcations occur much in the same way as the $1\frac{1}{4}$ Hopf at H except that as n gets larger, the V-shaped resonance region (Arnold tongue) emanating from the Hopf curve gets increasingly narrow and their width at a distance from the Hopf curve is approximately $d^{\frac{1}{3}(n-2)}$. These predictions were confirmed for point I in figure 2 where the 5/1 resonance horn terminates on the Hopf curve. This soft resonance represents the way in which all p/q resonance horns terminate for $p \geqslant 5$.

5. DISCUSSION

The excitation diagram in figure 2 represents a forcing phase diagram in which the numerically computed boundaries separate domains of qualitatively different phase portraits. In principal, one can use such a diagram to predict the surface reaction system behaviour for any given forcing amplitude and frequency. The diagram we have shown, however, is only representative of the systems response to forcing in the small region of the parameters (α_1, α_2, γ_1, γ_2) where the system had an autonomous limit cycle. There are many feature of the excitation diagram that should be the same for forcing in every autonomous oscillating region (i.e. resonance horns) as well as some which are conjectured to be the same for forcing near a Hopf bifurcation such as the curve CD of saddle nodes in the 1/1 resonance horn and the period doubling curve EJF in the 2/1 resonance horn. Further connection between features in the excitation diagram and features in the autonomous bifurcation diagram remain to be made.

The excitation diagram was found to contain saddle-node, Hopf, period doubling, and homoclinic bifurcations for the stroboscopic map. In addition, many of these co-dimension one bifurcation curves were found to meet at the following co-dimension two bifurcation points: Bogdanov points (double +1 multipliers), points with double -1 multipliers, points with multipliers at $\pm1\frac{1}{3}$ and $\pm1\frac{1}{4}$, metacritical period-doubling points, and saddle-node cusp points.

A feature that, to our knowledge when we discovered it, had not been seen before in forced oscillators (Marek and his co-workers have also observed it (M. Marek, personal communication)) is the folding that occurs in the left side of the 3/2 and 2/1 resonance horns. Within these folds there are two sets of stable nodes and two sets of saddles, so that bistability between the two sets exists. There are also cases of bistability between subharmonic responses of period 3 and a torus in the top of the period 3 resonance horns. In addition to the implication of bistability, the fold in the side of the 3/2 resonance horn may be of mathematical significance. Aronson *et al.* (1986) put forth the mathematical conjecture that if the period 3 resonance horn is a simple disc-like resonance region that contains no other period 3 saddle-node points and whose boundary (except for the tip at zero forcing amplitude) contains only simple saddle-node bifurcation points, then there must be a period 1 fixed point in the interior of the resonance horn that has Floquet multipliers equal to the cube root of unity. Clearly, the 3/2 resonance horn in our model has non-elementary saddle-node bifurcation points on its boundary (cusp points) and consequently, there are other period 3 saddle-node bifurcations that exist in the interior of the resonance horn. This violates the conditions of the above conjecture, yet, just as Aronson *et al.* proposed, we have a $1\frac{1}{3}$ point within the 3/2 resonance horn. Evidently then, the conjecture may provide sufficient but not necessary conditions and it may be possible to relax some of the conditions of the conjecture.

Another point of interest in the excitation diagram of figure 2*a* is the point F where it appears that the period doubling, Hopf, and period 2 saddle-node curves all meet. This is a structurally unstable situation and if the region is greatly enlarged, it can be seen that point F really consists of three points: a double -1 point, a period 2 double $+1$ point, and a metacritical period-doubling point. In addition, there is a period 2 Hopf curve connecting the first two points above. These are exactly the same as points E, J and F (figure 2*b*). Therefore, we believe that the apparent single point F would result from the three distinct points if one of the other parameters were perturbed. This would be a codimensional-three bifurcation, and by using the labels for the three bifurcation points in figure 2*b*, would involve points E and J sliding the period 2 saddle-node curve. The period 2 Hopf curve connecting E and K would shorten until finally the triangle EJK would condense into a single point and the period 2 Hopf curve would have disappeared completely. Because we know of no analytical studies on this co-dimension three bifurcation, the above scenario is only conjecture based upon our numerical observations and may be incomplete. Perhaps these conclusions will provide insight for someone to take a serious analytical look at the bifurcation.

We have mentioned several times that curves of global homoclinic bifurcations exist for our excitation diagram (see figure 9*c*–*e* for before, during, and after phase portraits). This bifurcation involves the stable and unstable manifolds of the saddle crossing, with the net result that a torus is created or destroyed. Although this bifurcation is very similar to homoclinic bifurcations in continuous ODE systems, there are very important differences for maps. Unlike continuous systems, maps cannot have a simple homoclinic saddle loop. Instead, the stable and unstable manifolds must become 'wiggly' and touch each other with quadratic tangency. As the parameter is perturbed further, the manifolds intersect transversely, and since the manifolds are invariant by definition, the intersection of the two sets must also be invariant and there must be an infinite number of transversal intersection points (one intersection point can be mapped forwards and backwards in time creating more intersection points). The intersections accumulate along both direction of approach to the saddle and the structure is a mess that is often called a homoclinic tangle. Although the phase portrait shown in figure 9*d* is very close to the homoclinic bifurcation, it is still too far away to see the homoclinic tangle.

The Smale–Birkoff Homoclinic Theorem for maps (Guckenheimer &

Holmes 1983) states that when the above transversal homoclinic intersection occurs, that there is a structurally stable invariant Cantor set like the one for the Horseshoe map. It has also been shown by Holmes (1982) that this invariant set contains a countable, dense set of saddles of all periods, an uncountable set of non-periodic trajectories and a dense orbit. If nothing else is clear from the above, it is at least certain that homoclinic bifurcations for maps are accompanied by some very unusual phase portraits. Even if homoclinic bifurcations are not necessarily accompanied by the formation of stable chaotic attractors, they lend themselves to extremely long chaotic like transients before settling down to a periodic motion. Because there are large numbers of saddles present, their stable manifolds divide up the phase plane into tiny stability regions and extreme sensitivity to perturbations is expected.

An example of an apparently chaotic attractor is shown in figure 10 for parameter values that lie in the period 2 resonance horn just to the left of the period 2 Hopf curve shown in figure 2*b*. Although we have made no attempt to calculate the Liapunov exponents of the time series (Wolf & Swift 1984), it is obvious from the time series and its stroboscopic phase portrait that we have a chaotic attractor. The time series has three different timescales of interest. At short time scales, the time series appears to be random or chaotic. At larger time scales such as the one shown in figure 10*b*, the time series appears to be periodic with a period of 21 times the forcing period (the repeating pattern is indicated by the arrows). At even larger timescales, however, the time series appears to be random again because the near period 21 part of the time series eventually drifts into a random pattern for a short period of time until the period 21 pattern reappears.

It was found that the near period 21 part of the time series was very close to a period 21 saddle, which is located around the perimeter of the choatic attractor as shown in figure 10*a*. The chaotic attractor can be thought of in three dimensions as an attracting sheet of infinitesimal thickness and infinite extent wadded up into a finite volume so that the attractor folds over on itself many times. The stroboscopic phase portrait shown in figure 10*a* is a section of the attractor. What we believe is happening then, is that a trajectory wandering on the chaotic attractor eventually gets folded back near the stable manifold of one of the period 21 saddles. The orbit of the map then slowly approaches the saddle along this stable manifold with successive iterates becoming closer together. Eventually, because the trajectory is not precisely on the stable manifold, the orbit will move away from the saddle along the unstable manifold at an increasing rate. The chaotic attractor will then fold the trajectory back near a stable manifold and the process repeats. This explains the long lengths of time in which the trajectory appears to be periodic with period 21 as well as the short bursts of randomness in between. All of the other fixed points that have been found are unstable and period 2 except for the unstable period 1, which is in the very middle. The unstable manifolds of the period 2 saddles that lie to the inside of the chaotic attractor all lead to the chaotic attractor. Although these homoclinic bifurcations are generally restricted to small regions of parameter space, they are very common and are expected to occur in nearly every forced oscillator.

Chaotic attractors can occupy large regions of parameter space as in the

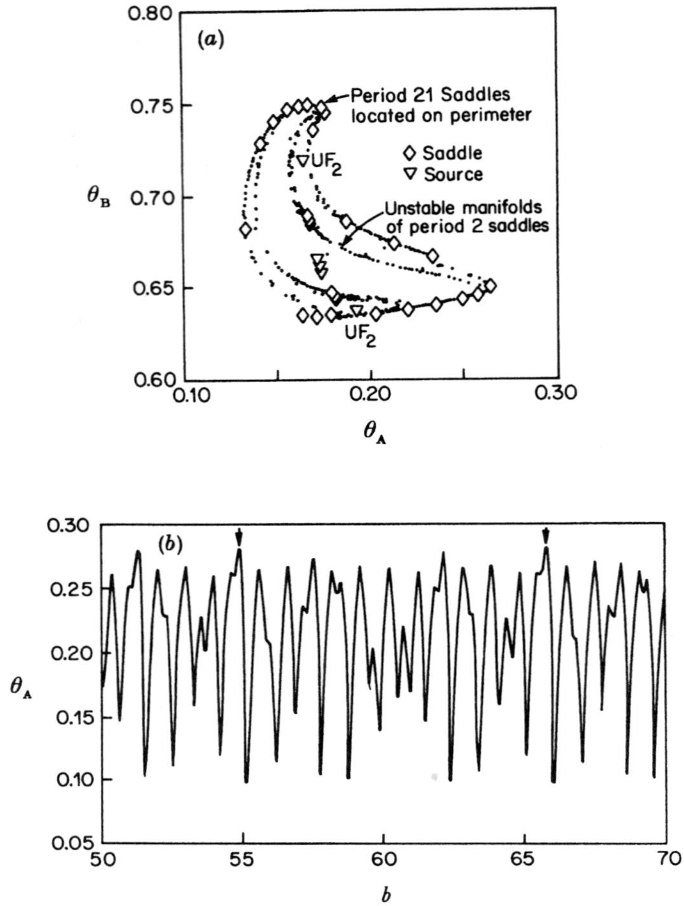

FIGURE 10 Example of chaos for A/A_0 1.45, $\omega/\omega_0 = 1.75$. (*a*) Stroboscopic phase protrait showing the intricate structure of the attractor. The diamonds are saddles and the triangles are sources, and no stable fixed points have been found. (*b*) The time series for a chaotic trajectory after 150 periods of forced oscillations. The arrows indicate a near periodic solution with period 21. The periodicity eventually slips into short random behaviour followed by long near period behaviour. This near periodicity reflects the fact that the chaotic attractor forces the trajectory to eventually pass near the stable manifolds of the period 21 saddle located around the perimeter of the chaotic attractor.

forced Brusselator model (Kai & Tomita 1979). They found a chaotic attractor inside a large elliptical region of the excitation diagram that was the limit of an infinite number of large concentric ovals of period doubling. This was a two-parameter manifestation of the period-doubling cascade to chaos studied by Feigenbaum (1983). Period-doubling cascades to chaos have also been experimentally observed in forced chemical reaction systems (for example, see Lamba & Hudson 1987). The large oval of period doublings in our model appears to be simple and no other period doublings were found within it. The similarity of this region with that of the Brusselator is so striking, that it may be possible for other period doublings to develop inside the region as another parameter, such as the centre of forcing, is varied.

ACKNOWLEDGMENTS

This research was sponsored by NSF under Grant No. CBT861542 and by the Minnesota Supercomputer Institute.

REFERENCES

Al-Taie, A. S. & Kershenbaum, L. S. 1978 Effect of periodic operation on the selectivity of catalytic reactions. *A.C.S. Symp. Ser.* **65,** 512–525.

Arnold, V. I. 1973 *Ordinary differential equations.* Cambridge: M.I.T. Press.

Arnold, V. I. 1977 Loss of stability of self-oscillations close to resonance and versal deformations of equivalent vector field. *Func. Anal. Applic.* **11**(2), 85–92.

Arnold, V. I. 1983 *Geometrical methods in the theory of ordinary differential equations.* New York: Springer-Verlag.

Aronson, D., McGehee, R., Kevrekidis, I. & Aris, R. 1986 Entrainment regions for periodically forced oscillators. *Phys. Rev.* A **33**(3), 2190–2192.

Bailey, J. E. 1973 Periodic operation of chemical reactors—a review. *Chem. Engng Commun.* **1,** 111–124.

Bailey, J. W. 1977 In *Chemical reactor theory. A review* (ed. L. Lapidus & N. R. Amundson). Englewood Cliffs, New Jersey: Prentice-Hall.

Bailey, J. E. & Horn, F. 1971 Improvement of the performance of a fixed bed catalytic reactor by relaxed steady-state operation. *A.I.Ch.E. Jl* **17,** 550.

Bailey, J. E. & Horn, F. 1972 Cyclic operation of reaction systems: the influence of diffusion on catalyst selectivity. *Chem. Engng Sci.* **27,** 109.

Bilimoria, M. R. & Bailey, J. E. 1978 Dynamic studies of acetylene hydrogenation on nickel catalyst. *A.C.S. Symp. Ser.* **65,** 526–536.

Bogdanov, R. I. 1975 Versal deformations of a singular point on the plane in the case of zero eigenvalues. *Func. Anal. Applic.* **9,** 144–145.

Bogdanov, R. I. 1981 Versal deformation of a singularity of a vector field on the plane in the case of zero eigenvalues. *Selecta Mathematica Sovietica* **1**(4), 389–421.

Capsaskis, S. C. & Kenney, C. N. 1986 Subharmonic response of a heterogeneous catalytic oscillator, the 'Cantabrator', to a periodic input. *J. Phys. Chem.* **90,** 4631–4637.

Chenciner, A. 1985 Hamiltonian-like phenomena in saddle-node bifurcations of invariant curves for plane diffeomorphisms. In *Singularities and dynamical systems* (ed. S. N. Puevmatikos). Amsterdam: Elsevier Science Publishers/North Holland.

Codell, R. B. & Engle, A. J. 1971 A theoretical study of a controlled-cycled stirred tank reactor. *A.I.ChE. Jl* **17**(1), 220.

Cutlip, M. B. 1979 Concentration forcing of catalytic surface rate processes. *A.I.ChE. Jl* **25**(3), 502.

Denis, G. H. & Kabel, R. L. 1970 The effect of temperature changes on a tubular heterogeneous catalytic reactor. *Chem. Engng Sci.* **25,** 1057–1071.

Dolnik, M., Schrieber, I. & Marek, M. 1986 Dynamic regimes in a periodically forced reaction cell with oscillatory chemical reaction. *Physica* **21**D, 78–92.

Dulos, E. & DeKepper, P. 1983 Experimental study of synchronization phenomena under periodic light irradiation of a nonlinear chemical system. *Biophys. Chem.* **18,** 211–223.

Feigenbaum, M. J. 1983 Universal behaviour in nonlinear systems. *Physica* **7**D, 16–39.

Fjeld, R. J. 1972 Asynchronous Quenching. In *Periodic optimization,* vol. 2 (ed. A. Marzollo). New York: Springer-Verlag/CISM.

Fjeld, R. J. 1974 Relaxed controls in asynchronous quenching and dynamical optimization *Chem. Engng Sci.* **29,** 921.

Gore, F. 1967 Performance of a chromatographic reactor in cyclic operation. *Ind. Engng Chem. Process Des. Develop.* **6,** 10.

Guckenheimer, J. & Holmes, P. 1983 *Nonlinear oscillations, dynamical systems, and bifurcations of vector fields.* New York: Springer-Verlag.

Guevara, M., Glass, L. & Shrier, A. 1981 Phase locking, period-doubling bifurcations and irregular dynamics in periodically stimulated cardiac cells. *Science, Wash.* **24,** 1350.

Hale, J. K. 1969 *Ordinary differential equations.* New York: Wiley.

Hall, G. R. 1984 Resonance zones in two-parameter families of circle homeomorphisms. *SIAM Jl Math. Anal.* **15**(6), 1075–1081.

Helmrich, A., Renken, A. & Schugerl, K. 1974 *Chemie Ing. Techn.* **46,** 647.

Herz, R. K. 1982 The dynamic behaviour of three way automotive catalyst. *A.C.S. Symp. Ser.* **178,** 59–78.

Hoffman, U. & Schadlich, H. K. 1986 The influence of reaction orders and of changes in the total number of moles on the conversion in a periodically operated CSTR. *Chem. Engng Sci.* **41**(11), 2733–2738.

Holmes, P. J. 1982 The dynamics of repeated impacts with a sinusoidally vibrating table. *J. Sound Vib.* **84,** 173–189.

Hudson, J. L., Lamba, P. & Mankin, J. L. 1986 Experiments on low amplitude forcing of a chemical oscillator. *J. Phys. Chem.* **90,** 3430–3434.

Iooss, G. & Joseph, D. 1980 *Elementary stability and bifurcation theory.* New York: Springer-Verlag.

Kai, T. & Tomita, K. 1979 Stroboscopic phase portrait of a forced nonlinear oscillator. *Prog. Theor. Phys.* **61**(1), 54–73.

Kareem, A. K., Jain, A., Silveston, P. L. & Hudgins, R. R. 1980 Harmonic behaviour of the rate of catalytic oxidation of CO under cycling conditions. *Chem. Engng Sci.* **35,** 273–282.

Kevrekidis, I. G., Schmidt, L. D. & Aris, R. 1984 On the dynamics of periodically forced chemical reactors. *Chem. Engng Commun.* **30,** 323–330.

Kevrekidis, I. G., Aris, R. & Schmidt, L. D. 1986 The stirred tank forced. *Chem. Engng Sci.* **41**(6), 1549–1560.

Kubicek, M. & Marcek, M. 1983 *Computational methods in bifurcation theory and dissipative structures.* New York: Springer-Verlag.

Lamba, P. & Hudson, J. L. 1987 Experiments on bifurcations to chaos in a forced chemical reactor. *Chem. Engng Sci.* **42,** 1–8.

Lee, C. & Bailey, J. E. 1974 Diffusion waves and selectivity modifications in cyclic operation of a porous catalyst. *Chem. Engng. Sci.* **29,** 1157.

McKarnin, M. A., Aris, R. & Schmidt, L. D. 1988 Autonomous bifurcations of a simple bimolecular surface reaction model. *Proc. R. Soc. Lond.* A **415,** 363–387.

Metcalfe, I. & Sundaressan, S. 1986 Oxygen storage in automobile exhaust catalyst. *Chem. Engng Sci.* **41**(4), 1109–1115.

Pugh, S. A., Schell, M. & Ross, J. 1986 Effects of periodic perturbations on the oscillatory combustion of acetaldehyde. *J. Chem. Phys.* **58**(2), 868–878.

Renken, A., Truffer, M. & Dettmer, M. 1984 The effect of cyclic operation on heterogeneous catalytic reaction with educt inhibition. In *ISCRE 8. 8th Inst. Symp. on Chem. React. Engng.* European Federation of Chemical Engineering: Pergamon Press.

Renkin, A., Mueller, M. & Wondrey, C. 1976 In *Proc. 4th Int., 16th Eur. Symp. on Chem. React. Engng.* Frankfurt: Dechma.

Sheintuch, M. & Schmitz, R. A. 1977 Oscillations in catalytic reactions. *Catal. Rev., Sci. Engng* **15,** 107–172.

Silveston, P. L., Hudgins, R. R., Adesina, A. A., Ross, G. S. & Feimer, J. L. 1986 Activity and selectivity control through periodic composition forcing over Fischer-Tropsch catalysts. *Chem. Engng Sci.* **41**(4), 923–928.

Slin'ko, M. G. & Slin'ko, M. M. 1978 Self-oscillations of heterogeneous catalytic reaction rates. *Catal. Rev. Sci. Engng* **17,** 119–153.

Takens, F. 1974 Singularities of vector fields. *Publs Math. Inst. ht. Étud. Scient.* **43,** 47–100.

Takoudis, C. G., Schmidt, L. D. & Aris, R. 1981 Isothermal sustained oscillations in a very simple surface reaction. *Surf. Sci.* **105,** 325.

Unni, M., Hudgins, R. & Silveston, P. 1973 The influence of cycling on the rate of oxidation of SO_2 over a vanadia catalyst. *Can J. Chem. Engng* **51,** 623.

Wandrey, C. & Renken, A. 1973 Zur Beeinflussung der Produktverkeiling durch periodisch Konzentrationsschwankungen bei der Oxidation von Hohlenwasserstoffen. *Chemie. Ing. Techn.* **45,** 854.

Watanabe, N., Kurimoto, H., Matsubara, M. & Onogi, K. 1982 Periodic control of continuous stirred tank reactors. II. Cases of a non-isothermal single reactor. *Chem. Engng Sci.* **37,** 745–752.

Wilson, H. & Rinker, R. 1981 Concentration forcing in ammonia synthesis. I. Controlled cyclic operation. *Chem. Engng Sci.* **37,** 343–355.

Wolf, A. & Swift, J. 1984 Progress in computing Lyapunov exponents from experimental data. In *Statistical physics and chaos in fusion plasmas.* New York: Wiley.

Zhou, X., Barshad, Y. & Gulari, E. 1986 CO oxidation on Pd/Al_2O_3. Transient response and rate enhancement through forced concentration cycling. *Chem. Engng Sci.* **41**(5), 1277–1284.

10
MASS AND HEAT TRANSFER

INTRODUCTION

Reprints M and N are representative of the questions in heat and mass transfer that I have considered at various times. A full list is contained in the index to the bibliography and it is pointless to repeat it here. M was written for a *festschrift* honoring Clifford Truesdell on one of his anniversaries; [156]. I felt it a pity that there should be no paper in Latin in such a circumstance, for as the founding editor of the *Archive for Rational Mechanics and Analysis,* he deliberately kept the door open for papers in the erstwhile *lingua franca* of natural philosophy.[1] No doubt publishing it in Latin is the best way of burying a paper, especially as the kindest critic could only have compared the rusticity of its Latin to that of a monastery in Lower Saxony in the Dark Ages. However, the problem is an interesting one, and I half-resurrected it as an appendix to my Hougen lectures at Madison. The tension I felt in doing the problem was that between the threeness of the semi-regular hexaga and the two-dimensionality of the plane they cover. The resolution comes about through the use of a generating function in two dummy variables, μ and ν, and in the way in which the three ciphers, μ, ν, and $\mu\nu$ play equal roles. In the limit in which the passages become more fine, a comparison can be made with the anisotropic continuum, and the two-by-two matrix of anisotropic diffusivities drops out in an expression symmetric in the three quantities defining the hexaga.

The second paper in this section, Reprint N, is typical of the happy collaboration with Ed Cussler that I have enjoyed since the late 1970s. His

[1] I recall Truesdell once referring to "that ignorance of Latin, which in our father's generation was taken for independence of mind, and which in our own generation may be taken for granted."

supervision of Jem's work [228], was over, and Jem himself had left when we noticed that we could wrap up the design of anisotropic membranes with a few simple graphs and formulae. The exponential dependence was the obvious case to use as a vehicle for the exposition, but it is shown how the method is not limited to this. The index may be used to locate other papers on membranes and mass transfer.

M.*

AN EXAMPLE OF THE RELATION BETWEEN DISCRETE AND CONTINUOUS MODELS†

Some bodies—*pace* Lucretius—appear to be continuous; others, discrete. Yet other bodies or structures, though discrete (i.e., formed with distinguishable parts) are so complicated by they demand equivalent continua for their description, and the properties of these equivalent continua are of interest. A structure of this kind is the catalytic monolith, consisting in a single piece of ceramic through which passes a multitude of narrow parallel holes. The most general shape of the cross-section is a hexagon, and a question of great interest concerns the transfer of heat in the monolith. For, if hot water were to be introduced into one passage and cold into all the others, the hot stream would emerge colder and the cold streams warmer, according to their proximity. An exact and general solution of this problem can be given, and in the limit of the passage size becoming very small this solution goes over into a solution of the equation of heat. This limit allows a matrix of conductivities of an equivalent continuum to be calculated.

THE GEOMETRY OF THE HEXAGA

Any three vectors, $\mathbf{a}$, $\mathbf{b}$, $\mathbf{c}$, restricted only by the conditions

$$\begin{aligned} \lambda(\mathbf{a} - \mathbf{b} + \mathbf{c}) &\neq \mu\mathbf{a} \\ \lambda(\mathbf{a} - \mathbf{b} + \mathbf{c}) &\neq \mathbf{a} - \mu\mathbf{b} \\ \lambda(\mathbf{a} - \mathbf{b} + \mathbf{c}) &\neq \mathbf{a} - \mathbf{b} + \mu\mathbf{c} \end{aligned} \tag{1}$$

for any numbers λ, μ, $0 < \lambda, \mu < 1$, define a hexagon whose vertices with respect to the center are $\pm \tfrac{1}{2}(-\mathbf{a} + \mathbf{b} + \mathbf{c})$, $\pm \tfrac{1}{2}(\mathbf{a} - \mathbf{b} + \mathbf{c})$, and $\pm \tfrac{1}{2}(\mathbf{a} + \mathbf{b} - \mathbf{c})$, and whose sides are the vectors, a, $-$b, c, $-$ a, b, $-$c (Fig. 1). Let ϵ^2 be the area of the hexagon

$$\epsilon^2 = |\mathbf{b} \wedge \mathbf{c}| + |\mathbf{c} \wedge \mathbf{a}| + |\mathbf{a} \wedge \mathbf{b}|. \tag{2}$$

and let us write $\mathbf{a} = \boldsymbol{\alpha}\epsilon$, $\mathbf{b} = \boldsymbol{\beta}\,\epsilon$, $\mathbf{c} = \boldsymbol{\gamma}\epsilon$. The vectors from one center to the centers of contiguous hexaga are

$$\pm\mathbf{A}\epsilon = \pm\,(\boldsymbol{\beta} - \boldsymbol{\gamma})\epsilon,\ \pm\mathbf{B}\epsilon = \pm\,(\boldsymbol{\gamma} - \boldsymbol{\alpha})\epsilon,\ \pm\,\boldsymbol{\Gamma}\epsilon = \pm\,(\boldsymbol{\alpha} - \boldsymbol{\beta})\epsilon. \tag{3}$$

* Reprinted with the permission of James Dumesic, Chair of the Department of Engineering at the University of Wisconsin, to whom the Regents deputed the copyright. From *Chemical Engineering in the University Context: Four Lectures in Honor of Olaf Hougen Given in the Department of Chemical Engineering, University of Wisconsin–Madison, 1979,* pp. 107–116. Madison, WI: Board of Regents of the University of Wisconsin System, 1982.

† This appendix is a translation of part of "De exemplo simulacrorum continuorum discritalumque," *Arch. Rat. Mech. Anal. 70,* 203–9 (1979).

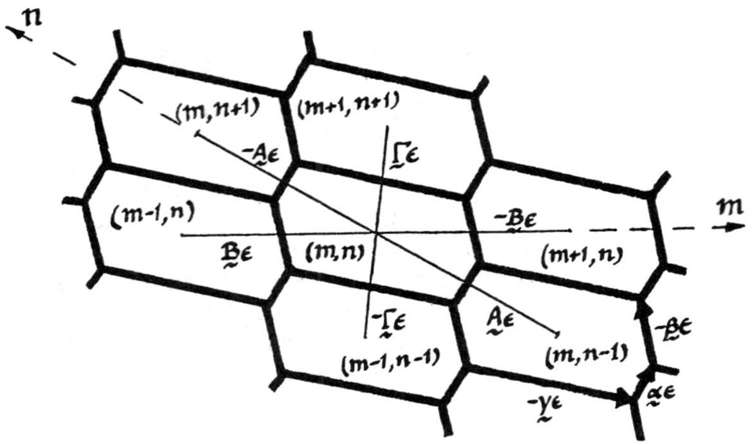

FIGURE 1 The disposition of the general hexagonal pattern.

Let (x_1, x_2) be rectangular coordinates in the transverse plane of the monolith, and r = (x_1, x_2) the vector and position of any point with respect to an origin which is the center of a particular hexagon. This hexagon is designated (0, 0), and when two of these six vectors $\pm$A, $\pm$B, $\pm\Gamma$ are chosen, the position of the centers of all the hexagons can be given. If, for example, $-$B, $-$A are chosen, the position of the center of the (m,n)th hexagon is

$$r_{m,n} = -m\mathbf{B}\epsilon - n\mathbf{A}\epsilon = m(\boldsymbol{\alpha} - \boldsymbol{\gamma})\epsilon + n(\boldsymbol{\gamma} - \boldsymbol{\beta})\epsilon. \tag{4}$$

Let the vectors

$$\mathbf{A}^+ = (\gamma_2 - \beta_2, \beta_1 - \gamma_1), \quad \mathbf{B}^+ = (\alpha_2 - \gamma_2, \gamma_1 - \alpha_1), \\ \boldsymbol{\Gamma}^+ = (\beta_2 - \alpha_2, \alpha_1 - \beta_1), \tag{5}$$

so that, if equation (2) holds,

$$\mathbf{A}^+ \cdot \mathbf{A} = \mathbf{B}^+ \cdot \mathbf{B} = \boldsymbol{\Gamma}^+ \cdot \boldsymbol{\Gamma} = 0, \\ \mathbf{B}^+ \cdot \boldsymbol{\Gamma} = \boldsymbol{\Gamma}^+ \cdot \mathbf{A} = \mathbf{A}^+ \cdot \mathbf{B} = -1. \\ \boldsymbol{\Gamma}^+ \cdot \mathbf{B} = \mathbf{A}^+ \cdot \boldsymbol{\Gamma} = \mathbf{B}^+ \cdot \mathbf{A} = 1 \tag{6}$$

and

$$m = \mathbf{A}^+ \cdot \mathbf{r}_{m,n}/\epsilon, \quad n = -\mathbf{B}^+ \cdot \mathbf{r}_{m,n}/\epsilon. \tag{7}$$

We shall use the matrix

$$\mathbf{N} = [\mathbf{A}, \mathbf{B}] = \begin{bmatrix} A_1 & B_1 \\ A_2 & B_2 \end{bmatrix} = \begin{bmatrix} \beta_1 - \gamma_1 & \gamma_1 - \alpha_1 \\ \beta_2 - \gamma_2 & \gamma_2 - \alpha_2 \end{bmatrix} \tag{8}$$

and its inverse

$$\mathbf{N}^{-1} = \begin{bmatrix} \mathbf{B}^+ \\ -\mathbf{A}^+ \end{bmatrix} = \begin{bmatrix} \alpha_2 - \gamma_2 & \gamma_1 - \alpha_1 \\ \beta_2 - \gamma_2 & \gamma_1 - \beta_1 \end{bmatrix} \tag{9}$$

If

$$\mathbf{1} = \begin{bmatrix} n \\ m \end{bmatrix},$$

equations (4) and (7) are

$$\mathbf{r}_{m,n} = -\epsilon\mathbf{N1}, \mathbf{1} = -\mathbf{N}^{1}\mathbf{r}_{m,n}/\epsilon. \tag{10}$$

For the arrangement of squares (Fig. 2.a),

$$\begin{gathered}
\boldsymbol{\alpha} = \begin{bmatrix} 1 \\ 0 \end{bmatrix}, \boldsymbol{\beta} = \begin{bmatrix} 0 \\ 0 \end{bmatrix}, \boldsymbol{\gamma} = \begin{bmatrix} 0 \\ 1 \end{bmatrix} \\
\mathbf{A} = \begin{bmatrix} 0 \\ -1 \end{bmatrix}, \mathbf{B} = \begin{bmatrix} -1 \\ 1 \end{bmatrix}, \boldsymbol{\Gamma} = \begin{bmatrix} 1 \\ 0 \end{bmatrix} \\
\mathbf{A}^{+} = [1, 0], \mathbf{B}^{+} = [-1, -1], \boldsymbol{\Gamma}^{+} = [0, 1].
\end{gathered} \tag{11}$$

For the array of regular hexaga (Fig. 2.b),

$$\begin{gathered}
\boldsymbol{\alpha} = \begin{bmatrix} \omega \\ 3^{1/2}\omega \end{bmatrix}, \boldsymbol{\beta} = \begin{bmatrix} 0 \\ -2\omega \end{bmatrix}, \boldsymbol{\gamma} = \begin{bmatrix} -\omega \\ 3^{1/2}\omega \end{bmatrix} \\
\mathbf{A} = \begin{bmatrix} \omega \\ \dfrac{-1}{2\omega} \end{bmatrix}, \mathbf{B} = \begin{bmatrix} -2\omega \\ 0 \end{bmatrix}, \boldsymbol{\Gamma} = \begin{bmatrix} \omega \\ \dfrac{1}{2\omega} \end{bmatrix} \\
\mathbf{A}^{+} = [\frac{1}{2\omega}, \omega], \mathbf{B}^{+} = [0, -2\omega], \boldsymbol{\Gamma}^{+} = [\frac{-1}{2\omega}, \omega]
\end{gathered} \tag{12}$$

where $\omega = 2(2 + 3^{1/2})^{-1/2}$. For the arrangement of Figure 2.c,

$$\begin{gathered}
\boldsymbol{\alpha} = \begin{bmatrix} a \\ 0 \end{bmatrix}, \boldsymbol{\beta} = \begin{bmatrix} 0 \\ -b \end{bmatrix}, \boldsymbol{\gamma} = \begin{bmatrix} -a \\ 0 \end{bmatrix} \\
\mathbf{A} = \begin{bmatrix} a \\ -b \end{bmatrix}, \mathbf{B} = \begin{bmatrix} -2a \\ 0 \end{bmatrix}, \boldsymbol{\Gamma} = \begin{bmatrix} a \\ b \end{bmatrix} \\
\mathbf{A}^{+} = [b, a], \mathbf{B}^{+} = [0, -2a], \boldsymbol{\Gamma}^{+} = [-b, a]
\end{gathered} \tag{13}$$

where $a = (\alpha/2\beta)^{1/2}$, $b = (\beta/2\alpha)^{1/2}$, $\alpha/\beta = |\alpha|/|\beta|$.

HEAT TRANSFER

The transfer of heat between two adjacent passages is a combination of transfer at the two surfaces of the wall between them and of conduction through the wall. The rate of transfer can be expressed in the form $h \times$ (area) $\times$ (difference of temperatures in the passages on one side and the other). We call h the overall heat transfer coefficient.

If the material at the junction of the walls is neglected, the temperature in the wall varies linearly with distance, and the addition of resistances gives

$$\frac{1}{h} = \frac{2}{h'} + \frac{\delta\epsilon}{k} \tag{14}$$

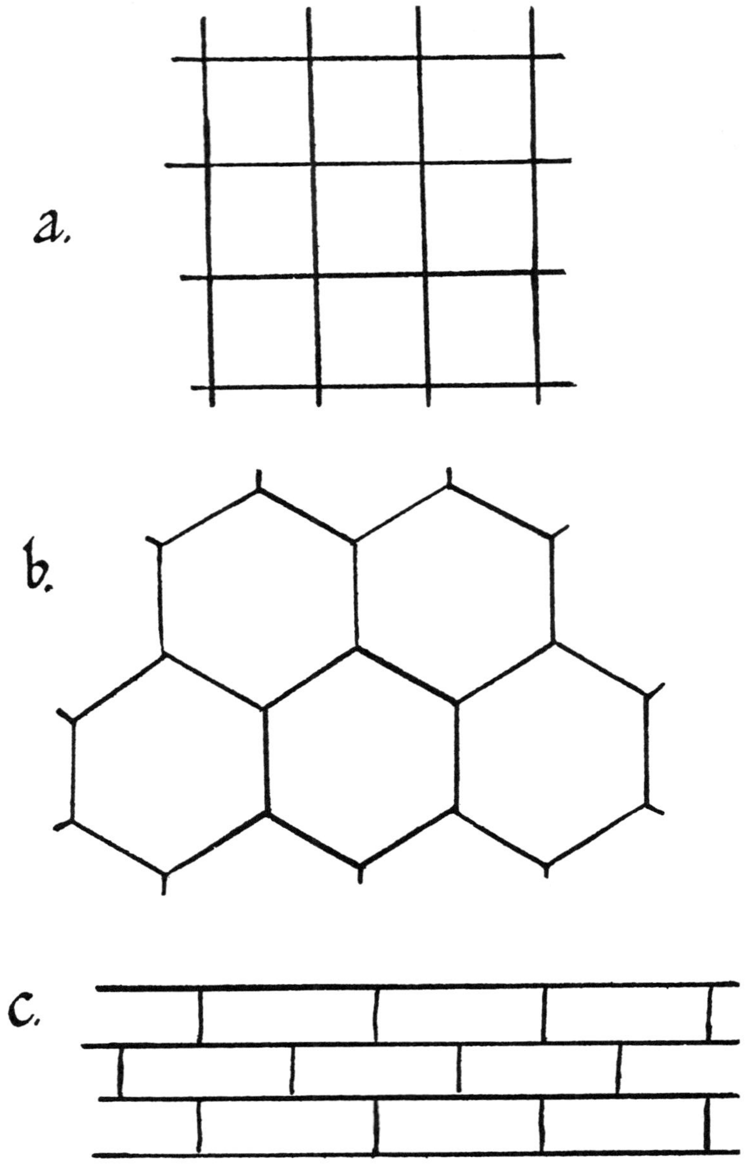

FIGURE 2 Special cases of hexagonal patterns.

where

h' = heat transfer coefficient at each surface
k = conductivity of the material of the walls
$\delta\epsilon$ = thickness of the wall.

Let

$$\kappa = h\epsilon = (k/\delta)\{1 + (2k/\delta h'\epsilon)\}. \tag{15}$$

In the limit $\epsilon \to 0$, $h'\epsilon$ does not approach zero; otherwise an infinite number of resistances would be found in a finite space. Equation (15) can be written

$$\kappa = \frac{k}{\delta} \Bigg/ \left\{ 1 + \frac{4}{Nu} \frac{k}{k'} \frac{1}{\delta(\alpha + \beta + \gamma)} \right\}, \tag{16}$$

where

k' = conductivity of the fluid which flows through the passages
Nu = Nusselt number (of magnitude 3–4 for many shapes of cross section)
α, β, γ = lengths of the sides in units of ϵ

If k/k' is large,

$$\kappa \doteqdot \frac{Nu}{4} k'(\alpha + \beta + \gamma). \tag{17}$$

THE DISCRETE MODEL

We shall use the vectors $-\mathbf{B}$, $-\mathbf{A}$ as axes of m,n (cf. Fig. C.1). Then, if we balance the transport of heat into and out of a cross-section of the (m,n)th passage, we have the equation

$$\begin{aligned} v\epsilon^2 \rho c_P \frac{d}{dx_3} T_{m,n} = h\alpha\epsilon\{T_{m,n+1} + T_{m,n-1} - 2T_{m,n}\} \\ + h\beta\epsilon\{T_{m+1,n} + T_{m-1,n} - 2T_{m,n}\} \\ + h\gamma\epsilon\{T_{m+1,n+1} + T_{m-1,n-1} - 2T_{m,n}\} \end{aligned} \tag{18}$$

where

$v\epsilon^2$ = volume flow rate of the fluid flowing through one of the passages
$T_{m,n}(x_3)$ = temperature in the $(m, n)^{\text{th}}$ passage at a distance x_3 from the inlet
ρ = density of the fluid
c_p = specific heat of the fluid

Let

$$z = \kappa y_3 / v\epsilon^2 \rho c_p \tag{19}$$

and

$$\mathscr{T}(\mu,\nu,z) = \sum_{-\infty}^{\infty} \sum_{-\infty}^{\infty} \mu^m \nu^n T_{m,n}(z); \tag{20}$$

then

$$\frac{1}{\mathscr{T}} \frac{d\mathscr{T}}{dz} = \alpha\left(\nu - 2 + \frac{1}{\nu}\right) + \beta\left(\mu - 2 + \frac{1}{\mu}\right) + \gamma\left(\mu\nu - 2 + \frac{1}{\mu\nu}\right). \tag{21}$$

Consider first the boundary conditions

$$T_{o,o}(0) = \frac{Q}{\epsilon^2}, \; T_{m,n}(0) = 0, m, n \neq 0 \tag{22}$$

when

$$Q = q/v\rho c_p$$
$$q = \text{excess heat which is introduced into the } (0, 0)\text{th passage}$$

If this is the case, $\mathcal{T}(\mu, \nu, 0) = Q/\epsilon^2$, and, as can be easily seen.

$$\begin{aligned}\mathcal{T}(\mu, \nu, z) &= \frac{Q}{\epsilon^2} e^{-2(\alpha+\beta+\gamma)z} e^{\alpha z(\nu+\frac{1}{\nu})} e^{\beta z(\mu+\frac{1}{\mu})} e^{\gamma z(\mu\nu+\frac{1}{\mu\nu})} \\ &= \frac{Q}{\epsilon^2} e^{-2(\alpha+\beta+\gamma)z} \sum_{-\infty}^{\infty} \nu^p I_p(2\alpha z) \sum_{-\infty}^{\infty} \mu^q I_q(2\beta z) \sum_{-\infty}^{\infty} (\mu\nu)^r I_r(2\gamma z) \\ &= \frac{Q}{\epsilon^2} e^{-2(\alpha+\beta+\gamma)z} \sum_{m=-\infty}^{\infty} \sum_{n=-\infty}^{\infty} \mu^m \nu^n \sum_{r=-\infty}^{\infty} I_{n-r}(2\alpha z) I_{m-r}(2\beta z) I_r(2\gamma z)\end{aligned}$$

Let

$$S_{m,n}(z) = e^{-2(\alpha+\beta+\gamma)z} \sum_{r=-\infty}^{\infty} I_{n-r}(2\alpha z) I_{m-r}(2\beta z) I_r(2\gamma z). \tag{23}$$

The solution of equations (18) and (22) is

$$T_{m,n}(z) = (Q/\epsilon^2) S_{m,n}(z). \tag{24}$$

Let

$$T_{m,n}(0) = T^0_{m,n}; \tag{25}$$

then

$$T_{m,n}(z) = \sum_{p=-\infty}^{\infty} \sum_{q=-\infty}^{\infty} T^0_{p,q} S_{m-p,n-q}(z). \tag{26}$$

These solutions are appropriate to an infinite array of hexagonal passages. For certain finite arrangements solutions can be found by the principle of reflection.

THE CONTINUOUS MODEL

In the continuous model we take v to be the velocity of the fluid and define a matrix K of thermal conductivities. This matrix has elements K_{ij} and the rate of transfer of heat in the ith direction ($i = 1,2$) is

$$J_i = -K_{i1} \frac{\partial T}{\partial x_1} - K_{i2} \frac{\partial T}{\partial x_2}. \tag{27}$$

Then we have the equation

$$v\rho c_p \frac{\partial T}{\partial x_3} = \nabla \cdot (\mathbf{K}\, \nabla T) = \sum_{i,j=1}^{2} K_{ij} \frac{\partial^2 T}{\partial x_i \partial x_j}. \tag{28}$$

The solution of this equation which takes the value $Q\delta(x_1)\delta(x_2)$ on $x_3 = 0$, is the continuous analogue of $(Q/\epsilon^2)S_{m,n}(z)$ and will be written as $QS(x_1, x_2, x_3)$.

This solution is

$$S(x_1,x_2,x_3) = \frac{\nu\rho c_p}{4\pi x_3 K^{1/2}} \quad \exp - \frac{\nu\rho c_p}{4x_3}(\mathbf{r}^T\mathbf{K}^{-1}\mathbf{r}) \tag{29}$$

where $K = \det \mathrm{K}$ and $\mathrm{r}^T = (x_1,x_2)$. If $\mathrm{T}(x_1,x_2,0) = \mathrm{T}_0(x_1,x_2)$ is the boundary condition, then the solution is

$$T(x_1,x_2,x_3) = \iint_{-\infty}^{\infty} T^0(\xi_1,\xi_2)S(x_1-\xi_1,x_2-\xi_2,x_3)\,d\xi_1 d\xi_2. \tag{30}$$

TWO LEMMAS

I wish to prove two lemmas

$$\text{I; } \operatorname*{Lim}_{\epsilon\to 0} \frac{1}{\epsilon} e^{-2y/\epsilon^2} I_{\lambda/\epsilon}\left(\frac{2y}{\epsilon^2}\right) = \frac{1}{(4\pi y)^{1/2}} e^{-\lambda^2/4y} \tag{31}$$

$$\text{II; } \operatorname*{Lim}_{\epsilon\to 0} \epsilon^{-2} S_{m,n}(z) = \frac{1}{4\pi y\chi} \exp - \frac{\alpha\mu^2 + \beta\nu^2 + \gamma(\mu-\nu)^2}{4\chi^2 y} \tag{32}$$

where $m = \mu/\epsilon$, $n = \nu/\epsilon$, $z = y/\epsilon^2$ and $\chi^2 = \beta\gamma + \gamma\alpha + \alpha\beta$.

The first limit follows from a formula of Olver (F. J. W. Olver, "Asymptotics and Special Functions," Academic Press, New York, 1974, p. 269).

$$\begin{aligned} e^{-z}I_\nu(z) &= (2\pi z)^{-1/2}\left\{\sum_{s=0}^{n-1} (-)^s \frac{A_s(\nu)}{z^s} + \delta_n\right\} \\ &\quad - i\, e^{-\nu\pi i}(2\pi z)^{-1/2} e^{-2z}\left\{\sum_{s=0}^{n-1} \frac{A_s(\nu)}{z^s} + \gamma_n\right\} \end{aligned} \tag{33}$$

where

$$\begin{aligned} &A_s(\nu) = (4\nu^2-1^2)(4\nu^2-3^2)\ldots(4\nu^2-\overline{2s-1}^2)/s!8^s \\ &|\gamma_n| \le 2\exp\{|4\nu^2-1|/4z\}|A_n(z)z^{-n}| \\ &|\delta_n| \le 2X_n \exp\{\pi|4\nu^2-1|/8z\}|A_n(z)z^{-n}| \\ &X_n = \pi^{1/2}\,\Gamma(\overline{n+2}/2)/\Gamma(\overline{n+1}/2) \end{aligned} \tag{34}$$

Let $\nu = \lambda/\epsilon$ and $z = 2y/\epsilon^2$; then

$$\operatorname*{Lim}_{\epsilon\to 0} \frac{A_s(\nu)}{z^2} = \frac{1}{s!}\left(\frac{\lambda^2}{4y}\right)^s.$$

It is possible to show that $\delta_n \to 0$ as $n \to \infty$ and that the second part of equation (33) can be neglected; the lemma follows from this.

In the second lemma, let $r = \rho\epsilon$ and notice that the limit of $\epsilon\Sigma_r$ is $\int d\rho$. Then

$$\operatorname*{Lim}_{\epsilon\to 0} \epsilon^2 S_{\frac{\mu}{\epsilon}\frac{\nu}{\epsilon}}\left(\frac{y}{\epsilon^2}\right) = \frac{1}{(4\pi y)^{3/2}}\frac{1}{(\alpha\beta\gamma)^{1/2}}\int_{-\infty}^{\infty} \exp - \frac{1}{4y}\left\{\frac{(\mu-\rho)^2}{\beta} + \frac{(\nu-\rho)^2}{\alpha} + \frac{\rho^2}{\gamma}\right\} d\rho$$

and lemma II follows when the integration is done.

EQUIVALENCE OF THE MODELS IN THE LIMIT $\epsilon \to 0$

Let $y = \kappa\gamma_3/\nu\rho c_p = z/\epsilon^2$.
The argument of the exponential function in equation (32) may be written

$$\frac{1}{4\chi^2 y}\left\{[\nu, \mu]\begin{bmatrix}\beta+\gamma & -\gamma \\ -\gamma & \gamma+\alpha\end{bmatrix}\begin{bmatrix}\nu \\ \mu\end{bmatrix}\right\} = \frac{1}{4y}[\mathbf{r}^T(\mathbf{N}^{-1})^T\mathbf{M}^{-1}\mathbf{N}^{-1}\mathbf{r}] \qquad (35)$$

where

$$\mathbf{M} = \begin{bmatrix}\alpha+\gamma & \gamma \\ \gamma & \beta+\gamma\end{bmatrix}, \det \mathbf{M} = \chi^2. \qquad (36)$$

If we compare equations (29) and (35), we see $\kappa^{-1}\mathbf{K} = \mathbf{NMN}^T$. Therefore

$$\mathbf{K} = \kappa\{\alpha \mathrm{AA}^T + \beta\mathbf{BB}^T + \gamma\boldsymbol{\Gamma\Gamma}^T\}. \qquad (37)$$

This formula shows the separation of the matrix K into two parts: κ is the physical part and $\{\alpha\mathbf{AA}^T + \beta\mathbf{BB}^T + \gamma\boldsymbol{\Gamma\Gamma}^T\}$ the geometrical.

The original paper [156] had the following Acknowledgments:

Confessiones

Solutio huius problematis nec ab ullo procuratione reipublicae nec beneficio privato subventa est, sed, gaudii mei causa, in tempore feriarum in Anglia Caledoniaque facta est. Multas gratias conlegae meo, G. PISTORI (W. MILLER, JR.), ago ob mentionem formulae Olveri. Haec formula formulae qua prius utebar praestat. Amica mea, E. S. LOTT, hoc opus vidit et ex multis ridiculis me eripuit. Noli tamen eam condemnare propter rusticum scribendi genus meum, redolentem provinciae septentrionalis longe remotioris quam Patavii. M. RIEBE, quae omnia bona facit, hac lingua aliena luctabatur et litteras claras machina quae scribit obtinuit.

N.*

A GENERAL THEORY OF ANISOTROPIC MEMBRANES†

R. ARIS and E. L. CUSSLER

Department of Chemical Engineering and Materials Science, University of Minnesota, Minneapolis, Minnesota 55455

(Received November 14, 1986; in final form March 12, 1987)

The theory of composite membranes with one "active" layer whose permeability varies with solute concentration is developed. The case of exponential dependence of permeability on concentration is treated in detail, but the case of a general dependence follows the same pattern and leads to the treatment of multi-layered membranes.

KEYWORDS Theory Anisotropic Composite Membranes Permeability

INTRODUCTION

A composite membrane, one of whose layers has a concentration dependent permeability, will behave anisotropically. If the permeability increases with the concentration, then putting the "active" layer on the high concentration side will result in a greater flux than would be obtained in the opposite direction if the concentrations were interchanged. The idea is not new, but in spite of its simplicity, a comprehensive theory has not been published. We do this in detail for an exponential dependence of permeability and show how arbitrary concentration dependence can be similarly treated.

This type of anisotropy has been studied as a possible separation method (Rogers *et al.*, 1957; Frisch, 1964; Rogers and Steinbert, 1968; Stannett *et al.*, 1968; Jem *et al.*, 1986). Curiously, it has not been as widely investigated as a strategy for packaging. For example, some bottles have an outer layer, made of poly co-vinylalcohol-vinylacetate, which is an oxygen barrier whose properties are compromised by swelling in water. The bottles have an inner layer, made of copolymers of vinyldine-chloride with other materials, which is a better barrier for water. If these layers were reversed, the bottle would be a much less effective barrier for both oxygen and water. Qualitative aspects of

* Reprinted with the permission of Gordon and Breach Science Publishers S. A. from *Chemical Engineering Communications,* volume **58,** pp. 3–16, 1987.

† This paper is dedicated to Neal Amundson on his 70th birthday.

this type of packaging have been discussed (Pike, 1986), but the fundamentals have been less completely covered.

EXPONENTIAL DEPENDENCE

We shall consider steady diffusion through a composite membrane, the notation and the arrangement of which are shown in Figure 1. There are a number of interesting phenomena, such as swelling, that enter into the transient case, but they will not be considered here. If the diffusivity in layer A depends on concentration exponentially

$$D_A(c) = D_{A0} \exp bc, \tag{1}$$

we have in b a natural characteristic quantity of dimensions (concentration)$^{-1}$. It is natural to define a dimensionless concentration

$$a = bc.$$

Let

$$u = bc_L, \qquad v = bc_R, \qquad w = b\bar{c} \tag{2}$$

The flux

$$N = -D_{A0}e^{bc}\frac{dc}{dx} = D_B\frac{\bar{c} - c_R}{l_B} \tag{3}$$

can be made dimensionless as

$$j = \frac{bNl_B}{D_B} \tag{4}$$

so that, if

$$\beta = \frac{D_{A0}l_B}{D_Bl_A} \tag{5}$$

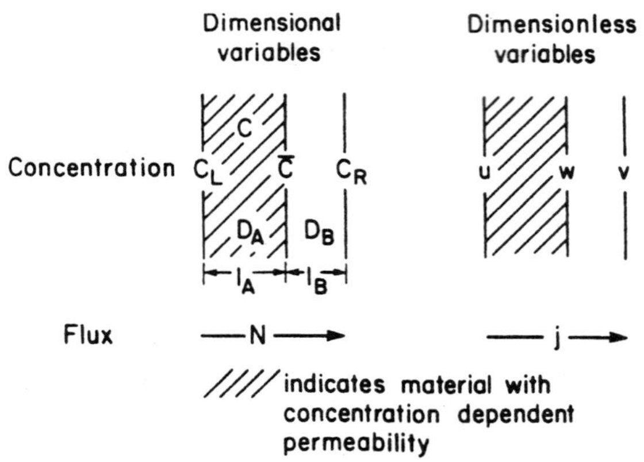

FIGURE 1 Composite membranes.

is the ratio of the mass transfer coefficients of the two layers

$$j = -\beta l_A e^a \frac{da}{dx} = w - v \tag{6}$$

Since j is independent of x, the second term is a constant and may be replaced by its average value, giving

$$j = \beta(e^u - e^w) = w - v \tag{7}$$

This equation, written as

$$\beta e^w + w = \beta e^u + v \tag{8}$$

has a unique solution $w = w(u, v; \beta)$ from which $j = j(u, v; \beta)$ can be calculated.

The exponential dependence, however, allows of a valuable simplification by absorbing β into the definition of the dimensionless concentrations. Let

$$U = u + \log\beta, \qquad V = v + \log\beta, \qquad W = w + \log\beta \tag{9}$$

Then Eq. (7) becomes

$$j = e^U - e^W = W - V. \tag{10}$$

Though Eq. (8), which now takes the form

$$e^W + W = e^U + V, \tag{11}$$

cannot be solved explicitly, we can rearrange Eq. (10) to give

$$U = \log[e^W + j], \qquad V = W - j. \tag{12}$$

These equations allow us to plot contours of constant W and constant j in the U, V-plane. It should be noted that, on account of the term $\log\beta$, U and V can be negative though u and v are always positive.

If $j > 0$, W can range from $-\infty$ to ∞, but if $j < 0$, e^W must be greater than $-j$ if the argument of the logarithm is to be positive. We can see from Eq. (12) that for fixed j and $W \to \infty$, $U \sim W$, $V = W - j$ so that the asymptotes, as V and U get large, are $V = U - j$. This is common-sense, since the layer B provides the only resistance of the film to mass transfer when the permeability of the layer A is so large. As $W \to -\infty$ for $j > 0$, $U \sim \log j$, $V \to -\infty$ and the asymptotes are the vertical lines $U = \log j$. If $j < 0$ and $W \to \log|j|$, $U \to -\infty$ and $V \to \log|j| + |j|$. Figure 2 shows the domain over which W and j may vary and Figure 3 shows the contours of constant j and may be used to determine the flux for any two layer (exponential-constant) membrane. As U and V become large, the dimensionless permeability becomes 1.

It is clear from the asymmetry of Figure 3 that the anisotropy becomes significant when β is small enough to make U and V negative. If we interchange u and v we change the sign of the flux and reduce its magnitude. Let the measure of anisotropy be

$$\sigma = -\frac{j(v, u; \beta)}{j(u, v; \beta)} \tag{13}$$

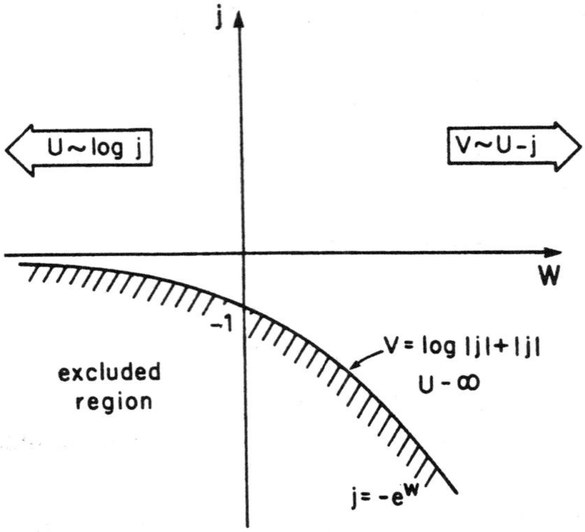

FIGURE 2 The *W*-*j* domain.

so that for $u \geq v$, $\sigma \leq 1$. If Eq. (10) and Eq. (12) are combined

$$j = e^{U} - e^{V+j} \tag{14}$$

then interchanging U and V and replacing j by $-\sigma j$ gives

$$-\sigma j = e^{V} - e^{U-\sigma j}. \tag{15}$$

Solving these two equations gives

$$U = \log\left[j\,\frac{\sigma e^{j} - 1}{e^{(1-\sigma)j} - 1}\right], \qquad V = \log\left[j\,\frac{\sigma - e^{-\sigma j}}{e^{(1-\sigma)j} - 1}\right]. \tag{16}$$

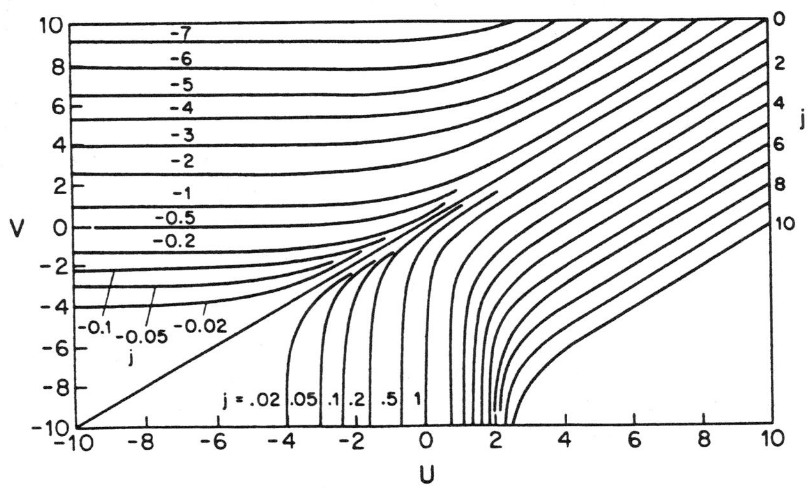

FIGURE 3 Contours of constant flux in *U*, *V*-plane.

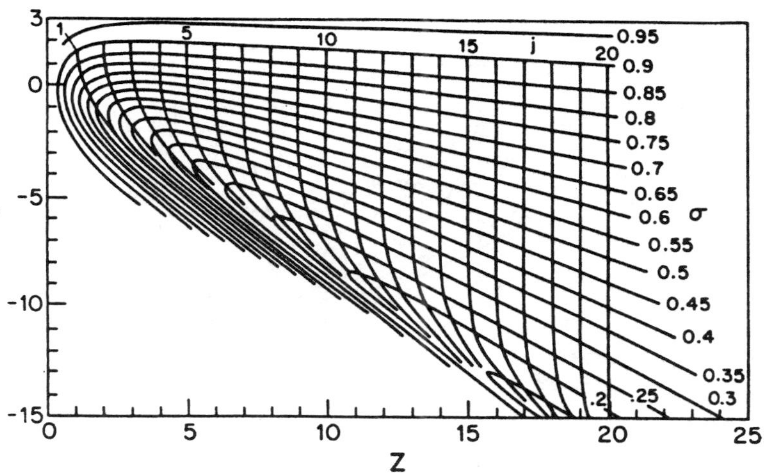

FIGURE 4 Contours of constant j and σ in Z, V-plane.

It is convenient to consider also the concentration difference across the membrane

$$Z = U - V = u - v = \log\left[\frac{\sigma e^{j} - 1}{\sigma - e^{\sigma j}}\right] \tag{17}$$

which, without loss of generality we can take to be positive. Thus the region within which σ and j may move is $\sigma^*(j) \le \sigma \le 1$, $j^*(\sigma) \le j \le \infty$, where $\sigma^*(j)$ is the inverse function of

$$j^*(\sigma) = \frac{1}{\sigma}\log\frac{1}{\sigma}. \tag{18}$$

From the loci of constant j and σ in the Z, V-plane in Figure 4, we determine the flux and concentration for any concentration difference, Z, and level, V. Figure 5 is an enlargement of the upper left part of Figure 4 which will be used later. We need only consider $Z > 0$ for results for negative Z can be obtained from the replacement of U, V, Z, σ and j by V, U, $-Z$, $1/\sigma$, and $-\sigma j$, respectively.

The asymptotic properties are best tabulated:

	Asymptotic Approach				
σ	j	U	V	Z	**Locus**
$\sigma^*(j) + 0$	fixed	$\log j$	$-\infty$	∞	$V + Z = \log j$
$1 - 0$	fixed	∞	∞	j	$Z = j$
fixed	$j^*(\sigma) + 0$	$\log j^*$	$-\infty$	∞	$V + Z = \log j^*$
fixed	∞	∞	$-\infty$	∞	$V \sim \log \sigma Z - (1 - \sigma)Z$

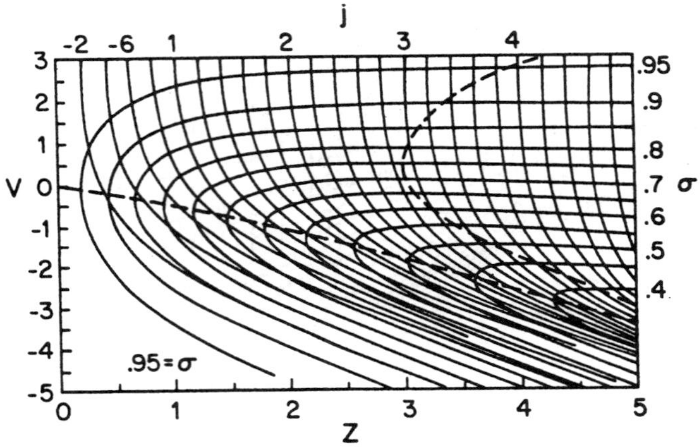

FIGURE 5 Enlargement of part of Figure 4.

DESIGNING FOR MAXIMUM ANISOTROPY

For any given pair of concentrations u and v, we can design a membrane to maximize the anisotropy. The given u, v define the difference

$$Z = u - v$$

and its is clear from Figure 4 that there is just one curve of constant σ that touches the vertical of constant Z. Let this be $\sigma = \hat{\sigma}(Z)$ and $j = \hat{j}(Z)$ be the corresponding value of j at the point of contact. By differentiating Eq. (17) with respect to j and setting the derivative to zero, the condition on $\hat{\sigma}$ and $\hat{j}$ is

$$\hat{\sigma} + \exp - (1 + \hat{\sigma})\hat{j} = (1 + \hat{\sigma})\exp - \hat{\sigma}\hat{j} \tag{19}$$

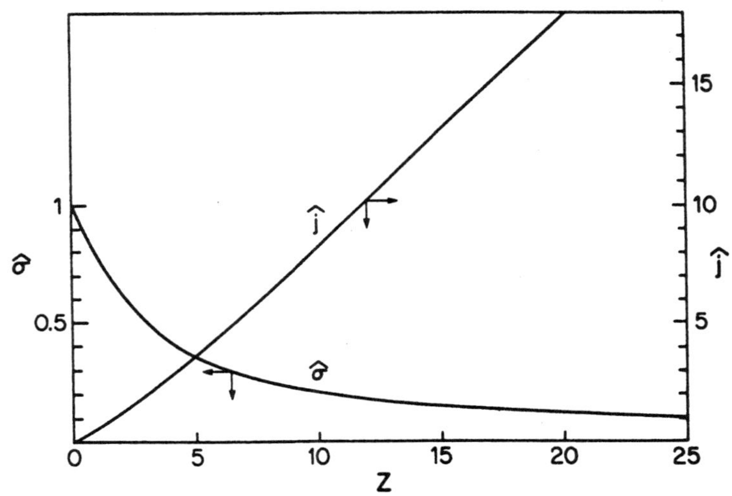

FIGURE 6 Maximum anisotropy for constant Z.

which, apart from the trivial $\hat{j} = 0$, has a unique solution in $j > 0$, $0 < \sigma < 1$. This relationship is shown in Figure 6. Using this solution in Eqs. (16) and (17) gives

$$Z = (1 + \hat{\sigma})\hat{j}, \qquad U = \left(\frac{Z}{1 - e^{-Z}}\right), \qquad V = \log\left(\frac{Z}{e^{Z} - 1}\right). \tag{20}$$

Thus

$$u = \log\left[\frac{Z}{1 - e^{-Z}}\frac{1}{\beta}\right], \qquad v = \log\left[\frac{Z}{e^{Z} - 1}\frac{1}{\beta}\right] \tag{21}$$

can be used to draw loci of constant β along which Z, running from 0 to $Z^*(\beta)$, $Z^*/(e^{Z^*} - 1) = \beta$, is the parameter. Some of these curves are shown in Figure 8.

It is sometimes useful to work with constant Z instead of treating it as a parameter. In this case $\hat{j} = Z/(1 + \hat{\sigma})$ and Eq. (19) becomes

$$\hat{\sigma} + e^{-Z} - (1 + \hat{\sigma})\exp\{-\hat{\sigma}Z/(1 + \hat{\sigma})\} = 0 \tag{22}$$

Numerical values are reported in Table I.

We shall also find it useful to know the locus of the minimal σ for fixed V, i.e., where a horizontal line in Figure 4 or 5 touches the curve of constant σ. The equation for this obtained by setting the derivative of V in Eq. (16) equal to zero is messier, being

$$\sigma + (\sigma j - 1)e^{-\sigma j} + \sigma\{(1 - \sigma)j - 1\}e^{(1-\sigma)j} + (1 - j)e^{(1-2\sigma)j} = 0. \tag{23}$$

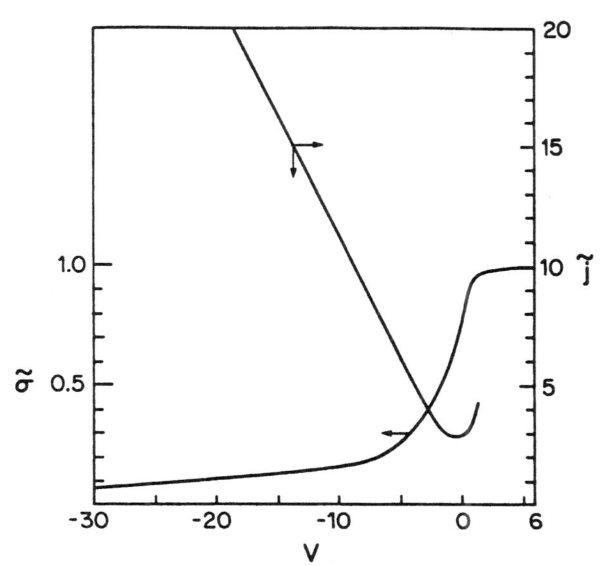

FIGURE 7 Maximum anisotropy for constant V.

TABLE 1 Conditions for Maximum Anisotropy for Given Concentration Difference

Z	$\hat{\sigma}$	$\hat{j}$	V	U
0.2	0.9512	0.1025	−0.1017	0.0983
0.4	0.9050	0.2100	−0.2067	0.1933
0.6	0.8612	0.3224	−0.3150	0.2850
0.8	0.8198	0.4396	−0.4265	0.3735
1.0	0.7808	0.5615	−0.5413	0.4587
1.2	0.7441	0.6880	−0.6593	0.5407
1.4	0.7096	0.8189	−0.7804	0.6196
1.6	0.6772	0.9540	−0.9045	0.6955
1.8	0.6468	1.0430	−1.0315	0.7685
2	0.6184	1.2358	−1.1614	0.8386
2.5	0.5550	1.6077	−1.4981	1.0019
3	0.5015	1.9980	−1.8503	1.1497
3.5	0.4562	2.4035	−2.2166	1.2834
4	0.4179	2.8211	−2.5952	1.4048
4.5	0.3852	3.2486	−2.9848	1.5152
5	0.3573	3.6839	−3.3838	1.6162
7.5	0.2637	5.9350	−5.4845	2.0155
10	0.2114	8.2546	−7.6974	2.3026
15	0.1547	12.9901	−12.2920	2.7080
20	0.1239	17.7950	−17.0043	2.9957
25	0.1042	22.6398	−21.7811	3.2189

Though we would like to find both σ and j as functions of V, say $\tilde{\sigma}(V)$ and $\tilde{j}(V)$ there is no simple way to eliminate either σ or j, as was possible for constant Z in Eq. (22). Figure 7 shows the curves of $\tilde{\sigma}(V)$ and $\tilde{j}(V)$ obtained by solving Eq. (23) for j as a function of σ and substituting in Eqs. (16) and (17). Numerical values are reported in Table II. It is useful to have numerical values as the extrema are very shallow and their exact positions are not easy to find.

TABLE 2 Conditions of Maximum Anisotropy for Given Membrane ($\beta = V$ fixed)

$\hat{\sigma}$	$\hat{j}$	U	V	Z	u(When $v = 0$) $= U + V$
0.1	24.1276	3.2935	−23.0943	26.3879	−19.8008
0.15	13.8233	2.8027	−12.8426	15.6453	−10.0399
0.2	9.3148	2.4848	−8.3261	10.8107	−5.8413
0.3	5.5007	2.1374	−4.3713	6.5087	−2.2339
0.4	3.9938	2.0188	−2.6330	4.6518	−0.5450
0.5	3.2854	2.0612	−1.6356	3.6968	0.3806
0.6	2.9472	2.2468	−0.9434	3.1902	1.3034
0.7	2.8355	2.5829	−0.3836	2.9665	2.1443
0.8	2.9232	3.1185	0.1364	2.9821	3.2549
0.9	3.3323	4.0569	0.7081	3.3498	4.7650
0.95	3.8778	4.9653	1.0833	3.8825	6.0491

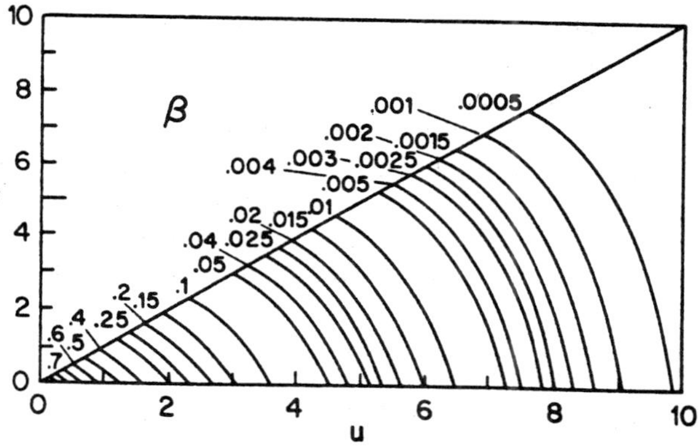

FIGURE 8 Design parameter for maximum anisotropy for given u, v.

APPLICATION

As an example of how this analysis can be used, we consider benzene diffusing through a composite membrane made of polypropylene on an inert microporous backing layer (Jem *et al.*, 1987). The permeability of benzene in polypropylene closely follows (Eq. (1), with $D_{A0} = 2.0 \cdot 10^{-5}$ cm^2/sec and $b = 5500$ cm^3/g. The backing layer has a constant $D_B = 1.3 \cdot 10^{-3}$ cm^2/sec; thus $D_B = 64D_{A0}$. As an illustration we assume that $C_L = 5 \cdot 10^{-4}$ and $C_R = 1.36 \cdot 10^{-4}$ g/cm^3 so that $u = bC_L = 2.75$, $v = bC_R = 0.75$. Thus $Z = 2$. From Figure 6 or Table I, we see that $\hat{\sigma}(2) = 0.6184$ with corresponding $\hat{j} = 1.2358$, $W = 0.8386$, $V = -1.1614$. For maximum anisotropy, we want $\log \beta = V - v = U - u = -1.9114$ or $\beta = 0.1479$. Thus we should set the ratio of thicknesses $l_\beta / l_A = \beta D_B / D_{A0} = 9.46$.

This design is notably robust in the following sense. Suppose we are content with a σ which is 5% greater than the minimum, i.e., $\sigma < 0.65$. Then from the enlarged figure (Figure 5) of the contours of constant σ, V can range

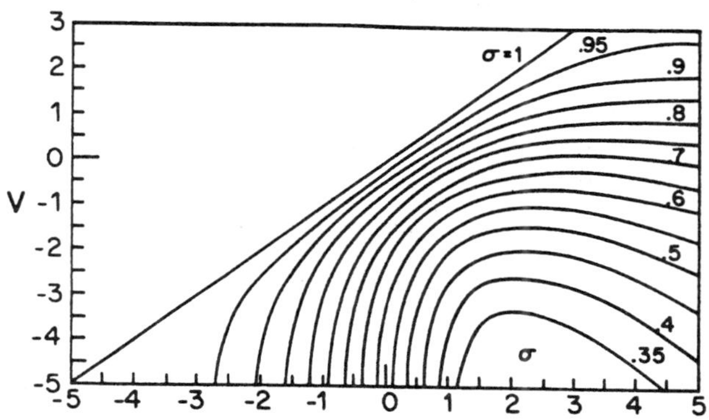

FIGURE 9 Loci of constant σ in U, V-plane.

from -0.5 to -1.7. But this means that β can be between 0.2865 and 0.0863, implying that the Celgard could be almost twice as thick (actually 194%) or nearly half as thick (actually 58%) with only 5% loss of anisotropy.

The range of conditions and maximum anisotropy of a membrane once it is made can best be seen in a figure such as Figure 9. If we take the membrane with $\beta = 0.148$, the range of positive concentrations is $U \geq -1.911$, $V \geq -1.911$. Now it is clear that the maximum anisotropy is where the maximum in V is just -1.911. This evidently occurs when $\sigma = 0.475$, $U = 2.05$ giving $u = 0.139$ (or $c = 0.25$ mg/cm^3) and $j = 3.6$.

ANISOTROPY WITH A GENERAL CONCENTRATION DEPENDENCE

We now abandon the exponential variation on which we have focussed for a more general concentration dependence. In particular, if

$$D_A = D_{A0}f(c), \qquad f(0) = 1, \tag{24}$$

then

$$N_1 = -D_{A0}f(c)\frac{dc}{dx} = \frac{D_B}{L_B}(c_i - c_l) = \frac{D_{A0}}{l_A}\{F(c_0) - F(c_j)\}, \tag{25}$$

where

$$F(c) = \int_0^c f(c^1)\, dc^1. \tag{26}$$

Let b be a parameter of the dimensions of $1/c$ derived from the dependence $f(c)$, e.g., $f'(0)$ as in the exponential case, and set

$$\begin{aligned} u = bc_0, \qquad V = bc_l, \qquad W = bc_i, \\ \Phi(x) = bF(x/b), \qquad j = bnl_B/D_B \end{aligned} \tag{27}$$

Then

$$j = \beta\{\Phi(u) - \Phi(w)\} = w - v. \tag{28}$$

Contours of constant j and w can again be drawn by the parametric formulas

$$u = \Phi^{-1}[j/\beta + \Phi(w)], \qquad v = w - j. \tag{29}$$

Again, as we should expect, $u = v$ for $j = 0$ and, since $\Phi \geq 0$, the domain $j + \beta\Phi(w) \geq 0$ is mapped onto the positive quadrant of the u, v-plane. The curves of constant w are again parallel and are given by

$$v = w + \beta\Phi(w) - \beta\Phi(u). \tag{30}$$

The curves of constant j can also be expressed as

$$v = -j + \Phi^{-1}[\Phi(u) - j/\beta] \tag{31}$$

To construct anisotropy diagrams such as Figures 4 and 5 the equations

$$\begin{aligned} \beta\Phi(u) - \beta\Phi(v + j) = j \\ \beta\Phi(u - \sigma j) - \beta\Phi(v) = \sigma j \end{aligned} \tag{32}$$

must be solved simultaneously for u and v as function of σ and j. In general this is not so easy as in the exponential case, nor is it usually possible to absorb the parameter β into the variables. From general properties of $f(c)$ various asymptotic properties of these curves may be derived. For example if $f(c)$ is bounded below as $c \to \infty$ then F, and so Φ, becomes unbounded and so, as $w \to \infty$, $u \sim w$ and $j \sim u - v$.

Stern and Saxena (1980, 1982) have used a total concentration of vapor, c_t, of the form

$$c_t = c(1 + K/(1 + \alpha c)) \tag{33}$$

and a concentration of mobile vapor

$$c_m = c(1 + FK/(1 + \alpha c)) \tag{34}$$

where c, as in the above discussion, is the ordinary dissolved concentration. They take the diffusion coefficient to be exponentially dependent on c_m and the flux to be

$$-J_A = -D_{A0}e^{bc_m}\frac{dc}{dx} \tag{35}$$

By working with c_m within the membrane and using Eq. (35) at the surface we can directly apply the results obtained above. While permeability minima may arise, we have chosen to work directly with the flux.

OTHER CONFIGURATIONS

The cylindrical geometry can be accommodated by a suitable definition of β. Consider a cylindrical membrane of material A between radii r_0 and r_1 ending a hollow cylinder of material B between r_1 and r_2. Then

$$\frac{N}{2\pi} = -rD_{A0}e^{bc}\frac{dc}{dr} = -rD_B\frac{dc}{dr}. \tag{36}$$

Again since these are constant, they can be integrated to give

$$\frac{N}{2\pi} = \frac{D_{A0}}{b}\frac{e^{bc_0} - e^{bc_1}}{\log(r_0/r_1)} = D_B\frac{c_1 - c_2}{\log(r_1/r_2)}. \tag{37}$$

Letting $u = bc_0$, $v = bc_2$, $w = bc_1$, and

$$\beta = \frac{D_{A0}}{D_B}\frac{\log(r_1/r_2)}{\log(r_0/r_1)}, \qquad j = \frac{Nb}{2\pi D_B}\log\left(\frac{r_1}{r_2}\right) \tag{38}$$

we again have the equations

$$\beta(e^u - e^w) = w - v = j. \tag{39}$$

If the layer B is composite, say planar and composed of layers of thickness l_{B1}, l_{C2} ... of diffusion coefficients D_B, D_C, ... then

$$\beta = \frac{D_{A0}}{l_A}\left\{\frac{l_B}{D_B} + \frac{l_C}{D_C} + \cdots\right\}. \tag{40}$$

The most general configuration of M layers requires a more general notation. In place of A, B, ... let the layers (in order from left to right) be A_1, $A_2, \ldots$ where l_i is the thickness of layer 1 and $D_{i0}f_i(c)$ its diffusion coefficient, $1 = 1, 2, \ldots, M$. Then

$$N = -D_{i0}f_i(c)\frac{dc_i}{dx} \tag{41}$$

Let c_{i-1} and c_i be the concentrations to the left and right faces of the 1 in layer, so that the membrane as a whole is functioning between c_0 on the left and C_M on the right. Then, with b chosen as before, and $u_i = bc_i$ and $\Phi_i(u)$ defined by Eq. (27), we have

$$j = \beta_i[\Phi_i(u_{i-1}) - \Phi_i(u_i)], \qquad 1 = 1, \ldots M, \tag{42}$$

where

$$\beta_i = \frac{D_{i0}}{l_i}\left\{\sum_{j=1}^{M}\frac{l_i}{D_{i0}}\right\}, \qquad j = NB\left\{\sum_{j=1}^{M}\frac{l_i}{D_{j0}}\right\}. \tag{43}$$

$(M - 1)$ of the M Eqs. (39) can be solved for $u_1, \ldots u_{m-1}$ and substitution made in the remaining equation to give

$$j = j(u_0, u_M). \tag{44}$$

Note that

$$\sum_{i=1}^{M}\frac{1}{\beta_i} = 1 \tag{45}$$

and that Eq. (42) implies that the lengths of the verticals A_iC_i in Figure 10 should be in the ratio of the $1/\beta_i$.

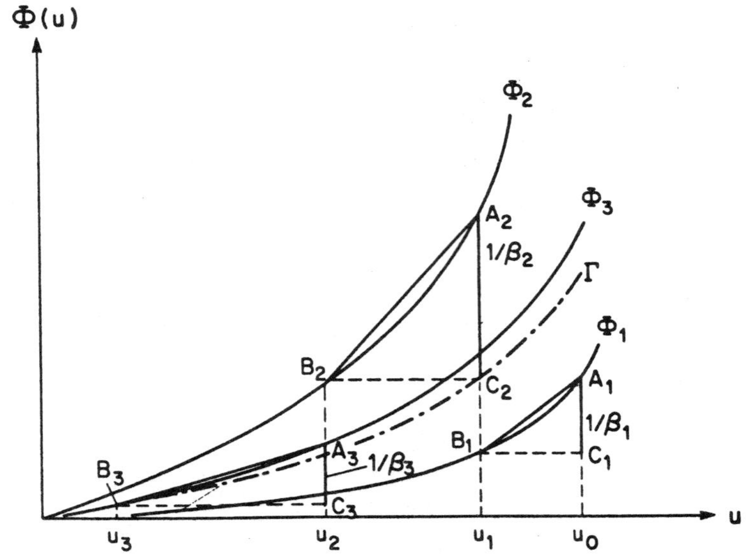

FIGURE 10 Graphical design for multilayered membrane.

It is easy to see that for a two layered membrane there is a unique solution for if we were to fix u_0 and u_2 in Figure 9 the ratio A_1C_1/A_2C_2 would range monotonically from 0 to ∞ as u_1 varies from u_0 to u_2. Hence there would be a unique point for u_1 at which this ratio was β_2/β_1. By doing this for various u_2 in the range (u_3, u_0) the locus of this point C_2 (shown as Γ) can be drawn. Then we can again work with only two curves Φ_3 and Φ_2 by always choosing a chord of Φ_2 for which C_2 lies on Γ. A graphical construction for M layers can be built up in this way.

ACKNOWLEDGMENTS

This work was partially supported by the National Science Foundation, grants 84-08999, and 86-11646, and by Questar, Inc.

NOMENCLATURE

A, B, C	names of layers in composite membrane
A_i	names of layers in composite membrane in general
a	dimensionless concentration
b	constant of dimensions [conc.]$^{-1}$; exponent in (1)
c	concentration
c_L, c_a	concentration at left and right of membrane
$\bar{c}$	concentration at A/B interface
c_i	concentration at right of 1 layer of general case
D_{A0}, D_{i0}	diffusion coefficient at zero concentration in A or A_i
D_B, D_c	diffusion coefficient at zero concentration in B, C ...
F	integral of f
f	general functional dependence of D on c
j	dimensionless flux
$\hat{j}$	dimensionless flux at maximum anisotropy for constant Z
l	thickness of layer; also with suffixes A, B, i.
N	flux of solute
r	radius in tube; also with suffixes 0, 1, 2
U	$u + \log \beta$
u, u_i	bc_L, bc_i
V	$v + \log \beta$
v	bc_R
W	$w + \log \beta$
w	$b\bar{c}$
Z	$U - V = u - v$
β	ratio of resistances of layers B and A
β_i	fraction of total resistance at zero conc. attributed to layer Ai
σ	anisotropy—$j(v, u)/j(u, v)$, $u > v$
$\hat{\sigma}(Z)$	maximum anisotropy (minimum σ) for constant Z
$\tilde{\sigma}(V)$	maximum anisotropy (minimum σ) for constant V
Φ	dimensionless F
ϕ	dimensionless f

REFERENCES

Frisch, H.L., "Multicomponent non-Fickian diffusion through inhomogeneous polymer membranes," *J. Polym. Sci.,* **A2** 1115 (1964).

Jem, K.M., Aris Rutherford, and Cussler, E.L., "Anisotropic Membrane Transport," *Chem. Engr. Comm.* (in press).

Pike, L., "Measurement of Oxygen Transmission Rates of Film and Thin Sheeting Under Humid Conditions," *TAPPI Proceedings,* Polymers Laminations and Coatings Conference (1986).

Rogers, C.E., Stannett, V., and Szwarc, M., "Permeability valves: permeability of gases and vapors through composite membranes," *Ind. Engr. Chem.,* **49,** 1933 (1957).

Rogers, C.E., and Sternberg, S., "Preparation and analysis of asymmetric polymer membranes," *J. Polym. Sci.,* **A1,** 1409 (1968).

Stannett, V., Williams, J.L., Gosness, A.B., and Gewasi, J.A., "Membranes with anisotropic flow properties," *J. Polym. Sci.,* **B6,** 1985 (1968).

Saxena, V., and Stern, S.A., "Concentration-Dependent Transport of Gases and Vapors in Glassy Polymers: II. Organic Vapors in Ethyl Cellulose," *J. Memb. Sci.,* **12,** 65–85 (1982).

Stern, S.A., and Saxena, V., "Concentration-Dependent Transport of Gases and Vapors in Glassy Polymers," *J. Memb. Sci.,* **7,** 47–59 (1980).

11 MODELING IN GENERAL

INTRODUCTION

I have always been fascinated by the ways in which scholars in different disciplines think and have tried making connections with other parts of the University. Indeed, Penn and I [168 = J′, App. 3] examined the notion of a model in the light of the relationship between the craftsman and his work. It is a practical viewpoint and emphasizes that the enjoyment is in the practice of the "craft or sullen art," not in the theorizing. It is not that I have anything profound to say about education, but I do feel it incumbent on an engineer who chooses to work in a university to be aware and concerned about the intellectual currents of the day and to be challenged by them, while resisting the temptation to go to extremes. We have had our minds opened in various directions over the second half of the 20th century, and it is important to keep them open, but that does not mean the rejection of traditional education. I have, from time to time, tried to formulate my thoughts on education in general, and Reprint O (K′, Ch. 1) is one such attempt.[1]

Reprint P (= [277]) is the last of three dialogue papers, the others being [145] = J′, App. 1., and [204] = J′, App. 2. There is some overlap in matter, and the description of the logistical equation here is much fuller than it was in the first dialogue simply because much more was known about it in 1993 than had been in 1977. A paper designed for a literary audience, having a

[1] Other relevant papers are: [77], which is concerned with graduate education; [80]; [127]; [134]; [142]; [143]; [154]; [201]; [202]; [243]; [256], the Phillips Lecture, Oklahoma State University, 1991; [283], the consequence of overdosing on literary criticism in the early 1990s; [284], a biographical article on that model scholar of the seventeenth century, Dom Jean Mabillon. He is referred to at the end of P.

similar structure but a more elementary example of modeling, was published in *New Literary History* edited by Ralph Cohen of Virginia [284].

Reprint Q (= [308]) was written for John Villadsen's *festschrift* and grew out of a series of seminars on aesthetic considerations in science and engineering. It has some overlap with others in the collection, thus betraying a lack of breadth in my reading, but I feel strongly on some of the points made and have used the strongest illustrations I know.

Chemical engineers rejoice in the use of dimensionless parameters almost more than any other type of engineer. It was therefore very meet and right to mention a few of them in fraternal greetings to the department at Princeton. As so often the verse of W. S. Gilbert was pillaged for its incomparable metrics and inventive vocabulary. The result is Reprint R.

O.*

OF CHEMICAL ENGINEERING AND THE LIBERAL ARTS: AN INAUGURAL FOR THE OLAF HOUGEN VISITING PROFESSORSHIP OCTOBER 3, 1979

Jeg tror det er bare høvelig at den først Olaf Hougen Professor skulle holde hans først forelesning på Norsk. Den var en passende anerkjennelse til en stor Amerikaner, en mann av den storste berømmelse i dette Universitet og i hele kemie ingenieurs fag. Olaf Hougen har ikke glemt eller forsømmte fedrelandets tradisjonene som er en levende del enda i vårt liv her i Amerikas hjerte. Men min kjennskap til Bjørnsons og Ibsens språk, till Welhavens og Wergelands mål, ja kansje endog til Thors og Odins tunge er altfor spinkel til å holde oppe en fullstendig forelesning og i dette departement av dyktige språkmenn det var bedre at jeg hold fast med mit fødespråk.

It would seem only fitting that the first lecture of the first Olaf Hougen Professor be given in Norwegian. It would be a great tribute to a great American, a man of the highest fame both in this university and in our profession, a man who has not forgotten or neglected the traditions of the old country, which are still a lively element in the life of this part of the country. But I fear my slight knowledge of the tongue of Bjørnson and Ibsen, of Welhaven and Wergeland, perhaps even of Thor and Odin, will not bear sustained discourse and in this department of formidable linguists I will do well to stick to my native tongue.

No doubt the Angelic Doctor, with his customary courtesy to the opposing viewpoint, would begin a discussion of the relations between chemical engineering and the liberal arts with *Videtur quod non sit necessarium inter artes utiliores et artes liberales ullum relationem haberi* ... , "It would seem that there is no relation between engineering and the liberal arts, for are not the latter those studies which are worthy of the free man, rather than the servile or mechanical, and is not engineering of all things the most mechanical?" But then he might go on, after laying out this and similar contentions, to answer each in turn and, having disposed of them one by one, to affirm the positive and gently put it into position, buttressed by suitable arguments. Admirable though this example be, I do not wish to emulate it here, and rather than attempting a tightly reasoned argument, I would like to reflect on the relation-

* Reprinted with the permission of James Dumesic, Chair of the Department of Engineering at the University of Wisconsin, to whom the Regents deputed the copyright. From *Chemical Engineering in the University Context: Four Lectures in Honor of Olaf Hougen Given in the Department of Chemical Engineering, University of Wisconsin—Madison, 1979,* pp. 3–26. Madison, WI: Board of Regents of the University of Wisconsin System, 1982.

ship of chemical engineering to the larger preoccupations of the university, and of its training to education in general.

Montaigne, in his essay "Of the Education of Children," tells us to let them have their "eye and ear in every corner; ... I have been present," he says, "when, whilst they at the upper end of the chamber have been only commending the beauty of the arras, or the flavour of the wine, many things that have been very finely said at the lower end of the table have been lost or thrown away." He goes on to say, "Let him examine every man's talents; a herdsman, a mason, a passing traveller: one may learn something from every one of these in their several capacities, and something will be picked out of their discourse whereof some use may be made at one time or another; nay, even the folly and impertinence of others will contribute to his instruction."

I am encouraged in this "folly and impertinence" by the fact that the many authors who have attempted to define a liberal education seem to be agreed on only one thing; namely, that it is not a question of subject matter but of attitude and development. It would seem, therefore, that though engineering would to be excluded from the liberal arts by the older senses of the word, it could provide the basis for a training as liberal as any when taken up in the true spirit of the university. Now it must be confessed that some of the earlier uses are, by the mores of our day, somewhat snobbish. Dr. Johnson goes so far as to refer it to those arts or sciences that are "becoming to a gentlemen" (a definition of little help to us in a country with no tradition of heraldry). The traditional liberal arts of the Middle Ages were of course seven: the trivium of grammar, logic, and rhetoric, and the quadrivium of arithmetic, geometry, music, and astronomy. It has always been popular to inveigh against scholasticism, forgetting perhaps that its insistence on the contingency of creation paved the way for experimental science, but it was scholasticism in its degenerate form of total reliance on Aristotle and disregard of all else that Montaigne was reacting against. He was certainly right, though, when he claimed that the test of education was how much we may make it our own and be truly informed by it. " 'Tis a sign of crudity and indigestion," he says, "to disgorge what we eat in the same condition it was swallowed; the stomach has not performed its office unless it has altered the form and condition of what we committed to it to concoct. Our minds work only on trust, when bound and compelled to follow the appetite of another's fancy, enslaved and captivated under the authority of another's instruction." And again it is certainly of the essence that we should examine and "thoroughly sift everything."

Thus, learning and assimilation are certainly the beginning of an education that will allow us to make free choices and hence be liberal in the true meaning of that word.

In a fascinating symposium volume entitled "The Meaning of the Humanities," Ralph Barton Perry defines the humanities as that which embraces all influences conducive to freedom. This he extends beyond mere curricula to institutions, activities, and relationships. He points out that the first condition of freedom is freedom from ignorance, which implies learning. That we should aspire to be as universal as possible is appropriate enough, though each must know his own limitations. A liberal education, however, requires some acquaintance with a wide range of subjects, for it does not follow that, having sharpened one's wits on mathematics, one can then speak penetratingly of politics, any more than a good classical scholar can get up "science" over the weekend, as the wife of a master of an Oxford college once averred on her husband's authority. Perry goes on to point out that imagination is a second condition of freedom of choice, for it allows the mind to entertain alternatives in a free way. Thirdly, there is a certain knowledge of sympathy which allows one to be assured that a choice is not entirely taken by default even though it involves leaving aside other possible courses of action.

There is a further role that education has to play, as the gluon for the up-and-down quarks of scientific abstraction and of the world of the "five and country senses," the beautiful quark that stains eternity's white radiance and the charmed quark of vision through those "casements opening on the foam of perilous seas in faery lands folorn." For disparate as the dancing molecules of the Eddingtonian table are from the adze-finished, rough-smooth, elbow-worn, age-darkened Jacobean board, the fact remains that they can be enjoyed by one and the same person and keep their place in the mind without confusion of substance. That this is possible is due to the way in which education makes us free within a tradition of thought. It gives us a self-critical awareness of what there is to explore and the realization that we are not necessarily the first to travel this way and that our age is not necessarily the one that has made the most progress. It gives us a sense of what is appropriate to the terrain so that we do not expect to scale the heights of physics without the oxygen of mathematics, nor do we wish to tear up the spring-softened alps of art with an engineering motorcycle.

What, then, are some of the aspects of liberal education that might engage us? I would like to discuss five of them without in any sense presuming to say that there are no others, nor equally important ones. The first is the value judgement that needs to be developed. Then there is a sense of relevance and appropriateness that marks the able mind. Thirdly, there is the ability to reason in the broadest sense. Fourthly, a sensitivity to language, and, fifthly, an innate courtesy which should be sought after and cultivated.

I believe that a liberal education would lead us to value that which has endured and to discern that which may be able to. Within the chemical engineering curriculum, for example, we value our knowledge of thermodynamics more than our acquaintance with the shape of a bubble cap. An educated outlook gives the lie to the phrase "the real world," used as if there were something more real about the world outside the university. Tolkien remarks: "The notion that motorcars are more 'alive' than, say, centaurs or dragons is curious; that they are more 'real' than, say, horses is pathetically absurd." My own favorite antithesis is that of the use of the word *quotation,* which in the business world is used every day for the most transitory of figures on the stock market. If one of these survives for 800 seconds it is doing well, while a quotation from humane letters may well survive 800 years and still be living and vibrant. Take, for example, the comfort that Peter the Venerable gave to Heloise, the abbess of the Paraclete. Heloise, you will recall, had taken the veil at the behest of her lover and husband, Peter Abailard, and, never fully reconciled to her vocation, was instructed by him from afar at an arm's length as rigid as once his embraces had been impassioned. Then she learned of his last days in the Cluniac priory of St. Marcel from Peter the Venerable, abbot of Cluny, who marvelled that "he who was known the world-over for his unique mastery of knowledge" should be so "steadfast in his own gentleness and humility." He closes his account with the incomparably lovely words: *Hunc, inquam, loco tui, vel ut te alteram in gremio suo confovet,* "Him, I say, in your place, or as another you, God cherishes in his bosom." The subdued restraint of the Latin scarce bears translation. It is the level light of the winter sun breaking out into clear sky between lowering clouds and western hilltops. As Etienne Gilson concludes: "In order that this particular human story may live anew in all its individual and concrete complexity, it is enough that we know it. In order that it may enrich us with its very substance, it is enough that we love it."

With value attached to enduring matters there goes then a certain gaiety in learning. If, as T. H. White makes Merlyn tell Wart in *The Sword and the Stone,* "the best thing for being sad is to learn something." then that something had better be durable. Montaigne says of philosophy, "There is nothing more airy, more gay, more frolic, and I had like to have said, more wanton. She preaches nothing but feasting and jollity; a melancholy and anxious look shows that she does not inhabit there." He concludes that "the most manifest sign of wisdom is a continual cheerfulness; her state is like that in the regions above the moon, always clear and serene."

If a sense of value puts a measure on what may be taken in the long term, a sense of relevance is the faculty by which we judge what is appropriate to the moment. Nothing is more hopelessly distracting than the introduction of the kitchen sink, Marxist dialectic, reincarnation, and "Uncle Tom Cobbley and all" into a discussion of who the "onlie begetter of these insuing sonnets" might be. To bring to bear on a given question only those considerations which most elegantly and economically may resolve it, is art of a high order and one that, for the most part, we must be content to admire in the truly great, but an elementary appreciation of what *not* to bring up is surely the beginning of the wisdom that we hope one day to approach.

Of course, irrelevance can be a wonderful weapon and a sure defense in certain situations. A footnote in Lord Wavell's *Other Men's Flowers* records how he and a friend routed a "lady of the night" in Piccadilly with the first stanza of Sir Henry Wotton's "Ode to Elizabeth of Bohemia":

> You meaner beauties of the night
> That poorly satisfy our eyes
> More by your number than your light,
> You common people of the skies:
> Where are you when the Moon shall rise?

(Lest the AlChE ever have a meeting in New York again, I can assure the reader that it works equally well there. By the third line they will turn away, clearly of the opinion that you are demented, if not dangerous.)

Now no one in his right mind would rate the poets by their dispersive effects on the underworld of our major cities and report that Blake terrifies them in Boston, Wordsworth bores 'em to death in Washington, while Lovelace is quite a turn-on in Los Angeles. However great the problems of the inner city, that would be an asinine approach to their solution, comparable to trying to judge the efficacy of impetrative prayer by longevity in the British royal family. Yet, aside from the extreme character of the example, comparable demands for inmediate relevance are being made on the humanities, both arts and sciences. "What use is it?" is asked of poetry: "dead languages," classical tongues are called; "How does it apply to social engineering?" is a demand made of psychology, or "to the energy crisis" of physical science.

This demand for a spurious relevance is being felt particularly keenly in the sciences, where the autonomy of the scientific discipline is coming under attack from both left and right, as both Marxist and capitalist agree in regarding science to be conditioned by economics. The Marxist has a doctrinaire position and must regard everything proceeds from the more pragmatic demand that science make its needed contribution to the good of mankind in a system designed to harness the enlightened self-interest of all to their mutual benefit. The tenor of research should thus be towards applicability, and the ultimate justification of pure research is its contribution to a fabric of knowledge that will be conducive to human prosperity on a cost-benefit basis.

There is a interesting passage in a thirteenth-century medieval bestiary which reads into the alleged natural history and reproduction of the elephant the biblical analogs of the story of Adam and Eve. It says that the elephant, when they want to conceive, go eastward toward paradise, where the female takes of a tree called the mandragora and gives to her spouse. It is on this basis that the female conceives and has her only baby elephant. For the learned author claims that "elephants remain pregnant for two years nor do they have babies more than once nor do they have several at a time, but only one." failing to observe that this would imply a decline in the elephant population by a factor of two in each generation. The account goes on to give a long description of what happens to an elephant if it falls down. At the fallen one's loud bellows, a large elephant appears but is unable to lift him up (this represents the law). He is followed by twelve more elephants (representing the band of prophets), but neither can they help until finally a most insignificant elephant appears who, putting his tusks under him, lifts him up without difficulty. This is held to represent the work of redemption.

Now it is all very well for us to smile at the fabulous nature of this kind of natural history, but we might remember that it is the product of the same temptation that we have today, namely, to render our studies relevant. This kind of forced relevance was to biblical story or theological truth in the thirteenth century, but we can make as many mistakes as a medieval monk by forcing our observations and studies into the relevance of contemporary needs and thought patterns. Relevancy, like pleasure, is not to be pursued for its own sake.

If relevance is not to be pursued for its own sake, reason is less elusive. It can and should be cultivated for itself, for it is the thread that binds together the several disciplines of the liberal arts. The methodologies of history and child psychology, or of political science and philology, may be very different—but they are at one in trying to draw together the threads of their evidence to form reasonable concepts that will give coherent and convincing pictures of their universes of discourse. Recent sophisticated psychological tests have shown that "students trained in the liberal arts are better able to formulate valid concepts, analyze arguments, define themselves and orient themselves maturely to their world" (Winter, Stewart, and McClelland in *Psychology Today,* September 1978, p. 69). I take this to be a psychologist's way of saying that the liberally educated student has learned to reason.

It would be unreasonable to try to discuss reason in a paragraph or two, and I wish to content myself with reminding you of the two aspects that Whitehead called attention to in his little book *The Function of Reason:* "There is Reason asserting itself as above the world, and there is Reason as one of many factors within the world. The Greeks have bequeathed to us two figures, whose real or mythical lives conform to these two notions—Plato and Ulysses. The one shares Reason with the gods, the other shares it with the foxes." The one is "Reason as seeking a complete understanding," the other, "Reason as seeking an immediate method of action." The danger of methodology is that we become captive to it and reject any evidence that does not harmonize with it. "Some of the major disasters of mankind," Whitehead concludes, "have been produced by the narrowness of men with a good methodology. Ulysses had no use for Plato, and the bones of his companions are strewn on many a reef and many an isle." This is an interesting statement which I confess I am far from understanding. Does he mean that there would have been less loss of life in the *Odyssey* if Plato rather than Odysseus had "roamed the wide world after the sack of Troy?" Would Plato have saved his men from the folly of eating Hyperion's cattle, or brought them out of Polyphemus' cave without blinding the giant and so incurring the undying hatred of Poseidon? Would the goddess of the flashing eyes, Pallas Athene, daughter of aegis-wearing Zeus, have protected Plato as she favored the cunning Ithacan and guided his path through the wind-dark sea? I leave you to work out this particular Whiteheadian dictum, but his distinction is well-framed for all that.

The reason of Plato might indeed be called reflection. Reflection, that is, in the sense of the happy occupation of the mind with a subject in which it is viewed from many angles, rolled around the intellectual palate, and savored as fully as possible. The very word suggests something which goes beyond the active mode of reason to a more contemplative approach that synthesizes what reason has analyzed and organized. It is "theory," not in the crude sense that opposes it to experiment, but in the sense of the Greek word for "to behold." "And so they tell us," says Aristotle in the Ethica Eudemia, "that Anaxagoras answered a man who was raising problems of this sort and asking why one should choose rather to be born than not—'for the sake of beholding the heavens and the whole order of the universe." He, then, thought the choice life for the sake of some sort of knowledge to be precious." For calling it a happy occupation we have the authority of the Common Doctor, Thomas Aquinas, who in one of those terse statements of his *Summa contra Gentiles* says, "The ultimate happiness of man lies in the contemplation of the truth." For this happiness there may indeed be no better preparation than the discipline of humane letters, but chemical engineers too, by virtue of their training, should have an appreciation of the scope of truth, the importance of context, and the vitality of the great themes of art and science.

In an isolated fact one can scarcely take any particular delight. For example, the freezing point of water under a pressure of X atmospheres is Y degrees. It would be a sign of strangeness, not to say insanity, if one spent much time contemplating tables of such data, and they must have our sympathy whose lot it is to compile them. In any case, the very units in which these numbers are expressed are arbitrary and in themselves afford no intellectual satisfaction. But the formula for the variation of the freezing point with pressure is another matter, for it wraps up in a simple expression all the data and links them with the whole of thermodynamic theory, and in this one can take some pleasure. Similarly, the fact that the nth digit of the number π is p_n is again of no great significance, though the whole number is of course of tremendous importance, and there is some interest in knowing whether the sequence of digits p_n is in any real sense random. What is interesting about π is the fact that it is neither rational nor algebraic but transcendental, for here we approach matter of true intellectual content, comparable to that of thermodynamics.

For practical applications the computation of π beyond a few digits is superfluous. For practical applications, the formula for the variation of freezing point with pressure may not always be needed; it is sufficient to know the sign of the derivative of freezing point with pressure. For it is this that allows the "skate's heel" to sweep "smooth upon a bow-bend." Hopkins' line evokes a totally different set of images in which one can take delight at a new level. The recollection that π is transcendental may have the same passionate force, but its overtones are of a much more abstract nature. With Hopkins we are led immediately to human, and indeed to spiritual, vistas which mathematics, for all its beauty, cannot afford. You will recall that the sonnet "The Windhover: To Christ our Lord" goes on:

> the hurl and gliding
> Rebuffed the big wind. My heart in hiding
> Stirred for a bird,—the achieve of, the mastery of the thing!
> Brute beauty and valour and act, oh, air, pride, plume here
> Buckle! AND the fire that breaks from thee then, a billion
> Times told lovelier, more dangerous. O my chevalier!
>
> No wonder of it: sheer plod makes plough down sillion
> Shine, and blue-bleak embers, ah my dear,
> Fall, gall themselves, and gash gold-vermilion.

This is no place to attempt a commentary or to tease out the echoes of Virgil and Herbert, but the word "Buckle" cannot be so easily passed over. It is highly ambiguous, and critics have argued over which of its three principal meanings has primacy. It is part of the poet's art to provide this polyvalency that stimulates thought and imagination. A critic may be perceptive or insensitive, but not necessarily right or wrong. It is the peculiar glory of mathematics that it is patient of definitive judgements of right and wrong, yet leaves room for choices governed by aesthetics. One of the particularly subtle problems of engineering education is the task of forming the student's mind so that he is prepared to leave the limited certitude of the schoolbook and launch out into the deep of genuine professional judgement. In this enlargement of the mind literary studies may well have a role to play.

Then again there are certain themes large and small that run, appearing and reappearing, through the fabric of a subject or of literature—the great conservation principles and their embodiment in balances over elementary volumes, for example, or the notion of a base case and variations on it, that is, of partial derivatives. In the arts what a theme the "juice and joy" of springtime has been to poet, artist, and musician! From the Song of Songs which is Solomon's we have the spring of lovers:

For, lo, the winter is past,
the rain is over and gone;
the flowers appear on the earth;
the time of the singing of birds is come;
and the voice of the turtle is heard in our land.

No one knows what moved Jerome to put Rachel's death in the springtime when he translated the Vulgate, for there appears to be no justification for it in the Hebrew. Perhaps a touch of poetry stirred in his querulous old bones, for he put it not only in the primary account in Genesis 35, but also in the patriarch's backward look (chap. 45): "As for me Rachel died by me in the way when yet there was but a little way to come unto Ephrath"—*eratque vernum tempus.**

With Chaucer there is the holiday mood, and every vein is bathed "in swich licour/Of which vertu engendred is the flour." And before him, in one of the oldest medieval love songs, surviving from the tenth century in three different musical settings, we read:

Ego fui sola in silva	Alone in the wood
et dilexi loca secreta	I have loved hidden places,
Frequenter effugi tumultum	Fled from the tumult
et vitavi populum multum	And crowding of faces.
Iam nix glaciesque liquescit	Now the snow's melting,
Folium et herba virescit	Out the leaves start,
Philomena iam cantat in alto.	The nightingale's singing,
Ardet amor cordis in antro	Love's in the heart.

(Translation by Helen Waddell, *Mediaeval Latin Lyrics* [Harmondsworth: Penguin, 1952].) It should appeal to the Wisconsin faculty who often flee the tumult of the throng to be alone in the woods and their hidden places. But to read it as invitation to go hiking would be ahistorical—for hiking is a modern sport—and might lead to a secular version of the error committed

* In connection with the remarks made above concerning the Vulgate-version of Rachel's death. I owe to my learned friend Reuel Shinnar the following observations. There is evidence that Saint Jerome was familiar with some of the rabbinical sayings and legends that were being collected together in his own time. The Genesis Raba records a Jewish tradition that Rachel died in the spring. Commenting on Gen. 35:16. Rabbi Elieser Ben Yaakov says, "At the time when the earth is pitted like a sieve and the wheat is standing, the rain has passed and the heat has not yet come ..." This defines the period just after the Passover and rests on a typical paronomasia with the Hebrew words for distance, sieve and growing wheat.

by one editor who, overlooking the frank come-hither of the last two verses, included it in his *Analecta Hymmica,* believing it "in the innocency of his heart to be a hymn to the Blessed Virgin with imagery drawn from the Song of Songs."

The contrast of spring and death is with us in Henrik Wergeland, the radical nineteenth-century Norwegian poet, whom spring could not save loved he it never so tenderly.

O Foraar! foraar! red mig!
Ingen har elsket dig ømmere end jeg
... Og anemonerne,
Dine blaaøiede børn, knaele og bede, at du skal
Redde mig—mig, der elsker dig saa ømt.

(O Spring! Spring! save me!
For none has loved you more tenderly than I
.... And the anenomes,
Your blue-eyed children kneel and pray you
To save me—me, who have loved you so tenderly.)

Indeed, to quote Hopkins from his sonnet "Spring,"

Nothing is so beautiful as spring—
When weeds, in wheels, shoot long and lovely and lush:
Thrush's eggs look little low heavens, and thrush
Through the echoing timber does so rinse and wring
The ear, it strikes like lightnings to hear him sing;
The glassy peartree leaves and blooms, they brush
The descending blue; that blue is all in a rush
With richness; the racing lambs too have fair their fling.

Turning now to the question of sensitivity to Language, I take it as axiomatic that educated people should seek to express themselves with clarity, economy, and elegance. This clearly involves a developed sensitivity to the meaning of words, their overtones and texture. I am not, of course, referring to the avoidance of sheer illiteracies such as abound in our day, diverting though it might be to illustrate these. Edwin Newman has regaled us with two books recently giving some of the stranger aberrations, and it is very humbling to find that academics are well represented in the perpetrated nonsense that is recorded there. Nor am I chiefly concerned with a sense of the origin of words, though I believe this is important. For example, take a word as colorful as *conspirator.* You can literally see the plotters in the word, those cloaked figures huddled together, their breaths comingling as a hazy cloud in the night air. Or take the morality implicit in the slighting use of the word *simple* or even more slighting *simpleton.* To be simple is indeed an honor, for it is the opposite of being double-folded or characterized by duplicity. Yet it is curious that though duplicity is not held in any great honor, simplicity is too often held in contempt. Richard Chevenix Trench in his *Study of Words* comments that "inasmuch as it is felt that a man without these folds will in a world like ours make himself a prey, and as most men, if obliged to choose between deceiving and being deceived, would choose the former, it has come to pass that 'simple,' which in the Kingdom of Righteousness would be a word of highest honor, carries with it in this world of ours something of contempt." He comments. "It is to the honor of the Latin tongue, and very characteristic of the best aspects of Roman life, that 'simplex' and 'simplicitas' never acquired this abusive signification." The word *silly* has gone through a similar transformation from its original Old English *selig,* which means "blessed."

Then there is the intrinsic poetry of the word, for as someone has said, "All language is in some sort a collection of faded metaphors." *Tribulation,* which we associate with affliction and sorrow, originally comes from the threshing floor. *Tribulum* was the threshing instrument or harrow whereby the Roman husbandman separated the corn from the husks, and *tribulatio* simply means this act of threshing. But someone picked up this meaning and invested it with a much deeper one, using the metaphor of the threshing floor for the deeper experiences of life. Coleridge claimed that "in order to get the full sense of a word, we should first present to our minds the visual image that forms its primary meaning." *Desultory* immediately brings to mind its origin from *de* and *salto* and the image of a circus rider jumping from one horse to another without spending much time on either; or *capricious,* from *capra,* "goat," immediately evokes the image of that animal suddenly, unexpectedly, and unaccountably leaping up and springing forward or sideways as whim may move it.

Think of the store of poetic legend which is still secreted in words. *Halcyon* wraps up in itself an allusion to one of the most beautiful and significant legends of heathen antiquity, according to which the sea preserved its perfect calmness for the "fourteen halcyon days" during which the halcyon, or kingfisher, was supposed to brood her nest on the seas. Or take the relics of customs that pervade the language. A "stipulation," or agreement, reminds us of the *stipula,* or straw, which two Romans making some mutual arrangement would break between them. *Thrall* reminds us of the ancient custom of thrilling, or drilling, a hole in the ear of a slave as he was taken into servitude, a custom that lasted from early Hebrew times right through to the Anglo-Saxon period. *Lumber* reminds us that the first bankers or pawnbrokers were Lombards and that in their "lombard" room they kept the pledges. Even *calculation* reminds us of the pebbles (*calculi*) that were used to help the practice of counting. We could wander in the beautiful garden of words, admiring here and there the flowers that grow so profusely. But it is not even this that I have in mind.

There is, of course, an importance to guarding the meaning and significance of words that can scarcely be overemphasized. Words are the vehicle of our thought and communication, and if they lose their exactness, there is neither hope for our generating anything significant in ourselves nor of enjoying the intellectual intercourse which is our birthright. C. S. Lewis has called attention to the crime of "verbicide," pointing out that inflation is one of the commonest murder weapons. We use *fantastic* for "unusual," *awfully* for "very," *tremendous* for "great," *unthinkable* for "undesirable," and so on. The greatest cause of verbicide is our anxiety to express our approval or disapproval of things rather than to describe them. Hence "the tendency of words to become less descriptive and more evaluative; then to become evaluative, while still retaining some hint of the sort of goodness or badness implied;

and to end up by being purely evaluative, useless synonyms for good and bad." This has happened to *villain,* which was originally an innocent enough name for a certain class of citizen, and even more markedly to *gentleman.* You have no need these days to be armigerous; you can earn the double distinction of becoming "a gentleman and a scholar" by pouring someone a cup of coffee! There is undoubtedly a need to preserve exact meanings to words if our thoughts are to have any exactness, just as there is an importance to being sensitive to their overtones if our expression is to have any richness. When the Duke of Wellington referred to the battle of Waterloo as a "damned nice thing," he used the word *nice* nicely. An argument or a distinction can be "nice." It is seldom that a lady is or needs to be.

Eliot ruminates on the difficulty of maintaining this precision in the fifth section of the second quartet ("East Coker"):

> So here I am, in the middle way, having had twenty years—
> Twenty years largely wasted, the years of *l'entre deux guerres*
> Trying to learn to use words, and every attempt
> Is a wholly new start, and a different kind of failure
> Because one has only learnt to get the better of words
> For the thing one no longer has to say, or the way in which
> One is no longer disposed to say it. And so each venture
> Is a new beginning, a raid on the inarticulate
> With shabby equipment always deteriorating
> In the general mess of imprecision of feeling.
> Undisciplined squads of emotion.

Even beyond the intellectual and argumentative use of words for scientific or analytic propositions, there is a need to be sensitive to their values and appropriateness for the matter on hand. Again T. S. Eliot (*Four Quartets,* "Burnt Norton," lines 149–58) says,

> Words stain,
> Crack and sometimes break, under the burden.
> Under the tension, slip, slide, perish,
> Decay with imprecision, will not stay in place,
> Will not stay still. Shrieking voices
> Scolding, mocking, or merely chattering.
> Always assail them. The Word in the desert
> Is most attacked by the voices of temptation,
> The crying shadow in the funeral dance,
> The loud lament of the disconsolate chimera.

It is important to be sensitive to the *texture* of words and sentences. The word *surely,* for example, has an open texture as well as a confidence and reassurance. So in that beautiful phrase of the second Isaiah which is so familiar to us through Handel's setting, "Surely he hath borne our griefs and carried our sorrows," *surely* conveys an overtone of "certainly it is true," but it conveys it in a open form, inviting contemplation and discussion rather than foreclosing it with a cocksureness or false and shallow security. It is indeed, as Saint Peter Damian calls it, "the reticence of Holy Writ wherein silence itself cries out that some greatness is at hand." Or, to take another example from an early author, consider Saint Paul's statement on the meaning of the death of Christ: "Now that he ascended, what is it but that he also descended first into the lower parts of the earth?" The interrogative form leaves an open texture that any affirmative statement would have closed off. Thus the reader is invited to consider the fact and its implications, to try and understand it, to contemplate the depth of a proposition of such awful magnitude that no affirmative statement could ever do it justice. Yet the very interrogative of this sentence provides a springboard for the affirmative statement of the next: "He that descended is the same also that ascended up far above all heavens, that he might fill all things." Again, the abounding triumph of the second statement takes some of its life from the question form of the previous one. The two interact and resonate.

Finally I come to the courtesy that I believe should be one of the products of a liberal education. I cannot do better than to illustrate it from the early history of palaeography. In the early seventeenth century the Bollandists, a group of Dutch Jesuits, began a massive compilation of the lives of the saints. In 1643 the first volume of the *Acta Sanctorum* was published; now more than 300 years later the Bollandists have reached the month of November. By 1675 part of the work was being done by an old scholar, Godefroid Henschen, and his younger colleague, Daniel Papebroche. The older man was sent for his health's sake to the Ardennes, and his younger colleague went along. Ever improving the shining hour. Papebroche examined some local charters and became very doubtful as to their authenticity. As a result, in the introduction to the second volume for April of the *Acta Santorum* Papebroche undertook to establish the general principles for determining the truth and falsity of documents—an effort, indeed, that he had been encouraged to make some years before by Pope Alexander VII. The observations of Papebroche were erratic, and he took occasion to cast some aspersions on the charters of the Abbey of Denys near Paris, where, he said, "although there are many very early documents, not one of them is an original, and few can be believed to be authentic."

This assertion drew the critical attention of Jean Mabillon, a Benedictine of the congregation of Saint Maur who in the 1670s was editing the works of Saint Bernard. He too was interested in the principles of distinguishing true and false documents and recognized the basic fallacy of Papebroche's method. Papebroche had examined far too few examples to be able to make any sound generalization, and Mabillon proceeded in five years to examine some thousands of documents and produced his masterpiece, modestly entitled *De Re Diplomatica.* It is a folio volume of 640 pages with 120 reproductions of hundreds of manuscript pages. The kernel of Mabillon's work was the recognition that certain scripts (he described them as ancient Roman, Gothic [i.e., Visigothic], Anglo-Saxon, Lombardic, and Merovingian) could be distinguished, and overly simple though this classification is, we recognize it as the beginnings of the science of palaeography. Mabillon's masterpiece was published in 1681, but what is interesting is the correspondence that passed between Papebroche and Mabillon some two years later. "Ever since I had a chance to go through your work *De Re Diplomatica,*" wrote Papebroche, "I have been wanting to tell you how valuable I have found it. The main result for me is that I am now totally unsatisfied with those few papges that I devoted to this subject, and I can take only slight comfort from the fact that

I provided the occasion for an outstanding and exhaustive treatise. And I intend to say precisely that in the preface to a venture of mine on the Papacy, which will go to the printer tomorrow. I have held up my manuscript until I could learn from your book what errors I had to correct as to the reputation of the Archives of Saint Denys, which I seem to have impugned. I am amazed that you have been able to produce such a masterpiece in so short a time." He goes on: "I must confess that when I first started to read your book, I felt rather bad; but, as I went on, my pleasure at the soundness of your whole argument became so great that, as I find your pages fairly gleaming with the bright light of truth, and most of all, became filled with admiration for the thousands of things that I had never known before, I had to find someone with whom to share my joy, and I hurried to my colleague Father Baerts to tell him of what I learned. If the occasion arises, I wish you would make completely clear to everyone how thoroughly I have come around to your view. And I ask you to continue to love me—as one who may not be very learned but who is always willing to be taught."

So beautiful a letter received an equally lovely reply. The following November Mabillion wrote: "I have received the books and the letter which you sent me in which you speak of my *De Re Diplomatica.* I am filled with admiration of the extraordinary modesty of so learned a man, of which, I am sure no finer an example could be found. Where would one find among learned men a scholar who not only admits that he has been defeated in a literary argument but totally repudiates his former stand and, what is more, wants that fact to be spread far and wide? For that is precisely what you have done, and in such a friendly fashion that if I didn't already know you, this very fact would fill me with an affection for you." He continues: "All I can say is, if my work is approved by the public as it is by you, the credit goes to the authority and resources of the documents available and not to any talent of mine. But whether or not it is of any value. I should much rather have been the author of a modest letter like yours than of any work of shallow ostentation." As Father Quain remarked in his lecture at St. John's University in 1974, "In those days there were giants in the earth."

Now you may argue that these are two men in holy orders who should indeed be demonstrating unquestionable humility, but that, alas, has not always been the spirit of clerical disputation. I would argue that, though indeed the attitudes of these two men may in this case have been primarily formed by their religion, there is in the pursuit of learning a certain innate courtesy and respect for the truth which should breed a like attitude to one another. If in a state university we are wisely interdicted from dilating too much upon the one, we can indeed fall back upon the other. In concluding I would quote to you from the dedicatory letter of Mabillion's own work of 1681. "In this work," he says, "I wish to propose some rules, not made up arbitrarily at whim, but rather, drawn out of the multitude of documents over a long period of time, so that students may be assisted in distinguishing truth from falsity. Far be it from me to assume the role of master in this matter. In the republic of letters we are all free."

Such indeed might be the motto of any university, of any discipline, or of any department. *Quippe in republica litteraria omnes liberi sumus.*

P.*

TWO EYES ARE BETTER THAN ONE: SOME REFLECTIONS ON THE IMPORTANCE OF HAVING MORE THAN ONE VIEWPOINT IN MATHEMATICAL MODELLING AND OTHER DISCIPLINES†

R. ARIS

Department of Chemical Engineering and Materials Science, University of Minnesota, Minneapolis, MN 55455, U.S.A.

(Received and accepted July 1993)

Abstract—*Written in the form of a conversation between a mathematician, an engineer, and a palæographer, this paper attempts to explore the value of complementary viewpoints: the physical and mathematical in modelling, the logical and the imaginative in mathematics, the historical and the artistic in palæography.*

It was one of those bright, brittle October mornings in southern Italy, when the slight ground fog is not a mollifying blanket, but has its own texture and brilliance and seems to be an extension of the hoar-frost that rims the puddles and feathers the grasses by the roadside. The road itself zig-zagged up to the Abbey of Montecassino, the mother-house of the Benedictine Order and, in a certain sense, of all western monasticism, twisting and turning up the mountainside through terrain so rough as to leave only an occasional short-cut for the hardier walker. But it was not busy, and the three friends who were making the climb that morning could impose their self-similar Chestertonian meanderings on it without risk of being run down.

American, English, and Italian, they formed an odd trio and, save that

* Reprinted with the permission of Elsevier Science from *Mathematical and Computer Modelling,* volume **18,** number 8, pp. 95–115, 1993.

† Invited paper for the Department "Modelling—A Personal Viewpoint."

I owe the title of this paper to a remark of the Prefect of the Vatican Library, Fr. L. E. Boyle, O.P., who drew attention to the complementary function of the *oculus palæographicus* and the *oculus calligraphicus* in the foreward he kindly wrote for my *Explicatio Formarum Litterarum.* Needless to say, he must not be held responsible for any heresies I have uttered in elaborating his comment. I am obliged to him also for identifying the last figure as membrane 4 of Barb. lat. 592, and for permitting me to use it.

I am indebted to the kindness of D. N. P. Murthy for permitting me to treat with such personal freedom his invitation to write on modelling from a personal viewpoint.

they were all wermen,[1] had between them two ends and the middle of the scales of almost all the characteristics of mankind. But these degrees—of height and weight, condition and speech, temperament and perceptiveness—were so curiously comingled that it is left to the reader to sort them out from the various clues that lie around in what follows. More interesting, but not susceptible of gradation in the same way, were their disciplines, for one was a palæographer, one a mathematician, and one an engineer. They had known each other since graduate school at an enlightened institution that housed all graduate students together, whatever their disciplinary affiliation, and had met by chance in the lobby of the hotel that was housing the participants of a conference that *B* and *C* were attending. *A* was doing "his own thing" during a single quarter leave that had taken him to several Italian libraries.

B, who was somewhat portly, had had to make a few extra tacks on the first stretch and was out of breath by the time he caught up with the other two at the first bend. "I seem to remember you at dinner, *A*," he puffed, "as always coming out with remarks like 'You should have seen the dragon I encountered this morning.' "

"Or what about the time," said *C*, "when he undertook to give us the low-down on the sex life of the elephant, or, at least, some thirteenth century monk's version of it."

"Est animal quod dicitur elephans in quo non est concupiscentia coitus,"[2] intoned *A*, who was not in the least out of breath and might have continued to quote the Latin text had he not been headed off by *C* with "They have a gestation period of two years, don't they?"

But *A* was not to be denied. "There's much more to it than that," he said. "When the time comes for the once-in-a-300-year-lifetime experience for this paragon of continence, the male takes his mate eastwards, near to the earthly paradise, where the mandragora tree grows. She picks a fruit from the tree and seduces him to eat. Whereupon they mate with their backs to one another and the female at once conceives. 'Pregnancy lasts two years, nor do they give birth more than once, and never to several young, but only one.' The female gives birth in a pond with the male standing guard against their enemy the dragon. The elephant is afraid of a mouse, though it strikes terror into the bull and tramples snakes. It is incapable of getting up by itself once it falls down so it trumpets loudly. A huge elephant comes to the rescue but is unable to lift the fallen one. They both trumpet, summoning twelve elephants, but still he cannot be moved. So they all trumpet, and a most insignificant elephant of great cunning immediately appears, puts his trunk under the fallen elephant, and lifts him easily. It provides an elaborate allegory of the Christian doctrine of the Fall and the Redemption, complete with the Tree of Knowledge, the impotence of the Law and Prophets, and the humility of the Incarnation, finally effective in raising up Mankind."

"Sounds like pure myth to me," said *B*.

[1] 'Werman,' as opposed to 'woman' or 'wiffman,' is a man of the male sex or gender, 'man' being a human being without reference to sex or age (see, OED). The revival of this useful word would have saved us from many of the risible neologisms that afflict contemporary English. Men may poke fun at their own sex with impunity, but run a certain risk in pulling the legs of the other.

[2] "There is an animal called an elephant, which has no desire to mate" [1].

A bristled at the popular misuse of the word 'myth,' but excused him as being only an engineer, and went on. "No, actually it's a fine farrago of truth and error. The manuscript has the gestation period about right,[3] but its lifespan is at most a quarter of the three centuries that the typical bestiary attributes to it. Some things in these bestiaries are way ahead of their time. The migration of the swallow, for instance, and its wintering overseas is clearly stated, though the most bizarre explanations of their disappearance in winter—that they swarmed like bees and dove into the mud at the bottom of a pond, for instance—persisted to Dr. Johnson's day. Some of the fantasies live on in our language. The Latin name of the bear, *ursus*, is said to come from *orsus* (beginning) because the bear has so short a gestation period that the cubs are born shapeless and have to be licked into shape. I'll bet you told your wife, *B*, that you couldn't take her out just before you left because you had to lick your paper into shape for this meeting at Cassino."

"Come to think of it, I believe those were my very words," replied *B*. "But if I remember rightly, you're a palæographer and not really concerned with anything beyond the shape of letters."

"Oh, I don't know about that. At the very least, you might give me credit for some concern with codicology—sometimes known as the archæology of the book—with abbreviations, punctuation, and scribal practice in general, the transmission of texts, traditions of the scriptorium, rise of the stationers, characters of the fifteenth century Renaissance—not to mention that of the twelfth—cultural history of Western Europe for the first fifteen centuries of the era, with particular reference to monastic history and that of the Universities. At the risk of offending any politically correct Latinasters, fortunately a small population, though the palæographer be the *ancilla ancillarum textus*, she is at least a maid-of-all-work and does service in a noble house."

Obviously, *A* had been accused of narrowness before and had some well-fletched arrows in his quiver. Indeed he might have launched into the introductory lecture of his survey course had not *B* caught up with him. "I didn't mean to imply you were narrow," he said. "How could anyone be narrow who had eaten as many dinners in Hall as we did! But what would you say were the most important aspects of your field—the directions of approach, the attitudes?"

"I can tell you better when we get to the top and I can show you some real, live manuscripts," he replied. "For now I'll turn the question on you and ask you to explain these mathematical models that you play with. You used to blether on about mathematical descriptions of nature in graduate school, do you still do that sort of thing?"

"Yes, the kind of engineering I do consists in taking a physical situation and trying to make a mathematical model of it. The aim may sometimes be design, that is, I have to show how to choose certain quantities to meet certain specifications, but sometimes to explore possible behaviour that would be hard, or dangerous, to achieve by experiments. You can't do a good experiment without first having some kind of theoretical model, nor can you have a good

[3] The gestation period of the elephant is approximately 645 days, and its average lifespan 40 years.

model without proper attention to the physical background. So I would say, generally, that you need the binocular vision of the physical and the mathematical eyes to construct a really satisfying mathematical model. Penn and Aris in the first paper of the journal *Mathematical Modelling* [2] argue that the model, when formulated, has a life of its own, and so it does, in the sense that mathematicians, like *C* here, can play with it and build purely mathematical structures without having to come down to earth. But the engineer finds satisfaction in the interplay of the physical and the mathematical."

"Give an instance I can understand," said *A*.

"Take your own example of the elephant population. Generations of monks copied that business about 'nor do they give birth more than once and never to several young but to only one' without making the slightest mathematical model of it. If two elephants produce only one, then even a perfectly balanced population with 100% survival will be halved in one generation, quartered in two, and so on. Even with a life time of 300 years and a mediæval estimate of the age of the world, there be a few left. Worse yet, the two that survived in the ark would not have been capable of propagating the species at all. As soon as you bring mathematics in (you asked for mathematics you could understand!), even in the simplicity of division by two, you can see that something's wrong."

"There might be a number of reasons for that," said *A*. "Not all scribes were particularly literate and copied slavishly, even introducing their own careless errors. You're flogging this idea of the need for multiple viewpoints, so I suppose that you would say that they were so obsessed with the moral or theological viewpoint that they overlooked the physical. There's something to that."

"In mathematical terms, *A*," said *C*, taking up the dialogue in an effort to slow *A* down a bit and give *B* a chance to catch up, "if N_0 is the original population, the first generation will only be $N_0/2$, the second $N_0/4$, and, if N_m is the population of the mth generation, $N_m = N_0/2m$. I think you can see that if the ratio of two generations is constant, say g, that is, $N_{m+1} = gN_m$, then $N_m = g^m N_0$, and the population will become extinct if $g < 1$. If $g > 1$, the population will grow without bound. The situation is either stable—and the population veers toward extinction—or unstable and it grows without bound. Only when $g = 1$ exactly, does the population remain finite and like the Platonic Beauty, which 'suffers neither increase nor diminution, nor any other change'.[4] But $g = 1$ is like a knife edge on which the system is poised between intemperate growth on the one hand and unhappy extinction on the other. (See Figure 1.) Now suppose that the growth rate g is a function of N, which, without loss of generality, we can take to be ... "

"Hold hard there, *C*. 'Without loss of generality,' my foot! You mathematicians love your clichés; you'll be trotting out 'It is obvious that' next. Besides, you're going far too fast for *A*. Let me take over here and set up the equations, then you can give the solution," *B* had caught up with the other two and had his wind back.

"One of the principal contributions that engineers have made to intellec-

[4] Plato, *Symposium*, 211.

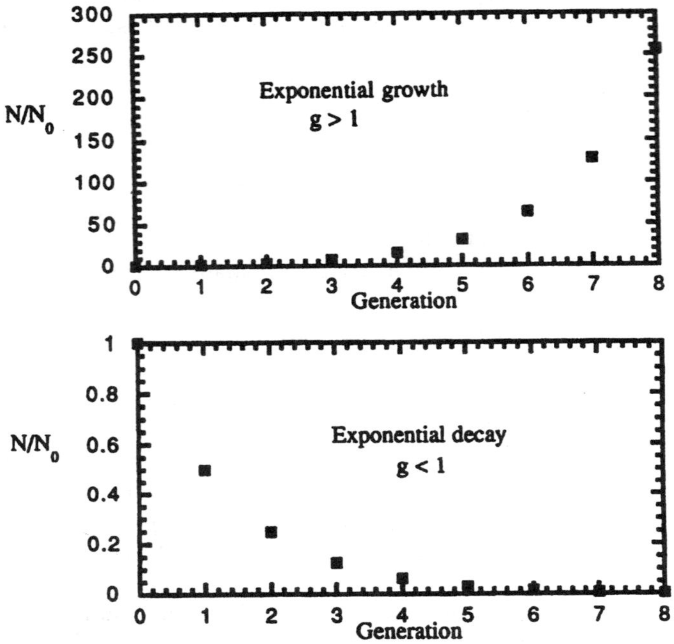

FIGURE 1 Exponential growth and decay.

tual history is their definition and use of dimensionless parameters." *B*, normally a man of hesitant speech was off and running, for evidently the subject was one that was dear to his heart. "As vivid imagery is to poetry, so are the dimensionless numbers to mathematical modelling, they establish magnitudes of entities within the context of the problem/poem. Take the opening stanza of Keats' poem, "The Eve of St. Agnes."

> St. Agnes' Eve—Ah, bitter chill it was!
> The owl, for all his feathers was a-cold;
> The hare limped trembling through the frozen grass,
> And silent was the flock in woolly fold:
> Numb were the Beadsman's fingers, while he told
> His rosary, and while his frosted breath,
> Like pious incense from a censer old,
> Seemed taking flight for heaven, without a death,
> Past the sweet Virgin's picture, while his prayers he saith.

It is so consummately, even if somewhat fulsomely, done that we shiver with the owl, we feel the sharp and sided grass, we press silently against our woolly neighbour, we see the cold in the persistence of the beadsman's breath and flex our fingers against the numbness of his. It establishes the coldness vividly in terms of the poetic environment. If we read: 'St. Agnes Eve—Ah, bitter chill it was! / The temp. was ten above, right on the line. / But everyone felt much, much colder 'cos, / The wind chill index made it minus nine. / The beadsman wanted both with minus sign, / So tried to change these temps. to

Centigrade / Which would have suited everybody fine, / If he had not a tragic error made, / Forgetting, in his haste, the part subtraction played,' we would know at once that we had a parody of Keats and a pretty poor one at that.

"But just as Keats' imagery makes us feel the magnitude of the cold, so, in a mathematical model, we form the ratios of the variables to naturally occurring combinations of quantities having the same dimensions."

"You've no need to explain that; palæographers do it all the time. We speak of the weight of a script in terms of the ratio of the height of a short letter such as *c*, *e*, *i*, *m* to the width of the pen nib it was written with. So a heavy Insular letter may be only two and a half pen widths high, but a light Rustic capital may be five or six. The absolute height of a letter is certainly a datum of the hand, but the way it looks, its character and identity, are more a matter of the letters' weight."

B looked so crest-fallen at having his grand exposition so preempted that *A* felt compelled to add, "I like your analogy to peotry. But I don't see a naturally occurring number to divide your population by. You can't divide by zero or infinity, if I remember rightly, and the initial population is quite arbitrary."

"Precisely," said *B*, "and that's because it's linear, $N_{m+1} = gN_m$, but suppose we made g a function of the population N_m, then the equation would become nonlinear and the fun begins. The simplest hypothesis would be to assume that there is a natural saturation level N_{sat}, at which the resources of the environment are so exhausted that the growth rate falls to zero, and the simplest way to do that would be to put $g_m = g(1 - N_m/N_{sat})$. The naturally occurring population is clearly N_{sat} and we can take the dimensionless population to be the ratio $x = N/N_{sat}$, or, for the mth generation $x_m = N_m/N_{sat}$, the dimensionless variable x where N had none. $N = 2000$ might be ridicuously large for a herd of elephants or hopelessly small for a colony of ants, but $x = 0.95$ (or 0.05, or 0.5) leaves you in no doublt that it is a large (or small, or middling) population for its environment. So $g_m = \gamma(1 - x_m)$ and, dividing $N_{m+1} = g_m N_m$ by N_{sat}, gives the defining equation

$$x_{m+1} = \gamma(1 - x_m)x_m.$$

All we need is the initial dimensionless population x_0 and we're off and running; with x_0 we use the equation to calculate x_1, and x_1 to calculate x_2 and so on."

By this time they had reached another bend in the road and were half sitting on a low wall. *B* pulled out a pad, "Here, let me show you a graphical construction."

"Before you do that," said *A*, "I want to ask you about this binocular vision business. You were making a big fuss, a few moments ago, about the need to keep both the physical and the mathematical viewpoints in balance. Haven't you gone overboard with the mathematics?"

"Heavens no!" said *B*. "I have not yet begun to mathematize. In fact I'll leave that to *C*. But you're right. I should have pointed out that we introduce ideas from the physics, or rather the biology, of the situation in positing a saturation population. The constant γ is also illuminated from the physical angle, for it is seen to be the maximum growth rate attained when the population is vanishingly small. In the assumption of a linearly decreasing growth rate, we were also trying to incorporate some biological realism. Had we been studying a particular population on which observations had been made, we might have used actual data or a function fitted to them. As it turned out, we just took the simplest function we could think of that satisfies $F(0) = 1$, $F'(x) < 0$ in $0 < x < 1$, $F(1) = 0$. The remarkable thing is that virtually any function that satisfies those conditions will give the same qualitative behaviour for our equation, which would be $x_{m+1} = \gamma F(x_m)x_m$. Of course, mathematicians like C worry about the weaker set of conditions that can be imposed on the function $F(x)$.

"We are still at the stage of encoding the problem, as John Casti calls it in his book *Reality Rules* [3]. Physical intuition plays a big role in setting up the model and it is often necessary to start with a simplified model and gain some experience on the road to building up a model, much as a sculptor or painter simplifies the masses or colour tones in first working out a satisfying composition, returning later to elaborate and finish the several parts. Once the physical system is encoded in the formal, mathematical one, it can be temporarily laid aside while the solution of the mathematical system is developed. The two viewpoints are still needed, however, for though the more that the model can be tied back to physical reality the better, this must not be allowed to restrict the free ranging of the mathematical imagination. I think this will come out more clearly if we let C take up the tale."

They had started up the hill again, and A and C stopped from time to time to let the puffing B catch up whilst C used his pad to explain the solution to A.

"The graphical solution that B was going to show you is really rather elegantly simple. Let's draw a unit square and, for any γ (see footnote[5]), draw the curve $y = f(x) = \gamma(1 - x)x$ and the diagonal $y = x$. If we start from a point x_0 on the base and go vertically up to the curve, we hit it at a point of height $y = f(x_0)$ and this, since $x_{m+1} = \gamma(1 - x_m)x_m$, is x_1. By going horizontally to the diagonal and dropping back vertically to the base again, we have come to the point x_1. Repeating the process, we get from x_1 to x_2, and so on. For example, if $\gamma < 1$, the curve $y = f(x)$ lies wholly underneath the diagonal (Figure 2, $\gamma = .75$). Successive steps take us quickly to extinction, for $\gamma < 1$ is like $\gamma < 1$ in the linear case, only more so since $x_{m+1} = \gamma(1 - x_m)x_m < \gamma x_m$."

They had stopped so that C could draw the descending staircase (Figure 3) of iterates on his paper and A, stepping back, swung an arm out in a vatic gesture:

> Facilis decensus Averni:
> Noctes atque dies patet atri ianua Ditis
> sed revocare gradum superasque evadere ad auras,
> Hoc opus, hic labor est.[6]

[5] We will keep $\gamma \leq 4$ so that (0, 1) is mapped onto itself. The figures have been redrawn from the very rough efforts of C.

[6] "Easy is the descent of Avernus! Night and day the portals of gloomy Dis stand open; but to reverse your step and to win again the daylight—that's labour, that's hard work," Virgil, *Aeneid*, vi, 126.

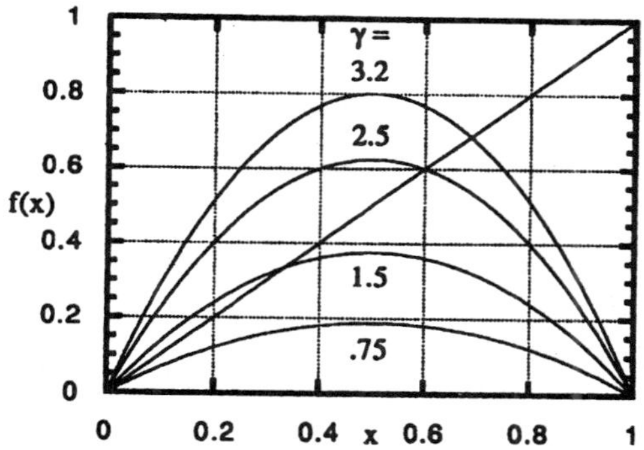

FIGURE 2 The curve $y = f(x)$ for several γ.

"What's that all about?" said B, ducking, for, puffing up from behind, he had almost run into A's outflung hand.

"Pardon me, I was just detonated by C's staircase," said A, making sure he had not actually hit his friend. "It reminded me of Aeneas' descent into the underworld; not far from here actually, a few miles south near Naples. But I see it can be dangerous to interrupt, so go on, C, before I get carried away again."

C had taken the opportunity of sketching the next figure whilst they were stopped and passed it over to A as they continued (Figure 4). "That descending staircase was for $\gamma < 1$. If $\gamma > 1$, the curve $y = f(x) = \gamma(1 - x)x$ has a nontrivial intersection with the diagonal at $x = 1 - (1/\gamma)$. The intersection at $x = 0$ is called trivial because it corresponds to the possibility of extinction which is always there, but not very interesting. But the intersection at $x = 1 - (1/\gamma)$ is interesting for it is a steady state of the population. If I start exactly at this point, I shall remain there for all succeeding generations, i.e., if $x_0 = 1 - (1/\gamma)$, then $x_m = 1 - (1/\gamma)$ for m, 1, 2, 3, ... If I start at a low

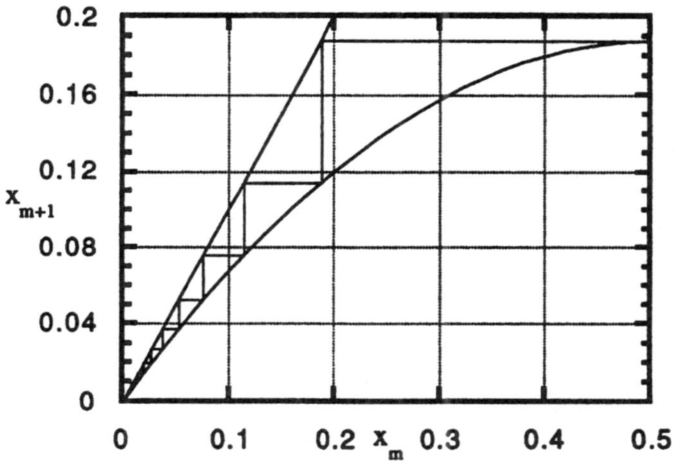

FIGURE 3 Staircase to extinction, $\gamma = 0.75$.

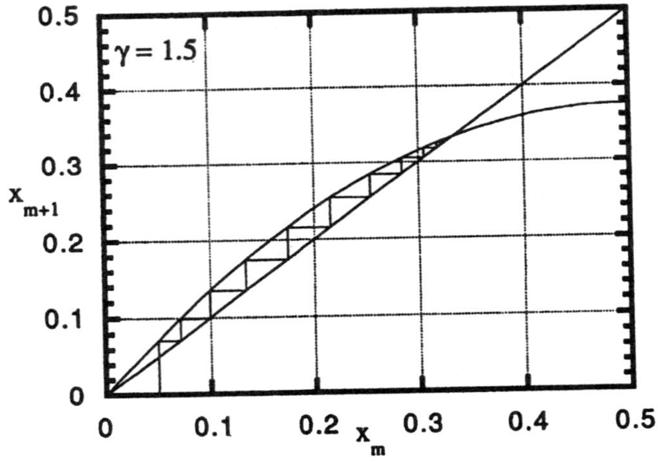

FIGURE 4 Monotonic approach to a steady state.

population and follow the construction I sketched just now (Figure 4), I get an ascending staircase that approaches the steady state from below. There's your labour, A, there's your work! If the population is greater than the steady state population to begin with, then the population decreases in successive generations and approaches the steady state from above. From the physical viewpoint, the steady state is clearly stable, since if some perturbation were to increase the population temporarily, it would come back down again, while, if the perturbation decreased the population below the steady state level, it would go back up again. Mathematically we need a criterion for this stability in terms of the parameter γ. We see that the reason that the staircase ascends from below is that the slope of the curve $y = f(x)$ is less than that of the diagonal at the point of intersection; i.e., $0 < f'(1 - 1/\gamma) = 2 - \gamma < 1$ or $1 < \gamma < 2$. When γ gets above 2, the slope $f'(x)$ at the steady state is negative, and the construction of the solution by successively going vertically to the curve and horizontally to the diagonal takes us into the steady state by spiraling inwards (Figure 5). If, however, the slope is less than -1, we have a spiral that goes outwards. The steady state is stable if the slope of $f(x)$ at the steady state is between -1 and 1, in other words, $1 < \gamma < 3$.

"We notice that the steady state exists beyond $\gamma = 3$, but it is unstable, and no sequence x_m generated by the equation will go to it. But no sequence of points generated by the equation will go to the trivial steady state either, for this is unstable when $\gamma > 1$. Where, then, does the trajectory, as this sequence of points is called, go when γ is slightly greater than 3? If we were to put a microscope on the steady state for $\gamma = 3$, i.e., $x = 2/3$, we would see two lines, one absolutely straight, the diagonal $y = x$ at 45 degrees, and the other as near straight as makes no matter, since the microscope enlarges the curve so that a little segment of it is virtually identical with its tangent. This tangent would be at -45 degrees and the two would be the diagonals of a small square around the steady state. What we are looking at is the birth of a period-two solution in which the population alternates between two values,

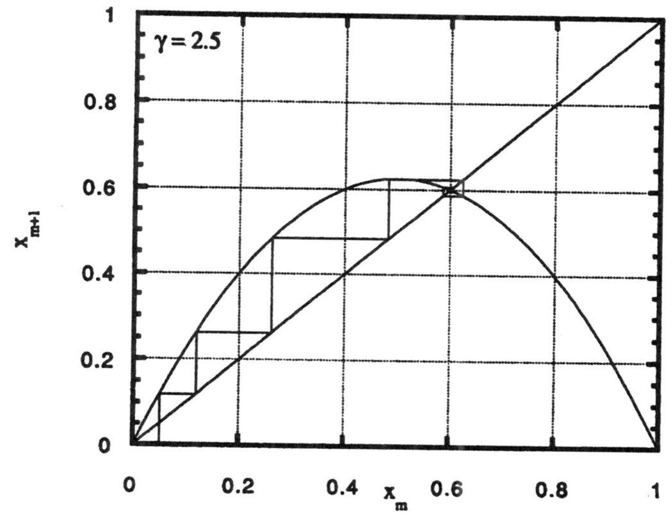

FIGURE 5 Stable steady approached spirally.

taking one in even years and the other in odd. Either of these two values will satisfy $x_{m+2} = x_m$, so they are the fixed points satisfying $x = f(f(x)) = \gamma^2 x(1 - x)\{1 - \gamma x(1 - x)\}$. (Figure 6). This is a curve of the fourth degree, and its four solutions give us back the trivial $x = 0$, the nontrivial, but unstable, steady state $x_1 = 1 - 1/\gamma$ (for a steady state cannot help but be a period-two solution as well), and the two values $x_{2,1}$ and $x_{2,2}$ between which the period-two solution alternates (in Figure 6 these are .513 and .799). It is almost impossible to draw the approach to the period-two solution freehand, but

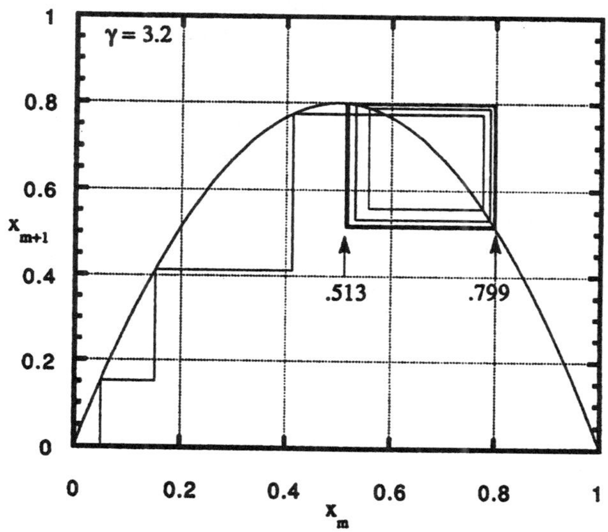

FIGURE 6 Period-two solution for $\gamma = 3.2$.

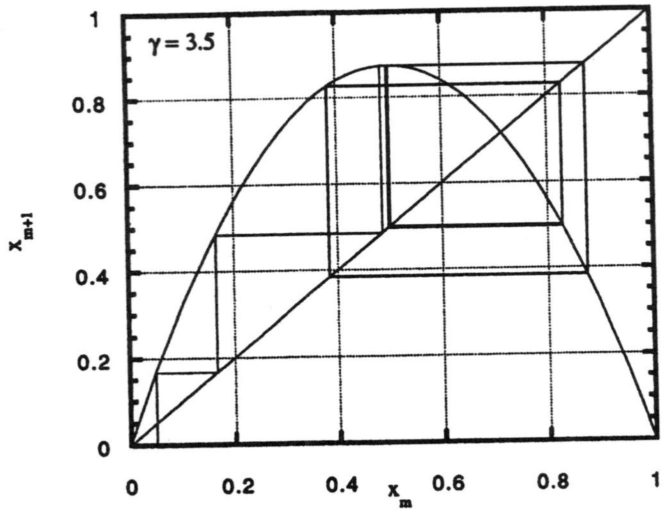

FIGURE 7 Period-four solution.

there is a test for the stability of this also, namely, $|f'(x_{2,1}) \cdot f'(x_{2,2})| < 1$. In the microscope's field at exactly $\gamma = 3$, the period-two solution is born as an infinitesimal square whose centre is the period-one solution or steady state. But as γ increases slightly above 3, the nonlinearity of the function $f(x)$ turns it into a small rectangle that grows rapidly as γ increases. It is at first stable, but at about $\gamma = 3.45$, it becomes unstable and gives birth to a period-four solution (Figure 7). The period-two solution still exists as do the period-one (the nontrivial steady state) and the trivial steady state (extinction), but they are all unstable. This means that you may start as close to any of them as you please, but you will find that you move away. From a physical viewpoint, they are unreachable, but the mathematical viewpoint comes to the rescue, for there is no reason why you cannot run the equation backwards and calculate x_m from x_{m+1}; f is a quadratic which may be solved to give $x_m = (1/2) \pm \{(1/4) - (x_{m+1}/\gamma)\}^{1/2}$. This is one of the virtues of the mathematical modelling approach, that you can study unstable things, as readily as stable, by reversing time.

"The period-four solution loses its stability even more rapidly than the period-two lost its, and gives birth to a period-8, which gives birth to a period-16, and so on *ad infinitum*. In fact, the distance between one transition and the next gets shorter by a factor of about 4.67 each time, and the sequence of critical values of γ at which the successive solutions of period-2^n are born, $\{\gamma_n\}$, converges on a finite $\gamma_\infty = 3.5699456$. More precisely, $(\gamma_n - \gamma_{n-1})/(\gamma_{n+1} - \gamma_n)$ approaches the Feigenbaum constant 4.6692016091029 ... , so named after its discoverer [4]. But what is so marvelous about Feigenbaum's constant is that it does not depend on the exact form of the function $f(x)$ used in the iteration. Any continuous function that maps [0,1] into [0,1] and has a single, quadratic maximum (i.e., $f''(x)$ is not zero there[7]) will exhibit the

[7] It must also have a negative Schwarzian derivative, $(f'''/f') - (3/2)(f''/f')^2 < 0$.

period doubling scenario in which the ratio of distances between successive transition values of the parameter converges to Feigenbaum's constant."

C paused for breath and *A* interjected, "But isn't this class of transformations a pretty unrealistic thing? You surely can't claim it has anything to do with the real world."

C was about to bridle at this, but *B*, who had been looking for an excuse to stop for a moment, saw his chance. "I can answer that in two ways," he puffed, planting himself in the way of the other two. "The period doubling scenario, as *C* called it, arises both in other much more sophisticated and realistic models and is actually observed in experiments. In the early '80s, a number of experiments were done by physicists and engineers in acoustics, laser physics, chemical reaction, electronics, and hydrodynamics with three, four, or five successive period doublings. Although you could not expect much agreement so early in the sequence (three transitions only gives you one Feigenbaum ratio), the ratios were all in the ballpark of the Feigenbaum constant and often within experimental error [5, Table, p. 618]. Then again, in quite realistic models of mechanical vibrations or chemical reactors, period doubling is quite common. Think of a chemical reactor in which two reactions are taking place. Because there are two reactions, we need two concentrations, together with the temperature, to specify its state (Figure 8a). It can be described, therefore, by a point moving around in three-dimensional space. If it goes to a stable steady state, the point will follow a curve that goes to the point representing the steady state and slow down as it approaches it until you can't see it move—it just sits there. It could go on a smooth, monotonic curve directly to the steady state point, but let us suppose that it spirals inwards. Now as we change the flow rate to the reactor (this is the tunable parameter like γ in the population model), we find that the spiral gets slower and slower until, at a critical flow rate, it starts to spiral out away from the steady state. This is because the steady state has become unstable, repelling, rather than attracting nearby states. If it is the only steady state there is, we again ask the question that arose in *C*'s logistic model when γ passed through the critical value 3; where does the trajectory go? The answer is similar (Figure 8b). Just as the discrete system developed a periodic oscillation between two states, so the point representing the state of the reactor spirals outwards and finally settles down on a closed loop. Every time it comes back to its starting point it has completed a cycle. It is at first (i.e., for flow rates not much different

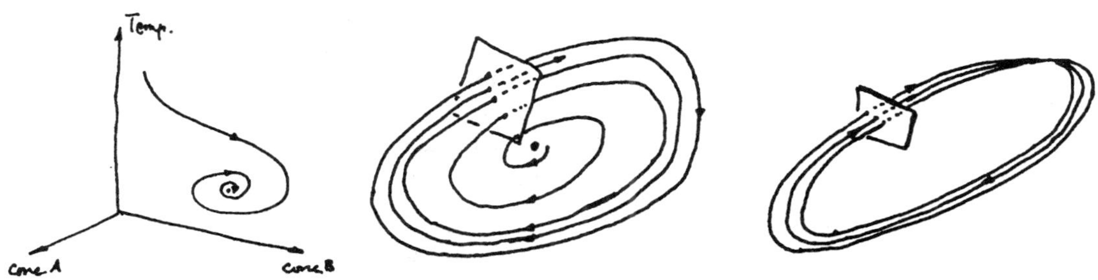

FIGURE 8

from the value at which the steady-state becomes unstable, corresponding to γ slightly greater than 3); you start the state on the loop, it will stay on the loop; if you start it nearby, it will approach the loop. Now imagine a little plane set up transverse to the loop. As the trajectory spirals its way onto the stable loop, it will pass through this plane every time it comes around. Since the loop is stable, these points of passage will converge on the point at which the loop itself penetrates the plane. This is called a Poincaré map, after the great French mathematician who initiated so much in the theory of dynamical systems around the turn of the century. The periodic solution, represented by the loop, will be unstable if the points of passage diverge rather than converge. The critical situation will be when a point starting very close to the loop on one side comes around to a point almost exactly the same distance from the loop but exactly on the other side (Figure 8c), for it follows that if it goes around again it will be virtually where it started. But this really makes it a periodic solution of twice the period of the original. So at a critical flow rate the period of the cycle in which the reactor is running doubles. It is as if the centre line of a narrow Möbius strip splits, and the halves move out the strip's edge."

B was just about to tear a strip off the edge of his pad and demonstrate when *A* surprised him with: "Oh, I know what a Möbius strip is. It's made by putting half a twist in a long strip before making it into a loop. It has only one side and only one edge. I once saw that well-known line from the Pervigilium Veneris

> Cras amet qui nunquam amavit,
> Quique amavit cras amet[8]

inscribed on a Möbius strip. The first two and the last two words being the same, they did not need to be repeated."

"Very interesting," said *C*, without much conviction, for he was a bit miffed at the interruption of his exposition, "but I hadn't quite finished with the logistical equation."

They resumed the ascent and *C* continued. "In this sketch, which I did whilst we were standing around, you'll see the period-two solution that is stable from $\gamma = 3$ to 3.44949 (Figure 9 for $\gamma = 3.2$), but this time I've shown it as the population against the generation number, n. It is clear that after ten iterations it has settled down to its periodic oscillation between .513 and .799. Similarly for $\gamma = 3.5$ (Figure 10), which is in the interval (3.44949, 3.54409) in which the period-four solution is stable, this sketch shows the periodic solution closely approximated after ten generations. But what happens when we get just beyond the limit $\gamma_\infty = 3.5699456\ldots$? Then the solution becomes chaotic (Figure 11), being neither periodic nor asymptotic to a periodic solution. You might think, to look at another presentation of the solution for $\gamma = 3.57$ (Figure 12), that it is completely random. But this is not so; there is a subtle structure built into the chaos that develops."

[8] In Parnell's rather free translation, "Let those now love, who never love'd before:/Let those who always lov'd, now love the more."

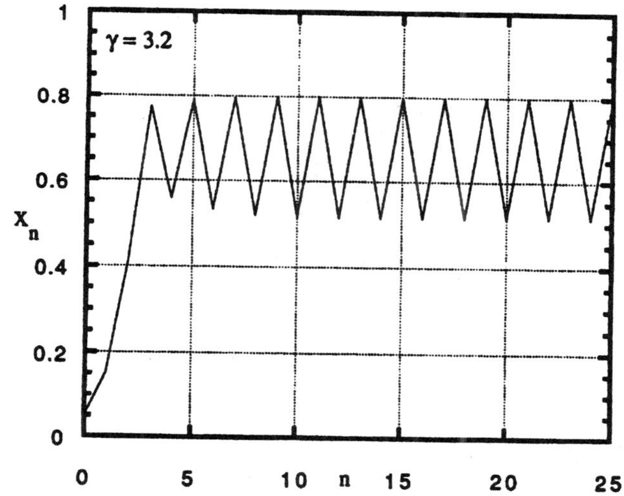

FIGURE 9 Period-two populations versus generation.

"You're losing me, my dear fellow," said A, not without a touch of irony.

"Yes, you'd better give over and let A have a chance to tell us about his manuscripts," said B, who felt that C was in danger of monopolizing the conversation. But C was not to be denied.

"Just give me two minutes," he said, "to sketch the way the subject develops.[9] Just look at this xerox summarizing the behaviour of this system

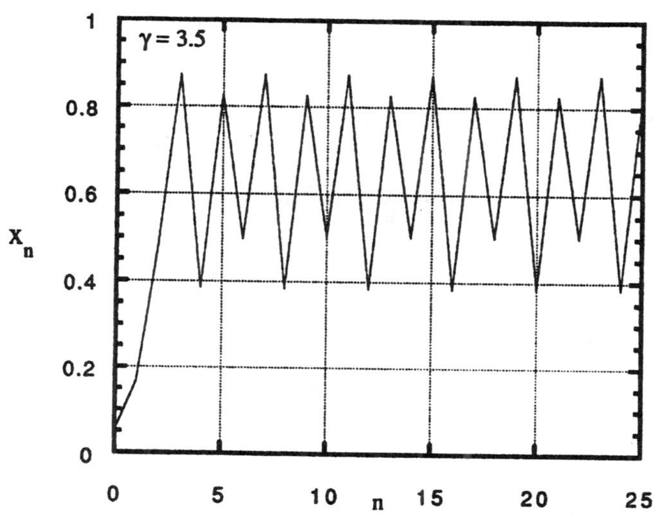

FIGURE 10 Period-four population as a function of generation.

[9] C's exposition follows on the lines of the excellent discussion of 'period doubling and its chaotic mirror' in [5, Chapter 11].

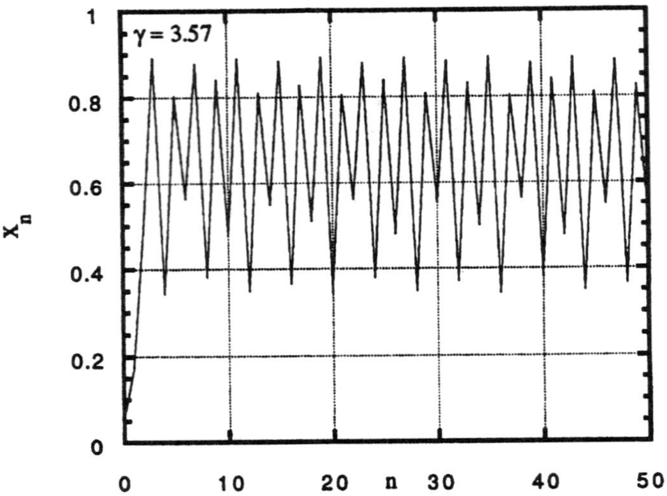

FIGURE 11 Time course of incipient chaos.

(Figure 13). It is made by running the iteration for fixed γ a couple of thousand times to get rid of the transients and then plotting the next couple of hundred values that x_m takes in the vertical line above γ. When γ is between 1 and 3, this will give a single point, for the stable steady will have been reached and all two hundred points will be on top of one another. When γ is in (3, 3.44949), the two branches of the stable period-two solution will emerge, since the points will alternate between the two populations. The branches keep on bifurcating at a succession of values of γ that closes down by a factor that

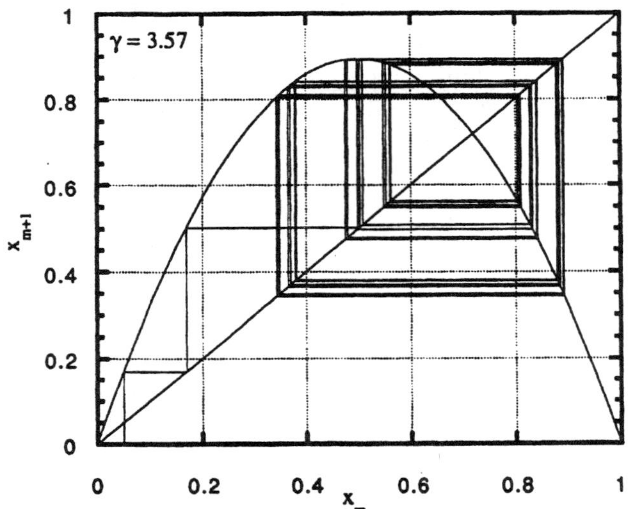

FIGURE 12 Incipient chaos.

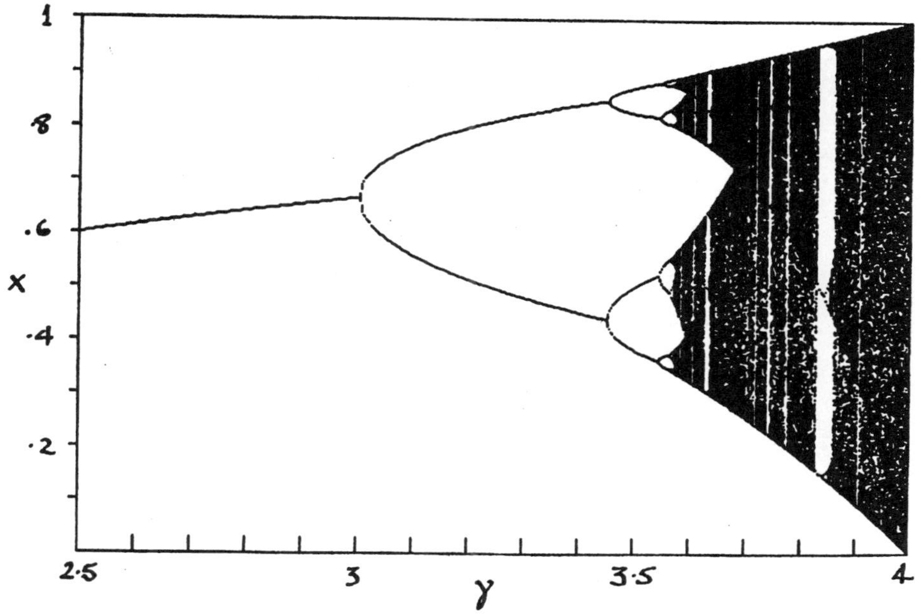

FIGURE 13 Ultimate states of the logistical population.

rapidly approaches the value of Feigenbaum's constant. This tree is in a loose sense self-similar. The upper and lower branches that come off the single stem are themselves trees of the same topological structure. The scale decreases by a factor that is asymptotically 4.669 in the γ direction and about 2.3 in the other direction. At the limiting value $\gamma = 3.5699$, there are infinitely many points which form a truly self-similar Cantor set. By studying the ways in which the periodic solution moves between the various upper and lower branches, we can get a reading on the dynamics on the Cantor set.

"Then, for γ greater than γ_∞, we find more grey areas, where it seems that the dynamics is chaotic, but there is also a succession of windows where stable periodic solutions are the only ones to be found. The last of these is from $\gamma = 3.828$ to $\gamma = 3.857$ where period-three solutons exist. Even the structure of this is beautifully self-similar. And then there's the Charkovsky sequence. ... But do just let me mention the sensitivity question.

"Chaotic solutions are those which are neither periodic nor asymptotic to a periodic solution but are characterized by extreme sensitivity to initial conditions. A solution that is asymptotic to a stable periodic solution is not sensitive to starting point, for, if we start from two nearby values, the trajectories will both converge on the same periodic solution and get closer and closer together. With a chaotic solution, the trajectories starting at two nearby points ultimately diverge no matter how close they may have been at the beginning. If $f_{(n)}(x)$ denotes the nth iterate,

$$f_{(n)}(x+h) - f_{(n)}(x) = h \cdot \left(\frac{d}{dx}\right) f_{(n)}(x) + O(h^2),$$

and it is natural to study the limiting behaviour of the derivative. In particular, the Liapounov exponent

$$l = l(a) = \lim\left(\frac{1}{n}\right)\log\left|\left(\frac{d}{dx}\right)f_{(n)}(x)\right|$$

can be found (it is independent of x), and this provides a measure of the rate of divergence of two nearby points, since

$$|f_{(n)}(x+h) - f_{(n)}(x)| \sim hAe^{n\lambda}.$$

Thus, if λ is negative, there is no divergence or sensitivity, whereas for positive λ there is. Just where chaotic solutions occur is often very difficult to determine, for direct calculation has to be finite in length, or rounding error would be overpowering, and so an apparently chaotic solution might be part of a solution of very large period, and it is only by checking the Liapounov exponent that we can be sure."

C trailed off into silence as it dawned on him that he was talking to himself. The other two had stopped at the turning to the Polish cemetery, a grim reminder of how bitterly this ground had been fought over as recently as fifty years before. They were some yards behind, but catching up.

"You were so intoxicated by the exuberance of your own verbosity," said *B*, "that we didn't have the heart to interrupt."

"You missed the good part about sensitivity to initial conditions," said *C*.

"I wouldn't have understood it anyway," admitted *A*.

"No, but *B* might have," said *C* loftily, "and it would have given him a good example of binocular vision, for it's where the intuitive, geometrical viewpoint needs to be checked by rigorous, analytical reasoning. But he evidently doesn't need the support of a pure mathematician."

"Let's get *A* to tell us something of the history of this place before we get up to the gate," said *B* by way of covering up *C*'s huffiness for which he felt a little responsible. "We can return to binocular vision when we've heard his spiel."

"To give you a thumbnail sketch of its history: the Abbey of Monte Cassino was founded in 529 by Benedict of Nursia, he being then about 49 years of age. Here he composed his Rule for Beginners—a document that was to become the foundation of western monasticism—and here he died at the age of 63. The complex was destroyed by the Lombards about 580 and the monks fled to Rome where the community spent the next 137 years. Under Abbot Petronax, the monastery was rebuilt in the early part of the eighth century and came to be regarded as the exemplar of Benedictine tradition and was visited by many abbots and others seeking to know more of how the Rule should be worked out in practice; Charlemagne himself visited in 787. At the turn of the ninth century, the buildings were considerably extended, but in 883 the Arabs sacked the place and killed the abbot and some of the monks, but some managed to escape to Teano. Even here misfortune dogged them, for a fire destroyed their treasured autograph of the Rule and drove them on to Capua, where they spent the next 67 years. However, the two centuries after their return in 950 were prosperous both in the public and the intellectual life. It was an important centre for the transmission of the classics,

Varro's *De Lingua Latina,* some of Tacitus' histories and Apuleius' *Golden Ass* are preserved only through copies made at Monte Cassino. Perhaps the zenith of the monastery's influence was reached under Desiderius, who was abbot from 1058–87 and went on to be Pope Victor III, and his successor Oderisius (1087–1105). The abbey took its present shape under Desiderius' building, for it was rebuilt faithfully after its destruction by earthquake in 1349 and, though it suffered some Renaissance embellishments, the basic layout of Old St. Peter's is still preserved. Its fourth destruction was by saturation bombing on the fifteenth of February 1944. Fortunately, the manuscripts and movable art had been taken to Rome and were preserved there until the rebuilding was complete. So you will see some new frescos, but the manuscripts will be those that Mabillon[10] saw during his visit in 1685.

"It was to study the transmission of the classics through Montecassino that Traube sent his doctoral student, the young E. A. Lowe, to the abbey in 1912. Lowe quickly realized that the key lay in the history of the peculiar script of the region and devoted himself to the study of this, producing that 'perfect model both of palæographic method and of style,' *The Beneventan Script: A History of the South Italian Minuscule* in 1914 and two folio volumes containing a hundred plates—the gift of Oxford University to the Abbey of Montecassino on their fourteen hundredth anniversary in 1929. I want to show you some of the manuscripts when we get up there, but I need a breather after that last zag, so let's sit before we zig."

"You won't forget that you promised to address my thesis that at least two viewpoints are needed to see anything in depth," said *B*.

"No," replied *A*. "In fact I can give you a brief example while we sit here." He rummaged in his backpack and brought out a xerox of a page of manuscript. "Why do you think this is called the script of Luxeuil? Luxeuil is a monastery in eastern France, founded in 590 by St. Columban."

"I suppose," said *B*, "that you've got some manuscript with a sign-off at the end: 'I, Gondohinus, finished this writing on the afternoon of January 26, 876. Now let's go for a drink!' "

"Not quite, though there are colophons—please, not 'sign-offs'—and marginalia even more racy than that in some manuscripts. As a matter of fact, no book written in the Luxeuil script can be securely placed at Luxeuil, though there is a manuscript in the Pierpont Morgan Library[11] that is dated and placed by its colophon (*apud coenubium Lussovium anno/duodecimo regis Chlothach*) which fixes it at Luxeuil in 669. Now this Ms. has a display script—one used for headings—that has several distinctive features in common with the display scripts of seven other manuscripts whose texts are written in the Luxeuil minuscule. There is the tendency to slimness and the bifurcation of finials in *T*, *N*, *V*, *E*, *X*, the *A* with a *v*-like cross bar, *G* with a spiral, *H* with the sinuous bar, the lozenge shaped *O*, and the unusual form of *X* (see

[10] Jean Mabillon (1643–1707), a Benedictine of the Congregation of St. Maur resident at St. Germain des Près, is reckoned to be the founder of modern palæography (the study of scripts) and diplomatic (the study of documents). He published his *De Re Diplomatica* in 1681 and made three major *voyages litteraires* in the next few years. One of these was to Italy and lasted eighteen months. He spent some time in Rome and went as far south as Naples.

[11] MS. 334, Plates V,1 and B,3 of *EFL*, J. P. Morgan Lib., New York. Figures 14–19 are 60% facsimile.

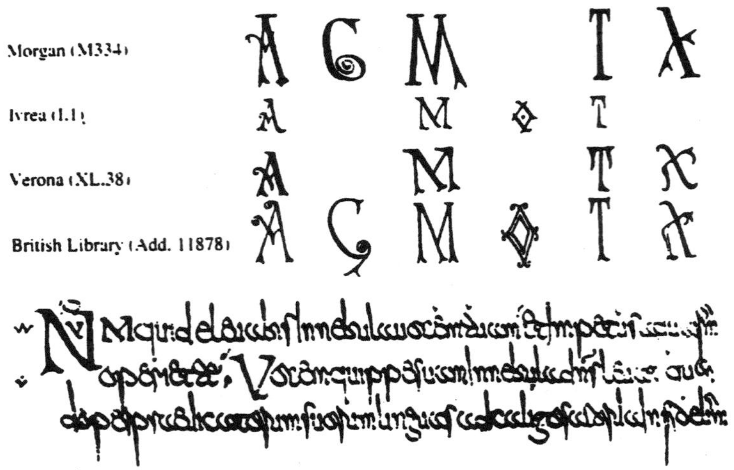

FIGURE 14 The display capitals in some Luxeuil manuscripts and a few lines of the minuscule.

footnote[12]). Here's a sampling of these similarities, given in tabular form to illustrate Lowe's argument that this constellation of common features cannot be accidental, but must be the product of a definite attempt to create a house style in the scriptorium (Figure 14). You see the artistic and historical viewpoints at work. Here he is using the *occulus calligraphicus* and viewing the stiuation from the scribal point of view, asking how a script type is created. 'Style,' writes Lowe, 'is invariably the creation of a single master. The master may use, and often does use, material familiar to many, but the stamp of his individuality, his particular use of his skill, gives it style.' It is this style that Lowe sees in the Luxeuil script and in the distinctiveness of the display script that allows him to associate the two. But just think of the ocular memory that allowed him to remember manuscripts as widely scattered as these, for they are to be found in London, Ivrea, Verona, Paris, Leningrad, and the Vatican. No wonder T. A. M. Bishop used to refer to Lowe, who was still alive when I went to Bishop's classes in 1964–1965, as 'the greatest of living scholars'—strictly speaking meaningless, but a pardonable exaggeration."

By now the three friends were on the move again and came rather quickly on the main gate. *A* took *B* and *C* for a quick you-can-come-back-and-see-it-later-if-you're-interested tour of the place. This was partly to see what manuscripts were on display in the museum, for he did not want to call for one that was already visible. There was a calendar with a worn footnote to the effect that Thomas Aquinus had been a novice at Monte Cassino before he joined the Order of Preachers. "Mabillon saw that very page," said *A*. "He comments on it in the account of his journey, the *Iter Italicum*." Then they went back to the Library, where *A* addressed the librarian in halting Italian, but with such an execrable accent that the librarian quickly said, "Linguam latinam loquamur," and they conversed slowly but surely in Latin. *A* was vastly pleased to find that Pomerius' *De Vita Contempletiva* was on

[12] A single example was given. Ivrea Bibl. Cap. I (1).

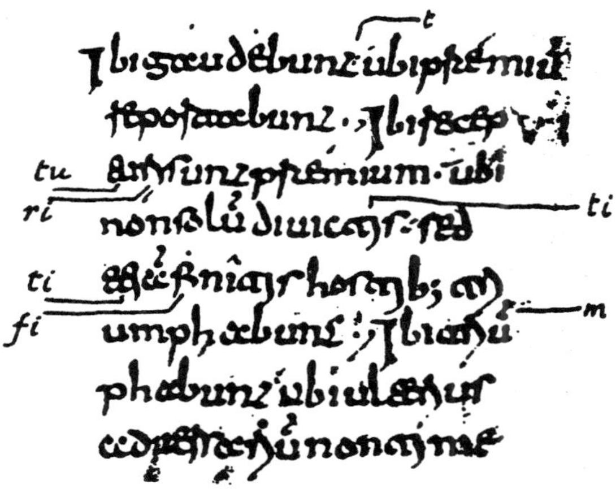

FIGURE 15 Beneventan script of the tentative period (s. VIIIex–IX).

loan from Naples for an up-coming exhibition, and he asked if he might show it to *B* and *C*.

"This manuscript (Figure 15) is dated between 817 and 835, and belongs to what Lowe calls the tentative period—from the end of the eighth through most of the ninth century—when the script was first emerging as a regularization of the kind of cursive charter hand that was familiar to scribes. The hand is bold but irregular and the rules of the script, which later become absolutely *de rigeur* are not yet formulated. In this example, however, I would draw your attention to the ligaturing of certain letters: *nt* and *ræ* in the first line; *turi* at the very beginning of the third; *etiam finitis hostibus* in the fifth. This last shows the distinction, which was later to become absolute, between assibilated and unassibilated *ti*. The former, occurring when *ti* comes before a vowel, but not after *s*, was like a backwards beta; the latter was a *j*-like stroke depending from a *t*. Stylized ligatures of *i* following *e*, *f*, *g*, *l*, *r* were also developed and *t* regularly stood on its head in ligature with a following *a*, *e*, *u*.

"By a careful examination of many manuscripts and with the use of that marvellous eye and vast memory, Lowe was able to chart the development of the script from these tentative beginnings through the formative period of the tenth century (Figure 16) when the script becomes more fluent and rhythmical through what seems to have been a conscious effort to regularize it. At this point, he is using his *oculus calligraphicus,* looking for regularity of alignment and word separation, noticing that the letters have taken a normalized form, yet observing that there is 'a certain freedom of form as opposed to later strictness and conventionality.' At the same time there is ample exercise of his supreme *oculus palæographicus* which sees that the ligatures with enclitic *i* have by this period become obligatory, the *ti*-distinction is strictly observed, and the use of *i*-longa is as definite as it ever becomes. The two eyeballs are comprised in the pair of eyes and work together so harmoniously in the living being that I risk having to kill off my specimen in thus anatomizing them.

"The high point of the script is the eleventh and twelfth centuries, in particular the period of Desiderius' abbacy (1058–1087), for it was Desiderius

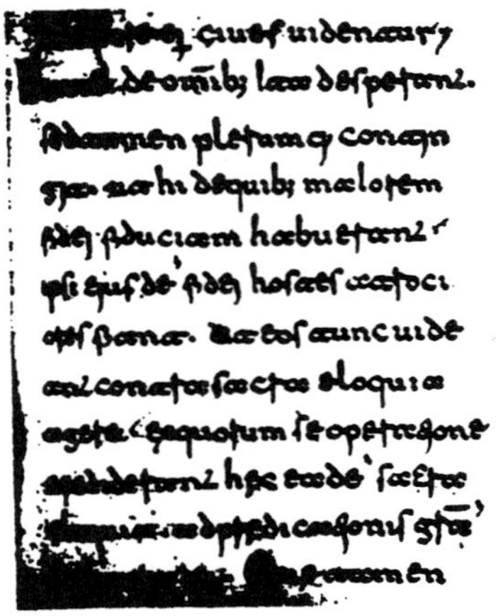

FIGURE 16 Beneventan script of the formative period (s. X).

who rebuilt the monastery and greatly enriched its art and library. At its best, Beneventan has immense coherence and strength, though it is not always easy to read. We're in luck, here's a Desiderian manuscript on loan from the Vatican and it's the one Lowe used to illustrate the period of maturity. The first three lines here (Figure 17) read *mente tranquil / la conuocatos / frs allocutus est*—where *frs* is an abbreviation for *fratres.* In *allocutus* you see how similar *a* and *oc* are. Yet the script has a stately rhythm. If you look at the whole page (from which the previous figure was taken), you see it is one of the magnificent products of the Desiderian period—a lectionary, written by Leo, prince of Beneventan scribes. Here is another of the books he wrote[13] which can be dated to 1062 by an inscription on f. 1v and a couplet

Huius scriptorem pie Christe *Leonem*
In libro uitae dignanter supplicio scribe.

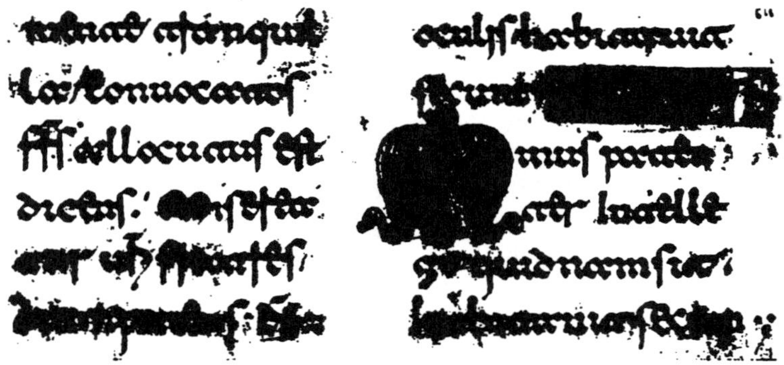

FIGURE 17 Beneventan script of the time of Desiderius (1058–87).

[13] Monte Cassino 99.

It contains this splendid opening to a homily on Isaiah 55, with a lavishly decorated *O* filling half the page. The other four letters of the opening word *Omnes* are written in the border at the four points of the compass and the second word *sitientes* in two golden tabs at either side. Look at it close-up and you'll see the *N* and the *S* and, rather less visible, the *SITI* of *sitientes* in the first tab. Leo's instructions to the illuminator can be seen in the margin. Then the text goes on in Leo's fine hand *uenite ad aquas*—'Ho, everyone that thirsteth, come ye to the waters, and he that hath no money; come ye, buy, and eat; yea, come buy wine and millk without money and without price. Wherefore do ye spend money for that which is not bread? And you labour for that which satisfieth not?' The thunder of Isaiah reverberates in the cumulus rhythms of the script, carrying the reader forward with its measured urgency. The reader is aided by a peculiar feature of Beneventan punctuation. The question mark has a short, angled upper stroke with two dots below it and comes, like our question mark, at the end of the question. But look back to the first line where a small bird-like *v* is flying over the *a* of *Quare.* This is a sign to the reader that a question is at hand and must be anticipated with the rising inflexion in the tone of voice.

"The ordinary Caroline minuscule began to rival the Beneventan in the 12th century. The prestige of Montecassino declined and so did the dominance of the Beneventan script. It loses its rhythm and becomes angular and fractured in appearance (Figure 18).

"There was a dialect of the script practiced in Bari and in the monasteries around the Adriatic. It is of lighter aspect, as judged by the height of the letters in terms of pen widths, and shows traces of Byzantine influence. Tall *s* and *f* do not go below the line except sometimes in the *f-i* ligature as in the last word of the first line of Figure 19, *fidelibus.* It had a fondness for the double decker *C* that looks like a Greek epsilon, and for the division sign as an abbreviation for est. The *C* can be seen in the word *benedic* on the second line just before the smudged word at the right hand end. The use of thin-bodied initials with pearls between the interlace, such as the *B* at the top left of plate *D* is also characteristic.

"A most interesting volume, also on loan from Naples, is a very large

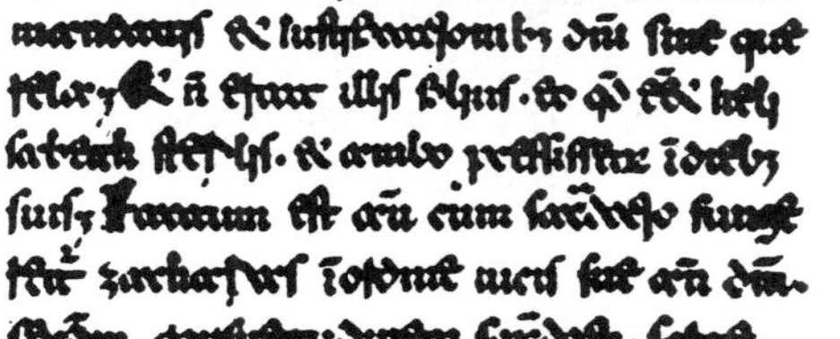

FIGURE 18 Beneventan script of the period of decline (s. XIII).

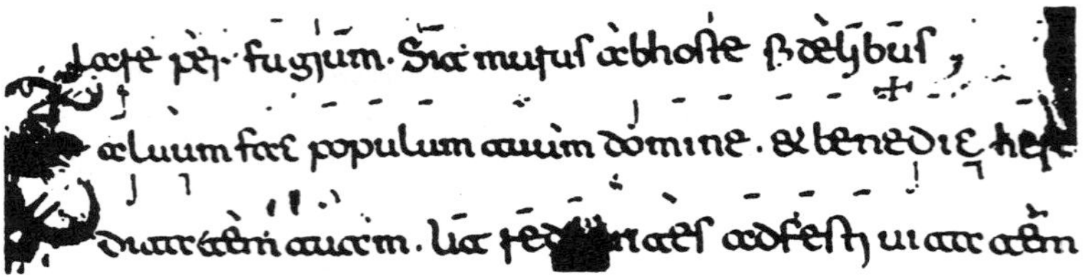

FIGURE 19 The Bari dialect of Beneventan script.

manuscript of Haymo's Commentary on St. Paul's Epistles.[14] This was copied by Ascarus of Carminiano who began his task on Thursday, 1 February 1145, being twenty years of age, as we learn from a note on the first page and an inscription on f. 219:

HIC LIBER FINITUS
ATQUE SCRIPTUS EST
DIGITIS ASCARI
ECCLESIE SANCTI
LAURENCII IN CARMI
NIANO SUBDIACONI
ETAS CUIUS ANNOS
VIGINTI OCCUPABAT.

Lowe at first thought, no doubt on the strength of 'occupabat,' that this meant that Ascarus spent twenty years of his life on the task, but later corrected the date.[15]

"So there," said *A*, "you have a short history of the Beneventan script. Now, if you'll excuse me, I really ought to get down to work. Perhaps we could have dinner together this evening."

"I've got an example of wildly imaginative viewpoint to try out on you," said *B* as they pushed back their chairs and took another cup of cappucino into the lounge after dinner that night. They had hashed over the viewpoint question throughout dinner and had agreed that there was always a need to keep the other eye open. The interaction of physical and mathematical insight when trying to build up a useful mathematical model had been declared by *B* to be "of the very essence"—this with great conviction when the second bottle of Frascati was running low! *C* chimed in with a reference to the importance of elegance, as well as rigour, in mathematics and quoted Croce on intuitive knowledge obtained through the imagination and logical knowledge obtained through the intellect.[16] *A* said he'd been amazed to read that even

[14] Naples, B. N. VI.B.3.

[15] In the description of Pl. LXXXVII in *Scriptura Beneventana*. See V. Brown's corrected edition of Lowe's "The Beneventan Script."

[16] The opening sentence of Croce's *Æsthetic* (tr. Douglas Ainslie), 1909.

physicists seemed to set great store by beauty as a criterion of truth.[17] The conversation was winding down when *B* offered his wild flight of imagination, so *A* said, "Okay, just one more model," in a resigned tone and *C* looked pained—or would have done had he not, in a very satisfying gesture, ordered cognacs all round.

"We are introduced fairly early in life to the platitudes of modern physics," began *B*, whose orotundity seemed to grow with his rotundity. "A gas we believe to be a very large number of very tiny molecules moving randomly at very high speeds colliding with each other and with the walls of its enclosure and so always capable of filling it. A solid, on the other hand, though it consist of the same tiny molecules, has them locked tightly together in an array that has certain regularities of structure which has its own boundaries independent of any enclosure; a liquid is in an intermediate state, conforming partially to its surroundings.

"An important interaction of a gas with the wall of the vessel that contains it is known as adsorption, a phenomenon in which there is the build up of molecules of the gas on the wall. This is not a static phenomenon, in which some of the molecules are 'wall molecules' and some 'gas molecules,' but a dynamic process in which molecules are constantly hitting each part of the wall, some of them staying for a certain time and then departing. If n molecules hit each square centimetre of surface per second and stay on it an average of t seconds, the number of molecules per square centimetre will, on the average, be

$$s = nt.$$

Now n is directly proportional to the pressure in the gas and inversely proportional to the square root of the molecular weight and the temperature. For oxygen at 20 degrees Celsius and 760 mm of mercury

$$n = 2.75 \times 10^{23} \quad \text{mols. per sq. cm per sec.}$$

"How can we get any realistic picture of what such a huge number means? J. H. de Boer in his book *The Dynamical Character of Adsorption* [7] has an ingenious way of visualizing this and I will follow him closely, adapting his model to the present situation. He calls for a linear magnification of 32 million so that one centimetre would become 320 kilometres and one cubic centimetre would become the volume of the space created by erecting a wall 110 kilometres high round the periphery of Italy and roofing it in. To make the volume more cubical, imagine a square with one corner in Venizia, another in Torino, the third in Terni, and the fourth somewhere off the western coast of Corsica, and let this be made cubical to reach 320 km into the sky—nearly three and a half times the height of Everest. On this scale of magnification de Boer points out that an oxygen molecule would be approximately the size of a bee.

[17] P. A. M. Dirac (quoted by Freeman Dyson, *From Eros to Gaia,* p. 305) said "the only basis for my work and the only reliable basis ... which was sufficiently general so as to secure me from making the same mistake again was to set up a principle of mathematical beauty; to say that we really don't know what the basic equations of physics are, but they have to have great mathematical beauty." See also [6].

FIGURE 20 An illustration of bee-keeping form an Exultet Roll in the Vatican Library.

"In this huge space there would be about 20 trillion bees (i.e., 2×10^{19}) and this number is equally unimaginable. But in a room 20 m long by 10 broad by 5 high there would be a million bees. A high vacuum of 0.001 mm of mercury would reduce this to one bee in the room.

"To make the speed of a bee the same as the velocity of the molecule, we must stretch the scale of time and make a second into a year. Then the million bees in this large room are flying around with an average speed of 1600 km per hour, colliding with each other about 150 times a second and travelling about three metres between each collision. Ten bees will strike each square centimetre of the wall every second, so that even in this slow-motion picture we can hardly count the impacts. This is a somewhat nightmare vision, but if we reduce the pressure to 0.1 mm of mercury, there is only one bee in 10 cubic metres and only ten bees will hit an area of one square metre each second, staying there about five minutes. This means that there will be about 3000 bees swarming over this square metre of surface hopping around and vibrating rapidly."

C was nodding off by this time, but *A* was still awake.

"Have I got a picture for you," he said. "There's a famous illustration (Figure 20) in an Exultet roll that was written and illustrated not far from here in the eleventh century. The script is the Bari dialect of Beneventan. The Exultet roll takes its name from the opening words of the Easter service in which the paschal candle is blessed. Naturally the bees come in for blessing as the providers of beeswax, and in this particular roll mediæval apiculture is vividly illustrated. I'll show you a picture of it tomorrow and you'll notice that the words and pictures are in opposite directions. That's because it was a mediæval multi-media presentation. The lector would feed the roll forward over the top of the lectern so that the faithful might see the picture whilst hearing the word.

"After all, two avenues, like two eyes, are better than one."[18]

REFERENCES

1. *Bestiary,* a translation of MS. Bodley 764, translated by R. Barber, Folio Society, London, (1992).
2. M. Penn and R. Aris, The mere notion of a model, *Mathematical Modelling* **1** (1), 1–12 (1980).
3. J.L. Casti, *Reality Rules. I, The Fundamentals; II, The Frontier,* John Wiley, New York, (1992).
4. M.J. Feigenbaum, Quantitative universality for a class of nonlinear transformations, *J. Stat. Phys.* **19,** 25–52. (1978).
5. H.-O. Peitgen, H. Jürgens and D. Saupe, *Chaos and Fractals. New Frontiers of Science,* Springer Verlag, New York, (1992).
6. S. Chandrasekar, *Truth and Beauty: Aesthetics and Motivations in Science,* University of Chicago Press, Chicago, (1987).
7. J.H. de Boer, *The Dynamical Character of Adsorption,* First edition, Clarendon Press, Oxford, (1953).
8. R. Aris, Re, k and π, *Appl. Mathematical Modelling* **1,** 386 (1977).
9. R. Aris, The jail of shape, *Chem. Eng. Commun.* **24,** 167 (1983).

[18] This is the third dialogue exploring various aspects of mathematical modelling. The first [8] looked at the hierarchical nature of modelling and the moulding of models. The second [9] treated the relationships between models and the usefulness of shape factors. There is a slight overlap in that the logistic equation is mentioned in the first and the measurement of letters by pen-widths in the second. An engineer is the common character in the three (Re, L, and B) and a mathematician appears in two (π and C); the others are a natural scientist (k), an art historian (B), a calligrapher (T), and a palæographer (A). The first two dialogues are scheduled to be reprinted as appendices to Dover's 1994 reprint of my *Mathematical Modelling Techniques,* originally published by Pitman in 1978.

Q.*

REFLECTIONS ON KEATS' EQUATION

RUTHERFORD ARIS

Department of Chemical Engineering and Materials Science, University of Minnesota, Minneapolis, Minnesota 55455, U.S.A.

(Accepted 15 May 1996)

Abstract—*Starting from the well known line of Keats' Ode on a Grecian Urn, the question of whether æsthetic considerations play any role in engineering is examined. Dirac claimed a place for them in physics and it is generally conceded that mathematicians recognize that elegance has a legitimate role in mathematics. Relating this to the engineer's craft of mathematical modelling brings in the power and beauty of the dimensionless parameter. Though beauty may not be a certain guide to truth, the exercise of mathematical modelling as a craft certainly leads to the enjoyment of it.* *© 1997 Elsevier Science Ltd*

Keywords: Mathematical modelling; dimensionless parameters; æsthetics.

KEATS' EQUATION

In his *Ode on a Grecian Urn*, John Keats apostrophises the silent form which has teased him out of thought in five stanzas concluding with the well-known lines:

> When old age shall this generation waste,
> Thou shalt remain, in midst of other woe
> Than ours, a friend to man, to whom thou say'st,
> "Beauty is truth, truth beauty," that is all
> Ye know on earth, and all ye need to know.

The complete identity which the poet puts into the 'mouth' of the Grecian urn would seem to be so extreme as to be written off at once by common sense. Indeed it has been so written off and by considerable critics too. Q (Sir Arthur Quiller-Couch) found it "an *uneducated* conclusion albeit most pardonable in one so young and ardent". To T.S. Eliot it seemed meaningless, though he conceded that "perhaps the fact that it is grammatically meaningless conceals another meaning from me". But one of the functions of poetry is surely to challenge common sense and to open up our understanding and imagination—like the nightingale's song, to charm "magic casements opening on the foam of perilous seas in faery lands forlorn".

Bate (1963) in his *John Keats* points out that the poet is not speaking for himself, but is putting the sentiment into the silent testimony of the urn as if to a passing stranger. He notes that, in his correspondence of the previous

* Reprinted with the permission of Elsevier Science from *Chemical Engineering Society,* volume **52,** pp. 2447–2455, 1997. Copyright © 1997 Elsevier Science.

year and a half, "Keats never comes close to anything as bald as the simple equation of these two abstractions, 'beauty' and 'truth', that he permits the urn to make here" (Bate, 1963), p. 517). Hence "the special and restricted character of the urn is stressed before its inscriptional message is permitted" (Bate, 1963, p. 518). It is aloof from the world in which age wastes the generations and only woe varies—an "Attic shape", a "fair attitude". Its message is to minds "teased out of thought" and so not accessible to immediate rationality. Bate even suggests that Keats' "irrepressible" humor may not have been fully reined in 'on even the most serious occasions'.

Dirac's Statements and Dyson's Analysis

I have not heard it suggested that Paul Dirac was less than serious in making the remark, so often quoted, that "it is more important to have beauty in one's equations than to have them fit the experiments". (Chandrasekhar, 1994). Or again, referring to his early work for which he sought a more general basis than Bohr orbits, "the only reliable basis I could think of which was sufficiently general ... was to set up a principle of mathematical beauty; to say that we don't really know what the basic equations of physics are, but they have to have great mathematical beauty. We must insist on this, and that is the only feature of the equations that we can have confidence in ..." This seems to me to be a generalization at least as sweeping as that of Keats' urn.

Dyson, in a letter to Chandrasekhar which the latter quotes, suggests that Dirac had arrived at that position when later work shewed that his physical logic had been flawed, but that his final equations were left intact, leading him to conclude that "one should not trust the physical logic but should only trust the mathematical beauty" (Chandrasekhar, 1994). Such, at any rate, is the analysis of Freeman Dyson whose reading of the matter is surely the most cogent and informed in the world today.

Uses of the Word Beautiful

First however I should say something about the way we use the adjective Beautiful. The dictionary entry can be summarized as follows (OED, 1928):

Beautiful—

(1) Excelling in grace of form, charm of colouring or other qualities that delight the eye and call forth admiration.

(a) of the human face or figure

(b) of other objects.

(2) Affording keen pleasure to the senses generally, and hearing in particular; delightful. In modern colloquial speech it is applied to anything a person likes very much. "Soup of the evening, beautiful soup, beautiful, beautiful soup".

(3) Impressing, with charm, the intellectual or moral sense, through *inherent fitness or grace or exact adaptation to a purpose*; hence applied sometimes to things otherwise repulsive, a beautiful operation in surgery.

Aside on Adventitious Beauty

The first two meanings are not prominent in scientific or engineering thought, though in the latter part of the last century there was a passion for decorating machinery. What might be called *adventitious* art is however not uncommon in scientific studies. Scientific societies have been known to make minor awards at meetings for the most artistic electron micrograph, for example. Such images were not produced for their artistic qualities and there need be no correlation between their scientific value and their artistic merit. It requires only a certain flair for composition in presentation and an eye for colour to produce spectacular and award-winning displays. Similarly the pictures of mixing flows that Ottino (1989) has produced excel in "grace of form and charm of coloring". The self-similarity of fractals has delighted the eyes and called forth the admiration of multitudes even of those who do not understand how they are generated. This is adventitious beauty in the sense that its beautiful qualities come as an extraneous addition; they form no part of the scientific motivation or internal structure of the problem. It is to be distinguished from the 'inadvertant art' of the craftsman (vide infra).

Understanding and Appreciation

The third sense would seem to be the one which we are most commonly using when we judge a piece of scientific work to be beautiful or choose one of several pathways because it has the most elegance. The notions of fitness and suitability are bound up with our appreciation of the beautiful and we would be inclined to say that any lack of meetness or incongruity of means and ends is a flaw or blemish. That the inward and outward aspects of a thing should be congruent is a necessary, but perhaps not sufficient, condition for a certain beauty to inhere within it. In this sense any understanding we have enhanced the beauty of the thing which we are contemplating and few scientists would endorse Wordsworth's lines (1798):

Sweet is the lore that Nature brings;
Our meddling intellect
Misshapes the beauteous form of things:
We murder to dissect.

The coexistence of two orders of appreciation was tellingly put in a famous passage of Eddington's Gifford Lectures (1929) on the nature of the physical world. He contrasted the section in Lamb's Hydrodynamics on the generation of waves on the surface of a liquid by wind blowing across it with Rupert Brooke's sestet:

There are waters blown by changing winds to laughter
And lit by the rich skies all day. And after
Frost, with a gesture, stays the waves that dance
And wandering loveliness. He leaves a white
Unbroken glory, a gathered radiance,
A width, a shining peace, under the night.

Eddington went on to say that we do not look back on the moments when the beauty of rippling water gradually freezing over has 'teased us out of thought' as moments when we fell below ourselves—"We do not look back on them and say, 'It was disgraceful for a man with six sober senses and a scientific understanding to let himself be deluded in that way. I will take Lamb's *Hydrodynamics* with me next time.' "

But we also have to take account of the relation between cognition and æsthetic experience. This raises important questions: is, for example, a work of artistic landscaping as beautiful as it was at first sight when we learn it is ecologically pernicious or has been created by exploited labour? Humans certainly have the ability to separate an experience into compartments by impermeable barriers or to become completely insensitive to certain aspects of it. Time and place enter into our appreciations. We do not admire a Gothic cathedral the less for knowing that it was built with safety measures so primitive that today they would be regarded as criminal.

Can it then be said that understanding changes, not the ability to see the beauty of things, but the taste that regards a thing as beautiful? *Quot homines tot sententiae, suo quoque mos* (Terence, -161), is a sentiment as old as the hills and would seem to militate against any rational discussion of æsthetics.* For David Hume (1757), "Beauty in things exists within the mind that contemplates them", and this, of course, raises all sorts of questions such as the notorious mind–body problem, over which many a philosopher has laboured.

There is obviously a sense in which æsthetic considerations play a role in science and engineering in that they motivate certain types of study on the part of an individual. I, personally, have no empathy at all with distillation columns, that being entirely the province of my *quondam alter ego*, Aris Rutherford (Who's Who, 1974–1975). The Macabe–Thiele diagram has a certain elegance, but it is the geometric elegance of a simple mapping. Acrivos and Amundson's distillation of continuous mixtures is interesting, but again its virtues are those of mathematics rather than of engineering. It is a personal distaste and I am thankful that I have never had to work in that area, but what if my taste were the taste of the age and everybody was as averse to distillation columns as I?

Birkhoff's Æsthetic Measure

G. D. Birkhoff tried to give a mathematical theory of æsthetics by recognizing the complexity, C, and the order, O (i.e., its harmony and symmetry) that an object could be said to have. He then restated Hemsterhuis' definition of the beautiful as "that which gives us the greatest number of ideas in the shortest time" as the relation:æsthetic measure

$$M = O/C.$$

* "There are as many opinions as there are persons, each has his own tradition". It is usually quoted as, *Quot homines tot sententiae, et de gustibus non est disputandum*, of which the second clause is often rendered as 'and there is no accounting for taste'.

It was then his task to find suitable ways of calculating O and C for a class of æsthetic objects. He posited certain indices of tension that matched the automatic adjustments made in the process of perception as when one runs the eye round the sides of a polygon. Presumably an irregular polygon would have greater indices of tension than a square, since one would have to register. The order was a weighted sum of the æsthetic associations, i.e. of elements of harmony, symmetry, alignment or balance counting positively and ambiguity, undue repetition or unnecessary imperfection counting negatively. A center of repose in a painting was a plus; a poem overburdened with alliteration or assonance a minus. It all sounds rather quaint now but it engaged the attention of one of the foremost American mathematicians of his day, the founder of ergodic theory and the theory of dynamical systems. One wonders what he would have made of Penrose tiles that fill space without ever having double periodicities.

Mathematical Beauty

Though it would be hard to find anyone today who would find the notion of æsthetic measure at all enlightening, there are few mathematicians who would not claim that there is such a thing as mathematical beauty and that it may even at times be the guide that Dirac thought it to be in physics. I remember vividly when I first went up to Edinburgh and was assigned to John Cossar as tutor. I was very much in love with epsilons and deltas at the time and burbled on to Artur Erdelyi about my need to learn rigour. His reply was simply that "if you are going to Dr. Cossar you will have the opportunity not only to learn rigour, but also, what is perhaps more important, elegance".

G. H. Hardy in his well-known 'A Mathematician's Apology' says, "Beauty is the first test: there is no permanent place in the world for ugly mathematics". Hardy was the quintessential pure mathematician, rejoicing in the fact that the theory of numbers had no application in the mundane world and regarding ballistics as 'repulsively ugly'. He would have been horrified to learn of some of the developments of number theory that make unbreakable codes possible and have had little sympathy with the trend that has made the dichotomy of 'pure' and 'applied' mathematics distinctly passé. In his apology he draws attention to the way in which theorems can be appreciated even though their proofs may be recondite. Fermat's two square theorem—that the primes greater than two are expressible as the sum of two squares if and only if they are congruent to one modulo four—is an example of this. What he would have made of the recent proof of Fermat's last theorem is an interesting speculation and raises a point made by Professor Schmalzried of Hanover in a lecture at Minnesota, where he shewed a slide of Albrecht Altdorfer's Alexanderschlacht [or The Battle of Issus; vide, e.g. Janson (1986)] and asked whether a painting loaded with such a vast amount of information could be regarded as truly beautiful. If a mathematical proof is so long and recherché that only a handful of people in the world can follow it, can it be said to be beautiful? Does a computer proof have any beauty? Not for me, but I suppose it might in the eye of a computer scientist. But I think most mathematicians would reckon the proof of Bieberbach's conjecture a greater achievement than that of the four colour theorem, though the latter can be stated in terms more easily understood.

In his essay on 'the Study of Mathematics', Bertrand Russell (1976) pays tribute to Plato's concept of mathematics as 'divine necessity' (Laws, 818) and goes on: "Mathematics, rightly viewed, possesses not only truth, but supreme beauty—a beauty cold and austere, like that of sculpture, without appeal to any part of our weaker nature, without the gorgeous trappings of painting or music, yet sublimely pure, and capable of a stern perfection such as only the greatest art can show. The true spirit of delight, the exaltation, the sense of being more than man, which is the touchstone of the highest excellence, is to be found in mathematics as surely as in nature. What is best in mathematics deserves not merely to be learned as a task, but to be assimilated as part of daily thought, and brought again and again before the mind with ever-renewed encouragement. Real life is, to most men, a long second-best, a perpetual compromise between the ideal and the possible; but the world of pure reason knows no compromise, no practical limitations, no barrier to the creative activity embodying in splendid edifices the passionate aspiration after the perfect from which all great work springs. Remote from human passions, remote even from the pitiful facts of nature, the generations have gradually created an ordered cosmos, where pure thought can dwell as in its natural home, where one at least of our noble impulses can escape from the dreary exile of the actual world".

Russell, like Hardy, espouses an exalted view of pure mathematics, and seems to be a trifle carried away by the exuberance of his own enthusiasm. But he had been a mathematician, though he confesses in his autobiography that the sustained concentration that he expended over the 'Principia Mathematica' was never again at his command.

Robert Bridges, was not a mathematician but a medical man as well as a classical scholar and poet, and took a more earthy, or less abstract, view. In the first book of his 'Testament of Beauty' he has a striking metaphor for scientific research and scholarly endeavor. (The unusual words are not misprints, but are made in accord with his personal theory of orthography (Bridges, 1924).).

As a man thru' a window into a darken'd house
peering vainly wil see, always and easily,
the glass surface and his own face mirror'd thereon
tho' looking from another angle, or hooding his eyes
he may discern some real objects within the room.

He adds an ironic, indeed prophetic, touch: "some say 'tis so with us, and also affirm that they/by study of their reflection have dicover'd in truth/ther is nothing but thatt same reflection inside the house". But he goes on:

Not emotion or imagination ethic or art
logic or science nor dialectic discourse,
not even thatt supersensuous sublimation of thought,
the euristic vision of mathematical trance
hath any other foundation than the common base
Of Nature's building.

Since it is generally conceded that mathematics is the touchstone of intellectual beauty we should ask what its characteristics are. *Economy of thought and terseness of style* are certainly of the essence, for in no discipline is Occam's razor whetted so well or used with a keener edge. In practice, say for purposes of exposition, some redundancy may be needed. (Some years ago I suggested that for teaching purposes the principle of Ior's Ears would complement that of Occam's Razor. Ior was so surprised at being greeted on his birthday that he wanted it again in the other ear—Inmitte istud in alteram auriculam, dixit. Age!) (Milne, 1958) But this must not obscure the integrity of the mathematical structure. *Integrity* is the requirement, common to all disciplines, of wholeness—that everything is there that needs to be. *Coherence*, or *consonance*, is a related quality which keeps everything in proportion and reconciles diverging trends and dissonant voices. Then there is the aspect of beauty which compels attention, not by self-advertisement, but by its *radiance* or *effulgence*. This may be an element of surprise that 'grabs' you. (Cf. the Norwegian idiom "*Det var* (not *er*) *pen*" for 'That *is* beautiful', referring back to the first impression.) In Aquinas these are the formal characteristics of all art—*integritas, consonantia et claritas.* Hardy adds the quality of *seriousness* to distinguish the beautiful pattern of a mathematical structure, say Cauchy's theorem, from an end game at chess, where the thought may be of comparable complexity but the seriousness of purpose is not.

Mathematical Modelling

In so far as mathematics is the language of science and engineering certain features of mathematical beauty might be expected to spill over into the world of applied mathematics. That master mathematical modeller, J. D. Murray, has suggested that the beauty might lie in the natural system being modelled rather than in the model itself. However a model might well be deemed ugly if it obscured or failed to reflect the natural beauty. The artistic heights that can be attained in computer visualization, go far toward the satisfaction of Poincaré's desire for an organ that expresses that taken in by the eye as the mouth serves the ear (Beddow, 1995). On the other hand, the integrity of Jolls' graphic presentation of thermodynamic properties and relations lies in a Gibbsian self-restraint that deliberately avoided the excitement of the cutting edge (Jolls, 1990).

Detailed comments from now on will be mainly concerned with mathematical modelling in the context of chemical engineering. It is not the aspect of our trade which is getting the most support these days, when application is at such a premium that more direct routes to actuality are preferred, but it is something that has to be done to a greater or smaller extent and a craft which engineers, in particular have cultivated. The mathematicians have recently discovered it under the guise of industrial mathematics; physicists have burst into a nonlinear world; the chemists are into super-computing and have immense computer models; the engineers stand around like bourgeois gentlemen muttering to themselves—*par ma foi! il y a p'us de quarante ans que j'ai dis de la prose, sans que j'en susse rien.* [Egad! I've been speaking prose for forty years without realizing it; Poquelin (1670).]

Dimensionless Parameters

Are there any distinctively engineering practices that have æsthetic appeal that is not mathematical in nature? I would suggest that the engineer's use of dimensionless parameters comports with a notion as profound and far-reaching as any in his repertoire. Strictly speaking all measurements are dimensionless, for to say that an object is two metres in length is to say that the ratio of its length to that of the standard metre laid up in Paris is the real number 2. A most elaborate and critical organization attends the propagation of the standard so that engineers and others who make measurements can be sure that others will understand and be able to compare them. But the dimensional units are arbitrary and only by chance do they fit the problem. Whereas, if we work with quantities taken from the problem itself and use them as units, we shall get meaningful numbers. The principle of economy is upheld by the fact that we shall get the minimum number of parameters by suitable nondimensionalization.

The beauty of dimensionless parameters is something of the same sort as the beauty of imagery in poetry. Here we can turn again to Keats:

St. Agnes' Eve—ah, bitter chill it was!
The owl for all his feathers was a-cold;
The hare limped trembling through the frozen grass,
And silent was the flock in woolly fold:
Numb were the Beadsman's fingers, while he told
His rosary, and while his frosted breath,
Like pious incense from a censer old,
Seemed taking flight for heaven, without a death,
Past the sweet Virgin's picture, while his prayer he saith.

You could not wish for richer imagery; a shudder goes through you as you see the owl, feathers slightly ruffled to trap the insulating layer of body-warmed air, hunched on a snowy branch, or the shivering hare limping warily through the 'sharp and sided' blades of frozen grass, or the sheep huddled together so that the fold itself seems woolly. They are all images taken from the context of the poem. Had you read

St. Agnes' Eve—ah, bitter chill it was!
The temp. was ten above, right on the line.
But everyone felt much, much colder 'cos
The wind chill index made it minus nine.
The Beadsman wanted both with minus sign,
So tried to turn these temps. to centigrade,
Which would have suited everybody fine,
If he had not a classic error made,
Forgetting, in his haste, the part subtraction played

you would rightly have suspected the work of a parodizing poetaster. Temperature is conveyed in arbitrary units quite foreign to the setting and the imagination has nothing to play with. The images of a poet work from within the context of the poem just as the dimensionless variables and parameters characterize and give shape to the model.

A MODEL OF ALGAL GROWTH

To illustrate the modelling process, let us consider a problem in the growth of algae in a stagnant pond which, by remaining in its simplest form, admits an easy and complete solution.† There are two independent variables, z, the depth of a point below the surface of the pond, and t, the time. $I(z, t)$ is the intensity of light at depth, z, and time, t, and $N(z, t)$ the density of algal growth there. The growth is exponential with a growth rate constant proportional to the intensity of light, which falls off with a Beer's Law constant proportional to the algal growth. Thus

$$(\partial/\partial z)I(z,t) = -aI(z,t)N(z,t) \tag{1}$$

$$(\partial/\partial t)N(z,t) = bI(z,t)N(z,t) \tag{2}$$

is a pair of first-order partial differential equations. We can specify the initial distribution of algae and the intensity of light at the surface,

$$I(0,t) = I_0(t) \quad \text{and} \quad N(z,0) = N_0(z). \tag{3}$$

This specifies the model, and it is well at this point to look back at the assumptions that have been made. They are (1) the immobility of the algae, which cuts out convective and diffusive terms, and (2,3) the validity of the growth and Beer's laws. No mention need be made of the depth of the pond if the immobility is conceded, for the solution for infinite depth is simply truncated at the finite depth. How adequate the model is depends on its purpose. For the present purpose of illustrating model building, it is admirable; for predicting the total growth, it may be less accurate and, for the details of distribution, still less. The latter purposes will demand a comparison with experiment, but since we are concerned with the first we can proceed with equanimity.

Here there are no obvious characteristic values of I and N; they are bound up in the boundary conditions. Let I^* and N^* be such values, to be determined later, then $(1/aN^*)$ is a characteristic length and $(1/bI^*)$ a characteristic time. Thus we put

$$u(x,y) = I/I^*, v(x,y) = N/N^*, x = aN^*z, y = bI^*t. \tag{4}$$

and obtain

$$-u_x = v_y = uv \tag{5}$$

$$u(0,y) = f(y), v(x,0) = g(x). \tag{6}$$

† Professor McCoy drew my attention to this case and I give my solution much as I gave it to him at the time. I should have recalled Bischoff's (1969) paper on a class of problems of which this is a particularly simple case, which I knew of, or even the text on first-order p.d.es (Amundson and Aris, 1972) but the problem looked so innocent that I solved it before thinking of it as a standard form. Ozawa (1970) has shown that the Legendre transformation can be applied and, I doubt not, there is a wealth of literature that should be consulted. Sometimes it is more fun to do it yourself.

It is possible to get a simple solution to these equations and indeed to much more general ones in which uv is replaced by $F(u) \cdot G(v)$. The fact that $-u_x = v_y$ suggests the use of a stream function $u = L_y$ and $v = -L_x$, so that $L_{xy} = L_x L_y$. The further substitution $L(x, y) = -\log(M(x, y))$ gives $M_{xy} = 0$, of which the general solution is

$$M(x, y) = P(x) + Q(y) + R \tag{7}$$

where, w.l.o.g. we can put $R = -1$.

But substituting back into the boundary conditions gives

$$v(x, 0) = P'(x)/\{P(x) + Q(0) + R\} = g(x) \tag{8}$$

$$u(0, y) = -Q'(y)/\{P(0) + Q(y) + R\} = f(y)\}. \tag{9}$$

Thus, if $F(y)$ and $G(x)$ denote the integrals of $f(y)$ and $g(x)$ from 0 to y and x respectively,

$$u(x, y) = f(y)/\{1 + \exp(F(y))[\exp(G(x)) - 1]\} \tag{10}$$

$$v(x, y) = g(x)/\{1 - \exp(-G(x))[1 - \exp(-F(y))]\}. \tag{11}$$

It is seldom that one can get a solution to the equations of a mathematical model in closed form and most usually a definite situation has to be considered and a solution computed. By way of illustration the figure shows the buildup of algae throughout an equinoctal day when the surface intensity is $f(y) = \sin(\pi y/12)$ (i.e. I^* is the intensity at noon and y the time in hours after 6 a.m.) and the initial distribution a Gamma distribution of index 1 (i.e. N^* is chosen to make $G(\infty) = 1$). It is clear that there is not much growth in the morning, but the algae in layers just beneath the surface grow rapidly and deprive the lower algae of light so that they grow scarcely at all (Fig. 1).

This model bears the imprint of its own limitations as a good model should. It is assumed that there is no movement, yet algae are growing in the upper layers of the water and presumably making the medium more dense. This suggest that sedimentation or convective roll-over should be considered. We shall not go into such generalizations here, but merely leave this example with the comment that the same equations turn up in the theory of a continuous reactor in which a belt of material is to be impregnated with a solution, whose solute is immobilized on sites in the belt.

Are there any æsthetic considerations at work here? Of course there are trivial ones, like the choice of notation. In the matter of nomenclature u and v, f and g, P and Q go together naturally in the English-speaking world as contiguous in the alphabet and commonly so used. Usage rather than contiguity governs the Greek triple ξ, η, ζ. We would be surprised, though not dismayed, at a choice of λ and α for u and v, the dimensionless intensity and density of algae, much as we might murmur 'de gustibus' to ourselves on seeing a man wearing one blue sock and one brown; φ and α might pass muster, much as one can get away with one black and one navy sock, at any rate in all male society. This, however, is trivial for the mathematics is the same whatever the notation. The solution could be given as a special case of Ozawa's treatment of $u_x = v_y = f(u)g(v)$. That would be legitimate enough

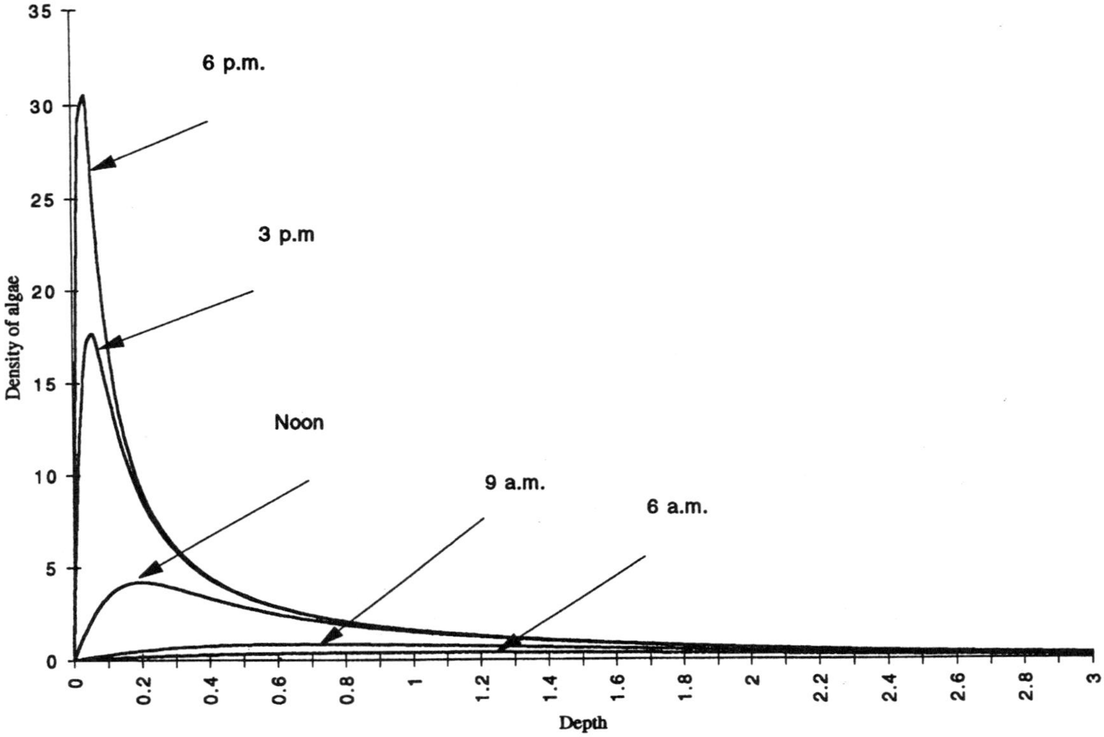

FIGURE 1 Algal growth at 6 a.m, 9, noon, 3 p.m, 6 p.m.

for Ozawa as illustrating his previous exposition of the Legendre transformation, but, in a presentation of a simple model, I submit that the solution should be as simple and direct as possible. To have reduced the problem to linear form $M_{xy} = 0$ and then to have solved this equation by, say, the Laplace transform, insead of recognizing the general solution of the equation, would, to my mind, have been æsthetically inferior.

Beauty as a Criterion of Truth

Are there any situations in which beauty is taken as a criterion of truth? Suppose we have some data that is very exactly fitted by an ugly formula—something that only the empiricist who thought it up could love. Take, for example, a formula for the rate of synthesis of ammonia

$$k\,\frac{p_{H_2}^{3.16} p_{N_2}}{p_{NH_3}^{1.65}} - k'\,\frac{p_{NH_3}^{2.35}}{p_{H_2}^{2.84} p_{N_2}}.$$

Nobody in their right mind would attach any significance to those powers to which the partial pressures are raised. They are just chosen (in a consistent manner, it is true, for the equilibrium relationship must correspond to the locus on which the rate is zero) to fit the data. Suppose this fit is near perfect. Then you could certainly use the formula with confidence to interpolate between data points, say for the purposes of integrating a differential equation, and you might even venture to extrapolate a little. Then someone comes along with

$$\frac{kp_{H_2}^3 p_{N_2} - k' p_{NH_3}^2}{1 + Kp_{NH_3}}.$$

Here is a formula that at every point is more elegant than the former. Whether one would lavish praise upon it is an altogether different question, but it strikes me as having a simplicity and directness. What is more they come with a suggested mechanism which is amenable to independent tests. Unfortunately it fits the data far less well. Nevertheless I should be inclined to accept it in preference to the former. Is this a case Dirac would have considered to be supportive of his position? The beauty is modest and particular and he was talking about the deeper mysteries of natural philosophy.

I am inclined to take a middle road and say that æsthetic considerations are indeed dominant, but the judgment based on them needs to be taken tentatively. Thus I would have no use for the first formula, but would also seek the best fit of a more general formula:

$$\frac{kp_{H_2}^3 p_{N_2} - k' p_{NH_3}^{21}}{\{1 + K_1 p_{H_2} + K_2 p_{N_2} + K_3 p_{NH_3}\}^m}.$$

Even this form, backed as it is by a theory of adsorbtion and catalysis, would have to be tested in its several parts, for as Michel Boudart once said, "It's all over with the overall".

Craftsmanship and the Enjoyment of Beauty

A few years ago Mischa Penn and I wrote a paper on the mere notion of a model, which opened the International Journal of Mathematical Modelling, (now of Mathematical and Computer Modelling). In it we put some emphasis on the craftsmanship of modelling as a *via media* between the instrumentalist and realist positions. The former would take models merely as a preparation for computation; whereas, the latter claims to be telling us something about the real world. Pye (1968), in a book on the nature of design, distinguishes between the workmanship of risk—of which he says that it is "an uncouth phrase, but at least descriptive"—and the workmanship of certainty; e.g. between calligraphy and printing, or between the hanging of a door on the Rolls-Royce and the robotic attachment of the doors of a lesser vehicle.

In craft there is what Polanyi (1962) calls a tacit knowledge; we know more than we can express. And so perhaps it is in the processes of engineering synthesis and analysis or in the lines of enquiry, theoretical or experimental, that æsthetic considerations have their place. Who can forget the simplicity and directness of some of G. I. Taylor's experiments? Astarita suggested that the utilitarian ends create their own æsthetic, but is not the process guided by it too? The well-known couplet from Horace (Flaccus, -13) comes to mind;

omne tulit punctum qui miscuit utile dulci
lectorem delectando pariterque monendo.‡

In this connection there is the incidental elegance, the ancillary beauty, of the small, but significant, idea or method that 'curves rightly to the hand'. Villadsen's use of differential inequalities in the diffusion-reaction problem comes to mind; or his discovery (with Michelsen) that, not just the end point, but *every* point in the numerical integration of this two point boundary value problem could be made to give an effectiveness factor for some problem with parameters on a straight line in the space of the parameters β and γ, the dimensionless heat of reaction and activation energy (Villadsen and Michelsen, 1972). Villadsen himself has written of "the primitive feeling of joy in having participated, albeit in a minor role (Nielsen and Villadsen, 1984), in the quest for the final truth about the mathematical structure of the simple second order boundary value problem for an nth order reaction in a pellet of uniform temperature" (Villadsen, 1988), and many a less accomplished worker will resonate to this sentiment.

The innate æsthetics of craftsmanship has never been better expressed than by Vita Sackville-West (1927) in her poem 'The Land':

All craftsmen share a knowledge,
They have held reality down fluttering to a bench;
Cut wood to their own purposes; compelled
The growth of pattern to the patient shuttle;
Drained acres to a trench. Control is theirs.
They have ignored the subtle
Release of the spirit from the jail of shape.
They have been concerned with prison, not escape;
Pinioned the fact, and let the rest go free,
And out of need made inadvertant art.
All things designed to play a faithful part
Build up their plain particular poetry.
Tools have their own integrity,
The sneath of scythe curves rightly to the hand,
The hammer knows its balance, knife its edge,
All tools inevitably planned,
Stout friends with pledge
Of service; with their crochets too
That masters understand,
And proper character, and separate heart,
But always to their chosen temper true.
So language, smithied at the common fire,
Grew to its use; as sneath and shank and haft,
Of well-grained wood, nice instruments of craft,
Curve to the simple mould the hands require,

‡ "He carries every vote who has mixed the useful with the beautiful by delighting and instructing the reader at one and the same time."

Born of the needs of man.
The poet like the artisan
Works lonely with his tools; picks up each one,
Blunt mallet knowing, and the quick thin blade,
And plane that travels when the hewing's done;
Rejects and chooses; scores a fresh faint line;
Sharpens intent upon his chiselling;
Bends lower to examine his design,
If it be truly made,
And brings perfection to so slight a thing.

ACKNOWLEDGMENTS

I wish to thank the reviewers for two of the most careful and considerate reviews that a paper of mine has ever received. They encourage me to think that this paper, though not in the usual subject area, may yet be stimulating to the readership of Chemical Engineering Science. Two larger issues that the reviewers bring up, but which I have not attempted to incorporate in the text concern the late section on kinetics and a point of psychological interest. Concerning the first, the question is raised as to whether the 'ugly' kinetic expression, by its superior fit with the data, does not contain more truth, and hence, perhaps, more beauty, than the Langmuirian laws. If it is purely empirical, I think not, for it is tied to a specific set of data: if however it is derived as the expression of some wider theoretical underpinning that is appropriate to the region of the data, then it might well be so. Secondly, one reviewer points out that our appreciation of beauty may go through an optimum as our knowledge increases the stages of unfamiliarity, recognition, understanding to the familiarity, which even though not breeding contempt, yet takes the beautiful so much for granted as to dull its radiance. The lively appreciation of beauty becomes a reward for the effort put into the work of understanding. It is certainly true that interest in something once thought beautiful can fade especially if it lies lightly on the mind '*som Sondags enriss på et skurat golv*'§, as Anders Østerling puts it. But I would have thought some things would be built so securely into the mind as to become canons for later judgments.

NOTATION

a	Beer's law constant
b	growth rate constant
f	dimensionless light intensity at surface
F	indefinite integral of f
g	dimensionless initial density of algae
$I(z, t)$	intensity of light at depth z and time t
$I_0(t)$	intensity of light at surface
I^*	characteristic intensity
$N(z, t)$	density of algae at depth z and time t
$N_0(z)$	initial density of algae at depth z
N^*	characteristic density
t	time
$u(x, y)$	dimensionless intensity of light
$v(x, y)$	dimensionless density of algae

§ "As the Sunday juniper on a scrubbed floor".

x	dimensionless depth
y	dimensionless time
z	depth

REFERENCES

Afer, P. T. (161 BC) *Phormio,* p. 454.

Amundson, N. R. and Aris, R. (1972) *Mathematical Methods in Chemical Engineering: First Order Partial Differential Equations with Applications.* Prentice-Hall, Englewood Cliffs, NJ.

Bate, W. J. (1963) *John Keats.* Belknap Press, Cambridge.

Beddow, J. (1995) private communication.

Bischoff, K. B. (1969) General solution of equations representing the effects of catalyst deactivation in fixed-bed reactors. *Ind. Engng Chem. Fundam.* **8,** 665–668.

Bridges, R. B. (1924) *The Testament of Beauty,* BK. I, pp. 350–370. Clarendon Press, Oxford.

Chandrasekhar, S. (1994) The perception of beauty and the pursuit of science. *Bull. Am. Acad. Arts Sci.* **XLVII,** 14–29.

Eddington, A. S. (1929) *The Nature of The Physical World.* Cambridge University Press, Cambridge.

Horace, see Flaccus.

Flaccus, Q.-H. (post 14–13 BC) *Ars Poetica,* 343. See Brink C. O., (1963) *Horace on Poetry.* Cambridge University Press, Cambridge.

Hume, D. (1757) III. Of tragedy. In *Four Dissertations.* Millar, London.

Janson, H. W. (1986) *History of Art: A Survey of the Major Visual Arts from the Dawn of History to the Present Day,* 3rd ed. Prentice-Hall, Englewood Cliffs, NJ.

Jolls, K. (1990) Gibbs and the art of thermodynamics, In *Proceedings of the Gibbs Symposium,* ed. G. D. Mostow and D. G. Caldi, pp. 293–321. American Mathematics Society, Providence.

Milne, A. A. (1958) *Winne ille Pu.* Methuen, London.

Molière, see Poquilin.

Nielsen, P. H. and Villadsen, J. (1984) An analysis of the multiplicity pattern of models for simultaneous diffusion, chemical reaction and adsorption. *Chem. Engng Sci.* **40,** 571–587.

Ottino, J. M. (1989) *The Kinematics of Mixing: Stretching, Chaos and Transfer.* Cambridge University Press, Cambridge.

Oxford English Dictionary, (1928) 1st edn. Vol. 1, p. 744, col. 1. Clarendon Press, Oxford.

Ozawa, Y. (1970) Application of the Legendre transformation to one-dimensional packed bed model. *Chem. Engng. Sci.* **25,** 529–533.

Polanyi, M. (1962) *Personal Knowledge: Towards a Post-Critical Philosophy,* Routledge & K. Paul, London.

Poquilin, J. B. (Molière) (1670) *Le Bourgeois Gentille-homme*, Act II, Sc. 4.

Pye, D. (1968) *On the Nature and Art of Design.* Cambridge University Press, Cambridge.

Russell, B. W. (1976) *A Free Man's Worship, and other Essays,* p. 63. Unwin Books, London.

Sackville-West, V. (1927) *The Land.* Doran, New York. Terence, see Afer.

Villadsen, J. (1988) Challenges and cul-de-sacs in reactor modelling. *Chem. Engng Sci.* **43,** 1725–1738.

Villadsen, J. and Michelsen, M. L. (1972) Diffusion and reaction in spherical catalyst pellets: steady state and local stability analysis. *Chem. Engng Sci.* **27,** 751.

Who's Who in America (1974–1975) Vol. 2, p. 2672, col. 1. St. Martin's Press, New York.

Wordsworth, W. (1798) *The Tables Turned. In The Poetical Works of William Wordsworth,* Vol. III, eds. E. de Selincourt and H. Darbishire. Clarendon Press, Oxford. 1947.

R.*

CHEMICAL ENGINEERING GREETINGS

UNIVERSITATIS MINNESOTENSIS
COLLEGIUM FABRORUM CHYMICORUM
SODALES PRINCETONIENSES,
NUNC ANNUM EORUM QUINQUAGESIMUM
FELICITER AGENTES,
GRATULATUR,
MAGNO CUM GAUDIO,
GRATUITO CUM CONSILIO,
VENIAMQUE W. S. GILBERTI POSCENS

EDITOR'S NOTE: *We are grateful to Prof. John Seinfeld (California Institute of Technology) for bringing our attention to the following poem, written by Prof. Rutherford Aris of the University of Minnesota to congratulate Princeton University's Chemical Engineering Department on their fiftieth anniversary.*

When you're taking a whack,
in your orange & black,
at a problem defying solution
In kinetics or flow,
or the ways that bugs grow,
or just how to get rid of pollution.
When you're wracking your brain
& you cannot obtain
a result that fulfills your intentions
Remember that Chem.E.s
do nothing by semis,†
and always get rid of dimensions
Recollect that p'raps Schmidt's
is the number that fits
and that Peclet is Reynolds times Prandtl.
Just remember your D's,
it'll come out with ease
when it's wrapped up in Damköhler's mantle.
If you wish you could borrow
relief from your sorrow,
the Ohnesorge number we'll mention,
Into Reynolds you shoot
the good Weber's square root—
it accounts for the surface's tension.
For tubes in the boiler,
use t'number of Euler
and for drug use the factor of Fanning,

* Reprinted with the permission of *Chemical Engineering Education,* volume **15,** number 1, p. 12, 1981.

† Except Centenaries, of course.

Which in matters of fiction
was used for the friction
'twixt Newman and Cardinal Manning.
But it's really quite rude
to rhyme Froude with St. Jude
or to mess up a grave patronymic,
To dress up in gingham
the number of Bingham,
or regard Boussinesq's as a gimmick,
While to think that Biot
has to do with the glow
that both heat and hard work often foster,
Or that Deborah's song
has the permanent pong
of an overly ripe double Gloucester.
To madden a Hatta
or exchange for the latter
the group that's been named Svant Arrhenius,
To mock at the balding
of Sherwood or Spalding;
these are all of 'em crimes hyperhein(i)ous.
If none of these fit,
just reflect for a bit,
devise one of your own—
it's a sine qua non—
for there's many a name
that has brought Princeton fame.
For size of reactor,
use great Wilhelm's factor;
for hard computation,
'tis Leon's equation;
and if that don't fit well
the number of Whitwell.
If you've pulled a slight boner,
be absolved by Dick Toner;
your problem's all garbled?
Why, Dean Elgin's not marbled;
While, if biking's your goal,
use the Johnson Control.
You've got Schowalter (Bill),
and there's Dudley Saville,
Andres, Russel and Hall
Prud'homme, Ollis et al.‡,
for from such you can scoop
a dimensionless group
and you'll out of your fix be
with some help from Ms. Bixby.
But away cerebration.
Now 'tis your celebration
and we all think it's nifty
that you flourish at 50.
So we wish you good cheer
for full many a year.

RUTHERFORD ARIS

‡ With due apologies to those not mentioned.

III

MISCELLANEA

ACKNOWLEDGMENTS

An Autobiographical Appendix with Asides

To write autobiographically is to essay the difficult passage between the Scylla of conceit and the Charybdis of false-modesty. True modesty is a rare and paradoxical virtue—those who have it don't claim it and those who claim it don't have it—and I must be content with an ironical view of myself and a recognition of my indebtedness to others as the two oars with which to keep the coracle of my recollection from dangerous imbalance as I rush between cave and whirlpool on the treacherous waters of uncertain memory. Looking back on 40 years of mathematical modeling, my chief reaction is to say what fun it has been and continues to be. For the exercise of a craft to the best of one's ability is great fun as well as being a great joy, and, although joy is a peculiar possession and the province of friends, fun is its public aspect and may be shared promiscuously. I have restricted myself to my mathematical activities and career as a chemical engineer except in the bibliography, where all my writings are listed, and in one paper [277], where a comparison with the methods of paleography was specifically asked for.

Mathematical modeling has its highs and lows. There are the high points where a long, technical manipulation comes out the same at the first and second attempts, instead of the third and fourth, or a final formula that is transparently correct is obtained. There are the low points when you meet an integral of the kind which cometh not out save by prayer and fasting (as G. H. Hardy is said to have quoted in that context), or you realize that some assumption made right at the beginning has led to an obviously false result at the end, but you cannot for the life of you see which one it can be. To share some of these highs and lows and so to give a context for the selection of papers is partly the reason for this appendix. As I wrote it, I realized how manifold was may indebtedness to others and I regard it as the "Acknowledgments" paragraph that quite properly goes at the end of a paper, although here drawn out into a narrative with asides.

EARLY EDUCATION, 1935–1943

It was at Canford School in the pleasant county of Dorset that I was converted from would-be chemist to would-be mathematician. Until then, my keenest pleasure had been the nitration of glycerin in a conical flask, a reaction that went, after a quiet induction period, with a fine flurry of frothing and boiling and evolution of brown vapor. Of course, I was a little scared of the stuff when I had made it and did not quite know what to do with it. Eventually, I poured it out on the coke pile that served the boiler plant of my father's photo-finishing works, where it seems to have done no harm. If this nitration ended with a whimper, it was quite otherwise with the nitration of whatever one nitrates to form picric acid. Picric could be evaportated to crystalline form, mixed with red lead, and put in a tin lid on a tripod in the open-air behind the self-same works. Quickly, one put a roaring Bunsen burner under it and even more quickly one retired, turning at 10 paces to witness a most satisfying bang that rent the tin lid in three and wrapped the pieces around the top of the tripod. Small wonder that for most of the rest of my career I have been content to react *A* and *B* to form *C* and, possibly, *D*. I did learn quite a bit of chemistry by reading and it stood me in good stead for the School and Higher School Certificate examinations. I developed a great admiration for the periodic table and was puzzled by the additivity of Sugden's parachor—ah! but there is a name to conjure with, like Harmon Killebrew or Emerson Fittipaldi. The chemistry master at Canford, Evelyn Underhill (known among us boys as "Pants"), in his report on my first year said I should pay more attention to fundamentals and be less occupied with recondite matters. However, by then my heart was not in it, and my occupation with the recondite was probably the show-off of a bright fifth former who was not very good at games.

I should perhaps interpolate a word about the peculiarities of the English educational system as it was in the late 1930s through the late 1940s. One started, on the private side, by attending a coeducational kindergarten from age 5 to about age 8. "All you needed to know" was taken for granted, and from the serious business of learning the alphabet and "times-tables" we passed quickly to reading, writing, and "doing sums." Reading led to French and arithmetic to algebra in the second year, whereas writing developed into parsing, analysis, and the mystery of dictation. I was very poor at dictation and still remember spelling out the punctuation in the first sentence we were given by Miss McTaggart: "Peter had a cold coma cogh and a runy nose ful stopp." In the third year, Latin and geometry were introduced, and one had the pleasure of explaining to one's parents, who obligingly feigned ignorance, just what was meant by that very satisfying conclusion to a theorem, QED. The school was run by two formidable spinsters, Miss MacTaggart and Miss Woodcock, who had named it after St. Martin of Tours, of whose proclivities in the matter of sharing we were frequently reminded.

From kindergarten one graduated to a Preparatory school; mine was dedicated to St. Wulfram, whose virtues were lost in Teutonic mists, and run by a Mr. Titley, whose glass eye and imposing stature made him the personification of authority. Here I passed a year in a form that began each day with a sum such as, "Convert 1,234,567 inches into miles, furlongs, poles, yards, feet, and inches," a tricky business when you remember that there are 5½ yards to a pole, which might also be disguised as a "rod" or a "perch." I was never much good at arithmetic and it was a relief to return to geometry again in the next form. We also began Latin again, which was not too bad, as I was a step ahead of the game. However, when 3 years later, I began Latin for the third time after another gap of a year, boredom and rebellion set in, and instead of using my background to excel in it, I stupidly paid it no attention, dreamed of chemistry in class, was careless in homework, and ultimately walked out of the School Certificate Examination that I sat at age fourteen after half an hour—an item I have studiously kept off my *curriculum vitae*. It was a mistake that I have never ceased to regret in later years when the recovery of some facility in Latin became as important as it was then difficult. *Facilis est decensus Averni, Noctes atque dies atri iuanua ditis, Sed revocare gradum, superasque evadere ad auras, Hoc opus, hic labor est.* (Easy is the descent to Avernus. Night and day its gloomy portals stand open. But to ascend the path and emerge again into sunlight, that is labor, that is work.)

CANFORD, 1943–1946

The preparation in a "preparatory school for boys" was for the Common Entrance and Scholarship Exams of the Headmasters' Conference Schools, or, as they were commonly called, the public schools. They are indeed public in the sense that they are not owned privately or run for profit, but in American usage they are fee-paying and private. Most of them are boarding schools, and because my parents would never have considered my being a boarder, it was only the lucky chance of Canford's starting a house for dayboys (to keep up their numbers in wartime) that allowed me to get the benefit of a public school education. It was a benefit that I appreciated the more for having spent the previous 3 years—age 10 to age 13—at Wimborne Grammar School, a public school in the American sense.

Founded by Lady Margaret Beaufort, the mother of Henry VII, and chartered by Elizabeth I, Wimborne Grammar had a very good chemistry mistress whose able teaching encouraged my passion for chemistry, but was otherwise of indifferent quality and struggling, like all others, to maintain itself under difficult wartime conditions. However, I pass over any further discussion of the English educational system and return to my starting point.

I entered Canford on the eve of my 14th birthday and was put in the Fifth Form. Needless to say, there is nothing as logical as the American K-12 about the English nomenclature for forms or grades. The lowest form at Canford was the Fourth, which was normally entered by 13-year-olds from their prep schools in either the Lower or Upper division. Then came the Shells, Removes, and Fifths, all in two divisions and each representing a year's sojourn, although one could be held back or put forward for good scholastic reason. The Fifth Form took the Oxford and Cambridge Joint Board School Certificate at the end of the school year in July, an exam roughly of the level of high school graduation. Those who stayed on, and most did, went into the Lower Sixth and the following year into the Sixth Form, which, by tradition, was always the top form in the school. Although one was in either the Lower or Upper division of one's form as a general assessment, one was in set A or set B for each individual subject. So, on the first day the masters came around, asking each new boy some questions designed to allow them to place one in the right set. A short, oldish master with a rather nasal voice and a magnificently leonine head visited me in the Lower Fifth's room, introduced himself as Mr. Piggott and put me through my mathematical paces. I had worked on my algebra that summer, solving a lot of quadratic and simultaneous linear equations and manipulating surds, but I had not done much geometry. Through a lucky guess at an answer to one of his questions, however, he concluded that I knew all about similar triangles and put me in set A, which he taught himself.

When, 15 years later, I was able to dedicate my *Discrete Dynamic Programming* to him with the words "*Ad H. E. Piggott, omnium magistrorum meorum facile principem*," I spoke no more than the plain truth, for the liveliness, enthusiasm, and care that he brought to his teaching were unparalleled in my experience. He had been an Organ Scholar at Clare College, Cambridge, and taken a first in the Tripos many years before and, in fact, had spent most of his career at the Royal Naval College. With many of the masters away in the Armed Forces, Canford was fortunate to have a group of devoted masters and staff who, at considerable personal sacrifice, kept the school together during the very difficult war years.

I took the School Certificate at the end of my first year at Canford and the Higher School Certificate Part IV [Maths, Physics, and Chemistry] at the end of the second. I finished up in Mr. Piggott's Sixth Form set, reading for Part III of the HSC, which consisted wholly of pure and applied mathematics papers. This extraordinary degree of specialization for a 16-year old was ameliorated by classes in English ("Yes, yes, we would judge a boy by his performance in a game of Rugby football and an English essay," was one of our favorite declamations when imitating the Headmaster on ribald occasions), current affairs (taught by a master who had entered the Army as a private and returned as a colonel), and Divinity (taught by the Headmaster himself, and redolent with unlikely combinations—one term we read the *Revelation of St. John the Divine* and *Plato's Republic*, another, *St. John's Gospel* and the *Phaedrus*). There were four of us in this special mathematics set, three of whom became professional mathematicians (John Morrison at A.T.&T. and John Reeve at Reading University), a tribute to the infectious enthusiasm of Piggott's teaching.

BILLINGHAM, 1946–1948

It was Mr. Piggott who got my my first job, for I finished school at the height of the demobilization and entry to the universities were difficult. A cousin of his, who was a director of I.C.I., happened to visit him at this time, and, being told about me, arranged an interview for me at Billingham-on-Tees. Thus, I found myself translated from the relatively privileged ambiance of a southern public school to the work-a-day world of the industrial northeast and, for the next 2 years, a lab boy with the Mechanical Engineering Department of the Research Labs. A nasty little 17-year old I must have been too, with great disdain for the slack work habits of the British workman, and without the tact to conceal it, for, although the department was primarily for research, its head was the authority on bursting disks for the whole of I.C.I., and most of these were made by a fitter and his mate in a workshop that I frequented. They were kind enough to warn me when they were about to test a disk to destruction—at least most of the time. For my part, I amused them greatly by drinking raspberry leaf tea (unrationed, in contrast to ordinary tea), not realizing it was a specific remedy for morning sickness. I made surface roughness measurements in a wear-testing program, and later learned some of the practicalities of attaching strain gauges to a variety of surfaces.

One evening and one afternoon a week was spent at Constantine Technical College in Middlesborough doing physics for the subsidiary subject (minor) of the B.Sc. (London, External). The External Division of the University of London provides an excellent way to get a degree, although perhaps not so good a way of getting an education. They publish a syllabus and set and mark the papers; the candidates go to London and sit the exams; nobody is concerned with how they have prepared for them. The scheme served a useful purpose when a new university (called a University College until it came of age in those days) started up in any quarter of Britain or the Empire, giving credibility to what would otherwise be a degree of unknown standard. Although, for some unknown reason, one could take a degree in mathematics without a subsidiary subject, I sat the Physics papers and took the practical (experiments with thick lenses were popular!) in a lab somewhere in the bowels of Imperial College. I had also been taking a correspondence course in mathematics and, with the physics out of the way, I doubled up on the maths courses and took a shot at the B.Sc. (Special)—as the Honours Degree was called—at the end of my second year with I.C.I. The 12 papers that I sat in June of 1948 covered a range of topics in pure and applied maths: algebra, conics and algebraic geometry, spherical trigonometry, analysis, statics, dynamics, and hydrostatics, but no functions of a complex variable beyond the platitudes of the Argand diagram. I was lucky enough to get a first, and I.C.I., who regularly sent off any employee who got a first for a year at a university, felt I was a bit wetter behind the ears than most and sent me to Edinburgh for two. I should thank the anonymous tutor of my correspondence course, for promptness in returning, and sometimes commenting on, the problems I had worked was of the essence. It cannot have been a very inspiring job to do this week after week and I can only hope that my good result boosted his standing on Wolsey Hall, the organizing body of these correspondence courses.

EDINBURGH, 1948–1950

Edinburgh is a marvelous city in which to live. "East windy" it certainly was, for the rush of air as one rode the tram platform would banish the megrims for a week. However, "west endy" I never found it for I was treated with the greatest kindness. Trams were still the city's transport in 1948 and "tuppence" the maximum fare. The Mathematical Institute was in Chambers Street and presided over by Professor A. C. Aitken, an algebraist, probabilist, and statistician by specialty and famed for his prodigious memory and calculating ability. He taught a course in numerical methods, and having demonstrated a method, such as Gauss-Seidel iteration, would assign a problem, for instance, three simultaneous equations with coefficients of three decimals, and, by the time we had copied it down he was iterating, two decimal places at a time, to give the answers to six decimal places. Of course he could have been doing it all from memory, but there was plenty of evidence in the course that he could have iterated 3×3 matrices in his head with no trouble.

Aitken had very kindly received me for post-graduate study, but wisely did not allow me to proceed to the Ph.D. I was assigned to Dr. John Cossar as tutor, a gentle, quiet man, greatly respected by those who knew him, but not widely known. When I told Artur Erdelyi (he was just back from Cal Tech, where he had been leading the Bateman Project, to which he returned the following year) that I felt the need of acquiring more mathematical rigor than I had got from my correspondence course, he replied in his perfect, almost courtly, English; "But, if you are going to Dr. Cossar, you will have the opportunity to learn, not only rigor, but, what is of more importance, elegance." Aitken was rather shocked that I had done my analysis from a translation of Goursat and suggested, even before sending me to Cossar, that I should read de la Vallé Poussin's *Cours d'Analyse* in the original to get the real flow of the subject, and I was not many pages into it before I saw his point. I went to all the courses I could and sat the exams, but my name was kept off the list of marks as it was held that had I been first, it would not have been fair to the others, because I had a first degree, and if I were not highly placed, it would not have been fair to me! There were some notable students in that class, or rather, students who went on to be notable; Lucas, who returned to teach in the sub-department of Technical Mathematics just as I left for the United States in 1958, and who showed great kindness to W. L. Edge in his later years; and McKay, who later left mathematics for the law and became Lord Chancellor. (The transition from mathematics to the law is a classic one, and more than one wrangler has finished on the bench. The last of the Lord Chief Barons—a Pollock, my grandmother's grand uncle—was Senior Wrangler, and confessed that he did it all on his knowledge of Euclid, of Mr. Newton and his fluxions, he knew nothing!)

W. L. Edge was Reader in Mathematics when I went up in 1948; later he was given an ad hominem chair. He was a great character; of booming voice and reactionary opinions with which he pulled the leg of the Astronomer Royal for Scotland when that worthy came in for the departmental tea at 4 o'clock on the days he taught in the Mathematical Institute. Edge's first publication had been a big blue Cambridge University Press book, appropriately enough entitled "Ruled Surfaces." He had been a pupil of H. F. Baker, and had gone straight from a Trinity prize fellowship to teach at Edinburgh in Sir Edmund Whittaker's day. A man of an unyielding standard of excellence, he could yet show the greatest patience with anyone, however slow, who was genuinely trying to learn. He showed great kindness to generations of students, helping and following their careers with enthusiasm or, sometimes, dismay. He was proud of the fact that the works cited in his papers were seldom later than the nineteenth century, and when I arrived in Edinburgh was enjoying afterglow of having published in *Acta Mathematica*, the Victorian amplitude of whose margins became the fullness of his description of the Klein group of order 96. Every one of his 91 papers is a model of clarity and succinctness. He was a great walker on the Highland and Lowland hills and in the Lake District. He retired in 1975, but was active until his death in 1997.

The Astronomer Royal for Scotland, W. M. H. Greaves, was a great character also. His repertoire of humorous stories was told with such gusto that one did not mind hearing it on a monthly cycle, although his subsequent approximation of aspect led one to appreciate it best from behind the first row of desks in the Research Room. His ritual for lighting his pipe was famous. Every pocket would be patted vigorously for each item of the paraphernalia, the last of which, a box of Swan Vestas, would be shaken so vigorously by the ear that it was a wonder that spontaneous combustion was not achieved. A dive into the armchair provided a base for the iterative process of lighting an old pipe, whose stem was so short that the bowl was almost under the left nostril. The iterations having converged, he would bounce back into the conversation just where he had left it 10 minutes before!

The level of the senior classes was that of graduate courses in the United States and there was only one post-graduate series of lectures each term. These were attended by most of the faculty. Sir Edmund Whittaker, who was still working on his history of the theories of ether and electricity, attended some of them. I arrived late for the term and had to catch up with Erdelyi's lectures on the Laplace transform. Cossar gave a course on Lebesque integration and we had occasional visitors, including Titchmarsh and Fréchet. However, the series that stands out in my memory was Edge on group theory in algebraic geometry. Starting from the simplest groups, he went on to those of higher order, showing how insight could be obtained by the skillful combination of geometric and algebraic methods. It was a virtuoso performance of symphonic dimensions in which the structure of the subgroups of a group of high order was beautifully laid out and their action on, for example, the 27 bitangents of the plane quartic curve described in loving detail. I still have the notes that I took in rough at the afternoon lecture and transcribed, excitedly, in the evening.

Rather unnecessarily, for I.C.I. had given me *carte blanche,* I registered for an External London M.Sc. to be taken in analysis, complex variables, and differential equations. Among the texts specified was the second volume of Courant and Hilbert, which had not yet been translated into English. With the scantiest of German that I had taken one term at Canford, I worked away at the translation with grammar and dictionary and finished it in the Long Vacation between the 2 years of my leave from I.C.I. In the spring of 1950 I went down to London and sat the four, 3-hour papers, but, unfortunately, failed them, just how badly I do not know. Cossar had covered the syllabus perfectly in his direction of my reading, so the fault was entirely my own. The fact that in 1949 I had not had to face a major exam for the first time in 6 years may have had a little to do with it, for there is a certain intellectual athleticism that clings to the British system of comprehensive examinations that requires one to "keep in training." At any rate, the experience, although humbling, enlarged my sympathy for those who find examinations more daunting than, up to that time, had I.

BILLINGHAM, 1950–1955

Returning to I.C.I. Billingham in the summer of 1950, I was put under C. H. Bosanquet's tutelage. This was something of a novelty, for he had never worked with an assistant before. He had been Dr. Lee's Reader in Physics at Oxford and was seconded to industry as part of the war effort. When the war ended, he simply stayed there, much as, when he was moved out of his office for it to be redecorated, he stayed in the new one so that no one would know where to find him. He had saved I.C.I. his weight in platinum many times over and he was allowed to do what he wanted. He was, therefore, a sort of in-house consultant to whom anyone could pose their insoluble problem and from whom, if it interested him, they would receive an inscrutable answer. He was a man of extraordinary physical and mathematical intuition and feeling for magnitude. When (a year or two after my apprenticeship) I was first beginning to read up the literature on diffusion and reaction in catalyst pellets, I asked him what the effectiveness of the catalyst in a water-gas-shift reactor might be, he got up from his desk, circumnavigated the otherwise empty office three times, followed by the two Rhodesian Ridgebacks that he brought into work, pressed his head against the wall twice and came up with the answer, "About two-thirds." Sure enough, a month later, when I had read Hougen and Watson and Wheeler *et ceteros,* I calculated an effectiveness factor of 0.63! He had no interest in writing a report and still less in writing a paper. He was quite capable of falling asleep while I was showing him my latest effort, and many times I left his office quietly without an answer to my question.

The first problem I looked at concerned the thermal conductivity of furnace bricks. This had been measured by a steady-state experiment in which a cube of these bricks was built with a heater inside and thermocouples in various places. To reach steady-state, the heater had to be turned on for a day or more before the measurements were made, and the whole experiment, requiring a number of different rates of heat generation, might take a week. The furnace people were, therefore, delighted to be told of Angstrom's method, in which a periodic heating of the end of a bar of the material and the measurement of the temperature at two points would give both the conductivity and the emissivity from the amplitude ratio and phase lag of the two temperature records. The only trouble was that they got very different values for the conductivity. The first suggestion was that they were using a square wave heating, but analyzing the experiment using formulae derived from sinusoidal heating. I worked for several weeks on a method of using the square wave data and learned a lot about evaluating solutions of the heat equation. The Fourier series solution was of little use if it took a hundred terms to converge, for you must remember that these were the days when the best you could hope for was an electrically powered calculating machine that did its most sophisticated operation, the division of two numbers, with a wonderful whirring and clacking that could be made rhythmical by choosing the right denominator. Square roots were looked up in the excellent tables of L. J. Comrie or obtained by iteration. Later I was to spend 3 months integrating a pair of ordinary differential equations from various starting points using a predictor-corrector method on this machine, a task that I could now do in an hour with Mathematica or some such program. However, to conclude my first assignment: when I finished a rather clumsy, but correct, analysis of the square wave, I still got conductivities that were distinctly different from the values obtained by the traditional method. While we were still confused by this, Bosanquet asked how the bricks were made and was told "by extrusion." Because the bricks were very porous, we had a material that was anisotropic, the pores being long and narrow and oriented in the with the long side of the brick. The two methods were measuring two different conductivities! Lesson #1: Learn all you can about the background of your problem. I also spent a week chasing an inconsistency between two of the formulae. The ratio of the two discordant results was 3.15, so I thought I had dropped a π somewhere. When I found it, however, it was the square root of 10.

Another problem I looked at in the year I spent as Bosey's assistant was the design of the venturi scrubber. To clean up a dusty gas stream it is made to flow past a water-spray and into a converging-diverging nozzle. As the gas accelerates in the converging section, the drops of water lag behind, and so move relative to the gas and pick up some of the dust, and later, in the diverging section, where the gas decelerates, they are moving faster than the gas and again have a scrubbing action. Of course, the water drops have not reached the speed of the gas in the throat, and the immediate deceleration of the gas means that there is a pocket of very low relative velocity, and, therefore, of low scrubbing efficiency. Bosey had suggested putting in a straight section between the converging and diverging sections to increase the efficiency of this pocket. He gave me a correction factor for Stokes' law, which he said he had found useful in the past, with which to calculate the motion of the drops relative to the gas. This gives a straightforward enough differential equation but it had to be solved on the clackety-clack calculating engine. In addition, the collection efficiency of the drop had to be determined, for the dust particles have their own inertia, and in the stagnation flow around a spherical water drop, the smaller ones can be swept around the drop. As the machine had no memory, virtually every intermediate number had to be remembered or written down. My accuracy with Runge–Kutta was so abysmal that it was a relief to discover the predictor-corrector formulae of Milne. After several months, I wrote a report on the design of the scrubber, which Bosey approved, and I thought we had heard the last of it. However, several months later the Patents Department took out a patent on the straight section. I cannot imagine it was ever asserted in any way, but you never know.

During the year that I worked under Bosey's supervision, I had several interesting problems, such as that of damping the vibrations of a centrifuge when, at the end of a batch, it spun down and passed through the critical frequency of its mounting (this one provided a nice introduction to Fresnel integrals), or the efficiency of silencers (I managed to get elliptic integrals into that). I learned a lot from him, although, unfortunately, his physical intuition did not rub off. However, I well remember the delight of seeing the conductivity problem written in dimensionless variables for the first time. I had some elementary dimensional analysis from as far back as Mr. Brentley's physics class at Canford, but had used it only to check the consistency of equations. For the rest, I had the mathematician's outlook that dismissed dimensions a phrase like *'in consistent units.'* Many stories could be told of him: How he came in one morning to borrow a pencil, and he had forgotten to change out of his goat-milking jacket when that daily chore had been done (a fact strongly confirmed by the olfactory sensibilities). Did I have any books on the theory of relativity? He had heard a talk on it on the wireless, had been working on it all week, and had got a formula for the background luminosity of the sky. The less unusual members of the department included Dorothy Annable, who was working on the design and control of ammonia and methanol converters, and Menna Jones (later Vincent), whom I pestered with a discussion of my latest reading or results. We once were talking about modeling a physiological problem and I started off with, "Consider your legs to be cylinders." Her "Thank you very much" was exquisitely dismissive, and I learned Lesson #2: Modeling may be abstract, but still demands tact. George Bell came with an Oxford D.Phil. in chemical physics during this time, and his pocket-patting and two-speed locomotion suggested that we might have another genuine eccentric in the making.

My first major assignment after I was promoted to the Technical Officer grade, was to help the Analytical Division interpret the data they were getting from their latest enthusiasm, the vapor phase chromatograph. These instruments were not yet available "off the shelf," but were made for the job. Consequently a bit of theory was needed, and I was called in to work out the effect of sample size and stationary phase loading and so forth. So I studied the literature from Martin and Synge (my first encounter with the theoretical plate), Cassidy and Onsager, Gluekauf and Weiss (whose polemics enliven a Faraday Society's discussion), to Amundson and Lapidus (whose inversion of Laplace transforms evoked my admiration), Goldstein and Murray, and Whitham and Lighthill (who came later, but whose flood waves and traffic patterns opened up the scope of kinematic waves). I soon realized that the combination of molecular diffusion and convective dispersion was an important factor in chromatography, and just at that time Taylor's seminal paper on dispersion came out. In it, he showed that the effect of diffusion across the parabolic profile of the flow in a circular tube could be represented by an apparent dispersion coefficient, $a^2U^2/48D$ (see pp. 109–120). In trying to understand Taylor's paper, I calculated the moments and showed that, asymptotically, the center of gravity of a dispersing cloud moved with the mean speed of the stream, and its variance increased linearly with an apparent dispersion coefficient of $D + a^2U^2/48D$. The technical aspects of these results, and others that used temporal moments, are dealt with previously. Here I want to record my indebtedness to Sir Geoffrey Taylor, who not only communicated the paper I wrote to the *Proceedings of the Royal Society,* but also corresponded with me, an unknown stripling, on several ancillary questions.

The writing and rewriting of the Taylor diffusion paper took most of my evenings in the first half of 1955, for by then I had been moved off chromatography and was in the thick of reactor design. The occasion of this transfer was the departure of Dorothy Annable, leaving a tremendous gap in the fundamental work being done at Billingham. (These were the days when industry still saw the wisdom of doing broad-based research and the bottom line had not become top dog). I was just getting really involved with a number of problems that had grown out of chromatography and I was very fed up with the summary transfer, and resolved to try to move to a university. I applied for several lectureships during the school year 1954–1955, but without success. Meanwhile, however, there was plenty to learn about reactors. Ammonia synthesis had been put in order by Annable, but new methanol plants were needed. Heaven forefend that I should break my vows of confidentiality by saying how it was made. For when, some months later, I asked permission to publish a paper on the graphical design of an adiabatic reactor for $2A + B = C$, it was refused on the grounds that a competitor might smell methanol and from the lines on my graph deduce the kinetics we were using! Actually, we did not have any kinetic law, and it was my job to find a formula like the Temkin law for ammonia. In doing so, I learned two valuable lessons. The first came from an attempt to get the activation energy of the forward reaction from existing plant data. I argued that the data from the first bed of a multibed adiabatic reactor would be dominated by the forward reaction, that the rate would be proportional to the temperature rise DT, and that the temperature to correlate with the rate should be close to the mean temperature in the bed. There were reams of data from another I.C.I. plant, and the correlation was perfect (corr. coeft. 0.9999), but the result was meaningless. I had failed to notice that the plant operators had kept the inlet temperature virtually constant and we had been correlating DT against a constant plus $DT/2$. The other lesson I learned some years later when chatting with my successor: never trust results that are obviously correct. In this project, I assumed that the equilibrium constants calculated for this synthesis under operating conditions would certainly be correct and wrote the rate with the forward reaction multiplying a factor depending only on equilibrium. The data fit I got with this expression was pitiful, but by freeing up the equilibrium term, my successor was able to devise a useful kinetic expression.

Whether I helped to increase the supply of methanol, I shall never know. Actually, I was far more interested in a chart I had devised while trying to see what things had to be chosen to get a definite design of a multi-bed adiabatic reactor. In the plane of concentration and temperature, the adiabatic beds were represented by straight lines or shallow curves of positive slope joined by horizontal lines for the heat exchangers between beds. If the interbed cooling were by cold shot, the horizontal lines would be replaced by a pencil of lines through the feed point. It was nothing very new, but it was new to me and it was getting at the mathematical structure of the system. In fact it was not new, for an M.S. student had been doing something rather similar at Minnesota, and although I.C.I. would not let me publish the report I wrote on it, it was to be the cardinal point of my career, for it favorably impressed a visitor to Billingham in the winter of 1955.

Neal Amundson was on sabbatical in Cambridge in the academic year of

1954–1955, and in the early part of 1955, he visited Billingham with Kenneth Denbigh. I had some acquaintance with the latter's Faraday Society papers, but I do not now recall when I first read Amundson and Bilous' papers in the first volume of the *AIChE Journal.* However, I knew that they were leaders in the field and put my best foot forward when they stopped by for a brief visit during their tour of the Research Department. Amundson did not give a talk at Billingham, but two lectures on reactor dynamics were announced in Cambridge and I asked permission to go down for them. For some reason, this was not possible, but I was sent down to visit with him for a day in April or early May. Now, I had not told my superiors at I.C.I. that I was trying to leave and get an academic job, nor had they yet told me or done anything about a suggestion that the Research Director had made during Amundson's visit, namely that I.C.I. should send me to Minneapolis for a year's study. This put Amundson in a bit of quandary when, lunch with Denbigh at the Senate House in Cambridge, I revealed that I was actively trying to get into academia. However, having assured himself that I had for some time been determined to leave I.C.I., he felt free to ask if I would be interested in a research fellowship at Minnesota. This seemed an excellent suggestion, the more so when, on my telling my superiors that I was going, they told me that they had plans to send me. Although this might have happened without any obligation to return (as had been the case with my 2 years in Edinburgh, at the end of which I nevertheless felt that I did have an obligation; but one that my 5 subsequent years of service had now fulfilled), it was clearly better way to do it. As soon as I had worked off my notice period at I.C.I., I sailed for America, arriving a little late for the Fall Quarter of 1955.

MINNESOTA, 1955–1956

The first problem that Amundson suggested was the problem of reaction in laminar flow. Why I did not see that limited, but significant, insight could be gained by extending the method of moments, which I was still using to explore the ramifications of Taylor diffusion, I do not know, but it was 25 years before I caught on to that approach [169]. Instead, I got a solution in terms of Kummer's hypergeometric function, whose zeros are notoriously hard to calculate and which formed a non-orthogonal set if longitudinal diffusion was included.

I should have said "were hard to calculate," for nowadays what were then recondite and untabulated functions are available at the click of a mouse on one's laptop. Then we had an old analog computer in the Experimental Engineering Department and a Univac Scientific. The former had performed wonders in its day and, under Olegh's gentle touch, had produced the diagrams in the classic Amundson and Bilous papers. For me, however, it would not even generate $A\exp(-E/RT)$ over a very limited temperature range. Admittedly, it was a year or two older than it was when Bilous communed with it in the stillness of the night, but I think it also, like an experienced hack, sensed that a dude was on its back. The Univac Scientific, however, was the latest word from the budding Minnesota computer industry and occupied most of the second floor of the Experimental Engineering building. Moreover, it had air-conditioning, a privilege shared with the animals in the hospital, but with hardly any other part of the University. The only trouble was that it had to be programmed in machine language and you had to know in which register you had stored each number of the calculation—all this in octal too! I went to the course on programming later in the year, but was getting thoroughly confused when Amundson came to the rescue by hiring a graduate student in computer science to work with me. Our problem went in as punched cards and came out as yellow tape that was then printed out as columns of numbers that had to be plotted by hand. What went on between the cards and the tape was the province of R. A. Wonderly, who was aptly named. I have often hoped I would have a chance to thank him properly, in the light of what has grown out of the work he helped me with. The experience showed me the advantage of vicarious computation, and I have been infinitely obliged to the abilities and generosity of my students that have allowed me to retain my computational virginity. I have constantly stressed that it should not be surrendered lightly, for rushing off to the computer before one has a real feel for the problem is a prescription for disaster. I cannot, alas, claim to have been so resistant to the blandishments of that other great seducer of our day, the Xerox machine. Not a few are the papers that I have copied with the best of intentions of reading, only to find them, months later, at the bottom of the pile, unread, unstudied, and unused.

These remarks on digital computing belong to the second problem that Amundson suggested when it became clear that the parabolic flow was leading nowhere. This was the problem of controlling a stirred tank reactor, which, without control, would have three steady states, so that it operated at the naturally unsteady state of the uncontrolled system. The steady-state problem was easy enough: the addition of a controller, whose action was proportional to the deviation of temperature from that of the desired steady-state would change the heat rejection line into a parabola, and the increase of the control parameter would move the two outlying steady-states into coincidence with the controlled state. However, what did the phase planes look like as the control tightened? We discussed this dynamic evolution at length and laid out the alternative transistions that were possible in a phase plane. These included a homoclinic orbit and other features that were only to be understood in their fullness some 20 years later. In particular, the controlled steady state could attain its stability either by the collapse of a stable limit-cycle or by shedding an unstable limit-cycle (i.e., a sub- or super-critical Hopf bifurcation). Some alternatives were resolved by computation, but the mode of the Hopf bifurcation had to be found using the formulae in Minorsky's *Introduction to Nonlinear Mechanics,*[1] an expansion in polar coordinates up to the cubic terms. I did it thrice and got the same answers from the last two performances; I need not add that I did not give it a fourth try. There was an element of luck in our landing on so interesting a niche of parameter space, but this kind of luck usually owes much to good insight, in this case Amundson's.

[1] Minorsky, N. *Introduction to Nonlinear Mechanics.* Ann Arbor: university microfilm (1947).

Amundson had kindly extended my Research Fellowship for a second year, but a letter arrived from Edinburgh telling of a vacancy caused by an appointee's last-minute withdrawal. It was a lectureship in the Technical Mathematics subdepartment, headed by L. M. Brown, a gentle geometer whom I knew from my earlier time in the department. As this was just the kind of position I had been seeking when I was still with I.C.I., I jumped at the chance and returned in October. This left Amundson rather in the lurch, but I promised to write up the work and to return for 3 months the next summer.

EDINBURGH, 1956–1958

My first task was to get our work on the control of the stirred tank written up. The resulting paper was rather long and Denbigh, the communicating editor of *Chemical Engineering Science,* asked me to divide it up. I did so, but Pergamon printed the three parts consecutively. I also wrote up the work I had done on shape factors at I.C.I. The steady state part of this [3a] showed that the linear dimension V_p/S_x (i.e., the pellet's volume divided by its external surface area) should be used in the Thiele modulus for the diffusion and reaction problem, as this brought all the effectiveness factors together for large and small values of the Thiele modulus and brought them closer to each other. If I recall correctly, the idea is nascent in Wheeler's work[2] and Hougen and Watson's[3] treatise has a good treatment of the topic, but my paper seemed to make it more generally known. It is interesting that the proof is very shaky and, as mentioned previously, I did not properly understand the Robin boundary conditions, shortcomings that I did not pubicly remove until I wrote the first volume of [I′].

It was at Edinburgh that I first learned a bit about teaching. One of the courses I had to teach to third-year engineers was on Fourier series. I thought I was getting on famously until, at about the third week into the term, the Servitor[4] took me aside and explained that I had been covering the material at such a rate that I had completely lost the class. Since that day, I have had a healthy respect for the staff that supports the efforts of the faculty. Much wisdom lies in this stratum of the university.

Throughout my 2 years on the faculty at Edinburgh I benefited greatly from the presence and interest of K. G. Denbigh, who, in 1955, assumed the revived chair of chemical technology. Not only did I have the opportunity of going to his lectures on thermodynamics, but he also encouraged me to publish

[2] Wheeler, A. Reaction rates and selectivity in catalyst pores. *Adv. Catal.* **3,** 249 (1951).

[3] Hougen, O. A. and K. M. Watson. Chemical process principles, Vol. III. *Kinetics and Catalysis.* New York: John Wiley, (1947).

[4] The name given to the factotum responsible for the good order of a building at the University of Edinburgh. Mr. Henderson, the Servitor of The Mathematical Institute, was a magnificent representative of this staff category, presiding in his top hat and silver-buttoned coat on formal occasions and keeping the coal fires in the offices and the blackboards in the classrooms in apple-pie order in the day-by-day routine. He had more letters behind his name than the Professor himself, for he was a semi-professional dancer.

my papers on shape factors in *Chemical Engineering Science,* of which he was then an editor. It seems to me that, perhaps because he was also a leading thermodynamicist, perhaps because he eventually turned to the natural philosophy of time and to academic administration, Denbigh's fundamental contributions to chemical reaction engineering have not been sufficiently recognized.[5] However, in his Faraday Society papers of 1944,[6] he anticipated the problems of multiplicity and optimality [A′, 16] that occupied others 20 years later. A brilliant expositor, as the users of his thermodynamics text know, his 1965 text on *Chemical Reactor Theory*[7] has been very ably kept up to date by J. C. R. Turner.

I returned for the summer of 1957 and worked on the statistical aspects of the stirred tank [8], but the most far-reaching development was nontechnical. One evening in August I became engaged to Claire Holman of Minneapolis and mentioned this to Amundson the next morning. His response was no feeble felicitation, but the very practical question "Do you want a job?" Because, at that time, the only way of living on one side of the Atlantic and keeping in touch with the other side was to live in America, my response was an instant affirmative. In those days, the process of appointment was much less complicated than it is today and a week later I had an offer of an assistant professorship, which I accepted immediately—a decision that I have never for a moment regretted.

MINNEAPOLIS, 1958–1964

I really had only the haziest notion of what was expected of a tyro assistant professor. Amundson had said that I should write papers on my own—i.e., in addition to those I was writing with him—and that it would be hard to get me promoted without a Ph.D. Fortunately, I had registered 3 years earlier with the University of London, which also gives external Ph.D. examinations for students who have their first degree from London. As with all their external degrees, they paid no heed to how one prepared for it and demanded only the appropriate registration, an outline of the proposed research after 3 years (at which time they appointed the examiners, who approved the scope of the work or suggested modifications), the submission of a dissertation, and an oral examination after the examiners had read and approved it. Amundson, who had given the Institute Lecture in 1958 on optimization—a field he had pioneered with Bilous—also suggested that Bellman's[8] method of Dynamic Programming was worth looking into. Because we discussed the results I got along the way, Amundson was my supervisor *de facto,* although I had none

[5] Thus, the omission of his name in the Founder awards at the International Symposium on Chemical Reaction Engineering in 1996 can only be accounted for by supposing the selection committee to have had a strict limitation of numbers.

[6] Denbigh, K. G. Velocity and yield in continuous reaction systems. *Trans. Faraday Soc.* **40,** 352 (1944).

[7] Denbigh, K. G. *Chemical Reactor Theory,* New York: Cambridge University Press (1965).

[8] Bellman, R. *Dynamic Programming.* Princeton University Press (1957).

de jure and I count myself at least a stepson in his distinguished academic family.[9] The suggestion proved fertile, and I wrote the dissertation in the summer of 1959 and defended it in July 1960. One of the examiners was W. G. Bickley, a blind mathematician at Imperial College, for whom the thesis had been translated into Braille. He had read and remembered it so thoroughly that, while waiting in his office for the other examiner, P. V. Danckwerts, to arrive, we discussed Ausonius and Carroll, two poets whom I had quoted in the dedication and in the chapter headings.

A revision of the dissertation [A′] was published by Academic Press in Bellman's series on Mathematics in Science and Engineering. Bellman had taken an interest in my work and Amundson and I had visited him at Rand Corporation, the Santa Monica think-tank. He was rather fed up with the shakiness of the economic models with which he had developed dynamic programming and took to the definiteness and reality of chemical engineering. A joint paper [14] immediately sprouted, for Dick was a prodigious publisher, and I returned to Santa Monica quite a few times in the following years, exchanging ideas with him and his colleague, Bob Kalaba. Bellman was one of the more remarkable characters I have known. Unrivaled for the speed with which he could generate ideas of his own or understand those of another, recalling instantly all he knew about the history of a problem and living life to the full, he always seemed to be a little larger than life. His autobiography is appropriately entitled "*Eye of the Hurricane.*"[10] His reputation would not have suffered had he published fewer than his 620 papers and 47 books and monographs, for there is some repetition and incompleteness (vide supra pp. 69–73), but that could be said of many a lesser mortal. He was a master at coining a name and investing it with his meaning, "neither more nor less"—"the menace of the expanding grid," "the curse of dimensionality," and "dynamic programming" itself. I last saw him in the mid-1970's, when I spent a week or two at USC on an immunologic model he was interested in [148]. He had just had the botched operation that was the beginning of the decline in his powers of communication that so darkened the last years of his life. He worked in so many different fields and with such a great variety of colleagues that I was surprised, as well as honored, to receive the first Bellman Control Heritage Award in 1992.

Optimization dominated my first years at Minnesota, although I did some things in residence time distributions and Taylor diffusion [12, 21]. Dynamic programming gave way to Pontryagin's method, for the curse of dimensionality was heavy upon it. It is interesting that it has had a revival in the last few years through Rein Luus' modification, Iterative Dynamic Programming, which brings large problems within the capacity of the computing engines of our day. An extension to countercurrent processes, which at first blush are not suitable for dynamic programming, was made by Yesberg [36] and conversations with two of the leading optimizers, Doug Wilde and George Nemhauser, led to a way of suitably dissecting networks [46]. Optimization also

[9] Aris, R. and A. Varma. (eds.). *The Mathematical Understanding of Chemical Engineering Systems: Selected Papers of Neal R. Amundson.* New York: Pergamon Press (1980).

[10] Published by World Scientific Publishing Co. Singapore. 1984.

brought me one of my stellar pupils, M. M. Denn. He took up the maximum principle and, among other things, overcame the difficulty presented by the two-point boundary conditions. Five substantial papers and two notes came out of his dissertation.

May I interpolate a word of consolation here? I have supervised a goodly number of graduate students, and if I fail to mention some it is not that I think any the less of them, but that a bare recitation of *A* did *B* would be intolerably wooden, and a properly crafted survey of all that each one did would be intolerably long. All are listed in the index of co-authors and many will have their own evaluations of the others. If I praise some specifically it is because I am very sensible of the fact that academics' reputations are bound up with the luster of their pupils. Just as a department will not prosper unless its faculty are prepared to attract junior colleagues who are brighter than themselves, so also the individual faculty members will get nowhere without taking on pupils more able than they. We tell our graduate students that there will come a point in their research, if it is to lead to a Ph.D. rather than an M.Sc., at which they begin to tell us the direction in which to go, and our efforts to keep up with, say, a Kevrekidis, will either burst the lungs or strengthen the heart. If there is anything I have tried to instill in those I have supervised, it is that, in intellectual matters, the subject under consideration is a little more important than their own egos—or, indeed, than mine.

To have such students as Chien, Denn, Gavalas, and Ray in one's initial years—not to mention post-doctoral workers like Richard Mah and Willi Regenass—is an incomparable stimulation to good work. Ray took up some adaptive control problems [60, 64, 71, 72] that grew out of Chien's work [44, 45]. Siebenthal got the continuity problem in the optimal batch problem straightened out (vide supra [42, 43]), and Schindler looked at another form of adaptive optimal questing control [68–70]. Adam Chou [72] looked at a problem of a reactor with catalyst decay and labored with it in a very complex form while I was away on sabbatical. On my return, I suggested some major simplifications that allowed him to solve the problem and, with some additions by Ray, we published it. Unfortunately, I had completely—and culpably—forgotten a conversation with Octave Levenspiel before I went on sabbatical in which he had shown me what Steve Szepe had done, arriving at the same conclusion with the same system that we had now approached by another route. Octave wrote, more in sorrow than in anger, to point this out, but with characteristic generosity, believed me when I confessed to having completely forgotten our conversation.[11] Fortunately, we were able to get a note added in proof acknowledging their priority. I committed the same misdemeanor much more recently when I published a note with a colleague in industry on the centrality of apparent second-order kinetics in cracking a mixture with an uncrackable component [229]. This time it was Fred Krambeck who had to

[11] Some years later we were guests, along with other ASEE Lectureship awardees, at Wonewok, 3M's conference lodge. There was a table with an inlayed chessboard in the gameroom, but no pieces. Octave said "Let's play; we can remember where we've moved." Of course, he beat me handily, but the looks on the faces of those passing through the room were a delight to behold.

remind us that he had mentioned the very same result in his keynote paper at the ISCRE meeting in Edinburgh, a lecture that both of us had attended. Thankfully, our condign apologies were accepted, and our relations did not suffer.

George Gavalas belongs to this period, and if ever I had an excelling student, it was he. Minnesota has a great tradition of Greek graduate students stretching from Acrivos to Zygourakis (alphabetically, that is; not chronologically!) and Gavalas was certainly *primus inter pares.* At his oral prelim he gave a 10-minute summary of the theory of Fourier series (up to and including summability theorems) in answer to a leading question on Fourier series that normally leads a candidate on an erratic path to $1/\pi^2$. (Hyun Ku Rhee was another student who gave a masterly performance on his oral prelim.) Gavalas laid the foundations of the theory of reactions in continuous mixtures in his dissertation, and writing the paper that grew from this was one of my first sabbatical tasks. Before turning to this, however, I should mention that even more important for the development of mathematical modeling was the monograph he wrote on *Nonlinear Differential Equations of Chemically Reacting Systems* (Springer, Heidelburg. 1968). It spoke the language of mathematicians, showed how apt an environment for the application of the latest mathematical advances chemical engineering could be, and sparked the interest of outstanding mathematicians such as D. Cohen, his colleague at Cal Tech.

To this period belongs my, admittedly slight, involvement with fluid mechanics. Amundson asked me to teach the graduate course when I first arrived, and I cobbled together the little I had done at Edinburgh 7 years before with a crash course of reading the standard texts and produced a fine farrago of basics and applications. It was James Serrin's course on fluid mechanics—he had just published his *Handbuch der Physik* article—that showed me the structure of the subject. I had sat in on his course on potential theory when I first came to Minnesota in 1955 and knew him to be a first-class lecturer, and his treatment of the subject was as stimulating as it was apt. Throughout the 1950s there had been a growing interest in rational mechanics and thermodynamics, and Serrin, together with Truesdell, Noll, Coleman, and others, was in the forefront of this. It was partly as an effort to make this line of thought more accessible to the engineering student that I wrote *Vectors, Tensors and the Basic Equations of Fluid Mechanics.* The other motivation for this book was the work on surface flows of L. E. Scriven, who had joined the department in 1959. This work demanded the full panoply of tensor calculus and gave me the chance to dig around in a field that I had not visited since Erdelyi's lectures in 1949. It seems to have been useful to others, and I am glad to say it is still available in the honorable company of Dover reprints.

CAMBRIDGE, 1964–1965

The sabbatical custom—that, after 6 years of regular duty, you can have a year at half pay, provided you have a coherent plan of study and your department can manage without you—is one of the most valuable fringe benefits of academic life. It allows one to get caught up and to look ahead as well as to visit new places and people. We spent the academic year 1964–1965 at Cambridge and, thanks to the great kindness of everyone, we had the time of our lives. The Shell Department of Chemical Engineering of 1964 was very different from the department that Amundson had come to 10 years before. In 1954–1955, it had been in temporary buildings in Tennis Court Road; now it was in its own building in Pembroke Street. Then it had been exporting its talent: Denbigh to Edinburgh, Danckwerts to Harwell, Sellers to Cardiff, and Gray to Leeds. By 1964–1965, Danckwerts had returned via Imperial College; Davidson had just been made a Reader, and with Harrison had made Cambridge a focal point for the fluid bed community; J. R. A. Pearson, an applied mathematician of great ability and catholic interests, was soon to take up and excel at the "real tennis of the resounding court"; J. C. R. Turner, a physical chemist by training, of whose quick wit and lively mind one might say (adapting the second half of Johnson's epitaph of Goldsmith) *nullum quod tangit non ornat*; and Nigel Kenney, of happy memory, who pleased me enormously with a paper, just completed, that demonstrated the additivity of the molecular diffusion coefficient and the Taylor dispersion coefficient. The immediate collegiality and consideration of these and others, whom we knew through other connections, ensured that we had the best of times. Pearson arranged for me to supervise a brace of undergraduates at Trinity Hall, which gave me a better feel for the Cambridge system and also the privilege of dining at high table once a week. Various people asked me to dine in their colleges and I was able to meet some of the people I knew only from afar, such as: Sir Geoffrey Taylor, who had so kindly communicated my early papers; and J. E. Littlewood, who had never had to bother with a Ph.D. and was that year given an honorary Sc.D. I attended Part III lectures on homological algebra, thinking this to be the key to abstract chemical mechanisms (which it was not), and T. A. M. Bishop's classes on paleography, thinking this subject would be engrossing (which it has been). I had the run of the University Library (but not borrowing privileges); books were still affordable and I could walk from Heffers to the Whim and enjoy my latest purchase over a cup of coffee and an open fire. As a small token of thanks for the hospitality of the Shell Department, I gave six graduate lectures on optimization. I also gave seminar talks at Imperial College, where Denbigh was now the Professor, and several other universities, including: Brussels, under the aegis of Professor Jottrand; Copenhagen, where I first made the acquaintance of John Villadsen; and Trondheim, the home of Olav Erga, whom I knew from Wisconsin. I attended meetings in Amsterdam and Frascati, and even visited my old stamping ground, I. C. I. Billingham. In May, we went to Vienna to represent the University of Minnesota at the celebration of the 600th anniversary of the founding of the University of Vienna. This was a delightful melange of processions, laureations, meals, masses, and speeches.

In spite of these distractions, I managed to get quite a lot of work done, both in writing up things of the past and in making some progress in what might be called *formal chemical kinetics*. Because this is a rather well-defined area, which I have touched at several points in the research career I am attempting to describe, I will depart from a strictly chronological mode and limn the history of this subject with a very broad brush. The picture will be a little lopsided, for I am setting it in the context of my own work, but I hope I have at least mentioned the important contributions. The area is ripe for some historian of the exact sciences to give a disinterested account of its development.

Aside on Formal Chemical Kinetics, 1963–1995

In the broadest sense, I found the analogy with fluid mechanics to be very helpful. Just as kinematics provides the geometrical framework of fluid mechanics by exploring the motions that are possible, so also stoicheiometry defines the possible reactions and the restrictions on them without saying whether or at what rate they may take place. When dynamic laws are imposed on kinematic principles, we arrive at equations of motion; so, also, when chemical kinetics is added to stoicheiometry, we can speak about reaction rates. In fluid mechanics different materials are distinguished by their constitutive relations and allow equations for the density and velocity to be formulated; thence, various flow situations are examined by adding appropriate boundary conditions. Similarly, the chemical kinetics of the reaction system allow the rates of reaction to be expressed in terms of concentrations, and the reactor is brought into the picture as these rates are incorporated into appropriate equations and their boundary conditions.

The attempt to discover the mathematical structure of chemical kinetics begins with a paper by Wei[12] on chemical kinetics, in which he gave a set of axioms. He and Prater[13] also gave a very satisfying analysis of systems of first-order reactions.[14] From a mathematical point of view, there was nothing surprising about their results, but apart from a paper of Franklin's, there was little appreciation in the chemical world for the eigen properties of matrices in this context.[15] The attempt I made to bring an algebraic structure of a very different sort to second-order systems [39] did not turn out to be very fruitful. Wei[16] was able to show that linear paths were very rare, but not many of the theorems that Marcus had proved when he first saw the connection to non-associative algebras seemed applicable.

Another idea I tried at this time was intended to set up the organic chemists' planar structural diagrams in a mathematical framework. A hybrid matrix, in which the diagonal elements are the names of the atoms and the i,j^{th} element is the number of bonds between the ith and the jth atoms, would

[12] Wei, J. Axiomatic treatment of chemical reaction systems. *J. Chem. Phys.* **36,** 1578 (1962).

[13] Wei, J., and C. D. Prater. The structure and analysis of complex reaction systems. *Adv. Cat.* **13,** 203 (1962).

[14] Wei's later work with J. C. Kuo on lumping and that of Li and Rabitz at Princeton [*Chem. Eng. Sci.* **44,** 1413, (1959)] also opened up an important area of formal kinetics. See also Bischoff and Coxson, Lumping strategy. *I.E.C. Res.* **26,** 1329 & 2151, (1983). J. E. Bailey [Lumping analysis of reactions in continuous mixtures. *Chem. Eng. J.* **3,** 5261 (1962)]. forged an important link between continuous mixtures and lumps.

[15] Of course, chemists understood the eigen-properties of linear systems, but largely in the context of vibrational spectra rather than in kinetics.

[16] Wei, J. Structure of complex chemical reaction systems. *I.E.C. Fund.* **4,** 161 (1965).

seem to capture the information in a planar diagram. Moreover, because reaction is a matter of breaking some bonds and forming others, it can be represented by matrix addition. There are several difficulties in taking this much farther and, as far as I know, it has not been. Of course there is much work, both beautiful and ugly, on chemical kinetics that has been done in recent years by computational chemists, but this is at a much deeper level of physics and chemistry than the formal kinetics I am concerned with.[17]

Stoicheiometry is linear algebra, and the independence of reactions is an important concept. I wrote a paper with Richard Mah [33] that showed how the equations for the concentrations of all species could be reduced in number to that of the independent reactions. A note [38] of about this time suggested that, if a convention were needed, reactions might be orthonormalized. This is a notion, risible to the chemist but obvious to the mathematician, illustrates the difference between formal and practical chemistry.[18]

The work that I had been doing in this area was in the spirit of, although not directly connected with, the school of rational mechanics, which revived the methods and outlook of the age of reason. I had first had contact when writing my book on the basic equations of fluid mechanics and had paid tribute to the clarity and insight to which it led. I therefore sent off a paper of *Prolegomena to the Rational Analysis of Systems of Chemical Reactions* to the *Archive for Rational Mechanics and Analysis,* with a note expressing the hope that something chemical would not be outwith the policies of the journal. Truesdell, the editor, kindly accepted it, replying that the journal was devoted to a method rather than a subject. It is pleasant to note that several of the key papers in formal kinetics by Feinberg, Horn, Jackson, and Krambeck were subsequently published in this Springer journal, whose typographical standards and design are, to my mind, unsurpassed and equaled only by the *Proceedings & Philosophical Transactions of the Royal Society* (both products of Cambridge University Press) and the *Acta Mathematatica.*

If one name of the rational mechanics school is to be singled out—and to pick more than one would be hazardous to a degree—that of Clifford Truesdell it would have to be. This is not because he has stood head and shoulders above all others, for there are a number of outstanding scholars in this field, but because of his example and the centripetal influence he has had. Not for naught does he labor under the family motto *STRUIMUS,* for his works include lengthly *Handbuch der Physik* articles, editions of the works of Euler and others with detailed introductions (in Latin), a host of papers and numerous reviews, not to mention the labor (shared by his wife, Charlotte) of founding and editing journals, notably the *Archives: for the History of*

[17] E. C. Titchmarsh, in the preface to his *Theory of Fourier Integrals,* remarks that, in doing problems on heat conduction, he has retained "such terms as heat and temperature" but that "the reader need not know that such things exist." He was a pupil of G. H. Hardy, and shared his mentor's aversion to applied mathematics. By contrast, it is important in mathematical modeling to see as many links between the physical and mathematical as possible.

[18] Happel's use of the stoicheiometric number (*Isotopic assessment of catalysts*. Academic Press. 1986) should also be mentioned. Arpád Pethö has related stoicheiometry to dimensional analysis in *Three Lectures on Linear Algebraic Methods in Chemical Engineering* [see also *Acta. Math. Acad. Sci. Hung.* **18** 19, (1967)].

the Exact Sciences and *Archives for Rational Mechanics and Analysis.* I first encountered him in the stacks of the library at the University of Edinburgh in the form of an orange paperback on a unified theory of the special functions, a work that obtained certain forms of all the Special Functions from a single difference—differential equation and aimed to bring order to the multitudinous inter-relating formulae by deriving them by uniform procedures. It was written while (or soon after) he was an undergraduate at Cal Tech.—a remarkable achievement. Truesdell is a man of massive erudition and mordant eloquence, who not only knows the meaning of the word 'idiot'—a thing sufficiently rare in our day—but, rarer yet, he has the courage to use it correctly. His *An Idiot's Fugitive Essays on Science*[19] is full of provocative, but deeply considered, opinions couched in a prose of a standard that one seldom encounters today.[20]

I have often deplored, however, the indifference on the one hand and the animosity on the other that has existed between the school of rational continuum mechanics and what might be called the Cambridge school of fluid mechanics. Suspicion of sterility on one side and abhorrence of empiricism on the other have, at times, taken matters beyond the point at which a certain scholarly asperity merely serves as the whetstone for an epigram to the level at which neither can learn from the other.[21]

Although there are not many people with deep commitments to both the formal and the empirical approach to kinetics, there has fortunately been no divisive misunderstanding. Thus an expert on catalysis, both in the laboratory and the chemical plant, Michel Boudart, pointed out to me that my proof of the consistency of equilibrium and the kinetic expressions assumed too many degrees of freedom. This was corrected in a paper of *Some Addenda to the Prolegomena* [76].

The really fruitful development came from F. J. M. Horn and R. Jackson,[22] who introduced the concept of a complex, namely, that combination of species that occurs on one side of the chemical equation as normally written. Building on this and on Krambeck's paper on the structure of chemical kinetics in single-phase systems[23] and working with Horn while he was still at Rochester [see, for example, Horn's papers in *Arch. Rat. Mech. Anal.* **49,** 172, (1972); and *Proc. Roy. Soc.* **A334, 299 & 313,** (1973), and their joint *Chem. Eng. Sci.* **29,** 775, (1974)], Martin Feinberg has produced a corpus of outstanding work on networks of reactions with mass action kinetics. With great courage—for

[19] New York: Springer Verlag (1984).

[20] I once labored long in composing an introduction of him when he came to Minnesota to give a talk, and was rewarded by his *sotto voce*-remark as I left the stage and he came onto it, "Thank you for that introduction; you know I'm very seldom introduced in English." Truesdell was present when the only person to do so took umbrage at being the subject of my occasional verse. He told me that he would not take offense at being lampooned, but that I should remember what Suetonius had to say about the Emperor Vespasian. I could not find anything that appeared to be particularly relevant, but sent him a pseudo-Loeb tract on the life of the deified Truesdell. He replied with his thanks and a set of corrections to the Latin pages!

[21] Ever ironical, I refrain from giving examples of the exchanges or even from disentangling this rather convoluted paragraph!

[22] *Arch. Rat. Mech. Anal.* **47,** 81 (1972).

[23] *Arch Rat. Mech. Anal.* **38,** 317 (1970).

we still, to some extent, live in a "publish-or-perish" world—he produced on average about one paper a year, but saw to it that each one was of real substance. The goal of formal kinetics is to produce results of generality; that certain class of reaction networks is, for example, incapable of oscillatory behavior *whatever the kinetic constants may be.* Feinberg's results are theorems of this character, but he has also been concerned to make them available to those who do not have the patience to go through all the intricacies of his proofs, and has produced a disk whereby a network can be fully analyzed on the desk-top computer. His most recent work on attainable regions and their convexity properties shows the same generality with respect to optimal reactor configurations.

Gavalas' work on continuous mixtures, which I wrote up at the beginning of my sabbatical, laid the foundation for the branch of kinetics that has concerned itself with continuous reactions. The idea of a mixture of components that is so complex that it has to be described by cuts of a continuum, $A(x)dx$, rather than by membership of a discrete set, A_j, goes back in thermodynamics to De Donder, whose book *L'Affinité* (Paris: Gautier-Villars, 1931; Ch. III) was given to me by Professor Defay when I lectured on continuous reactions in Brussels in 1965. In the engineering literature, the equations of the simplest case of parallel first order reactions, $A(x) \rightarrow B(x)$ with the rate constant dependent on the index variable, x, arose, not in the context of cracking kinetics, but of froth flotation.[24] The idea for Gavalas' doctoral work was a spin-off of consulting work with Standard Oil of Indiana at Whiting, now Amoco of Naperville. Keith McHenry was my supervisor and he thought it strange that cracking data was often better fitted by a second-order reaction than by a first. We quickly saw that an initial distribution that was exponential in the rate constant would, as a whole, appear to disappear as if by a second-order reaction. This result was later generalized [76] by showing that an initial gamma distribution of parameter q would, as a lump, crack as if of the pth order, where $p = (q + 2)/(q + 1)$.

Continuous mixture theory found application in industry and was, as I understand it, incorporated into some of the models the chemical and oil companies were developing. Some of the work on polymerization and coal devolatization also used the notions of continuous mixtures, but there was little development on the formal side until the work of Astarita and Ocone in the latter eighties. Their paper (AIChE J. **34** 1299, 1988) introduced the idea of "uniform" kinetics. This allowed the time scale to be warped and results to be obtained when the underlying reactions were not of the first order. Indeed, it was shown that intrinsic kinetics (i.e. rate laws for A(x)dx) could be found that would mimic any kinetic law for the lump as a whole (AIChE J. **35** 529, 1989; [248, 249, 259]).

I will not attempt to summarize the present state of the theory of reactions in continuous mixtures. Progress has also been made in the understanding of nonlinear intrinsic kinetics by T. C. Ho (v. e.g., his papers in *Kinetic and Thermodynamic Lumping of Multicomponent Mixtures*. Ed. G. Astarita and S. I. Sandler. Elsevier. Amsterdam 1991) and in the applications in other areas

[24] I learned of this reference only in the late '80s when I took up the subject again.

such as coal gasification (Whittman) and polymerization (B. McCoy, AIChE J. **39** 1827, 1993). There are ramifications of lumping theory that are related and the work of Wei and Kuo and of Rabitz and his coworker Li at Princeton (v. sup. fn. 14, p. 438). The whole subject is ripe for a monograph.

MINNEAPOLIS, 1965–1971

The period between my first and second sabbaticals was extremely productive. F. G. Heineken, whom I had undertaken to supervise jointly with Henry Tsuchiya on a topic in enzyme kinetics, had the enterprise to come over to England and work at University College, London during my year of absence. Enzyme kinetics abounds with pseudo-steady-state hypotheses and these are treated in strange ways in biochemical textbooks. Heineken treated the so-called Michalis-Menten (a.k.a. Briggs-Haldane, Hougen-Watson) kinetics by singular perturbation methods.[25] This was published in the first volume of *Mathematical Biosciences,* a journal that Bellman had just launched [73]. I am not sure that we did justice to the priority of Acrivos, Bowen, and Oppenheim, who had first used the singular perturbation technique in 1963,[26] nor to Amundson, who suggested it to us, but I learned later that the first rigorous treatment of the pseudo—steady-state hypothesis was to be found in a paper by Fowler and Lock[27]; cf. [82, 96]. At this time, I was riding the second wave of first rate students, one of whom, J. M. Wheeler, even had the fortitude to do a purely experimental project in adaptive control [94]. Henry Tsuchiya again joined me in the supervision, and a photograph of us in front of the chemostat was taken in which I was to court the verisimilitude of practicality by holding a screwdriver. The result, however, was said to be unconvincing, because, although right-handed, I was holding the screwdriver in the left hand. The cooperation with Arnie Fredrickson and Henry Tsuhiya was very interesting and one of several that I have enjoyed over the years. Working with Amundson went back, of course, to my very first days and was pleasantly resumed in the last with a joint paper for John Davidson's retirement [280]; but I have also shared supervision or published with Carr, Cussler, Davis, Derby, Fredrickson, Keller, Schmidt, Stephanopoulos, and Tsuchiya. The ease of working with others, either officially or unofficially, as co-investigators or in joint supervision has been one of the strengths of the department.

Copelowitz, Drott, Farina, Hyun, Jackman, Jouven, Knapp, Mehta, Mercer, Pell, Rester, Rouyer, Schruben, and Viswanathan all belong to this period, and, although they were not officially working with me, I had, to my great benefit, quite a bit to do with Luss, Rhee, Schneider, Schuler, and Varma. Strieder came from Notre Dame for a year and we did some interesting things with nonlinear variational principles [103] that were later written up as a tract

[25] We used Vasil'eva's approach, which is a little ponderous. The clearest presentation is in C. C. Lin and L. A. Segal's *Mathematics Applied to Deterministic Problems in the Natural Sciences.* Philadelphia: SIAM.

[26] *Chem. Eng. Sci.* **18,** 177 (1963).

[27] *Proc. Lond. Math. Soc.* **29,** 127 (1922).

[H′]. Rhee, the doyen of our outstanding group of Korean students,[28] was doing a fine analysis of multicomponent chromatography [95, 97]. The revision of the book on first-order PDEs that Amundson and I had written in 1971 [G′] in the two volumes of 1986 and 1988 [M′] are largely Rhee's work, and contain many of the comprehensive results that he obtained at this time.

During this period between the first and second sabbaticals, I returned to the question of diffusion and reaction in a porous particle that had occupied me somewhat at I.C.I. and in Edinburgh. It must have been in 1960 that K. M. Watson came to Minnesota on an accreditation visit and, knowing of my shape-factor paper [3a], remarked to me that it would be helpful if someone would find a way to bring various kinetic expressions together. The penny did not drop then; indeed I do not remember giving it any thought, although the materials were all at hand—in fact, in the trial version of the famous Birdfoot that circulated at about this time there was a problem that could have led anyone straight to the normalization. I do not have any clear recollection of what prompted me to look at it again, but the time must have been just before going on sabbatical, for it appeared in *I.E.C. Fundamentals* in 1965 [52], four or more years after Watson's remark. Remarkably enough, Gene Petersen[29] and Ken Bischoff[30] published essentially the same result entirely independently within the space of a month or two and evidently with no overlap of referees. This was not the first time that independent results on a diffusion–reaction problem had sprouted at the same time. Amundson and Schilson,[31] Jim Carberry,[32] and Weiss and Hicks[33] published three different approaches to the nonisothermal catalyst pellet almost simultaneously, although in this case Amundson and Schilson's work had been done 4 or 5 years before it was published. A third triad lies at the root of the subject for Zeldovitch in Russia, Damköhler in Germany, and Thiele in the United States all contributed independently to the fundamentals of the problem and all were ignorant of Jüttner's work 20 or more years earlier. Perhaps time does ripen.

CAMBRIDGE, 1971–1972

I took the writing of a monograph on "the mathematical theory of diffusion and reaction in porous catalysts" [I′] as my sabbatical project when we went to Cambridge for the second time in 1971–1972. It is questionable whether

[28] Many of these students were leaders in the Korean community in the Twin Cities and many now occupy positions of responsibility. I must not try to list them all as I shall do someone an injustice, but J. C. Hyun should be mentioned as another outstanding leader; as also, later, Byong Cho and Ing He Lee. Like the Greek students, they form a subset whose ability and dedication have been and are exemplary. In dedicating the first volume of M′ to the Korean students who have done their graduate studies here, we spoke no more than the truth when we said, "They have set a standard that will not be surpassed."

[29] *Chem. Eng. Sci.* **20,** 587 (1965).

[30] *A.I.Ch.E.J.* **11,** 351 (1965).

[31] *Chem. Eng. Sci.* **13,** 226 (1961).

[32] *A.I.Ch.E.J.* **7,** 350 (1961).

[33] *Chem. Eng. Sci.* **17,** 265 (1962).

it is wise, in the long run, to undertake the writing of a lengthy book on a sabbatical. Perhaps a sabbatical should be a Sabbath, full of hard work to be sure, but a time of playing with new ideas, dreaming dreams, and seeing visions. We had an enormously good time, of course—how could we not with so many good friends in Cambridge and nearby—but the book was not finished until 1973 and then had to be sprinkled with additional bibliographical comments. The structure of the first volume on the steady state is sound enough and it could be brought up to date by adding the work since the late 1970s, notably that of Varma and Morbidelli on nonuniform distributions of activity. The second volume, on questions of uniqueness, stability, and transient behavior, may have served to stimulate some other work at the time it appeared, but there was so much relevant work going on that it was seriously dated soon after it appeared. In fact, the Clarendon Press wanted me to rework it into a more general text, but relented when we were able to find a subsidy through the generosity of Jim Carberry and of the Guggenheim Foundation.

The Guggenheim Foundation had kindly subvented my sabbatical with one of their Fellowships, and this additional subvention was almost certainly due to the kind interest of Marc Kacs, who combined a life of individual scholarship of the highest order with a far-reaching concern for the development of younger scholars of all kinds. He co-opted me for the board that oversaw the formation and development of the Center for Nonlinear Studies at Los Alamos, a group of very able physicists and mathematicians. This assignment took me to Los Alamos a number of times during the later 1970s and early 1980s for committee and scientific meetings. It was not only interesting to have contact with this fabled site, although, of course, I had no clearance to take me within the fence, but it also gave me the opportunity to meet some remarkable people. Although I only spent half an hour at the same table with Chandrasekhar, the memory of his presence has remained with me as that of the most totally recollected man I have ever met.

MINNEAPOLIS, 1972–1974

Apart from getting the diffusion–reaction volumes through the press and working with several graduate students, I became an editor of *Chemical Engineering Science,* succeeding Neal Amundson, who had held the United States' editorship for 25 years. As it turned out, I held it for only a couple of years because I passed the ball to Arnie Fredrickson when I became Head in 1974, but of the 150 papers that passed across my desk, none sticks in my mind more than that by Harmon Ray with his student Uppal and his colleague, Aubrey Poore.[34] The last, a student of Don Cohen's at Cal Tech, had been advised by Gavalas to make contact with Ray when he got to Buffalo—an illustration of the coincidences that lead to a propitious ferment. This work was one of the first fruits of a rather general revival of interest in Hopf bifurcation. Amundson and I had found limit cycles in the controlled stirred tank (a subcritical Hopf bifurcation in fact), but somehow it never occurred

[34] *Chem. Eng. Sci.* **29,** 967 (1974). See pp. 79–81.

to us to look for them systematically. Of course, computing facilities had greatly advanced in the intervening 16 years, and I, for one, mistakenly thought that they were artifacts of the controller. A meeting of the New York Academy of Sciences about this time brought together scientists of many different sorts [155]; catastrophes were in the air; singularity theory was aborning. Golubitsky found the stirred tank of interest while at the Institute of Mathematics and its Applications at the University of Minnesota and showed us how to do it properly. Luss and Balakotaiah did it properly and opened up new vistas in the landscape of chemical reaction engineering.

I must not let myself be sidetracked into random historical observations that come to me as I write, for they are merely that and I have not attempted to be systematic. Let none feel slighted by omission. It is a reflection of my limited recollection rather than of their importance. The field of chemical reaction engineering is ripe for harvesting by a good historian of science with sufficient technical expertise. One general remark I would make, however, is that I have profited greatly from consulting visits and have often been surprised at the way in which a problem I have been considering turns out to have some application in an industrial setting and, contrariwise, how fruitful an industrial problem proves to be when suitably abstracted. Moreover, my consulting led to friendship with such people as the late Keith McHenry of Amoco and Louis Hegedus, *olim* of General Motors, and Grace, *nunc* ELF, and I count this as a great benefit.

Aside on Neal Amundson and the Department

In 1974, Neal Amundson felt that he should step down from the departmental headship and, although we all felt that this would be a great loss, who could say him nay when he had nurtured us all for the previous 25 years? His building of the department was a solid achievement, and the atmosphere of mutually supportive striving for excellence his legacy. He was at home in the simpler times of the 1950s and 1960s, when a full professor could be hired by making a couple of telephone calls, and had built up the department by his keen nose for quality in whatever area of chemical engineering science he felt would be of importance. He protected us early in our careers from too much administrative palaver, gauged the opinion of the department from the lunchtime conversation, formulated his decision, and acted with benevolent autocracy as a good Head should. Tsuchiya came the year before I did; Ranz, who contributed so much to the system of team-teaching the undergraduates, Dahler, and Frederickson the same year; Scriven, who was "post-docing" at Shell Development, in 1959. John Dahler met Ted Davis and convinced him there was a future for chemical physicists in chemical engineering; Keller had the perceptiveness to enquire whether there was an opening and was promptly hired; Schmidt came from surface physics at Chicago; and Carr from Kistiakowski's lab at Harvard. Later came Chris Macosko (1970) and George Stephanopoulos (1974). Going back to the earlier 1950s and providing a strong engineering spine for these active limbs were Amundson himself; Isbin, the country's leading authority on nuclear reactor safety and design; and Ceaglske, a pioneer in process control. Preckshott went to head the department at

Columbia, Missouri, and, sadly, Madden and Stoppel had died. In addition, we had some of the members of the deceased School of Mines and Metallurgy, whose faculty had been dispersed among several departments by Dean Swalin: Hutchinson, Nicholson, Sivertsen, Toth, and Gerberich, who had come after this diaspora.

Such was the department when I was appointed Head[35] in 1974, full of vigor and with a tremendous variety of ongoing research.[36] I was probably the right person for the job, for it was clearly a transitional time and soon to be fraught with freezes and cuts. If a head can take credit for the successful hires of his administration,[37] I can hold mine up with the best, for Matt Tirrell came in 1977 and is now himself Head.[38]

It was during this period that Neal Amundson left us to go to Houston, a move that we, at the time, thought a tragedy, but which, in retrospect, seems for the best. For, aside from missing him personally, the continued progress of the department at the University of Minnesota shows how soundly he built, while his presence at UH greatly helped Dan Luss in his successful effort to build up that department. Both universities have benefited from his vision, from his insistence on the highest standards, and his generosity of concern and support. He has returned many times to Minnesota, and two of these have been formal occasions. The first was when the renovated Chemical Engineering Building was renamed Amundson Hall (1980); the second when he was given a Doctorate of Science, *honoris causa.* (1984). Because this is a volume of reprints, I venture to reproduce the *oratiuncula* that I made in presenting him to President Keller.

> "Mr. President,
>
> It is the custom in certain branches of engineering science to seek out the most singular point in the domain of a function. Such a point serves as an organizing center in its neighborhood all possible forms of behavior flourish and may be apprehended coherently. Though it smack a little of the vulgar solecism "very unique," I venture to commend
>
> Neal Russell Amundson
>
> to you as the most singular point in the domain of chemical engineering. He has long been recognized as its organizing center and in his neighborhood all forms of the discipline flourish. Thirty years ago, he was on sabbatical at Cambridge and it was

[35] Actually, "*acting head*" (in the hope that Amundson would change his mind), but I could never understand the difference.

[36] I should also mention here that we have benefited greatly in chemical engineering from the proximity of a mathematics department containing people who were not above talking to engineers, nor indeed of working, and sometimes publishing, with them. The names of Aronson, Friedman, McGehee, Marcus, Sell, Serrin, and Weinberger come immediately to mind, and there are others, I am sure, who should be mentioned. The Institute of Mathematics and Its Applications, which got off to a splendid start in under Hans Weinberger, has also been a great asset, bringing in, for short or longer stays many of the most active mathematicians in the world and organizing meetings around questions of central importance [c].

[37] George Stephanopoulos came at about the time I became head, but Amundson did all the negotiation.

[38] A photograph of Amundson, Aris, Kellor, Davis, and Tirrell, taken in 1996, was labeled "The Grateful Heads."

there that his thoughts on the directions of development in chemical engineering crystallized. They grew in a rich mother liquor of fifteen years of teaching, a decade of research and publication and a lustrum of administration under the deanship of Athelstan Spilhaus. They anticipated many of the trends that have since become popular—not in the pejorative sense, but because they have been seen to be important: the infusion of engineering by science, in particular by chemical physics and physical chemistry; the now-blossoming link with biology and medicine; and the perceptive use of mathematical models. To come later was the composite with materials science, but the monomer was already initiated and was soon to branch and grow.

"In his own scholarly work, Amundson has shown the power of mathematical methods, not so much in calculation or design (though even here he pioneered the use of computers in distillation calculations—and that a time when the Univac Scientific was the most advanced computer and Fortran was yet to be invented) but mathematics in its penetrating powers of analysis, its ability to show the essence of a problem, "the euristic vision of mathematical trance," as Bridges puts it. Neither has his eye for a ripe problem grown dim nor has his skill in using the tools of analysis abated. The last decade has seen a stream of papers on coal combustion that "mix the useful with the beautiful" as powerfully as those of the previous thirty years, occupied as they were with polymerization, tubular reactor behavior, reactor dynamics and control, chromatography and distillation and many other topics.

"In teaching, Amundson has an enviable reputation and generations of graduate students have flocked to his early Tuesday/Thursday lectures on mathematical methods for chemical engineers. It must be admitted, however, that he is not entirely free of that climatological machismo that afflicts Minnesotans, for at eight of the clock the lecture started whether in the Indian summer of September or the icy blast of January. What is more, he demanded a like attention to teaching from his faculty, establishing the tradition of team teaching and taking his share of recitations in courses taught by junior colleagues. It was not for nothing that the old title "Chief," used by heads of the divisions of the School of Chemistry where ChemE was born, lived on as a title of respect and affection for Neal Amundson.

"Mr. President, these virtues are a matter of public record, supported in the petition that was made to the All-University Honors Committee by the testimony of those who know the chemical engineering scene in the whole country and throughout the world. That you and I have privileged to experience a great part of this record at first hand, as have many others here tonight, adds a greatly valued touch of personal warmth to this the public recognition of how much the University of Minnesota owes to Neal Amundson. It is my privilege to present him to you now, Mr. President, for the degree of Doctor of Science *honoris causa tantum.*"

So much for the public record.[39] The reader will be very dense who does not realize that the depth and breadth of my obligations to the Chief are so great that they cannot properly be expressed within the decorum of the printed word. The hinges of the door of my little office at I.C.I. were indeed cardinal! Our joint publications have continued over the years, and it was a pleasant happenstance that our latest should have been dedicated to one of the very greatest chemical engineers of this century, John Davidson.

MINNEAPOLIS, 1974–1996

The period of my headship was not one of expansion, but I muddled through in the true British fashion. Mistakes in administration are not the subject of these confessions, for although I made plenty, they would not be very instruc-

[39] See also two articles in *Chem. Eng. Educ.* **V,** 104–106, (1971), and **XVI,** 50–54, (1982) [185].

tive even if I could remember them all. I tried to like administration enough to do it as well as I could, but it was a far cry from Keller's enormous ability in public affairs or Ted's vision and energy. So it was greatly to the benefit of the department when it discovered this and relieved me of the responsibility. The occasion of it was Princeton's attempt in 1978 to get me to move there, an effort that I have always regarded as a great compliment. In asking me to stay, my colleagues not only relieved me of the headship, but also (at Keller's imaginative suggestion) said I might be half-time in the department and persuaded in Academic Vice-President, Henry Koffler, to support the other half of me to pursue the study of paleography.[40] In the face of such generosity I could hardly have left, and I have not regretted my decision, nor, they are kind enough to tell me, have my colleagues. Some of the fruits of this freedom that bear on technical matters are to be found in the Phillips Lecture [256], the lectures given while Hougen Professor at Wisconsin in 1979 [K′], and several other papers, e.g., [202, 250, 281].

MINNESOTA AND SABBATICALS, 1978–1996

This is not the place to give a detailed history of the department, but I will make a couple of comments before resuming an account of the context of the papers dating from the second half of my career. Under Keller's leadership, the department got under way, but he was called to higher office only 2 years later and, in 1980, Davis assumed the headship. For the next 15 years we had the benefit of his vision, energy, and leadership. He brought the department into the computer age, scored an early success by bringing Cussler and Evans to Minnesota, and gradually built up the department, knitting it together. He has done for the Department of Chemical Engineering and Materials Science what Amundson did for the Department of Chemical Engineering and, also like Amundson, has maintained a high level of research output without neglecting teaching. In the early years of his presidency, Keller led a very successful fund drive that greatly increased the number of endowed chairs. The way in which Davis distributed the special chairs that came into being in this way, and others that came later, has made them objects of emulation rather than of envy. Two of the special chairs are of value to the whole department. The Piercy Professorship (the gift of George T. Piercy, a 1938 graduate who had a distinguished career with Exxon and has been an extraordinarily generous alumnus) supports a visitor for 3 months and allows a nontrivial contact with the department. The Amundson Professorship (built around a major gift from the Amundsons) is to supplement sabbatical semi-salaries.

I have "gone on" about the state of the department because it provides the context for my research during this period. There is nothing like an active

[40] This interest would generate a lengthy list of acknowledgements in its own right, but would be too much of a diversion here. Suffice it to say that at Tom Kraabel's invitation I started to teach a survey course in Latin paleography in the Classics Department, which served students in History, Art History, and Medieval Studies as well as Classics. My bibliography in humanistic studies is quite limited: see Subject Index to Papers in Bibliography.

group of colleagues with opportunities for interaction and cooperation to provide a propitious ferment for one's own research. Amundson and I shared the supervision of Georgakis, but with students of his caliber, this was a light task. Sometimes this may be formalized in a joint proposal or even grow into a center, as with the Center for Interfacial Engineering so successfully directed by Evans. Some times, it is more individual and *ad hoc.* Thus, the paper with Keller on a stable gradient across a membrane [116] was born of a lunch-time conversation, whereas a group of papers on chemotaxis [157, 192, 206] grew out of our joint supervision of Lauffenburger, if the light rein that we used with so brilliant a student can be called "supervision." It was a chance remark that suggested to Davis that a problem in surface diffusion to active sites would be a good way to get his new postdoc up to speed on finite element methods; this led to three papers [198, 199, 209]. The roots of the work with Frederickson and Stephanopoulos [158, 164] on commensal populations in a chemostat lie in the paper with Humphrey [144]. Let no one say that there is no engineering science at the annual meeting of the NAE! Bored by a symposium on taxation, Art and I sought refuge in the corridor, where he told me of the difficulty he was having in controlling the steady state of a chemostat with two microbial populations growing in it. The resolution of this problem, of which I have spoken previously (p. 45–46), linked immediately with work being done in Fredrickson's group, and Stephanopoulos showed that under the fluctuations of feed rate, the state would drift to one extreme or the other, and he later showed that forced oscillations of the feed rate could be used to stabilize the mean steady state.

Two problems related to the earlier work on diffusion and reaction occupied me in the late 1970s. One was derived from the observation that the equations for impregnating a catalyst are the same as those for chromatography. This was taken up by Sheng Yi Lee [197, 205], not to be confused with Sheng Tai Lee, who worked on the second topic, the catalytic monolith. So also did Zygourakis, yet another of the great Greeks. The old adage *Timeo Danaos et dona ferentis* will not wash at Minnesota; we would rather say that "we do not fear the Greeks, they come formidably gifted" (the paronomasia on gifts and gifted will not work in Latin). Svoronos was another. He came with a background in chemistry and took some of the senior courses. I had something to do with him at the masters' level, but, much as I would like to claim him, he was really one of George Stephanopoulos' students.

I must not fail to mention the enormous stimulus afforded me by the hospitality of Cal Tech through their Fairchild Distinguished Scholar program. When the Fairchild Foundation issued a call for proposals in the mid 1970s, most institutions asked for buildings; Cal Tech asked for people. Under Neal Pings' direction and Lee Sterett's consummate organization, this scheme brought in a stream of visitors, made them feel at home instantly, and allowed them to interact as little or as much as suited their work. I was privileged to have two spells as a Fairchild, for a quarter in the fall of 1977 and for a sabbatical year in 1980–1981. In the first I wrote "Mathematical Modelling Techniques" [J′]; the second was divided between palæography and chemical engineering, and even in the former area Cal Tech provided stimulus in the person of (the late) John Benson, a noted medieval historian. I greatly benefited from the friendship of Richard Rouse of UCLA and the proximity of the Huntingdon Library, and spent a lot of time of Mabillon's travels [285] and the mortuary roll of the Blessed Vitalis, but managed to get some engineering done as well, including papers with two of the faculty [152, 195]. The discerning reader will easily recognize the stream on the Cal Tech campus as the Ilissus of the first of the three dialogue papers [145, J′ App. 1]. The only problem that such a program as the Fairchild has is that the product must almost inevitably fall short of the opportunity. "Ah, but a man's reach should exceed his grasp, or what's a heaven for!"

The dynamics of reactors and their behavior under external forcing was one of the three main lines of my research in the 1980s and 1990s. It was greatly helped by having, students of high mathematical ability, including two of the very first water, Bill Farr and Iannis Kevrekidis. I struggled to keep up with the precision and depth of Farr's analysis and the indefatigable energy and multifarious activity of Kevrekidis. The latter was one of a group of very able students of which Lanny Schmidt and I shared the supervision. Almost all of them did an appropriate amount of experimental work and benefited from contact with students working on other projects of Schmidt's. Many of them had a large theoretical component in their work, and I naturally saw more of these than of the pure experimentalists.

One great benefit that was connected with this research was my getting to know Peter Gray and his research group at Leeds. Gray was in Cambridge when I went to visit Amundson there in 1955, but I did not meet him until the International Symposium on Chemical Reaction Engineering in Edinburgh some 30 years later. He and Steve Scott had devised a reaction system with an autocatalytic step $A + 2B \rightarrow 3B$ and a degenerative step $B \rightarrow C$, which, although it has certain limitations, gives the simplest form of the classic stirred tank problem without the computational complexity of the Arrhenius expression. It has a richer behavior than the Brusselator and is slightly more tractable algebraically than a system devised by Takoudis and Schmidt using vacant sites as the autocatalytic element [O′, 239, 240, 242]. We only wrote one paper together [242], but we had some stimulating exchanges and visits, including my spending a week or two in Leeds as a Brotherton Professor in 1985.

It was Peter Gray who asked me to give a paper in Brussels at a symposium held in honor of the 10th anniversary of Prigogine's Nobel Prize. I was to follow *senex ipse* on the first morning, and it must have been the prospect of my Parkinsonian stumblings clattering around the stage immediately following the Terpsichorean nimbleness of the Great Man's "light fantastic toe"—an offering to Irreversibility, *ni fallor*—that led me to ascribe my review solely to myself. True, I had written it all and it covered an extensive range, but it was so much a review of the work Schmidt and I had done, and were doing, collaboratively, that I should have put his name at the masthead. True, the acknowledgments at the end of the paper make the position clear (and name the students who had worked with us: Chang, Cordonier, Kevrekidis, McKarnin, Song, and Takoudis), but nothing short of co-authorship would have been really condign. The oversight, for there was nothing intentional about it, was compounded by the fact that I could have corrected the omission at any time within the next 18 months, during which Gray and his coeditors had to wait for Prigogine's overture.

Over the years the collaboration with Schmidt was very fruitful and we gained a good understanding of the natural and forced oscillations of stirred reactors and, later, of catalytic clusters through the work of another outstanding Greek, Dionysios Vlachos. I will not attempt to quantify that statement by totaling the number of students we served, of papers we published, of dollars we received, of citations we garnered, or the number of hours of Cray-time we used. Such libations to the deities of management science (which, some seem to think, have "brought us up out of the land of Egypt") have their place in the scheme of things, but it is not here. The statistics that can be derived from them have not been fully exploited; a "scientific" ranking would surely have need of the citations per dollar of support or the number of students that can dance on the head of a Cray-second. I have never been a good "rainmaker," although I have been able to support most of my students most of the time. All kidding aside, I am grateful to the agencies and funds that have supported my work since the late 1950s: the Research Corporation; the National Science Foundation; the Department of Energy; the Guggenheim Foundation; the Fairchild Foundation; the Petroleum Research Fund of the American Chemical Society; the Exxon Corporation (*Ad*–T. C. Ho *gratias ago*); and the Graduate School of the University of Minnesota.

A second fruitful collaboration has been with Bob Carr on chromatographic and countercurrent reactors. Again, the practical side of the research was Carr's work and I supervised the mathematical side. We had a very good stream of students and good support from the Department of Energy. Byong Cho, now with General Motors, was one of the first students on this project; others were Ted Petroulas, now a power on Wall Street; Barrie Fish, with Union Carbide; Anna Tonkavich, now with Batelle; Ajay Ray, now a professor at the University of Singapore; and Alan Wardwell, now with Conoco. As the work progressed, it became clear that moving beds of catalyst particles were not practical, and we passed to countercurrency, simulated by switching the feed between fixed columns. Tonkovich and Carr (for by this time the theoretical component was small) had considerable success in enhancing the C_2 yield from methane oxidative coupling over a Samarium catalyst, which was written up in *Science.* They put my name on the paper, as being a co-principal investigator, although by this time the theoretical component was buried deep in work that was wholly experimental. The very good increase in conversion that they obtained was, quite appropriately written up in C&ENews. Unfortunately the reporter, presumably thinking my name the better known, attributed the work to "Aris' Laboratory," an ascription that would be risible to my friends but misleading to the general public. Worst of all, it effectively denied proper credit to Bob Carr. I was in Australia at the time, but prompt action by Scriven and Davis and a flurry of trans-Pacific faxes got a partial correction in the next issue. Such corrections are usually ineffective, and it is quite clear that I should have kept my name off the paper in the first place. As I remarked to someone at the time, I thought that Samarium was the singular form of a well-known biblical place-name!

Besides the work with Carr and Schmidt, I had a very pleasant interaction with Ed Cussler over some problems in mass transfer through membranes with barriers [238], through anisotropic membranes [232], and by facilitated diffusion [247]. A severely mathematical problem of diffusion through a slot could be solved by a Schwartz-Christoffel transformation,[41] and provided an asymptotic formula that proved useful in the barrier problem [223]. Cussler was delightful to work with because he refused to accept any mathematical solution that could not be unpacked and shown to be physically sensible (cf. [244]). Next to Bosanquet, he is the most intuitive problem solver with whom I have worked. Chapter 19 [232] is reprinted here.

Why is there no word for '*festschrift*' in English? We are constantly having them and they are the sincerest forms of professional congratulations. I appreciated greatly the issue of *Chemical Engineering Science*[42] that was given me for my 60th birthday, with 25 papers by 50 authors, edited by Fredrickson, Gavalas, Ray, and Varma. But for *festschrift,* almost any word you can think of is clumsy, and "festival writing" is downright ridiculous. The elliptic function problem provided a paper for a *festschrift* for James Serrin [223], which, though not entirely worthy, provided a pleasant link with his course on potential theory that I had taken 30 years before. A dozen others have been called for: Amundson [220, 276]; Bellman [216]; Brenner [310]; Churchill [176]; Davidson [280]; Gray [306]; Jackson [313]; Marcus [281]; Ruckenstein [307]; Sharma [309]; Truesdell [156]; Villadsen [308]; and Wicke [161]. It was only meet that the contribution to Truesdell's *festschrift* [156] should be in Latin, but it is translated in Appendix C of the Hougen lectures [K′] and reprinted here as M, p. 337. These last papers, if I remember rightly, were requested for the year before the actual anniversary—a wonderful ploy for getting them in on time and confirmation of the well-known inability of the mathematician to do arithmetic!

The topic of reactions in continuous mixtures is one that has occupied my attention since learning of second-order cracking. With the revival of interest that Astarita[43] and Ocone's paper generated, I returned to it and got some further results in collaboration with Paolo Cicarelli.

I took my final sabbatical in 1993–1994, and spent a very pleasant term at the Institute for Advanced Study in Princeton, the All Souls of the New World. I was in the School of Historical Studies and working on the Mortuary Roll of the Blessed Vitalis. However, I must not digress into this or I shall have another long list of acknowledgements to recognize. I had been tapering

[41] The solution was not entirely original, as the paper made clear, but the application and asymptotics were. Moreover, it was great fun to work with elliptic functions again. I had not encountered them since first studying them in Edinburgh.

[42] *Chem. Eng. Sci.* **44,** 1751–2086 (1989).

[43] News of his untimely death reached me just as I was writing this section of the appendix. Doyen of Italian chemical engineers, Gianni was a man of broad interests, great accomplishments, and lively intellect. We wrote a couple of papers [248, 249] together, faxing each day's work to the other overnight. His warning about the potholes in the "information highway" [*Chem. Engng. Sci.* **51,** 4353–4354 (1996)] deserves to be heeded.

off on supervising so that no one would be left after my retirement.[44] Alhumaizi's was the last of 48 Ph.D. dissertations (21 of them jointly) and 20 M.S. that I have supervised. With the addition of an introductory chapter, it was published as an attempt to survey a dynamic system that has too many parameters to allow it to be exhaustively mapped [O]. So the phase plane of 1955 had become a three-dimensional phase space of 1995. The calculating engines had become incomparably more powerful, but the same pleasantly painful[45] thought was needed to tease out the wonderful variousness of its behavior.

I have referred to several ethical questions in Chapter 6 and this appendix, and I close with a more personal one. Were it not that someone in the Legislature might revive the ancient cry of the ultra-Protestant, "Would you have celibacy practiced in the open streets?" a case could be made for academic celibacy. As C. S. Lewis remarks somewhere, academics do not work hard, they just work all the time, and this can put an enormous burden on a spouse, particularly if he or she is not an academic. No doubt there are some who, like Euler, can write their papers in the midst of the domestic hubbub, but most of us get increasingly irritable and obsessed as a deadline approaches or a book is nearly finished (or abandoned). I must admit that I have, at times, been oblivious to the distress I have caused my wife, whose loyalty and support have ever been beyond my desserts. It is one of the blessings of retirement that, busier than ever although one may be, there are not the pressures that one feels in the thick of a career and no excuses for neglecting a relationship without which the career itself would be little worth.

Concluding with this reflection allows me the pleasant symmetry of finishing this, my last chemical engineering book, with the last four lines of the poem, the first four of which I used in the dedication of my first.

Nestore sim quamvis provectior aemulaque annis
vincas Cumanam tu quoque Deiphoben,
nos ignoremus quid sit matura senectus,
scire aevi meritum, non numerare decet.[46]

[44] My retirement was celebrated in 1996 with such extraordinary gusto that it quickly transcended any merit of mine and became a celebration of the large-heartedness of my friends. I enjoyed it enormously.

[45] I use the word in its older meaning 'painstaking'.

[46] Ausonius. Epigr. xviii. MGH p. 200. In her *Mediæval Latin Lyrics* (Penguin Classics, Harmondsworth. 1952). Helen Waddell translates the whole poem: Love, let us live as we have lived, nor lose/the little names that were he first night's grace,/and never come the day that sees us old,/I still your lad, and you my little lass./Let me be older than old Nestor's years,/and you the Sibyl, if we heed it not./What should we know, we two, of ripe old age?/We'll have its richness, and the years forgot.

BIBLIOGRAPHY

BOOKS

A′. *The Optimal Design of Chemical Reactors: A Study in Dynamic Programming.* New York: Academic Press, 1961.

B′. *Vectors, Tensors, and the Basic Equations of Fluid Mechanics.* Englewood Cliffs, NJ: Prentice-Hall, 1962. Reprinted, New York: Dover, 1989.

C′. *Discrete Dynamic Programming: An Introduction to the Optimization of Staged Processes.* New York: Blaisdell, 1964.

D′. *An Introduction to the Analysis of Chemical Reactors.* Englewood Cliffs, NJ: Prentice-Hall, 1965.

E′. *Dynamic Programming.* Dallas: Socony Mobil Oil Co., 1965.

F′. *Elementary Chemical Reactor Analysis.* Englewood Cliffs, NJ: Prentice-Hall, 1969. Reprinted, Boston: Butterworth, 1990.

G′. *First-Order Partial Differential Equations with Applications* (with N.R. Amundson). Englewood Cliffs, NJ: Prentice-Hall, 1972.

H′. *Variational Methods Applied to Problems of Diffusion and Reaction* (with W. Strieder). New York: Springer, 1973.

I′. *The Mathematical Theory of Diffusion and Reaction in Permeable Catalysts* (2 vols.). *Vol. 1: The Theory of the Steady State. Vol. 2: Questions of Uniqueness, Stability, and Transient Behaviour.* Oxford: Clarendon Press, 1975.

J′. *Mathematical Modelling Techniques.* San Francisco: Pitman, 1978. Reprinted, with 136, 145, 168, 204, 245, and 254 as appendices, New York: Dover, 1994.

K′. *Chemical Engineering in the University Context: Four Lectures in Honor of Olaf Hougen Given in the Department of Chemical Engineering, University of Wisconsin–Madison, 1979.* Madison: Board of Regents of the University of Wisconsin System, 1982.

L′. *An Index of Scripts in E. A. Lowe's 'Codices Latini Antiquores,' Pts. I–XI & Supplement.* Osnabruck: Otto Zeller Verlag, 1982.

M′. *First-Order Partial Differential Equations* (with H.-K. Rhee and N.R. Amundson) (2 vols.). *Vol. 1: Theory and Application of Single Equations. Vol. 2: Theory and Application of Hyperbolic Systems of Quasilinear Equations.* Englewood Cliffs, NJ: Prentice-Hall, 1986 & 1988.

N′. *Explicatio Formarum Litterarum = The Unfolding of Letterforms from the First Century to the Fifteenth.* St. Paul: Calligraphy Connection, 1990.

O′. *Surveying a Dynamical System: A Study of the Gray–Scott Reaction in a Two-Phase Reactor (with K. Alhumaizi).* London: Longman, 1995.

P′. *Mathematical Modeling: A Chemical Engineer's Perspective.* New York. Academic Press, 1999.

EDITED BOOKS

a. Aris, R., and A. Varma (eds.). *The Mathematical Understanding of Chemical Engineering Systems–Selected Papers of Neal R. Amundson*. New York: Pergamon Press, 1980.
b. Aris, R., H.T. Davis, and R.H. Stuewer (eds.). *Springs of Scientific Creativity: Essays on Founders of Modern Science*. Minneapolis: University of Minnesota Press, 1983.
c. Aris, R., D.G. Aronson, and H.S. Swinney (eds.). *Patterns and Dynamics in Reactive Media*. New York: Springer-Verlag, 1991.

CHAPTERS IN BOOKS EDITED BY OTHERS

Several papers in the General Bibliography are chapters in books. These two were omitted and could not be worked in without playing havoc with the numeration or doing conspicuous violence to chronology.

i. "Stirred pots and empty tubes" (with A. Varma). In (N.R. Amundson and L. Lapidus, eds.). *Chemical Reactor Theory*. Englewood Cliffs, NJ: Prentice-Hall, 1977 (pp. 79–155).
ii. "Reactor Steady-State Multiplicity and Stability " (with A. Varma and M. Morbidelli). In (J.J. Carberry and A. Varma, eds.). *Chemical Reactor and Reaction Engineering*. New York: Marcel Dekker, 1987 (Ch. 15).

JOURNAL PAPERS

An asterisk denotes that the paper has been reprinted in Part II, and its corresponding chapter number is given. Numbers in brackets represent

1. "On the application of Ångström's method of measuring thermal conductivity" (with C.H. Bosanquet) *Br. J. Appl. Phys.* **5,** 252–255 (1954).
2.* "On the dispersion of a solute in a fluid flowing through a tube" *Proc. Roy. Soc.,* **A235,** 67–77 (1956). (Reprint A)
3a. "On shape factors for irregular particles–I: The steady state problem. Diffusion and reaction." *Chem. Eng. Sci.* **6,** 262–268 (1957). Reprinted in "Classic papers from *Chemical Engineering Science.*" *Chem. Eng. Sci.* **50,** 3897–3903 (1995).
3b. "On shape factors for irregular particles–II: The transient problem. Heat transfer to a packed bed." *Chem. Eng. Sci.* **7,** 8–14 (1957).
4. "Some remarks on longitudinal mixing or diffusion in fixed beds" (with N.R. Amundson). *A.I.Ch.E.J.* **3,** 280–282 (1957).
5. "Stability of some chemical systems under control" (with N.R. Amundson). *Chem. Eng. Prog.* **53,** 227–230 (1957).
6. "An analysis of chemical reactor stability and conrol, parts I–III" (with N.R. Amundson). *Chem. Eng. Sci.* **7,** 121–155 (1958).
7.* "On the dispersion of linear kinematic waves." *Proc. Roy. Soc.* **A245,** 268–277 (1958). (Reprint C)
8. "Statistical analysis of a reactor: Linear theory" (with N.R. Amundson). *Chem. Eng. Sci.* **9,** 250–262 (1958).
9. "Diffusion and reaction in flow system of Turner's structures." *Chem. Eng. Sci.* **10,** 80–87 (1959).
10.* "On the dispersion of a solute by diffusion, convection and exchange between phases." *Proc. Roy. Soc.* **A252,** 538–550 (1959). (Reprint B).
11. "Notes on the diffusion-type model for longitudinal mixing in flow (Levenspiel, Smith and Van der Laan)." *Chem. Eng. Sci.* **10,** 266–267 (1959).
12. "The longitudinal diffusion coefficient in flow through a tube with stagnant pockets." *Chem. Eng. Sci.* **11,** 194–198 (1959).
13. "An analysis of chemical reactor stability and control–IV: Mixed derivative and proportional control." (with D.J. Nemanic, J.W. Tierney, and N. R. Amundson). *Chem. Eng. Sci.* **11,** 199–206 (1959).
14. "Some optimization problems in chemical engineering" (with R. Bellman and R. Kalaba). *Chem. Eng. Symp. Ser.* **56,** 95–100 (1960).
15. "On optimum cross-current extraction" (with D.F. Rudd and N.R. Amundson). *Chem. Eng. Sci.* **12,** 88–97 (1959).
16. "On Denbigh's optimum temperature sequence." *Chem. Eng. Sci.* **12,** 56–64 (1960).
17. "Studies in optimization–I: The optimum design of adiabatic reactors with several beds." *Chem. Eng. Sci.,* **12,** 243–252 (1960).

18. "Sudies in optimization–II: Optimum temperature gradients in tubular reactors." *Chem. Eng. Sci.* **13,** 18–29 (1960).
19. "Studies in optimization–III: The optimum operating conditions in sequences of stirred tank reactors." *Chem. Eng. Sci.* **13,** 75–81 (1960).
20. "The optimal design of stagewise adiabatic reactors." Paper presented at the AIChE/ORSA Symposium on Optimization in Chemical Engineering, New York, 1960.
21. "On the dispersion of a solute in pulsating flow through a tube." *Proc. Roy. Soc.* **A259,** 370–376 (1960).
22. "Chemical reactor design: For homogeneous flow reactions, a digital computer can determine the optimum combination of reactor type and operating conditions" (with G.T. Westbrook). *Ind. Eng. Chem.* **53,** 181–186 (1961).
23. "Studies in optimization–IV: The optimum conditions for a single reaction." *Chem. Eng. Sci.* **13,** 197–206 (1961).
24. "The determination of optimum operating conditions by the methods of dynamic programming." *Z. Elektrochem.* **65,** 229–244 (1961).
25. "Tubular reactor sensitivity" (with J. Coste and N.R. Amundson). *A.I.Ch.E.J.* **7,** 124–128 (1961).
26. "A study of iterative optimization" (with D. F. Rudd and N.R. Amundson) *A.I.Ch.E.J.* **7,** 376–384 (1961).
27. "Optimal bypass rates for sequences of stirred tank reactors" *Can. J. Chem. Eng.* **39,** 121–126 (1961).
28. "Heat transfer in fluidised and moving beds" (with N. R. Amundson). Paper presented at the Proceedings of the Symposium on the Interaction between Fluids and Particles, London, 20–22 June 1962.
29. "Studies in optimization–V: The bang-bang control of a batch reactor" (with N. Blakemore). *Chem. Eng. Sci.* **17,** 591–598 (1962).
30. "On optimal adiabatic reactors of combined types." *Can. J. Chem. Eng.* **40,** 87–92 (1962).
31. "Adaptive control" *Br. Chem. Eng.* **7,** 896–900 (1962).
32. "Stability of nonadiabatic packed bed reactors" (with S.L. Liu and N.R. Amundson). *Ind. Eng. Chem. Fund.* **2,** 12–20 (1963).
33. "Independence of chemical reactions" (with H. S. Mah). *Ind. Eng. Chem. Fund.* **2,** 90–94 (1963).
34. "Optimal adiabatic bed reactors for sulfur dioxide with cold shot cooling" (with K.-Y. Lee). *Ind. Eng. Chem. Proc. Des.-Dev.* **2,** 300–306 (1963).
35. "Control of a cold shot adiabatic bed reactor with a decaying catalyst" (with K.-Y. Lee). *Ind. Eng. Chem. Proc. Des.-Dev.* **2,** 306–309 (1963).
36. "Dynamic programming in countercurrent systems" (with D. Yesberg). *J. Math. Anal. Appl.* **7,** 421–424 (1963).
37. "Review of progress in control engineering." *Br. Chem. Eng.* **8,** 432–441 (1963).
38. "The fundamental arbitrariness in stoichiometry." *Chem. Eng. Sci.* **18,** 554–555 (1963).
39. "The algebra of systems of second-order reactions." *Ind. Eng. Chem. Fund.* **3,** 28–37 (1964).
40. "An analysis of chemical reactor stability and control–VIII: The direct method of Lyapunov. Introduction and applications to simple reactions in stirred vessels" (with R.B. Warden and N.R. Amundson). *Chem. Eng. Sci.* **19,** 149–172 (1963).
41. "An analysis of chemical reactor stability and control–IX: Further investigations into the direct method of Lyapunov" (with R.B. Warden and N.R. Amundson). *Chem. Eng. Sci.* **19,** 173–190 (1964).
42. "Studies in optimization–VI: The application of Pontryagin's method to the control of a stirred reactor" (with C.D. Siebenthal). *Chem. Eng. Sci.* **19,** 729–746 (1964).
43. "Studies in optimization–VII: The application of Pontryagin's methods to the control of batch and tubular reactors" (with C.D. Siebenthal). *Chem. Eng. Sci.* **19,** 747–761 (1964).
44. "An adaptive control of the batch reactor–I: Identification of kinetics" (with H.H.-Y. Chien). *Automatica* **2,** 41–58 (1964).
45. "The adaptive control of a batch reactor–II: Optimal path control" (with H.H.-Y. Chien). *Automatica* **2,** 59–71 (1964).
46. "Optimization of multistage cyclic and branching systems by serial procedures" (with G.L. Nemhauser and D.J. Wilde). *A.I.Ch.E.J.* **10,** 913–919 (1964).
47. "N-segment least-squares approximation" (with M.M. Denn). *A.I.A.A.J.* **8,** 432 (1964).
48. "Optimal policies for first-order consecutive reversible reactions" (with R. S. H. Mah). *Chem. Eng. Sci.* **19,** 541–553 (1964).
49. "Chemical reactor analysis: A morphological approach." *Ind. Eng. Chem.* **56,** 23–27 (1964).
50. "An elementary derivation of the maximum principle" (with M. M. Denn). *A.I.Ch.E.J.* **11,** 367–368 (1965).
51. "The dynamics of reactors of mixed type: I. The nature of the steady state" (with C.E. Gall). *Can. J. Chem. Eng.* **43,** 16–22 (1965).
52. "A normalization for the Thiele modulus (letter with correction in November issue)." *Ind. Eng. Chem. Fund.* **4,** 227–229 (1965).
53. "Stability estimates for the stirred tank reactor" (with W. Regenass). *Chem. Eng. Sci.* **20,** 60–66 (1964).

54. "Second order variational equations and the strong maximum principle" (with M.M. Denn). *Chem. Eng. Sci.* **20,** 373–384 (1965).
55. "Generalized Euler equations" (with M.M. Denn). *ZAMP* **16,** 290–295 (1965).
56. "Green's functions and optimal systems: Necessary conditions and an iterative technique" (with M.M. Denn). *Ind. Eng. Chem. Fund.* **4,** 7–16 (1965).
57. "Green's functions and optimal systems: The gradient direction in decision space" (with M.M. Denn). *Ind. Eng. Chem. Fund.* **4,** 213–222 (1965).
58. "Green's functions and optimal systems: Complex interconnected structures" (with M.M. Denn). *Ind. Eng. Chem. Fund.* **4,** 248–257 (1965).
59.* "Prolegomena to the rational analysis of systems of chemical reactions." *Arch. Ration. Mech. Anal.* **19,** 81–99 (1965). (Reprint D)
60. "An adaptive control of the batch racton–III: Simplified parameter estimation" (with W.H. Ray). *Automatica* **3,** 53–71 (1965).
61. "On simple exchange waves in fixed beds" (with N.R. Amundson and R. Swanson). *Proc. Roy. Soc.* **A286,** 129–139 (1965).
62. "On the theory of reactions in continuous mixtures" (with G.R. Gavalas). *Phil. Trans. Roy. Soc.* **A 260,** 351–393 (1966).
63. "Is sophistication really necessary? *Ind. Eng. Chem.* **58,** 32–37 (1966).
64. "Rationale for optimal reactor design. (with W.H. Ray). *Ind. Eng. Chem. Fund.* **5,** 478–483 (1966).
65. "Questing control of chemical reactors." Paper presented at the *Annual Meeting of the American Institute of Chemical Engineers,* Mexico City, October 1966.
66. "Bacterial growth as an optimal process" (with C.H. Swanson, A.G. Fredrickson, and H.M. Tsuchiya). *J. Theor. Biol.* **12,** 228–250 (1966).
67. "Compartmental analysis and the theory of residence time distributions." In K.B. Warren, (ed.,). *Intracellular Tranport,* (pp. 167–197). New York: Academic Press, 1966.
68. "Studies in optimization–VIII: Questing control of a stirred tank reactor" (with R.N. Schindler). *Chem. Eng. Sci.* **22,** 319–336 (1967).
69. "Studies in optimization–IX.: The questing control of a two phase reactor" (with R.N. Schindler). *Chem. Eng. Sci.* **22,** 337–344 (1967).
70. "Studies in optimization–X: Questing control with an economic criterion" (with R.N. Schindler). *Chem. Eng. Sci.* **22,** 345–352 (1967).
71. "An adaptive control of the batch reactor–IV: A more sophisticated controller" (with W.H. Ray). *Automatica* **4,** 139–161 (1967).
72. "Simple control policies for reactors with catalyst decay" (with A. Chou and W.H. Ray). *Trans. Instn Chem. Engrs.* **45,** 153–159 (1967).
73. "On the mathematical status of the pseudo-steady state hypothesis of biochemical kinetics" (with F.G. Heineken and H.M. Tsuchiya). *Math. Biosci.* **1,** 95–113 (1967).
74. "Transition between regimes in gas-solid reactions." *Ind. Eng. Chem. Fund.* **6,** 315–318 (1967).
75. "On the accuracy of determining rate constants in enzymatic reactions" (with F.G. Heineken and H.M. Tsuchiya). *Math. Biosci.* **1,** 115–141 (1967).
76.* "Prolegomena to the rational analysis of systems of chemical reactions–II: Some addenda. *Arch. Ration. Mech. Anal.* **27,** 356–364 (1968). (Reprint E)
77. "On what sort of place, if any, theoretical and mathematical studies should have in graduate chemical engineering research." *Chem. Eng. Educ.* **2,** 36–39 (1967).
78. "Sufficient conditions for the uniqueness of the steady state." *Chem. Eng. Sci.* **23,** 1501 (1968).
79. "Optimal control for pyrolytic reactors" (with A.P. Jackman). Paper presented at the Fourth European Symposium, Brussels, 1968.
80. "Canon and method in the arts and sciences (Olin Lecture, Yale University)." *Chem. Eng. Educ.* **3,** 48–52 (1969).
81. "On stability criteria of chemical reaction engineering." *Chem. Eng. Sci* **24,** 149–169 (1968).
82. "A note on mechanism and memory in the kinetics of biochemical reactions." *Math. Biosci.* **3,** 421–429 (1968).
83. "Communications on the theory of diffusion and reaction–I: A complete parametric study of the first-order, irreversible exothermic reaction in a flat slab of catalyst" (with D.W. Drott). *Chem. Eng. Sci.* **24,** 541–551 (1969).
84. "Communications on the theory of diffusion and reaction–II: The effect of shape on the effectiveness factor" (with S. Rester). *Chem. Eng. Sci.* **24,** 793–795 (1969).
85. "Communications on the theory of diffusion and reaction–III: The simulation of shape effects" (with S. Rester and J. Jouven). *Chem. Eng. Sci.* **24,** 1019–1022 (1968).
86. "Communications on the theory on diffusion and reaction–IV: Combined effects of internal and external diffusion in the non-isothermal case" (with B. Hatfield). *Chem. Eng. Sci.* **24,** 1213–1222 (1969).
87. "Mathematical aspects of chemical reaction" *Ind. Eng. Chem.* **61,** 17–29 (1969).
88. "Some problems in chemical reactor analysis with stochastic features: Linear systems with

fluctuating coefficients" (with T.M. Pell, Jr.). *Ind. Eng. Chem. Fund.* **8,** 339-345 (1969).

89. "A remark on the equilibrium theory of the parametric pump." *Ind. Eng. Chem. Fund.* **8,** 603 (1969).
90. "Problems in chemical reactor analysis with stochastic features: Control of linearized distributed systems on discrete and corrupted observations" (with T.M. Pell, Jr.). *Ind. Eng. Chem. Fund.* **9,** 15–20 (1970).
91. "Chemical kinetics and the ecology of mathematics." *Am. Scient.* **58,** 419–428 (1970).
92. "Communications on the theory of diffusion and reaction–V: Findings and conjectures concerning the multiplicity of solutions." (with I. Copelowitz). *Chem. Eng. Sci.* **25,** 906–909 (1970).
93. "Communications on the theory of diffusion and reaction–VI: The effectiveness of spherical catalyst particles in steep external gradients" (with I. Copelowitz). *Chem. Eng. Sci.* **25,** 885–896 (1970).
94. "Studies in optimization–XI: An experimental test of questing controller" (with J.M. Wheeler). *Chem. Eng. Sci.* **25,** 445–462 (1970).
95. "On the theory of multicomponent chromatography" (with H.-K. Rhee and N.R. Amundson). *Phil. Trans. Roy. Soc.* **A 267,** 419–455 (1970).
96. "Algebraic aspects of formal chemical kinetics." In M. Bunge (ed.), *Studies in the Foundations, Methodology and Philosophy of Science,* (Vol. 4, pp. 119–129). New York: Springer-Verlag, 1971.
97. "Multicomponent adsorption in continuous countercurrent exchangers" (with H.-K. Rhee and N.R. Amundson). *Phil. Trans. Roy. Soc.* **A 269,** 187–215 (1971).
98. "A note on a form of the Emden-Fowler Equation" (with B.N. Mehta). *J. Math. Anal. Appl.* **36,** 611–621 (1971).
99. "A note on the structure of the transient behavior of chemical reactors." *Chem. Eng. J.* **2,** 140–141 (1970).
100. "Communications on the theory of diffusion and reaction–VII: The isothermal *p*th order reaction" (with B.N. Mehta). *Chem. Eng. Sci.* **26,** 1699–1712 (1971).
101. "Transients in distributed chemical reactors part 1: A simplified model" (with D.L. Schruben). *Chem. Eng. J.* **2,** 179–188 (1970).
102. "Surface diffusion and reaction at widely separated sites" *J. Catal.* **22,** 282–284 (1971).
103. "Variational bounds for problems in diffusion and reaction" (with W. Strieder). *J. Inst. Maths Appl.* **8,** 328–334 (1971).
104. "On the realistic and interesting parameter ranges in the theory of diffusion and reaction" (with M.C. Mercer). *Latin Am. J. Chem. Eng. Appl. Chem.* **2,** 149–162 (1971).
105. *Some problems in the analysis of transient behavior and stability of chemical reactors.* First International Symposium on Chemical Reaction Engineering no. 109. Washington, D.C.: American Chemical Society, 1972.
106. "Communications on the theory of diffusion and reaction–VIII: Variational bounds on the effectiveness factor" (with S. Rester). *Chem. Eng. Sci.* **27,** 347–360 (1972).
107. "On a mechanism for autocatalysis" (with D.R. Schneider and N.R. Amundson). *Chem. Eng. Sci.* **27,** 895–905 (1972).
108. "Mobility, permeability, and the pseudo–steady-state hypothesis." *Math. Biosci.* **13,** 1–8 (1972).
109. "Diffusive and electrostatic effects with insolubilized enzymes" (with M.L. Schuler and H.M. Tsuchiya). *J. Theor. Biol.* **35,** 67–76 (1972).
110. "A method of representing the nonisothermal effectiveness factor for fixed bed calculations" (with J.G. Jouven). *A.I.Ch.E.J.* **18,** 402–408 (1972).
111. "The control of a stirred tank reactor with hysteresis in the control element–I: Phase space analysis" (with J.C. Hyun). *Chem. Eng. Sci.* **27,** 1341–1359 (1972).
112. "The control of a stirred tank reactor with hysteresis in the control element–II: Describing function analysis" (with J.C. Hyun). *Chem. Eng. Sci.* **27,** 1361–1370 (1972).
113. "On the equations for the movement and deformation of a reaction front" (with R.H. Knapp). *Arch. Ration. Mech. Anal.* **44,** 165–177 (1972).

114a. "Some problems common to chemical engineering and the biological sciences." Paper presented at the Proceedings of the Scandinavian Congress of Chemical Engineering, November 1971.

114b. "Some interactions between problems in chemical engineering and the biological sciences" In G. Lindner and K. Nyberg (eds.), *Environmental Engineering,* (pp. 215–225). Dordrecht-Holland: D. Reidel Publishing Co., 1973.

115. "Transients in distributed chemical reactors. Part 2: Influence of diffusion in the simplified model" (with I.H. Farina). *Chem. Eng. J.* **4,** 149–170 (1972).
116. "Asymmetries generated by diffusion and reaction, and their bearing on active transport through membranes" (with K.H. Keller). *Proc. Natl. Acad. Sci. USA* **69,** 777–779 (1972).
117. "An analysis of chemical reactor stability and sensitivity–XIV: The effect of the steady state

hypothesis" (with D.R. Schneider and N.R. Amundson). *Chem. Eng. Sci.* **28,** 885–896 (1973).

118. "Hydrodynamic focusing and electronic cell-sizing techniques" (with M.L. Shuler and H.M. Tsuchiya). *Appl. Microbiol.* **24,** 384–388 (1972).
119. "Diffusive and electrostatic effects with insolubilized enzymes subject to substrate inhibition" (with M.L. Shuler and H.M. Tsuchiya). *J. Theor. Biol.* **41,** 347–356 (1973).
120. "Communications on the theory of diffusion and reaction–IX: Internal pressure and forced flow for reactions with volume change" (with J.P.G. Kehoe). *Chem. Eng. Sci.* **28,** 2094–2098 (1973).
121. "Communications on the theory of diffusion and reaction–X. A generalization of Wei's bounds on the maximum temperature" (with C. Georgakis). *Chem. Eng. Sci.* **29,** 291–293 (1974).
122. "The theory of diffusion and reaction: A chemical engineering symphony (1973 ASEE Award lecture)." *Chem. Eng. Educ.* **8,** 20–40 (1973).
123. *Counter-current moving bed chromatographic reactors* (with S. Viswanathan). ACS Symposium Series no. 133. Washington, D.C.: American Chemical Society, 1974.
124. "An analysis of the countercurrent moving bed reactor" (with S. Viswanathan). *SIAM-AMS Proc.* **8,** 99–124 (1974).
125. "Phenomena of multiplicity, stability, and symmetry." *Ann. NY Acad. Sci.* **231,** 86–98 (1974).
126. "On the ostensible steady state of a dynamical system." *Rend. Lincei Ser. VIII* **57,** 1–9 (1974).
127. "What's the use of a Ph.D. anyway?" *AIChE Stud. Memb. Bull.* **15,** 5–7 (1974).
128. "On the concept of the steady state in chemical reactor analysis" (with S. Viswanathan). *Chem. Eng. Commun.* **2,** 1–4 (1975).
129. "The design of stirred reactors with hollow fiber catalysts for Michaelis-Menten kinetics" (with C. Georgakis and P.C.-H. Chan). *Biotechnol. Bioeng.* **57,** 99–106 (1975).
130. "Diffusion, reaction and the pseudo-steady-state hypothesis" (with C. Georgakis). *Math. Biosci.* **25,** 237–258 (1975).
131. "Diffusion and first order reaction in a general multilayered membrane" (with B. Bunow). *Math. Biosci.* **26,** 157–174 (1975).
132. "Carberry's ultimate paper." *Chem. Eng. Educ.* **9,** 118–119 (1975).
133. "Diffusion and reaction in mycelial pellets." *J. Ferment. Technol.* **53,** 899–901 (1975).
134. "Some thoughts on the nature of academic research in chemical engineering." *Chem. Eng. Educ.* **10,** 2–5 (1976).
135. "Computational methods for the tubular chemical reactor" (with A. Varma, C. Georgakis, and N.R. Amundson). *Comput. Meth. Appl. Mech. Eng.* **8,** 319–330 (1976).
136. "How to get the most out of an equation without really trying." *Chem. Eng. Educ.* **10,** 114–124 (1976).
137. "Geometric correction factors for the Weisz diffusivity cell" (with W.W. Meyer and L.L. Hegedus). *J. Catal.* **42,** 135–138 (1976).
138. "Modeling the monolith: Some methodological considerations" (with S.T. Lee). Paper presented at the Fourth International–Sixth European Symposium on Chemical Reaction Engineering, Heidelberg, 6–8 April 1976.
139. "Studies in the control of tubular reactors–I: General considerations" (with C. Georgakis and N.R. Amundson). *Chem. Eng. Sci.* **32,** 1359–1369 (1977).
140. "Studies in the control of tubular reactors–II: Stabilization by modal control" (with C. Georgakis and N.R. Amundson). *Chem. Eng. Sci.* **32,** 1371–1379 (1977).
141. "Studies in the control of tubular reactors–III: Stabilization by observer design" (with C. Georgakis and N.R. Amundson). *Chem. Eng. Sci.* **32,** 1381–1387 (1977).
142. "Academic chemical engineering in an historical perspective." *Ind. Eng. Chem. Fund.* **16,** 1–5 (1977).
143. "The sciences and the humanities" (with M. Penn). *Chem. Eng. Educ.* **11,** 68–73, 85 (1977).
144. "Dynamics of a chemostat in which two organisms compete for a common substrate" (with A.E. Humphrey). *Biotechnol. Bioeng.* **19.** 1375–1386 (1977).
145. "Re, k and π: A conversation on some aspects of mathematical modelling." *Appl. Math. Modelling* **1,** 386–394 (1977).
146. "Art and craft in the modelling of chemical processes." Paper presented at the Proceedings of the First International Conference on Mathematics Modelling, St. Louis, MO, 29 August–1 September 1977.
147. "On the effects of radiative heat transfer in monoliths" (with S.-T. Lee). *Chem. Eng. Sci.* **32,** 827–837 (1977).
148. "The infiltration of lymphocytes and macrophages into a site of infection." In A.I. Bell, A.S. Perelson, and G.H. Pimbley (eds.), *Theoretical Immunology,* (Vol. 8). New York: Marcel Dekker, 1978.
149. "Finite stability regions for large-scale systems with stable and unstable subsystems" (with M. Morari and G.S. Stephanopoulos). *Int. J. Cont.* **26,** 805–815 (1977).
150. "Models of the catalytic monolithic." Paper presented at the Levich Conference, Oxford, 1978.
151. *Poisoning in monolithic catalysts.* American

Chemical Society Symposium Series no. 65. Washington, D.C.: American Chemical Society, 1978.
152. "An analysis of thermal desorption mass spectra. I" (with C.-M. Chan and W.H. Weinberg). *Appl. Surf. Sci.* **1,** 360–376 (1978).
153. "Temperature gradients in porous catalyst pellets." *Ind. Eng. Chem. Fund.* **17,** 309–313 (1978).
154. "Horses of other colors—Some notes on seminars in a chemical engineering department." *Chem. Eng. Educ.* **12,** 148–151 (1978).
155. "Chemical reactors and some bifurcation phenomena." *Ann. NY Acad. Sci.* **316,** 314–331 (1979).
156.* "De exemplo simulacrorum continuorum discretalumque." *Arch. Rat. Mech. Anal.* **70,** 203–209 (1979). (Translation in K′ and P′) (Reprint M)
157. "Measurement of leukocyte motility and chemotaxis parameters using a quantitative analysis of the under-agarose migration assay" (with D. Lauffenburger). *Math. Biosci.* **44,** 121–138 (1978).
158. "A stochastic analysis of the growth of competing microbial populations in a continuous biochemical reactor" (with G.S. Stephanopoulos and A.G. Frederickson). *Math. Biosci.* **45,** 99–135 (1979).
159. "Stability analysis of structured chemical engineering systems via decomposition" (with M. Morari and G.S. Stephanopoulos). *Chem. Eng. Sci.* **34,** 11–15 (1979).
160. "The role of dimensionless parameters in the Briggs–Haldane and Michaelis–Menten approximations" (with P.S. Crooke and R.D. Tanner). *Chem. Eng. Sci.* **34,** 1354–1357 (1979).
161. "Creeping fronts and travelling waves (E. Wicke festschrift)." *Chem. Eng. Tech.* **51,** 767–771 (1979).
162. "Effect of catalyst loading on the simultaneous reactions of NO, CO, and O_2" (with L.L. Hegedus, R.K. Herz and S.H. Oh). *J. Catal.* **57,** 513–515 (1979).
163. "Method in the modeling of chemical engineering systems." In C.T. Leondes (ed.), *Control and Dynamic Systems: Advances in Theory and Application,* (Ch. 20). New York: Academic Press, 1979.
164. "The growth of competing microbial populations in a CSTR with periodically varying inputs" (with G.S. Stephanopoulos and A.G. Fredrickson). *A.I.Ch.E.J.* **25,** 863–872 (1979).
165. "Traveling waves in a simple population model involving growth and death" (with C.R. Kennedy). *Bull. Math. Biol.* **42,** 397–429 (1980).
166. "Bifurcations of a model diffusion–reaction system" (with C.R. Kennedy). In P. Holmes (ed.), *New Approaches to Nonlinear Problems in Dynamics* (pp. 211–233). Philadelphia: SIAM, 1980.
167. "A continuous chromatographic reactor" (with B.K. Cho and R.W. Carr, Jr.). *Chem. Eng. Sci.* **35,** 74–81 (1980).
168. "The mere notion of a model" (with M. Penn). *Math. Model.* **1,** 1–12 (1980).
169. "Hierarchies of models in reactive systems." In W. E. Stewart, W.H. Ray, and J. Conway (eds.), *Dynamics and Modelling of Reactive Systems,* (pp. 1–35). New York: Academic Press, 1980.
170. "This number omitted."
171. "A note on the Beneventan script." *Soc. Scribes Illum. Newsl.* **18** (1980).
172. "Bilinear approximation of general non-linear dynamic systems with linear inputs" (with S. Svoronos and G.S. Stephanopoulos). *Int. J. Cont.* **31,** 109–126 (1980).
173. "Observations on fixed-bed dispersion models: The role of the interstitial fluid" (with S. Sundaresan and N.R. Amundson). *A.I.Ch.E.J.* **26,** 529–536 (1980).
174. "A new continuous flow reactor for simultaneous reaction and separation" (with B.K. Cho and R.W. Carr). *Sep. Sci. Technol.* **15,** 679–696 (1980).
175. "Effects of random motility on growth of bacterial populations" (with D. Lauffenburger and K.H. Keller). *Microb. Ecol.* **7,** 207–227 (1981).
176. "Some canonical chemical engineering catastrophes (Churchill festschrift)." *Chem. Eng. Commun.* **9,** 51 (1981).
177. "Isothermal sustained oscillations in a very simple surface reaction" (with C.G. Takoudis and L.D. Schmidt). *Surf. Sci.* **105,** 325–333 (1981).
178. "Multiple steady states in reaction controlled surface catalysed reactions" (with C.G. Takoudis and L.D. Schmidt). *Chem. Eng. Sci.* **36,** 377–386 (1981).
179. "Steady state multiplicity in surface reactions with coverage dependent parameters" (with C.G. Takoudis and L.D. Schmidt). *Chem. Eng. Sci.* **36,** 1795–1802 (1981).
180. "The notions of uniqueness and multiplicity of steady states in the development of chemical reactor analysis." In W.F. Furler (ed.), *A Century of Chemical Engineering,* (pp. 389–404). New York: Plenum Press, 1982.
181. "On the behaviour of two stirred tanks in series" (with S. Svoronos and G.S. Stephanopoulos). *Chem. Eng. Sci.* **3,** 357–366 (1982).
182. "Weakly coupled systems of nonlinear elliptic boundary value problems" (with K. Zygourakis). *Nonlin. Anal.* **6,** 555–569 (1982).
183. "The intangible tints of dawn." *Thor. Quart.* **1,** 52–60 (1982).

184. "The mathematical theory of a countercurrent catalytic reactor" (with B.K. Cho and R.W. Carr). *Proc. Roy. Soc.* **A383,** 147–189 (1982).
185. "Chemical engineering at the University of Minnesota." *Chem. Eng. Educ.* **16,** 50–54 (1982).
186. Continuous reaction gas chromatography: The dehydrogenation of cyclohexane over Pt/g-Al_2O_3 (with A.W. Wardwell and R.W. Carr, Jr.). American Chemical Society Symposium Series no. 196. Washington, D.C.: American Chemical Society (1982).
187. "Isothermal oscillations in surface reactions with coverage independent parameters" (with C.G. Takoudis and L.D. Schmidt). *Chem. Eng. Sci.* **37,** 69–76 (1982).
188. "The scope of R.T.D. theory." In A. Pethös and R.D. Noble (eds.), *Residence time distribution theory in chemical engineering,* (pp. 1–21). Weinheim: Verlag Chemie GmbH, 1982.
189. "Residence time distribution with many reactions and in several environments." In A. Pethö and R.D. Noble (eds.), *Residence time distribution theory in chemical engineering,* (pp. 24–40). Weinheim: Verlag Chemie GmbH, 1982.
190. Review of *Insights into Chemical Engineering,* by P.V. Danckwerts, *Chem. Eng. Sci.* **37,** 1123 (1982).
191. "Some characteristic nonlinearities of reacting systems." In A. Bishop, D. Campbell, and B. Nicolaenko (eds.), *Nonlinear Problems: Present and Future,* North-Holland, Amsterdam, 1982.
192. "Effects of cell motility and chemotaxis on microbial population growth" (with D. Lauffenburger and K. Keller). *Biophys. J.* **40,** 209–219 (1982).
193. "Multiple oxidation reactions and diffusion in the catalytic layer of monolith reactors" (with K. Zygourakis). *Chem. Eng. Sci.* **38,** 733–744 (1983).
194. "Monotone iteration methods for solving coupled systems of nonlinear boundary value problems" (with K. Zygourakis). *Comput. Chem. Eng.* **7,** 183–193 (1983).
195. "A two-layer model of the atmosphere indicating the effects of mixing between the surface layer and the air aloft" (with D.D. Reible and F.H. Shair). *Atmos. Environ.* **17,** 25–33 (1983).
196. "The interpretation of sorption and diffusion data in porous solids." *Ind. Eng. Chem. Fund.* **22,** 150–151 (1983).
197. "Theoretical and experimental aspects of catalyst impregnation" (with S.-Y. Lee). In G. Poncelet, P. Grange, and P.A. Jacobs (eds.), *Studies in Surface Science and Catalysis,* (Vol. 16; pp. 35–45). Amsterdam: Elsevier, 1983.
198. "Effectiveness of catalytic archipelagos–I: Regular arrays of regular islands" (with D.-Y. Kuan and H.T. Davis). *Chem. Eng. Sci.* **38,** 719–732 (1983).
199. "Effectiveness of catalytic archipelagos–II: Random arrays of random islands" (with D.-Y. Kuan and H.T. Davis). *Chem. Eng. Sci.* **38,** 1569–1579 (1983).
200. "On the dynamics of a stirred tank with consecutive reactions" (with D.V. Jorgensen). *Chem. Eng. Sci.* **38,** 45–53 (1983).
201. *Chemical reaction engineering as an intellectual discipline.* American Chemical Society Symposium Series no. 226. Washington, D.C.: American Chemical Society 1983.
202. "R.H. Wilhelm's influence on the development of chemical reaction engineering (from the Wilhelm lectures at Princeton)." *Chem. Eng. Educ.,* **17,** 10–41 (1983).
203. "Discrete cell model of pore-mouth poisoning of fixed-bed reactors" (with B.K. Cho and L.L. Hegedus). *A.I.Ch. E.J.* **29,** 289–297 (1983).
204. "The jail of shape." *Chem. Eng. Commun.* **24,** 167–181 (1983).
205. "The distribution of active ingredients in supported catalysts prepared by impregnation" (with S.-Y. Lee). *Catal. Rev.* **27,** 207–340 (1984).
206. "Traveling bands of chemotactic bacteria in the context of population growth" (with D. Lauffenburger and C.R. Kennedy). *Bull. Math. Biol.* **46,** 19–40 (1984).
207. "Rate multiplicity and oscillations in single-species surface reactions" (with I. Kevrekidis and L.D. Schmidt). *Surf. Sci.* **137,** 151–166 (1984).
208. "Problems in the dynamics of chemical reactors." Paper presented at the International Chemical Reaction Engineering Conference, Pune, 1984.
209. "Estimation of fin efficiencies of regular tubes arrayed in circumferential fins" (with D.-Y. Kuan and H.T. Davis). *Int. J. Heat Mass Trans.* **27,** 148–151 (1984).
210. "Mathematical analysis for a chromatographic reactor" (with T. Petroulas and R.W. Carr). In R. Vichnevetsky and R.S. Stepleman (eds.), *Advances in Computer Methods for Partial Differential Equations–V: Proceedings of the Fifth IMACS International Symposium on Computer Methods for Partial Differential Equations,* New Brunswick: IMACS, Dept. of Computer Science, Rutgers University, 1984.
211. "On the dynamics of periodically forced chemical reactors" (with I.G. Kevrekidis and L.D. Schmidt). *Chem. Eng. Commun.* **30,** 323–330 (1984).
212. "More on the dynamics of a stirred tank with consecutive reactions" (with D.V. Jorgensen and W.W. Farr). *Chem. Eng. Sci.* **39,** 1741–1752 (1984).
213. "Thermodynamic limitations on the dynamic

behaviour of heterogeneous reacting systems" (with I. Kevrekidis and L.D. Schmidt). *Inst. Chem. Eng. Symp. Ser.* **87,** 109–115 (1984).

214. "Analysis of the counter-current moving-bed chromatographic reactor" (with T. Petroulas and R.W. Carr). *Comp. Maths. Appls.* **11,** 5–34 (1985).

215. "Numerical computation of invariant circles of maps" (with I.G. Kevrekidis, L.D. Schmidt, and S. Pelikan). *Physica* **16D,** 243–251 (1985).

216. "On the permeability of membranes with parallel, but interconnected, pathways." *Math. Biosci. (Bellman Memorial Issue)* **77,** 5–16 (1985).

217. "Analysis and performance of a countercurrent moving-bed chromatographic reactor" (with T. Petroulas and R.W. Carr, Jr.). *Chem. Eng. Sci.* **40,** 2233–2240 (1985).

218.* "Some common features of periodically forced reacting systems" (with I.G. Kevrekidis and L.D. Schmidt). *Chem. Eng. Sci.* **41,** 1263–1276 (1986). (Reprint I)

219. "The stirred tank forced" (with I.G. Kevrekidis and L.D. Schmidt). *Chem. Eng. Sci.* **41,** 1549–1560 (1986).

220.* "Yet who would have thought the old man to have had so much blood in him?'—Reflections on the multiplicity of steady states of the stirred rank reactor" (with W.W. Farr). *Chem. Eng. Sci.* **41,** 1385–1402 (1986). (Reprint J)

221. "Resonance in periodically forced processes" (with I.G. Kevrekidis and L.D. Schmidt). *Chem. Eng. Sci.* **41,** 905–911 (1986).

222. "Entrainment regions for periodically forced oscillators" (with D.G. Aronson, R.P. McGehee, and I.G. Kevrekidis). *Phys. Rev. A,* **33,** 2190–2192 (1986).

223. "On a problem in hindered diffusion (Serrin festschrift)." *Arch. Rat. Mech. Anal.* **95,** 83–91 (1986).

224. "The mathematical background of chemical reactor analysis–I. Preliminaries, batch reactors." *Physica* **D20,** 82–90 (1986).

225. "The mathematical background of chemical reactor analysis–II. The stirred tank reactor." In G.S.S. Ludford (ed.), *Reacting Flows: Combustion and Chemical Reactors, Lectures in Applied Mathematics,* (Vol. 24, pp. 75–107). Providence: American Mathematical Society, 1986.

226. "The continuous countercurrent moving bed chromatographic reactor" (with B. Fish and R.W. Carr). *Chem. Eng. Sci.* **41,** 661 (1986).

227. "An analysis of the counter-current adsorber" (with D. Altshuller, G. Vazquez, and R.W. Carr). *Chem. Eng. Commun.* **52,** 311 (1987).

228. "Anisotropic membrane transport" (with K.-M. Jem and E.L. Cussler). *Chem. Eng. Commun.* **55,** 5–17 (1987).

229. "On apparent second-order kinetics" (with T.C. Ho). *A.I.Ch.E.J.* **33,** 1050–1051 (1987).

230. "A global study of Kondepudi's pitchfork" (with X.-H. Song). *Sadhana* **10,** 1–12 (1987).

231. "Degenerate Hopf bifurcations in the CSTR with reactions $A \rightarrow B \rightarrow C$" (with W.W. Farr). *Can. Math. Soc. Conf. Proc.* **8,** 397–418 (1987).

232.* "A general theory of anisotropic membranes" (with E.L. Cussler) *Chem. Eng. Commun.* **58,** 3–16 (1987). (Reprint N).

233. "Nonlinear dynamics and strange attractors" (with K.S. Chang). *Kor. J. Chem. Eng.* **4,** 95–104 (1987).

234. "A sequence of scripts." *Scribe* **41,** 7–12 (1987).

235. "Ann Hechle's 'In the Beginning.' *Call. Revs* **4,** 41–46 (1987).

235a. "Painting with words: The art of Donald Jackson." *Call. Revs* **6,** 18–28 (1987).

236. "Forced oscillations of chemical reactors." In P. Gray et al. (eds.), *Spatial Inhomogeneities and Transient Behaviour in Chemical Kinetics.,* Manchester: Manchester University Press, 1990.

237.* "Autonomous bifurcations of a simple bimolecular surface-reaction model" (with M.A. McKarnin and L.D. Schmidt). *Proc. Roy. Soc.* **A415,** 363–387 (1988). (Reprint K)

238. "Barrier membranes" (with E.L. Cussler, S.E. Hughes, and W.J. Ward, III). *J. Membrane Sci.* **38,** 161–174 (1988).

239.* "Forced oscillations of a self-oscillating bimolecular surface reaction model" (with M.A. McKarnin and L.D. Schmidt). *Proc. Roy. Soc.* **A417,** 363–388 (1988). (Reprint L)

240. "Response of nonlinear oscillators to forced oscillations: Three chemical reaction case studies" (with M.A. McKarnin and L.D. Schmidt). *Chem. Eng. Sci.* **43,** 2833–2844 (1988).

241. "Chaotic behaviour of two counter-currently cooled reactors" (with K.S. Chang). *Lat. Am. Appl. Res.* **18,** 1 (1988).

242.* "Modelling cubic autocatalysis by successive bimolecular steps" (with P. Gray and S.K. Scott). *Chem. Eng. Sci.* **43,** 207–211 (1988). (Reprint F)

243. "Ut Simulacrum, Poesis." *New Literary Hist.* **20,** 323–340 (1988–1989).

244. "Theories of precipitation induced by dissolution" (with J. Kopinsky and E.L. Cussler) *A.I.Ch.E.J.* **34,** 2005–2010 (1988).

245. "Effects of velocity on homogeneous-heterogeneous ignition and extinction" (with R.J. Olsen and L.D. Schmidt). *Combust. Sci. Tech.,* **99** (1995).

246.* "On reactions in continuous mixtures." *A.I.Ch.E.J.* **35,** 539–548 (1989). (Reprint G)

247. "On the limits of facilitated diffusion" (with E.L. Cussler and A. Bhown). *J. Memb. Sci.* **43,** 149–164 (1989).

248. "Continuous lumping of nonlinear chemical kinetics" (with G. Astarita). *Chem. Eng. Proc.* **26,** 63–69 (1989).
249. "On aliases of differential equations" (with G. Astarita). *Rend. Lincei Ser. VIII* **83,** 7–11 (1989).
250. "Military technology and garrison organization—some observations on Anglo-Saxon military thinking in light of the burghal hidage" (with B.S. Bachrach). *Tech. Cul.* **31,** 1–17 (1990).
251. "Forced oscillations of chemical reactors with multiple steady states" (with G.A. Cordonier and L.D. Schmidt). *Chem. Eng. Sci.* **45,** 1659–1675 (1990).
252. "The simulated countercurrent moving bed chromatographic reactor" (with A.J. Ray, A.L. Tonkovich, and R.W. Carr). *Chem. Eng. Sci.* **45,** 2431–2437 (1990).
253. "The effects of phase transitions, surface diffusions, and defects on surface catalyzed reactions: Fluctuations and oscillations" (with D.G. Vlachos and L.D. Schmidt). *J. Chem. Phys.* **93,** 8306–8313 (1990).
254. "Manners Makyth Modellers (Fifth Danckwerts Lecture)." *Trans. I.Ch.E.* **69,** 165–174 (1991).
255. "The ignition criteria for stagnation-point flow: Semenov-Frank-Kamenetski or van't Hoff" (with X. Song and L.D. Schmidt). *Combust. Sci. Tech.* **75,** 311–331 (1991).
256. "Chemical engineering and the liberal education today." Paper presented at the 25th Phillips Lecture, Oklahoma State University, 26 April 1991.
257. "Steady states and oscillations in homogeneous-heterogeneous reaction systems" (with X. Song and L.D. Schmidt). *Chem. Eng. Sci.* **46,** 1203–1215 (1991).
258. "Diffusion and reaction in a Mandlebrot lung." *Chaos, Sol. Fract.* **1,** 583–593 (1991).
259. "Multiple indices, simple lumps and duplicitous kinetics." In A.V. Sapre and F.J. Krambeck (eds.), *Chemical Reactions in Complex Mixtures,* (pp. 25–41). New York: Van Nostrand Reinhold, 1991.
259a. "The mathematics of continuous mixtures." In G. Astarita and S.F. Sandler (eds.), *Kinetics and Thermodynamic Lumping of Multicomponent Mixtures.* Elsevier, 1991.
260. "The effect of phase transitions, surface diffusion, and defects on heterogeneous reactions: Multiplicities and fluctuations" (with D.G. Vlachos and L.D. Schmidt). *Surface Sci.* **249,** 248–264 (1991).
261. "Buoyancy-driven flows of a radiatively participating fluid in a vertical cylinder heated from below" (with A.G. Salinger, S. Brandon, and J.J. Derby). *Proc. Roy. Soc.* **A442,** 313–341 (1992).
262. "Dynamics of homogeneous-heterogeneous reactors" (with R.J. Olsen, W.R. Williams, and L.D. Schmidt). *Chem. Eng. Sci.* **47,** 2505–2510 (1992).
263. "Kinetics of facet formation during growth and etching of crystals" (with D.G. Vlachos and L.D. Schmidt). In K.S. Liang, M.P. Anderson, R.F. Bruisma, and G.G. Scoles (eds.), *Interface Dynamics and Growth,* (Vol. 237, pp. 145–150). 1992.
264. "Structures of small metal clusters–II: Phase transitions and isomerization" (with D.G. Vlachos and L.D. Schmidt). *J. Chem. Phys.* **96,** 6891–6901 (1992).
265. "Structure of small catalyst particles" (with D.G. Vlachos and L.D. Schmidt). *Chem. Eng. Sci.* **47,** 2769–2774 (1992).
266. "Structures of small metal clusters–I: Low temperature behavior" (with D.G. Vlachos and L.D. Schmidt). *J. Chem. Phys.* **96,** 6880–6890 (1992).
267. "Bifurcation and global stability in surface catalyzed reactions using the Monte Carlo method" (with D.G. Vlachos and L.D. Schmidt). In H. Swinney, R. Aris, and D. Aronson (eds.), *Patterns and Dynamics in Reactive Media,* (Vol. 37, pp. 187–206). New York: Springer-Verlag, 1991.
268. "Spatial and temporal patterns in catalytic oscillations" (with D.G. Vlachos, F. Smith, and L.D. Schmidt). *Physica A* **188,** 302–321 (1992).
268a. "Comments on mitigation of backmixing via catalyst dilution." *Chem. Eng. Sci.* **47,** 507–508 (1992).
269. "Modeling the spontaneous ignition of coal stockpiles" (with A.G. Salinger and J.J. Derby). *A.I.Ch.E.J.* **40,** 991–1004 (1993).
270. "Ends and beginnings in the mathematical modelling of chemical engineering systems." *Chem. Eng. Sci.* **48,** 2507–2517 (1993).
271. "Products in methane combustion near surfaces" (with D.G. Vlachos and L.D. Schmidt). *A.I.Ch.E.J.* **40,** 1018–1025 (1994).
272. "Ignition and extinction of flames near surfaces: Combustion of CH in air" (with D.G. Vlachos and L.D. Schmidt). *A.I.Ch.E.J.* **40,** 1005–1007 (1994).
273. "Ignition and extinction of flames near surfaces: Combustion of H_2 in air" (with D.G. Vlachos and L.D. Schmidt). *Comb. Flame* **95,** 313–335 (1993).
274. "Kinetics of faceting of crystals in growth, etching, and equilibrium" (with D.G. Vlachos and L.D. Schmidt). *Phys. Rev.* **47,** 4896–4909 (1993).
275. "Continuous reactions in a non-isothermal CSTR–I. Multiplicity of steady states" (with P. Cicarelli). *Chem. Eng. Sci.* **49,** 621–631 (1994).
276. " 'Almost discrete' Γ-distributed chemical spe-

cies and reactions." *Chem. Eng. Sci.* **49,** 581–588 (1994).

277.* "Two eyes are better than one: Some reflections on the importance of having more than one viewpoint in mathematical modelling and other disciplines." *Mathl. Comput. Model.* **18,** 95–115 (1993). (Reprint P)

278. "Enhanced C_2 yields from methane oxidative coupling by means of a separative chemical reactor" (with A.L. Tonkovich and R.W. Carr). *Science* **262,** 221–223 (1993).

279. "Design and performance of a simulated countercurrent moving-bed separator" (with B.B. Fish and R.W. Carr). *A.I.Ch.E.J.* **39,** 1783–1790 (1993).

280.* "Reaction of a continuous mixture in a bubbling fluidized bed" (with N.R. Amundson). *Trans I. Chem. E* **71,** 611–617 (1993).

281. "De Motu Arietum" (with B.S. Bachrach). In *A Festschrift for Professor Lawrence Marcus* (pp. 1–13). New York: Marcel Dekker, 1993.

282. "The simulated countercurrent moving bed chromatographic reactor: A novel reactor–separator" (with A.K. Ray and R.W. Carr). *Chem. Eng. Sci.* **49,** 69–480 (1994).

283. "An essay on contemporary criticism." *New Lit. Hist.* **25,** 35–46 (1994).

284. "Complementary viewpoints: Some thoughts on binocular vision in mathematical modelling and Latin paleography." *New Lit. Hist.* **26,** 395–417 (1995).

285. "Jean Mabillon." In H. Damico and J.B. Zavadil (eds.), *Medieval Scholarship: Biographical Studies on the Formation of a Discipline. Vol. 1: History,* New York: Garland Publishing, Inc., 1995.

286. "Chaos in a simple two-phase reactor" (with K. Alhumaizi). *Chaos, Sol. Fract.* **4,** 1985–2014 (1994).

287. "Adsorption kinetics for the case of step and S-shaped isotherms" (with A.V. Kruglov). *A.I.Ch.E.J.* **41,** 2393–2398 (1995).

288. "Mathematical models in catalyst design." In L.L. Hegedus (ed.), *Catalyst Design: Progress and Perspectives,* (pp. 213–244). New York: Wiley, 1987.

289. "Shooting method for bifurcation analysis of boundary value problems" (with X. Song and L.D. Schmidt). *Chem. Eng. Commun.* **84,** 217–229 (1989).

290. "Bifurcation behavior in homogeneous-heterogeneous combustion: II. Computations for stagnation-point flow" (with X. Song, W.R. Williams, and L.D. Schmidt). *Comb. Flame,* 292–311 (1991).

291. "Ignition and extinction of homogeneous-heterogeneous combustion: CH_4 and C_3H_8 on Pt" (with X. Song, W.R. Williams, and L.D. Schmidt). Paper presented at the 23rd International Combustion Institute, 1990.

292. "Reaction of a continuous mixture in a bubbling fluidized bed" (with N.R. Amundson). In N.P. Cheremisinoff (ed.), *Advances in Engineering Fluid Mechanics,* (pp. 105–117). New York, 1996.

293. "Comparison of small metal clusters: Ni, Pd, Pt, Cu, Ag, and Au" (with D.G. Vlachos and L.D. Schmidt). *Z. Phys. D.* **26,** 156–158 (1993).

294. "Dynamics of catalytic reactions on metal surfaces" (with L.D. Schmidt). In *Unsteady State Processes in Catalysis,* (pp. 203–216). Utrecht, Holland: VSP Press, 1990.

295. "Computer-aided experimentation in countercurrent reaction chromatography and simulated countercurrent chromatography" (with B.B. Fish and R.W. Carr). *Chem. Eng. Sci.* **43,** 1867–1873 (1988).

296. "Determination of Arrhenius constants by linear and nonlinear fitting" (with N.H. Chen). *A.I.Ch.E.J.* **38,** 626–628 (1992).

297. "Effect of pressure on the combustion of methane near inert surfaces" (with A. Balakrishna, D.G. Vlachos and L.D. Schmidt). *Comb. Flame,* (to be published).

298. "Forcing an entire bifurcation diagram: Case studies in chemical oscillators" (with I.G. Kevrekidis and L.D. Schmidt). *Physica* **23D,** 391 (1986).

299. "Finite element formulations for large-scale, coupled flows in adjacent porous and open fluid domains" (with A.G. Salinger and J.J. Derby). *Int. J. Num. Meth. Fluids* **18,** 1185–1209 (1994).

300. "Continuous countercurrent moving-bed separator" (with B.B. Fish and R.W. Carr). *A.I.Ch.E.J.* **35,** 737–745 (1989).

301. "Optimization of the countercurrent moving-bed chromatographic separator" (with B.B. Fish and R.W. Carr). *A.I.Ch.E.J.* **39,** 1621–1627 (1993).

302. "Autocatalytic continuous reactions in a stirred tank: I. Multiplicity of steady states" (with P. Cicarelli). *Chem. Eng. Sci.* **49,** 5307–5313 (1994).

303. "Pt-catalyzed combustion of CH4-C3-C8 mixtures" (with A. Balakrishna and L.D. Schmidt). *Chem. Eng. Sci.* **49,** 11–18 (1994).

304. "Ignition and extinction of methane–propane flames in air" (with A. Balakrishna and L.D. Schmidt). *Chem. Eng. Sci.* (Submitted).

305. "Steady-state flow transitions in the radiative Rayleigh–Bénard problem: Visualizing a bifurcation diagram" (with A.G. Salinger and J.J. Derby). *Vid. J. Eng. Res.* **3,** 97–109 (1993).

306. "Parallel Gray/Scott reactions in a stirred vessel

(Peter Gray festschrift)." *Faraday-Trans.* **92,** 2839–2842 (1996).

307. "Analysis of a continuous immobilization reactor." (Eli Ruckenstein festschrift) (with K. Roenigk). *Ind. Eng. Chem. Res.* **35,** 2889–2899 [12.10]

308.* "Reflections on Keats' equation (J. Villadsen festschrift)." *Chem. Eng. Sci.* **52,** 2447–2455 (1997). (Reprint Q).

309. "Mass transfer from small ascending bubbles" (M.M. Sharma festschrift). *Chem. Eng. Sci.* **53.**

310. "Model reduction in a class of multi-phase systems" (H. Brenner festschrift). *Chem. Eng. Comm.* **148–150,** 285–289 (1996).

311. "On some diagrams of chemical reaction engineering. *Chaos* (1999).

312. "Dissections, transgressions and perilous paths." *Mathl. Comp. Model.* **28,** 91–101 (1998).

313. "The beauty of self-adjoint symmetry" (Jackson festschrift) (with D. Ramkrishna). *Ind. Eng. Chem. Res.* (submitted)

314. "A fine flurry of fudge factors" (AIChE Institute Lecture, 1997) *Chem. Eng. Prog.* (Being revised for publication).

INDEX OF GRADUATE STUDENTS AND CO-AUTHORS

An asterisk denotes joint supervision of graduate work. The numbers refer to the enumeration in the Bibliography. The lowercase letters refer to the Edited Books section of the Bibliography. Those listed under my own name are those for which I am solely responsible, with the exception of 132, whose author is, or rather *was*, Aris Rutherford.

SUBJECT INDEX TO THE PAPERS IN THE BIBLIOGRAPHY

The numbers in this index are the numbers of the papers in the main bibliography. The square brackets used elsewhere, [ ], are not needed here. Papers reprinted in Chapters 7–11 are identified by the appropriate roman capital, as are the six reprinted in J′. References to the books are sparing and only included when the title might not suggest looking there for a particular topic. The Roman capitals denoting the books are primed to distinguish them from the papers reprinted here.

INDEX

Page numbers followed by an *f* indicate a footnote.